GEOSTATISTICAL ERROR MANAGEMENT

GEOSTATISTICAL ERROR MANAGEMENT

Quantifying Uncertainty for Environmental Sampling and Mapping

Jeffrey C. Myers

VNR VAN NOSTRAND REINHOLD
I(T)P® A Division of International Thomson Publishing Inc.

New York • Albany • Bonn • Boston • Detroit • London • Madrid • Melbourne
Mexico City • Paris • San Francisco • Singapore • Tokyo • Toronto

I(T)P® An International Thomson Publishing Company
The ITP logo is a registered trademark used herein under license

Printed in the United States of America

For more information, contact:

Van Nostrand Reinhold
115 Fifth Avenue
New York, NY 10003

Chapman & Hall
2–6 Boundary Row
London
SE1 8HN
United Kingdom

Thomas Nelson Australia
102 Dodds Street
South Melbourne, 3205
Victoria, Australia

Nelson Canada
1120 Birchmount Road
Scarborough, Ontario
Canada M1K 5G4

Chapman & Hall GmbH
Pappelallee 3
69469 Weinheim
Germany

International Thomson Publishing Asia
221 Henderson Road #05–10
Henderson Building
Singapore 0315

International Thomson Publishing Japan
Hirakawacho Kyowa Building, 3F
2-2-1 Hirakawacho
Chiyoda-ku, 102 Tokyo
Japan

International Thomson Editores
Seneca 53
Col. Polanco
11560 Mexico D.F. Mexico

1 2 3 4 5 6 7 8 9 10 BBR 01 00 99 98 97

Library of Congress Cataloging-in-Publication Data

Myers, Jeffrey C.
Geostatistical error management: quantifying uncertainty for environmental sampling and mapping / Jeffrey C. Myers.
p. cm.
Includes bibliographical references and index.
ISBN 0–442–01429–5 (hardcover)
1. Environmental sampling—Statistical methods. 2. Environmental mapping—Statistical methods. 3. Geology—Statistical methods. 4. Environmental geotechnology—Statistical methods. 5. Errors, Scientific—Statistical methods. I. Title.
GE45.S25M49 1996
363.7′0028′7—dc21

96-37272
CIP

http://www.vnr.com
product discounts • free email newsletters
software demos • online resources

email: info@vnr.com

A service of I(T)P®

To Deborah, the love of my life

Contents

PART 5. DATA QUALITY OBJECTIVES

Foreword

Having authored the first book on geostatistics in the English language and contributed to the growth of this field of work, I feel privileged to introduce this first book on environmental geostatistics. Since 1977, when *Geostatistical Ore Reserve Estimation* was published, geostatistical modeling concepts and techniques have become daily practice in mining, the basis for the characterization and management of petroleum reservoirs and, not surprisingly, a main quantitative tool in assessing uncertainty in environmental sciences and engineering.

The qualitative–descriptive and explanatory study of physical and chemical processes and human activities is fundamental to understanding environmental problems. However, decision making and actions relevant to detection, monitoring, assessment, and remediation are based as well on the quantification of both the information available and the uncertainties involved. This generates the need for the tools that can deliver reliable environmental assessments, including the quantification of inherent uncertainties.

This book introduces what its author terms *geostatistical error management,* or GEM. GEM can be seen as a framework that considers objectives and guidelines, while intelligently and effectively integrating the concepts and pragmatic tools needed to deal with today's problems in assessing hazardous waste and environmentally impacted sites. The major issues addressed can be summarized as (1) thorough site assessment, and (2) objective decision making. To address these broad issues, GEM integrates three major components. The first, *data quality objectives,* may be seen as a general management mechanism and constitutes the planning tool that identifies needs and assesses resources for environmental data collection activities. The second component includes Gy's *sampling theory and practice* adopted in the context of high-quality and high-reliability environmental data sampling. This component guarantees the integrity of assessment by analyzing sources of errors in sampling and is prerequisite for reliable quantification. The third and final component is the *geostatistical appraisal* which, based on the previous components, provides the actual quantification of the environmentally critical attributes at the investigated site. The integration of these three components provides a complete collection of tools for site assessment and decision making.

The introduction of an innovative framework and related methodology is only one of several attributes of this book. The development of the topic is logical and comprehensive. Concepts and their specifics are carefully articulated and serve as a strong foundation for examples that assist the practitioner in real-life situations.

The book describes what can be done today in characterizing and evaluating sites needing remediation, based on specific goals and objectives. In addition, it opens the way to expanding GEM to (1) the areas of environmental work beyond site evaluation/characterization, and (2) the geostatistical techniques of the future. Finally, I believe that the most important contribution of this book is the practical, pragmatic integration of topics and tools for site remediation. After all, geostatistics was always meant to be a tool for the practitioner and pragmatic problem solver.

Michel David

Preface

By design, this book differs markedly from other technical or statistical textbooks. It started with the idea that existing scientific, engineering, and economic tools could be viewed as a unit for advanced environmental site assessment. This idea spawned the concept of geostatistical error management, or GEM; the results are contained within.

This book takes several new perspectives, including:

- An environmental focus: To date, numerous books have been written on the subject of geostatistics. However, their focus has primarily been mining applications. GEM focuses primarily on the environmental applications of geostatistical techniques and how they fit into the regulatory, legal, and engineering aspects.
- Integration of related fields. The fields of Data Quality Objectives (DQOs), Sampling Theory & Practice (STP), and Geostatistical Appraisal (GA) have traditionally been viewed and applied as separate entities. Each has valuable error management tools that can be applied at appropriate stages of an environmental site assessment. By using them in an integrated fashion, many benefits can be obtained.
- An explanatory style. The subjects discussed herein are technical and mathematical in nature. Such topics discourage many readers immediately. This book attempts to address these technical areas in an explanatory way. Yes, there are formulas, but unlike most books on statistics, none appear until Chapter 3 and they don't become numerous until Chapter 6. In addition, an attempt has been made to explain what the formulas mean, realizing that the implications of an equation are not self-evident to those not mathematically inclined. As one reviewer put it, the text is written in a "friendly and chatty" style that makes it distinctly different. Hopefully this will take the edge off the mathematics and help readers to apply the GEM concepts.

The book is organized into sections of related material. Part 1 introduces the concept of GEM and the integration of the various fields. In Part 2 the concept of error is introduced, followed by an introduction to statistical methods as they relate to environmental situations. A new topic to most, Sampling Theory & Practice (STP) is discussed in Part 3, followed by chapters on correlated data and geostatistics in Part 4. Once the statistical tools have been presented, Data Quality Objectives (DQOs) are introduced in Part 5 and integrated with STP and GA.

GEM presents tools for quality environmental site assessment: Good planning, good sampling, good statistical analysis, and (hopefully) good decision making. GEM addresses three very common (and often difficult) questions: (1) How many samples should we take?; (2) Where do we put them?; and (3) How well do we know what's out there when we're done?

Decision making in the face of uncertainty is a difficult but unavoidable task in environmental remediation. GEM provides ways to manage uncertainty and errors at many phases of a project. The mining experience forced the development of new tools (STP, GA) because the old ones couldn't quantify the uncertainties involved in sampling and ore reserve estimation. Miners, a practical, no-nonsense, and profit-driven bunch, embraced these techniques for one reason: They worked.

Successful mining depends upon correct classification. Miners take samples to develop an ore inventory and locate areas of ore versus waste. Decisions are made daily about which material will be processed as ore and which material will be discarded as waste. Ore discarded as waste is an opportunity lost forever, penalizing the economic return expected. Conversely, waste processed as ore is an error that also reduces profitability. While errors and uncertainties are daily occurrences in a mining operation, geostatistical approaches assist in minimizing unprofitable decisions.

Geostatistical techniques have made their way into the environmental sciences in the last decade. For the same reason as in mining, they have been accepted by regulators, attorneys, engineers, and scientists. As in mining, environmental remediation and decision making are largely matters of classification. Based on an inventory derived from sampling, remedial decisions are made and, inevitably, errors. "Clean" material treated as "dirty" poses an inappropriate financial burden on the party funding the remediation, dirty material left untreated poses an unacceptable health risk. The goal of GEM is to minimize such errors so that all parties may profit.

GEM offers practical methods for site assessment. It is my hope that this book will provide readers with a unified and meaningful way to assess the ever-present question "how well do I think I know the problem" on environmentally impacted sites.

Acknowledgments

I would like to thank the following people for their assistance in the production of this book. Their contributions have made this book possible and their comments and suggestions have greatly enhanced the manuscript. Thank you to: Michel David, Francis Pitard, Charles Ramsey, Evan Englund, George Flatman, Kadri Dagdelen, George Robinson, Ed Perkins, Robert Gerzoff, Pat Madison, Doug Keckler, Gene Zika, Ron Candelaria, Bill Hartley, and Ruben Santos. I would also like to thank my family for their love, encouragement, and unwavering patience: Deborah, Katie, Libby, Megan, and Hunter.

PART 1 Introduction to Geostatistical Error Management

Chapter 1

Foundations of Geostatistical Error Management

SYNOPSIS: This chapter begins the discussion on the concept of geostatistical error management (GEM), including the purpose and goals. The relationship of GEM to the principles of total quality management (TQM) and quality by design (QBD) is presented. The chapter concludes with a discussion of each of the GEM components.

KEY WORDS: Geostatistical Error Management, Sampling Theory & Practice, Data Quality Objectives, Sampling, Site Characterization, Mapping, Total Quality Management, Quality By Design, Regionalized Variable, Variogram, Kriging.

1.1 Introduction to GEM

Geostatistical error management (GEM) is a gestalt of several fields of data design, sample collection, and statistical interpretation procedures. GEM combines existing techniques of *geostatistical appraisal (GA)* and *sampling theory & practice (STP)*, along with an environmentally specific adaptation of the scientific method called *data quality objectives (DQOs)*, to establish and achieve necessary goals in hazardous waste site remediation. The purpose of GEM is to provide objective, practical, understandable, and technically sound information for characterizing the extent and levels of contamination at hazardous waste or other environmentally impacted sites.

Geostatistical error management represents the integrated combination of the three separate but closely related fields (Fig. 1-1). GEM begins by using data quality objectives (DQOs) to limit error and uncertainty on sample data and subsequent decisions. Sound data form the basis of sound decision making and, as such, should not be compromised. DQOs also form the basis of a defensible strategy for site assessment and remediation, given that the data type, quantity, and quality may vary for different decision stages of a project. DQOs could have been appropriately called decision quality objectives; in either case, the acronym DQOs will suffice.

The second GEM component establishes the procedures necessary to implement correct sampling for environmental data. Sampling correctness is part of sampling theory & practice (STP) as developed by Pierre Gy. Sampling theory & practice will demonstrate that, although samples are collected frequently, they are frequently taken incorrectly. The effects of incorrect sampling are very detrimental, causing samples to be unrepresentative and leading to decision errors during remediation. Samples are our window on the world, and as Figure 1-2 shows, we ask a lot from them: A small volume of sample material determines the fate of much larger volumes.

The final GEM component addresses the issue of quality interpretation and mapping of the sample data. Geostatistical appraisal (GA) techniques offer many tools for data analysis, interpretation, and mapping. Mapping and visual displays of data play critical roles in site assessment. Geostatistical appraisal methods provide the ability to reduce the uncertainties and errors inherent in assessing

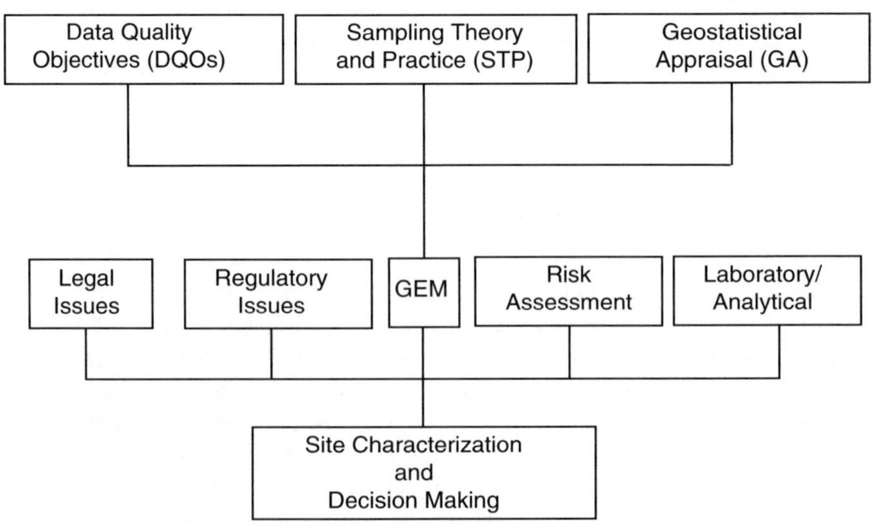

Figure 1-1. GEM components.

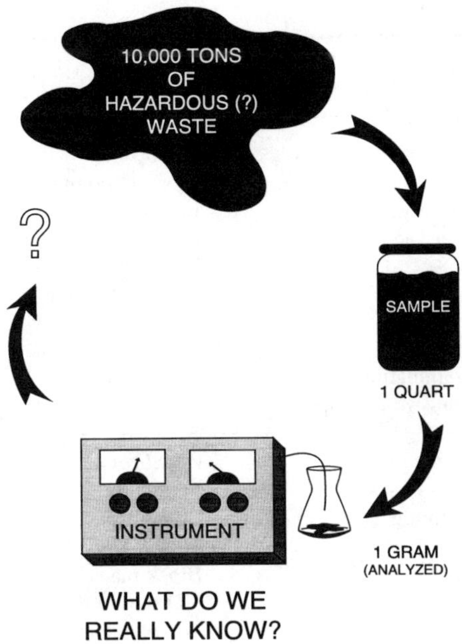

Figure 1-2. Small analytical volumes representing large remedial volumes raise questions as to the certainty of characterization.

environmental sites. Table 1-1 shows some of the uses of GEM in the site evaluation process.

Consequences of poor site characterization can be serious. Unremediated contamination poses health and safety risks to residents or workers and liability to site owners. Conversely, GEM will show how sites with small amounts of significant contamination can be overstated, resulting in large amounts of unnecessary remediation. Because of extremely high costs and liabilities associated with site cleanups, neither situation can be tolerated.

The GEM process provides a map, both figuratively and literally, that will be used in a *goal-oriented* manner to complete a successful remediation. GEM establishes the "road map," the project plan, that enables stakeholders to proceed through the site characterization phases in a systematic and unobstructed manner. GEM defines the starting point and end point of each project phase. Perhaps more importantly, it also shows critical decision points and objective sampling goals at various stages in the project. It is a truism that, without a road map or a goal, it is difficult to know where the end is or how to get there. If the destination is unknown or uncertain, it is highly likely that it also will not be attained.

GEM can be used whenever data are to be collected. For small sites, the GEM

Table 1-1. Uses of GEM.

Determine purpose of sampling:	*Incorporate site factors:*
Presence or absence Average or proportion Compare to threshold Compare to another population Search sampling Prove PRP Risk assessment Reportable quantity Cost recovery Determine extent of contamination Ascertain change in plume size or shape	Geology Site history Property boundaries Engineering constraints Physical conditions Regulatory issues Temporal factors Spatial continuity
Quantification of uncertainty:	*Visualization:*
Representativeness of sampling Global precision Spatial trends in variability Local uncertainty Local misclassification Mitigate effects of errors Improvement on classical statistics Select appropriate computer interpolation method (kriging, etc.) Probability of missing hotspots	Contour maps Mesh maps Block maps 3-D representations Wide audience ▪ Regulators ▪ Public interest groups ▪ PRPs ▪ Consultants ▪ Attorneys ▪ Judges/Juries ▪ Other stakeholders
Volumetric determination:	*Audit trail:*
Classification: clean vs. dirty Volumes: clean vs. dirty Level of effort determination Volume-variance influence Scheduling issues	DQO documentation Sampling plan Computer processing Graphics generation Link to legal admissibility

effort may be quite minimal. For large or complicated sites, the GEM process may be extensive. The final result of a complete GEM process is a series of maps displaying the extents and levels of the site contamination. Important and potentially costly decisions result from the interpretation of maps; a good initial "road map" will help ensure a high-quality final physical map that will enable precise decision making and reduce decision errors.

1.2 Goals of Geostatistical Error Management

Each GEM component has individual goals that it endeavors to achieve during the assessment process. Simultaneously, the goals of each GEM component contribute to the attainment of overall project goals. The individual component goals will be addressed first, followed by the overall GEM goals.

1.2.1 Component Goals

Component goals are presented below for data quality objectives, sampling theory & practice, and geostatistical appraisal.

1.2.1.1 Data Quality Objectives

An optimal project design needs to balance the uncertainties and their associated costs with the resources available for the project. Data quality objectives provide a valuable planning tool for environmental data collection activities. The major goals of the DQO process reflect EPA's seven-step DQO process and are set forth below:

1. State and clearly define the problem so that the focus of the study will be unambiguous.
2. Identify the decision to be made, and formulate a decision statement that you will attempt to resolve through the study.
3. Identify the informational inputs that will be required to resolve the decision statement, and determine which inputs require environmental measurements.
4. Define the spatial and temporal boundaries that are covered by the decision statement.
5. Develop a decision rule that defines the parameter of interest, specifies the action level and action support, and integrates previous DQO outputs into a single statement that describes a logical basis for choosing among alternative actions.
6. Specify tolerable limits on decision errors.
7. Optimize the design for obtaining data by identifying a resource-effective data collection design for generating data that are expected to satisfy the DQOs.

1.2.1.2 Sampling Theory & Practice

Quality site characterization is based on quality samples. However, it is often extremely difficult to obtain quality and reproducible samples. Sampling theory

& practice addresses the root causes of sampling error. The principal goals are listed below:

1. Define and separate the structural vs. the circumstantial properties of sampling.
2. Analyze the heterogeneities of sample media.
3. Select the appropriate dimensional module of observation.
4. Separate sampling errors into the correct categories.
5. Assess the correctness of sampling devices.
6. Develop sampling protocols to minimize sampling error.
7. Reduce variability between duplicate and replicate samples.

1.2.1.3 Geostatistical Appraisal

Geostatistical appraisal (GA) provides the statistical and visual portion of GEM. Whereas a picture is worth a thousand words, an inaccurate picture can badly mislead. Geostatistical appraisal endeavors to bring not only pictures, but quality and risk-qualified pictures, empowering decision makers to make informed decisions. Here is a listing of GA goals:

1. Obtain an assessment of the statistical nature of the sample data.
2. Acquire an understanding of the spatial relationship among the samples at the site.
3. Select an appropriate estimation method for interpolating the data.
4. Minimize the errors and uncertainty involved in the estimation process.
5. Map the levels and extents of the contaminants of interest at the site.
6. Minimize the cost of decision errors.

1.2.2 Overall GEM Goals

A superior site characterization is built on a balanced and focused environmental data collection program, quality samples, and correct visualization of data. Listed below are the overall and combined goals of the GEM process:

1. Integrate available assessment tools, putting special emphasis on techniques that minimize the uncertainties and eliminate errors.
2. Develop a defensible and logical decision and implementation plan for the site.
3. Generate a plan that will maximize the utility of the resources of time and money that will be expended during the project.
4. Provide goal-oriented checkpoints for decision making.
5. Minimize costs and liabilities on a projectwide scale.

1.3 Relationship of GEM to Total Quality Management

Many parallels exist between GEM and the more widely known concept of total quality management (TQM) developed and refined by Dr. W. Edwards Deming. The TQM doctrine emphasizes a systematic approach to productivity improvement using objective methods and all employees to continuously improve the quality of all products and services produced. The fundamental concepts of TQM are:

- Do the right things.
- Do things right the first time.
- Focus on the process, not just the results.
- Take a long-term focus.
- Quality cannot be inspected in.
- Strive for continuous improvement.
- Satisfy the customer.

From the GEM perspective, the most important, relevant, and applicable principle is that quality cannot be inspected in. This is the TQM principle of *quality by design (QBD)*. QBD applies to GEM in that it is not possible to inspect quality into a sampling or mapping program. Quality results from a quality design and implementation, based on a thorough understanding of the problem and related issues.

Each GEM component provides TQM analogues. DQOs stress the goals and planning aspect. Sampling theory and practice assures that quality data will be obtained, circumventing the numerous pitfalls awaiting the sampler. Geostatistical analysis provides quality interpolation and mapping so that appropriate decisions can be made.

The EPA itself embraced a TQM type of philosophy when it developed its data quality objectives (DQO) guidance document. To EPA's credit, this was part of an initial endeavor to build quality into sampling designs. Although TQM is not specifically mentioned in the DQO guidance documents, other publications by EPA personnel and others have discussed the objective of total quality as related to the Superfund, RCRA, and other environmental authorities.

1.4 The Role of Sampling and Data Interpretation in Site Assessment

Site characterization is a crucial step in the environmental cleanup process. It must be performed carefully so that subsequent remedial decisions can be made using both accurate and pertinent information. Decisions regarding which (if any) portions of the site will undergo remedial activities and the particular remedies to be used will be made. Remediation has a great price in time and money; the proper use of these resources should always be a primary goal in any project.

Sampling plays a critical role in site assessment. It is used to determine the presence or absence of hazardous or toxic chemicals. If harmful chemicals are found at the site, an appropriate risk assessment model is typically applied. An acceptable risk level, for example 10^{-6}, is converted to a tolerable dose standard. The dose standard is then used to derive the allowable concentration at the site. Finally, appropriate numbers of samples are taken to characterize the contamination at the site (Fig. 1-3).

Quality sampling, analysis, and data interpretation are essential to ensure that good decisions are possible. Decisions are easier to make when a high degree of certainty is present regarding the possible outcomes. Unfortunately, uncertainties abound in environmental sampling and assessment. Therefore, tools must be employed to minimize the uncertainties involved and maximize the confidence in decision making.

With remedial costs ranging from millions to hundreds of millions of dollars (and, in a few cases, billions of dollars), economic penalties are high for poorly conceived remedial plans and remedies. Lack of quality sampling and analysis may alter remedial decisions and planning so as to impact negatively both the economics and the quality of the site cleanup. At many sites, this has been an unfortunate but recurring theme.

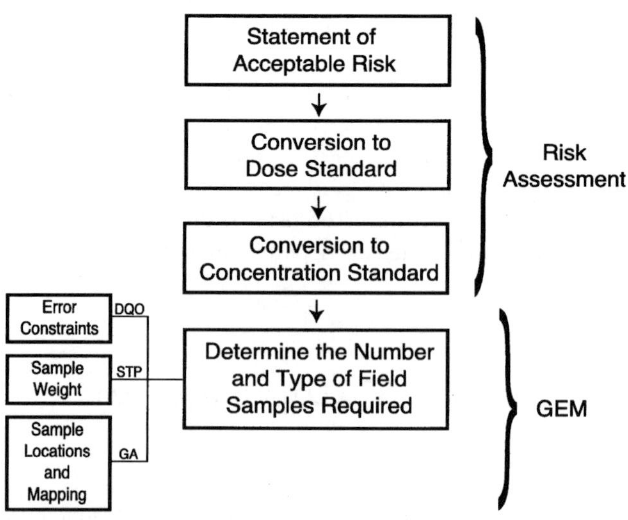

Figure 1-3. Relationship between environmental risk assessment and the GEM process.

Along with quality sampling, sufficient sampling is also necessary. For mapping tasks, this can often require a significant number of samples. Through sampling, one hopes to outline accurately the extent of the contamination. Figure 1-4 provides some insight into the potential magnitude of this task.

Starting with a few samples, two distinguishable areas appear. These are land and sea or, analogously, "clean" and "dirty" for environmental applications. As more samples are taken, undiscovered land appears with an increasingly complex boundary as the number of samples continues to rise. Even with 300 samples, the exact boundaries cannot be determined, opening the door for decision making errors.

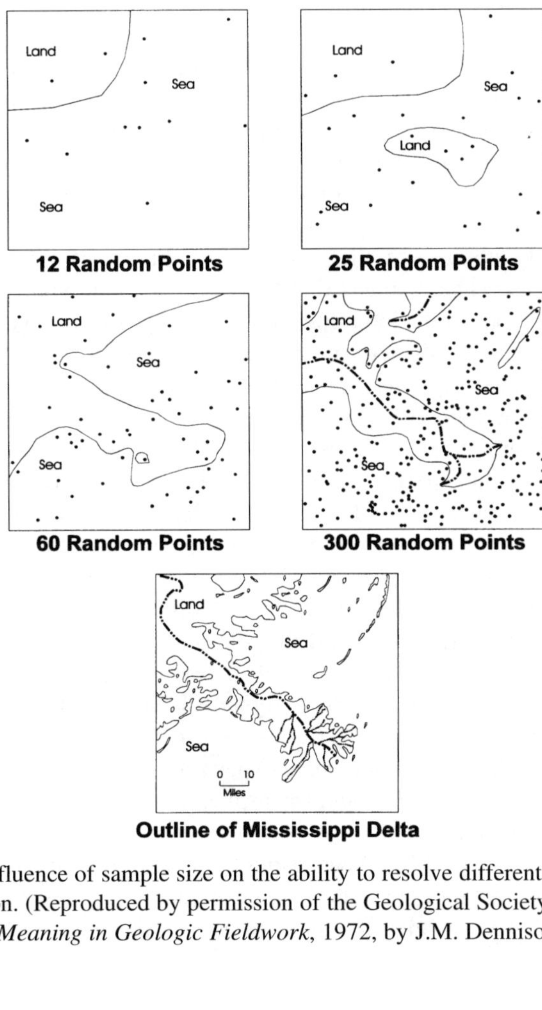

Figure 1-4. Influence of sample size on the ability to resolve different areas of a spatial population. (Reproduced by permission of the Geological Society of America from *Statistical Meaning in Geologic Fieldwork*, 1972, by J.M. Dennison.)

The reverse can also be true, however. An old sampling adage states:

Samples are like potato chips.
You're never satisfied with just one.
Every one you take makes you want more.
You're never sure you've had enough until you've had too many.

Part of the role of GEM is to show that the last statement can be avoided.

1.5 Legal Implications

Environmental cleanup activities commonly involve litigation. During litigation, the sampling and subsequent characterization of the site is exposed to intense scrutiny from government, industry, and public interest groups. The objective of this scrutiny is to discredit the sampling performed at the site. Discrediting the sampling may reduce liability for some parties which, in turn, may reduce liability and cleanup costs.

Unfortunately, not all sampling and analysis programs are designed to meet such intense review. The result is that these sampling programs are not legally defensible. In fact, one of the intentions of the GEM process is to demonstrate how effortless it would be to discredit many sampling programs. On a more positive and proactive note, GEM also presents techniques that allow one to design defensible sampling plans from the start.

It is also important to distinguish between quality scientific data and quality legal data; a good sample to the scientist may not be a good piece of evidence to an attorney. Table 1-2 shows a list of common attributes required for scientific validity. On the left are attributes dealing with quality parameters, laboratory procedures, and statistics. On the right are attributes required for legal defensibility.

While related, the attributes in column 1 differ substantially from those in column 2. Consequently, it is critical that the scientist understand the needs and requirements of the attorney, and vice versa. Good communication is needed on both sides so that a sample can be designed to meet both scientific and legal scrutiny. The arrow between the two sides represents the narrow bridge that links the two worlds. Both scientific and legal experts must work together to ensure that data may cross this bridge. Without proper planning, good scientific data may not be usable in legal venues.

The discussion of GEM will be restricted to the development of high-quality, scientifically valid data. Legal issues such as relevance, foundation, authenticity, and rules of evidence for scientific data are beyond the scope of this text.

Table 1-2. Elements of scientific validity contrasted with elements of legal admissibility.

Scientific Validity	Legal Defensibility
Media Collection	*Admissible Evidence*
Precision	Admissibility
Accuracy	Evidence
Linearity	Authenticity
Representativeness	Relevance
Completeness	Integrity
Comparability	Documentation
Matrix spikes	Chain of custody
Duplicates	Frye test
Selectivity	Rules of evidence
Limit of detection	Due process
Sample	Split sample
Reliability	
QA/QC	
Heterogeneities	
Sample selection mode	

1.6 GEM Components: A Detailed Introduction

As seen in Figure 1-1, GEM is composed of three basic and sequential components. Each component brings a special set of insights and tools to the process. An introductory discussion of each component is set forth below.

1.6.1 Data Quality Objectives

Data quality objectives (DQOs) were developed by the United States Environmental Protection Agency (USEPA) and published by the Office of Emergency Response and the Office of Waste Programs Enforcement. The first DQO guidance document was issued in March 1987 and represented an Agency effort of approximately three and a half years. In May 1984, the deputy administrator of the EPA requested that the assistant administrators provide active participation in the development of DQOs for site sampling and assessment. Updated and revised DQO documents were released by EPA's Quality Assurance Management Staff (QAMS) in September 1994.

The DQO process brings a needed focus to the site investigation and characterization effort. The DQO process, as outlined by the guidance document, is intended to be applicable to all data collection activities for site characterization. This includes, but is not limited to, activities under the Comprehensive Environmental Resource Conservation and Liability Act (CERCLA) of 1980, the Superfund Amendments and Reauthorization Act of 1986 (SARA), the Resource Conservation and Recovery Act (RCRA), and others. DQO applicable data collection activities run the gamut of environmental studies. Examples are a preliminary assessment/site investigation (PA/SI), a remedial investigation (RI), a feasibility study (FS), a RCRA facility investigation (RFI), a remedial design (RD), and a remedial action (RA).

The DQO process encompasses several functions, as follows:

1. Assesses any existing data for the site
2. Identifies the types of necessary decisions
3. Specifies the data uses and needs as well as data users and types of data
4. Develops a conceptual model for the site
5. Delineates specific objectives and goals
6. Analyzes the need for additional data
7. Sets specific goals for data quality, maps, and other graphical products developed from the site data

The goal is to have authoritative and accurate assessment of the operable units at the site, a set of sampling protocols that can procure accurate and defensible data, and a data analysis plan that can meet specified standards of performance during the mapping process.

Whereas the initial intent for DQOs was to provide a general framework that can be implemented during all cleanup phases of a Superfund site cleanup, DQOs can also be applied to sites that are not listed on the National Priorities List (NPL). Although these sites are not subject to the entire Superfund process, many aspects of the data planning and collection are often identical and should be conducted with equal care.

The seven-step DQO process addresses some issues related to managing uncertainty in decision making. Despite this, critics have charged that insufficient attention has been given to assuring that the statistical levels of error are properly considered in other phases of the process, including: (1) the design of sampling strategies, (2) the consideration of sample size (known as "support"), and (3) the use of statistical goals in data processing and mapping of sample data collected. Not coincidentally, these three issues are fundamental to the other GEM components.

For example, the 1987 DQO document recommended the use of PARCC parameters (precision, accuracy, representativeness, completeness, and comparability) to obtain quality data. However, PARCC parameters ignored many of the basic sampling errors addressed in STP. PARCC parameters applied a symptomatic approach rather than basic rules of sampling theory & practice.

Another notable inadequacy is that the PARCC approach addresses only one of the seven basic errors inherent in STP. It is impossible to obtain useful PARCC parameters if all aspects of sampling design are not incorporated. By understanding the fundamental problems involved in sampling and incorporating sampling protocols and devices to minimize these errors, acceptable quality data can be obtained. The 1994 DQO document no longer mentions PARCC parameters; rather, EPA has written a guidance document that incorporates STP techniques (USEPA 1992).

Whereas DQOs occur as a first step for regulators, potentially responsible parties (PRPs), and other stakeholders in the site characterization, they are presented last in this book. The reason is that a thorough understanding of STP and GA is beneficial as a prerequisite to establishing meaningful and defensible DQOs.

As an analogy, one can consider an architect who wishes to design and construct a house. His first step will be to design the structure. In order to do this, he must first have the knowledge of the materials and tools that he will use as part of the design. These include the types of wood, bricks, nails, windows, carpets, and so forth, along with the strengths and weaknesses of the materials, both in a structural and aesthetic sense. Once the design and material have been selected, a construction schedule to assemble the material components will be implemented. This will take into consideration such critical path components as assembly order, manpower availability, and weather. If both the design and construction phases are carefully executed, a solid and dependable structure will result.

In the GEM process, the regulators, PRPs, and other stakeholders perform the task of the architect in designing the site assessment plan. Once again, a thorough knowledge of the materials and tools available for the task will be essential. A primary "material" available for site assessment is the sample. STP brings quality sample material to the project; geostatistical methods provide a type of "tool" for using the acquired sample materials.

DQOs force all parties to take an in-depth look at the problems, goals, and objectives for the site characterization. It is important to keep in mind that often many stakeholders are involved in a remedial process, including those who are neither PRPs nor regulatory agencies. DQOs and the GEM process can help structure site characterization so that needs of multiple stakeholders are met.

1.6.2 Sampling Theory & Practice

The second component of the GEM process consists of formalized *sampling theory & practice (STP)*. This relatively new science was developed by Pierre Gy in the 1960s and 1970s to assist in the search and extraction of mineral resources. The objective of STP is to obtain sample data that reflect as accurately as possible the true in-situ contaminant concentrations. Anyone who has been involved with sampling and analysis programs recognizes that this is a formidable task. Until the development of STP, it was not possible to identify the sources and causes of the inevitable sampling errors.

An important issue in STP is sampling *correctness*. Because the requirements to achieve sampling correctness are not widely known, major sampling errors often occur. STP will discuss sampling correctness in detail, along with the repercussions of incorrect sampling.

STP will also demonstrate the dramatic difference between sampling and analytical errors. Samples errors produce sample data that are often poor approximations of the existing contaminant problem. These inferior data are then mapped or otherwise processed in order to make important remedial decisions. Most maps are based on the assumption that the data used are reliable. This is often not the case, so that one is simply mapping an illusion provided by the available but inaccurate data values.

Sampling has connotations far beyond simply acquiring a volume of material that is carefully prepared and analyzed. The inherent nature of the material to be sampled, the sampling devices selected or designed to procure the sample, the manner in which the sample is collected, and the subsampling stages at the laboratory are all critical to sampling success. These components are required to achieve sampling correctness. Besides misrepresenting contaminant concentrations, incorrect samples can lead to other problems.

One frustration that commonly arises during site assessment work involves duplicate or split samples. If a sample is split or if two samples are taken in what is considered to be the same location, each sample is analyzed separately. The two analyses may be done by the same laboratory or by separate laboratories, depending on the objectives. Experience has shown that the analytical concentrations will differ, often dramatically.

The discrepancy observed between values often triggers a series of data validation inspections. The chain of custody is examined, the preparation methods are evaluated, the analytical procedure and the operator influence are examined. This scrutiny may not disclose any problems. The problem may well have started before the sample was even taken. Thus, as TQM warns, the inspection process is partly just an academic exercise. STP will show how to build quality in, minimize the sampling errors that cause lab discrepancies, and focus inspection toward appropriate areas.

Sampling theory incorporates a logical approach to the problem of obtaining quality samples and respects the QBD philosophy. The first step is to recognize that errors are an inherent part of the sampling process. The reality is stark: Errors cannot be eliminated, only minimized. Not coincidentally, the acceptance of estimation error is one of the key steps in the geostatistical portion of GEM where, once again, errors can be minimized but not eradicated. Second, it is necessary to identify all possible causes of sampling errors. It will be shown that this does not mean assigning blame by vaguely accusing the common demons of sample matrix irregularities, heterogeneities, collection errors, bias, or accuracy. Sampling errors (1) are identifiable, (2) have causes, and (3) are separate and distinct from analytical errors. This critical distinction will be explained and emphasized in Chapters 7 and 8.

Sampling theory & practice has identified and organized errors in a systematic manner. Figure 1-5 shows the seven basic sampling errors. These are:

1. The Fundamental Error (FE)
2. The Grouping and Segregation Error (GE)
3. The Long-Range Heterogeneity Fluctuation Error (CE2)
4. The Periodic Heterogeneity Fluctuation Error (CE3),
5. The Increment Delimitation Error (DE)

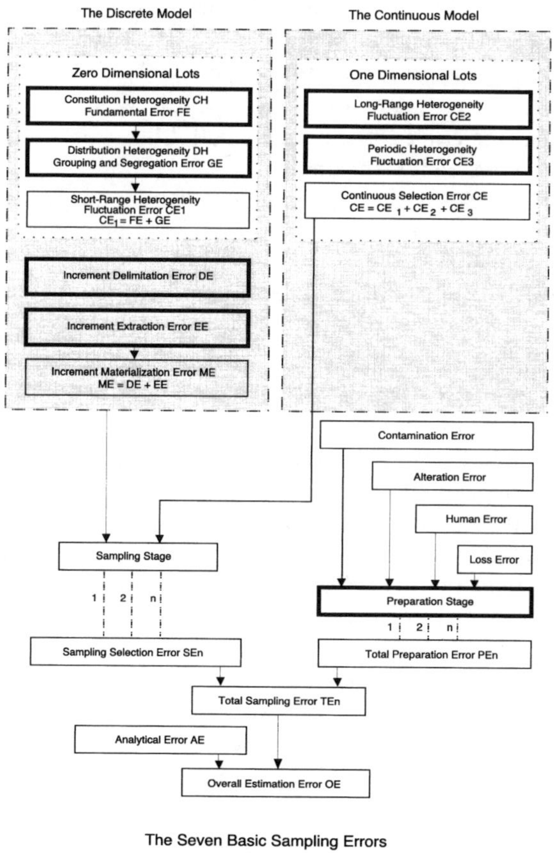

Figure 1-5. The seven basic sampling errors. (From *Pierre Gy's Sampling Theory and Sampling Practice: Heterogeneity, Sampling Correctness, and Statistical Process Control*, 2nd ed., by F.F. Pitard. Copyright © 1993 CRC Press, Boca Raton, FL. Courtesy of Francis Pitard.)

6. The Increment Extraction Error (EE)
7. The Preparation Error (PE)

Many of these errors may be unfamiliar to the reader. Details of each type of error will be presented in Chapters 7 and 8. In general, they address the nature of the material being sampled, local and regional differences in sample material, and the correct way to collect or "materialize" a sample.

Sampling concepts and objectives will be classified in a new frame of reference based on their nature and properties, either circumstantial or structural. For example, sampling theory will demonstrate that accuracy, an important objective, is only a *circumstantial* property. Circumstantial properties are aspects of a sampling problem over which we have no control. This is in contrast to *sampling correctness,* a *structural property,* over which we do have control.

The respected environmentalist Barry Commoner stated, "The first rule of ecology is that everything is related to everything else." While not meeting the definition of an ecosystem, sampling theory addresses another type of system in which many things are related to the final result. Like ecosystems, these interrelationships may not be readily apparent.

Chaos theory (Gleick 1987) has postulated that seemingly trivial alterations to an ecosystem can have significant and unexpected impacts, typically harmful ones. Research meteorologist Edward Lorenz coined the now well-known term *the Butterfly Effect.* In a dynamic system, such as weather or weather modeling, the almost imperceptible perturbation of a butterfly beating its wings could conceivably lead to a chain of unexpectedly turbulent events, from dust devils and squalls to continent-sized systems. Minuscule changes in initial conditions may indeed cause major effects.

Similarly, sampling theory shows that seemingly harmless assumptions may result in substantial sampling errors. In turn, sampling errors may cause serious decision errors and mapping errors. One simple example will be shown in the parable of the red beads, which follows this section.

Once the available tools of STP and GA have been presented, data quality objectives (DQOs) will provide a general framework for reducing errors and producing quality data. Sampling theory will first detail many of the specifics needed to actually obtain the objectives. Once component errors have been established and identified, mathematical definitions can be developed for such DQO concepts as bias, accuracy, precision, representativeness, and equitable sampling. Using sampling theory and geostatistics, comparability and completeness may also be objectively assessed.

1.6.2.1 The Parable of the Red Beads

To the untrained observer, the process of procuring a sample appears to be a fairly straightforward and simple task. One must first decide what medium is to be sampled, select a sampling device, and then use that device to collect a bit of material from whatever medium is being sampled. Depending on the nature of the sample, the result may be known once the sample is taken (a count or measurement); in other cases, chemical or physical analysis may be required.

Although the physical collection of a sample is often trivial, obtaining a sample that will accurately represent the critical concentrations in the medium being sampled can be extremely difficult. Often, it is falsely assumed that a sample accurately represents the contaminant concentration in the medium. For numerous reasons, the sample may, in fact, be quite different from the material it is intended to represent. It can even be a type of red herring that leads the sampler and data analyzer astray. Sampling theory & practice will provide a practical solution.

Through his work in TQM, Dr. Deming has developed a useful and wonderfully "stupid experiment" as a teaching tool for his introductory seminars. The exercise, while simple and understandable, clearly shows that a "simple" sampling design can have numerous inherent hazards. It will also serve to demonstrate some of the inherent relationships between TQM and STP. Later, STP will present techniques that can eliminate or reduce these hazards.

The experiment begins with (1) a plastic box containing 800 red and 3200 white wooden beads the size of peas; (2) "sampling devices," which are 6 paddles with 50 bead-size holes arranged in 5 rows of 10 each (representing one day's work and with the design carefully specified in a DQO type of plan); (3) a second box large enough for dipping the paddle; and (4) 6 volunteer workers or samplers.

For the TQM demonstration, the "job" of each sampler is to produce white beads. The imaginary customer (or regulator) will not accept red beads since they are considered to be contaminants. Before taking a sample, each sampler must mix the beads using a very rigid and well-defined procedure. The mixing process mimics the analogous operation of homogenization at the laboratory during sample preparation. Once the beads have been randomized, the first sampler dips the paddle into the box containing the beads and allows all the holes in the paddle to become filled.

The number of red beads—that is, the amount of contamination—resulting from the sampling event is then tallied and all the beads are returned to the box. Next, after another homogenization process, the second worker repeats the sampling process and a new tally of red beads is obtained. This process continues until all 6 workers have taken 4 samples each for a total of 24 samples.

The results of the samples taken by each worker appear in Table 1-3. The number of red beads obtained by each sampler for each sampling event is listed along with the event and worker averages. The data indicate that a significant amount of variability occurred during the sampling events. In three samples, a low count of 5 beads was obtained during the sampling event. This compares with a single value of 17 as the highest number of red beads obtained. Overall, a variance of 8.5, with a corresponding standard deviation of 2.9, was obtained.

Table 1-3. Sample data and statistics for Deming's red bead experiment.

	Day					
Sampler	**1**	**2**	**3**	**4**	**All 4**	**Average**
A	14	10	9	10	43	10.75
B	17	5	8	5	35	8.75
C	11	6	5	9	31	7.75
D	8	8	9	6	31	7.75
E	12	11	12	8	43	10.75
F	9	11	7	10	37	9.50
All 6	71	51	50	48	220	9.17
Average	11.8	8.5	8.3	8.0	9.2	9.17

Note that the overall average of the experiments is equal to 9.2. Since there were 4000 beads of all types (800 red and 3200 white), one might expect that an average value much closer to 10.0 would have resulted. After all, 800 red beads represent 20% of the total population and the beads were mixed to homogenize the distribution before each experiment. An average of 10 red beads would be 20% of the 50 total beads collected by the paddle during each sampling event. Could it be that an insufficient number of sampling events was performed?

Most statistical textbooks state that 30 samples are sufficient to appropriately estimate the average of a population. Since 24 is 80% of 30, the sample size of 24 already taken is not far from the magic number of 30. Also, consider that, if the remaining 6 experiments were actually performed, it would require (on average) that approximately 13.3 beads be obtained during each sampling event in order to achieve an average of 10.0.

Note that, in the first 24 experiments, only two events (8.5%) came up with a tally of red beads of 13 or greater! To expect the next 6 samples to be greater than 13 appears to be quite optimistic.

It is also interesting to note the average daily red bead totals of all samplers. The first day's average of 11.8 is far higher than both the overall average and the other three daily averages. In fact, the other three vary in total by only 0.5. Was there some error in the first day's sampling?

With regard to the averages for individual workers over four days, two had averages of 10.75 (relatively far from the overall average of 9.2), two had averages of 7.75 (relatively low), and two had averages relatively close to the overall average. Were these workers somehow biasing the data high, low, or toward the average?

Reflecting on these results, one wonders how such a simple experiment (or perhaps stupid experiment!) terminated with such enigmatic results. How could a homogeneous collection of red and white beads produce a high count more than three times the size of the low count? How did workers with identical tools, tasks, and talents come up with such surprising data? The questions raised by these results are not uncommon for sampling programs, whether simple or complex. In fact, some sampling programs raise more questions than they answer. First, some additional attention will be given to the problem. Then, STP will provide some insight into the assumptions of the experiment.

One logical next step would be to ask if any of the extreme data (5 or 17) exceeded the expected limits of variability. To calculate the average proportion of red beads drawn each day, *p*-bar ($\overline{P}$), it is necessary to determine the ratio of red beads in the count to the total number of beads drawn. The result is shown in equation 1.1 below.

$$\overline{P} = \frac{220}{(6)(4)(50)} = 0.18 \tag{1.1}$$

Given this, one can compute the upper control limits (UCL) and lower control limits (LCL) by means of the following formula:

$$\text{UCL / LCL} = \overline{x} \pm 3\sqrt{\overline{x}(1 - \overline{p})} \tag{1.2}$$

or

$$\text{UCL / LCL} = 9.2 \pm 3\sqrt{(9.2)(0.82)} \tag{1.3}$$

Figure 1-6 shows the plot of the number of red beads and the individual experiment, along with the upper and lower control limits. This figure shows that no experiment exceeded the control limits although one came very close, indicating that a moderately good state of statistical control exists for the experiment. In other words, no samples were out of bounds, nor was any real pattern observed that might indicate bias or other problems. For example, a run of perhaps seven or eight points in a row either above or below the average did not materialize during the experiment.

The stability of the system is denoted by the fact that none of the data were outside the control limits. This stability indicates that one might reasonably predict future limits of variation, at least in the short term. Given that only 24 data

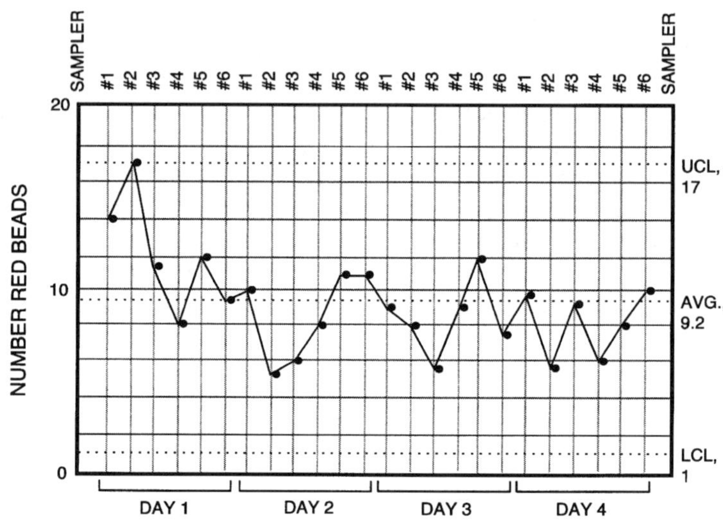

Figure 1-6. Control limits: red bead experiment. (Reprinted by permission of The Putnam Publishing Group from *The Deming Management Method* by Mary Walton. Copyright © 1986 by Mary Walton.)

were observed, it would be expected that, at some point, a datum would be outside the control limits.

If more data were taken, they could be combined with existing data to compute new limits for the future. New data, however, would most likely reflect the basic condition currently observed. As for the future, no one really knows because empirical evidence is always incomplete. This means that barring any change in the system that allows red beads to be captured, the number of red beads will fluctuate between the upper and lower control limits but not will exceed them. So, although initially the variability seemed a bit high, it actually was within the expected realm.

This, however, still does not explain why the observed average value did not match the anticipated average value. The reason is more subtle: More factors are involved in the system than just red and white beads. As it turns out, the red beads are coated with an added layer of pigment. This structural difference makes them slightly bigger and heavier. As such they "feel" different to the paddle and are more difficult to pick up. This accounts for the average of 9.2 instead of 10.0. This difference will also affect the homogenization process.

In addition, the paddles themselves are quite important. Tracking paddle experiments over the course of years, Deming learned that individual paddles have different averages. One gave 11.3 as an average number of red beads over hundreds of experiments. Another paddle yielded 9.6, another 9.4. Thus, there is a paddle bias present in the sampling process, specifically, a bias due to the sampling device.

The red bead experiment has identified two important concepts of STP. The first is the illusion of homogeneity. Homogeneity is, in fact, a myth, and the study of its counterpart, heterogeneity, constitutes a fundamental part of formalized sampling theory. The second is the importance of the sampling device and how biases can be introduced unintentionally.

In summary, even a simple sampling design can have complications. These are harbingers of the problems and assumptions yet to come. A few of the components of sampling theory and practice have been touched on here; a more detailed and complete description will be set forth in Chapters 7 and 8.

1.6.3 Geostatistical Appraisal

The final component of the GEM process is the geostatistical appraisal (GA) of the acquired sample data. At this point in the GEM process, it is expected that quality and representative data have been obtained and chemically analyzed using acceptable methods. As will be demonstrated in the chapters on geostatistics, having quality data at this point does not assure that the most accurate maps will be produced.

Geostatistical techniques have been used extensively for analyzing and mapping many types of phenomena. The approach was originally developed for the appraisal of ore reserves in the mining industry. In mining, samples are taken in or around a volume of material that is scheduled for extraction. Based on the acquired sample data, the concentration of metal in the block to be mined is estimated. If the estimated metal concentration is above an economic cutoff, the block volume is treated as ore; if below, it is treated as waste material.

Once mined, the block contents are assayed. This task is performed at the mill or processing facility to determine the "true" concentration of the ore. This gives ore estimators a unique opportunity: the chance to conduct a reality check on the estimated value. The estimated concentration is compared to the "true" concentration to assess the accuracy of the estimate. Geostatistical estimators have proved reliable in estimating true concentrations of economic commodities in the extractive industries.

Geostatistics has proved to be a tool with wide utility. Applications of the geostatistical estimation process used to date include determining the concentrations of hazardous compounds in soils and groundwater, rainfall, isopach thicknesses of geologic units, quantities of aquatic life, forestry, fishery and agricultural yields, soil compaction and stability parameters, mapping rock fractures, and demonstrating the effects of acid rain.

1.6.3.1 Introduction to Regionalized Variables

At this point it is necessary to make the distinction between GEM techniques and geostatistical appraisal techniques. GEM refers to a broad range of "geo"-statistical (geology or earth science related) techniques, including sampling and analytical methods and data quality planning, as well as "geostatistical" (regionalized variable) techniques. *Geostatistical* methods refer specifically to the application of the theory of regionalized variables (RVs).

Regionalized variables (RVs) is a new class of statistical analysis that differs significantly from classical statistical techniques. The fundamental assumption of independent data made in classical statistics is no longer valid in the realm of regionalized variables. Before the development of the theory of RVs, this was an obstacle to be overcome because all existing theory and practice was strictly geared for situations where data were independent. Mastery of the spatial correlation component of regionalized data empowered geostatistical techniques with abilities far beyond those available in the classical realm. This is not to say that classical techniques are no longer of use; they remain valuable for the initial assessment of data structures.

Statistical theory has a mathematical definition of independence based on probability theory. Until more detailed statistical theory is introduced, it is more important for the reader to have an intuitive understanding of the concept of independence. For now, a rudimentary working definition will be offered that can be expanded later. Simply put, two data points can be considered independent if knowing the value of one does not provide any information regarding the value of the second.

Independence can be easily demonstrated using an example familiar to most people, that of tossing a coin. Imagine that a coin is tossed and the result is heads. Knowing this result does not assist you in determining the outcome of the next toss of the coin, assuming that the coin is unbiased.

If a coin has been tossed 10 times and 10 heads have resulted, your intuition might mistakenly project that the next toss will almost certainly be tails. After all, the chance of tossing 10 heads in a row is less than 0.001 or less than 1 in 1000. The reality, however, is that each toss of the coin is a separate and independent event. That the previous 10 results were heads does nothing to imply that a tails outcome is imminent.

Despite the recent history of 10 consecutive heads, there is still a 50/50 chance that the eleventh toss will be heads. Many have been tricked by this concept, which has also earned the moniker "the gambler's ruin." It may be easy to explain, but it is also easy to see how one can fall into the logic trap.

The assumption and condition of independence is mandatory for the use of classical statistical methods. Statistical approaches, including STP, often rely on the additivity of variances for different error components. Problems occur when data are not independent and when unrecognized covariances sneak into the picture, adding a complicating correlation component.

Earth scientists have long realized that an entirely different set of independence conditions apply to geological data. Correlation between samples, a state of nonindependence, exists as the most frequent condition. The theory of regionalized variables and geostatistical techniques have quantified the relationships and made it possible to account for the presence of correlation. Advances in computer technology have accelerated the use of RVs.

If a field measurement or sample is taken, the value obtained may provide some information about the value of a sample taken nearby. For example, the thickness of an aquifer is a fairly consistent phenomenon. If a measurement made at a particular location yields a thickness of 50 ft, then another measurement made a few feet away would not be expected to vary by more than a relatively small amount, perhaps a few inches or a foot.

Similarly, if an arsenic sample taken near a smelter yields a relatively high concentration, a high arsenic concentration would be expected at other nearby locations. Conversely, if a sample at a greater distance from the smelter displayed a relatively low concentration of arsenic, other nearby samples would also be expected to exhibit low concentrations.

A simple but intriguing example can assist in illustrating the idea of independence and correlation. It will also demonstrate the inability of classical statistics to distinguish between independent and correlated data. The two data sets presented below will serve as the independent and correlated data set respectively (Royle 1979). For the example, it will be assumed that the sets of numbers represent sample concentrations taken along a given direction.

Sample data set #1 1 7 3 6 2 9 4 8 5

Sample data set #2 1 3 5 7 9 8 6 4 2

From a classical statistics point of view, the data sets are identical. The samples have the same average value, 5.0, and the same variance, 6.67. In a spatial sense, however, the first set can be described as highly erratic. Returning to the definition of independence, knowing the value at any location does not provide insight as to the next or previous value.

The second data set can be described as much more uniform spatially. Knowing the value at any location offers information as to the values nearby. This information is not available from the statistical analysis of the mean and variance. Therefore, one looks to the concept of regionalization to distinguish between these similar yet unequal quantities.

Similarity between nearby samples implies the concept of a neighborhood. In geologic terms, a neighborhood is a localized area within which conditions remain relatively constant. A sample is often assumed to represent or influence some area beyond the actual sample location. It becomes less reliable as a predictor of unsampled points or areas as the distance from the sample location increases. This concept can be seen in the two data sets just presented.

One can even extend the analogy to a real-life neighborhood. Knowing your own household income, you may be able to estimate the income levels of your neighbors. Trying to estimate the income in a more upscale neighborhood a mile away may be more difficult.

In the aquifer and arsenic examples, knowing the outcome of a sampling event at one location provided information about the concentrations at other nearby but unsampled locations. In the case of the coin toss, knowing one outcome gave no information regarding the next one. Clearly, two different situations exist: Coin tosses are independent whereas, in the earth sciences, data are correlated. Correlation is very typical for spatial data that exist in a "region," hence the term *regionalized variables*.

1.6.3.2 Introduction to Geostatistical Tools

Although the presence of correlation was long known, no way existed to quantify the amount of spatial correlation until the development of geostatistical methods, most specifically the variogram. The *variogram* determines the relationship between the distance separating nearby samples and the amount of correlation present. Through the process of variogram analysis, one-, two-, or three-dimensional spatial correlation structures of the variable of interest can be identified and quantified.

The variogram, departing from classical statistics, demonstrates that all samples are not equal for estimation purposes. Rather, the usefulness of a sample for prediction purposes is related to its spatial location. It will be shown mathematically that samples close to an unsampled location being estimated are better estimators than samples farther away.

This relationship can be shown graphically. Figure 1-7 shows an ideal

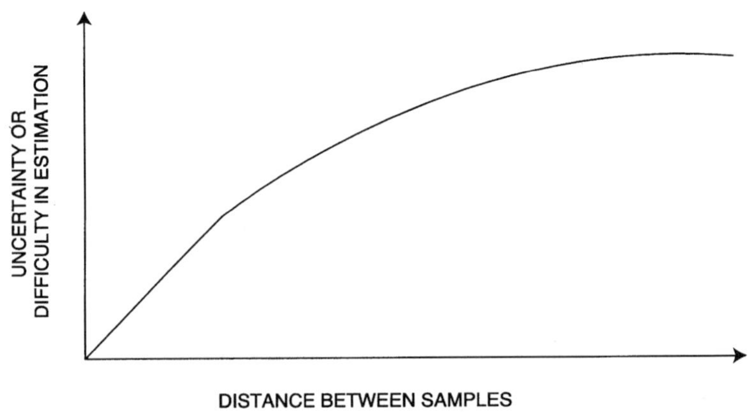

Figure 1-7. Idealized variogram.

variogram graph. The x axis represents the distance between sample points; the y axis shows the variance, which is related to the uncertainty or difficulty in estimating unsampled locations. The graph shows that, as the distance from known sample locations to unsampled points increases, the uncertainty or difficulty in estimating these unsampled locations also increases. At a certain distance, the difficulty reaches a maximum and remains constant.

Whereas variograms provide the assessment of the spatial correlation structure present, the technique of *kriging* provides the machinery that enables the geostatistics to use more fully the information derived from the variogram. Kriging, named for South African mining engineer Dannie Krige, uses sample values at known locations to estimate the value of unsampled points, areas, or volumes. Kriging assesses the usefulness of the information provided by each sample and derives a weighted estimate of the unknown quantity. When this technique is applied systematically to estimate points on a regular grid, contour and other maps may be produced.

Figure 1-8 illustrates one way in which kriging is commonly used. A square block is to be estimated using the three available sample data. Both intuition and the concept of regionalization suggest that the samples closer to the block should influence the estimated concentration to a greater extent than the more distant samples. In other words, the closer samples should receive more weight during the estimation process. Using the information provided in the variogram, kriging does assign more weight to the closer samples.

The estimate of the unsampled location produced by kriging has been designed to minimize the error associated with the estimate. Error minimization is a very desirable attribute and a fairly common technique in the statistical world.

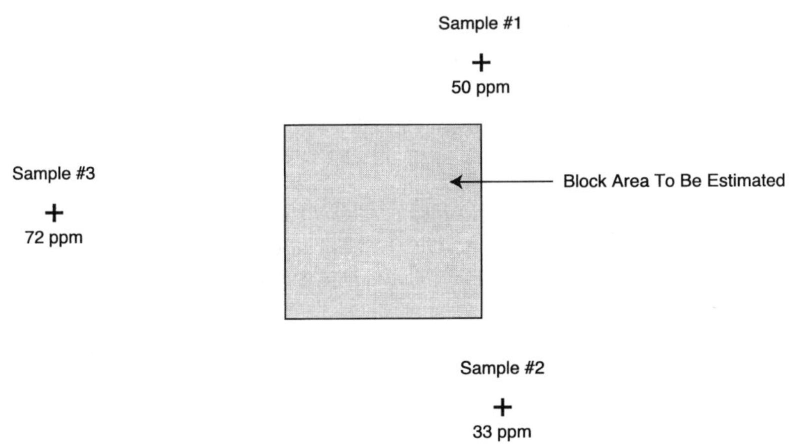

Figure 1-8. Example of kriging using three sample locations to estimate the concentration of a two-dimensional square block.

This puts kriging in the class of "best" estimators. The situation is complicated, however, by the existence of many types of kriging. Since all types of kriging qualify as "best" estimators, determining which type of kriging is the "best of the best" is not a trivial problem. It is one of the functions of GEM to provide methods that enable the scientist to chose the best of the best.

The end products of a complete GEM process (DQOs, STP, and GA) are maps on which important decisions will be made. One of the fundamental tenets of TQM is that quality cannot be inspected into a product, whether that product is a widget, a service, or a map. Quality must be built in by means of a process that removes obstacles to quality. Without the necessary underpinnings and quality foundation provided by the GEM process, the expected result will be maps and decisions of less than acceptable quality.

CONCLUSION

Chapter 1 has set forth the three major components of the GEM process: data quality objectives (DQOs), sampling theory and practice (STP), and geostatistical appraisal (GA). The goals of each component have been stated, as well as an integrated view of the goals from individual components. This chapter has shown that, whereas site characterization is highly dependent on sampling processes, three objectives must be achieved:

1. The sampling must be well planned to obtain the proper types and amounts of data through the DQO process.
2. The sampling must be performed in a correct manner so that quality and representative data are acquired (STP process).
3. The implications of the sample data must be distilled by statistical and mapping procedures (geostatistical appraisal) so that quality and informed decisions can be made for subsequent remedial activities.

The three components of GEM can be used effectively to produce a complete, integrated, and defensible site characterization plan.

Chapter 2

GEM Perspectives

SYNOPSIS: While GEM is an integration of three significant areas of study, it remains only one part of the total picture in environmental assessment and cleanup, as charted in Figure 1-1. It is helpful to understand where GEM fits into this larger picture. This introductory chapter is nontechnical and is intended to provide a broad view of the issues and concepts surrounding the GEM process. Discussion includes a brief history of the environmental movement in the United States; a regulatory perspective; insights into maps including their functions and problems; and the relationship of GEM to Chaos. This chapter will incorporate insights from all three GEM fields, plus a few others, using analogies to help the reader visualize the interconnections.

KEY WORDS and CONCEPTS: Environmental Impact; National Priorities List; Hazardous Ranking System; Comprehensive Environmental Response, Compensation, and Liability Act; Environmental Regulations; Visual Data Display; Mapping Errors; Decision Making; Kriging; Best Linear Unbiased Estimator; Interpolator; Public Domain Software; Chaos.

2.1 A Planet Enters Middle Age

In the last century, man has made a tremendous impact on the environment as a result of an accelerating industrial and technological base. A planet that had easily accommodated small populations impacting few resources was somehow assumed to be able to accommodate large populations with ever-increasing per capita demands. The inevitable result of this acceleration in technology and population is that man has done considerable localized and perhaps global damage to the environment.

Accelerated consumption has also tremendously increased the volume of waste created in the production of consumer, industrial, military, and other modern-day items. Realistic management of these wastes began only within the last generation, with an organized effort limited to the last 20 years. As a result, most countries have many locations that have suffered significant environmental damage. The United States is no exception and is among the world "leaders" in the number of hazardous waste sites.

The United States has placed over 1200 environmentally damaged sites on the National Priorities List (NPL). NPL sites are the "worst of the worst," having qualified for listing by scoring sufficiently high on EPA's Hazardous Ranking System (HRS) to be considered a significant risk to human health. Other equally damaged but less highly ranked sites abound and add to the total environmental cleanup required. Also of importance are active sites under the authority of the Resource Conservation and Recovery Act (RCRA).

Whether listed or not, active or inactive, each hazardous waste site is assumed to pose a risk to the surrounding populations and, consequently, must be cleaned to an acceptable level. Determining associated risk levels is a difficult task beyond the scope of this book. However, the methods and techniques described here are used to determine when appropriate cleanup has been achieved.

2.2 A Call for Action

Superfund's foundations were laid in the mid- to late 1970s. This snapshot in time was an environmentally tumultuous period. Widespread activism on the local and national level resulted in a new people's institution known as Earth Day. Sensational stories, such as those concerning the Love Canal site, made national headlines and became environmental icons. A daily barrage of media coverage highlighted polluted water supplies, leaking landfills, midnight dumping, children sick from chemical pollutants, conservation agency personnel taken hostage by irate citizens, and chronic releases of hazardous chemicals into streams, rivers, lakes, and groundwater. Such portrayals suggested that the entire nation was hemorrhaging hazardous waste.

Although emotion often clouded reality, this tidal wave of attention did bring to light many legitimate problems. Most important, it exposed many weaknesses in environmental regulations and the inability of the existing regulatory system to deal with serious environmental problems.

For example, the term *hazardous waste* did not even carry a precise definition. Often, any chemical discharged was assumed to be hazardous, regardless of its actual toxicity. This presented opportunities for sensational media coverage but not for effective cleanup. Another gap in the regulatory system was a mechanism for identifying waste sources or generators of waste. As a result, no way existed to track chemicals from the cradle to the grave. A void existed even in the definition of what an appropriate grave should be and how such a grave should function.

Ill-equipped governmental agencies only exacerbated the problem. Legislators had enacted few laws to provide the authorities necessary to initiate and complete cleanup actions. Even when available, funding mechanisms were inadequate to handle the escalating volume of complaints that bombarded environmental agencies. To further complicate matters, hazardous waste cleanup was a relatively new area for the agencies, and most lacked the technical knowledge required to complete an effective cleanup.

In the glaring light of the media, regulatory agencies were often exposed as impotent, penniless, and devoid of management and technical remedial tools. Citizens claiming damage were demanding immediate redress from a disorganized, underresourced, and embryonic bureaucracy. When they did not get satisfaction, some citizens went so far as to place agency officials under house arrest (as at Love Canal). Some state agencies responded by providing guns to so-called conservation officers.

Given the turmoil that existed in the 1970s, the social and regulatory pendulum began an understandably dramatic swing toward increased environmental protection and cleanup. Today, numerous legislative acts control waste discharges into air, water, and soil and provide the means to clean up the environmental sins of the past.

2.3 Regulatory Overview

The circumstances of the late 1970s sent a powerful message to Congress: The top priority was to stop the hemorrhaging of hazardous pollutants into the environment. Federal agencies and lawmakers reacted by hurriedly drafting rules, definitions, and standards. Attention quickly focused on the easy prey; industry "villains" of corporate America. Although there is no doubt that many corporations had been using inappropriate practices, they were not the only culprits, and many corporations were in compliance with the regulations of the time. Despite this myopic view of the problem, much of the "hemorrhaging" has been halted.

Most companies have become proactive in their efforts to manage environmental obligations, often developing new technologies en route.

In 1980, the United States Congress passed one of the most influential legislative acts in the history of the country. The Comprehensive Environmental Response, Compensation, and Liability Act (CERCLA), better known as Superfund, filled a gaping environmental regulatory void and provided authority to respond to the most seriously contaminated hazardous waste sites throughout the country. A brief review of the circumstances surrounding the birth of Superfund provides insight into the intent of the Congress and the structure of the law that is now a well-known model for hazardous waste cleanup.

The Superfund legislation is arguably the most pervasive, costly, and far-reaching environmental law ever passed. CERCLA liability is structured so that it reaches back in time to impose today's stringent environmental standards on a more ignorant and lackadaisical era. The United States Environmental Protection Agency (USEPA) can seek all sources for the recovery of site remedial costs.

When a site is placed on the NPL, a next step is to determine those responsible for the environmental damage. A search is conducted to discover any past or present owners of the site. Owners may then be declared potentially responsible parties (PRPs). Nonowners, such as small firms that may have sent waste to the site for disposal, may also be named as PRPs. Under the joint and severable liability provisions of CERCLA, each owner or PRP is wholly responsible for the cleanup costs at the site.

Once implicated as a PRP, the wealthier corporations must, in turn, initiate cost recovery suits against other firms to recover the portion of the cleanup costs owed by smaller PRP firms. This adds to the burden placed on the more affluent firms. Corporations are frustrated by frequent penalties for previous actions that, at the time the actions were taken, were in compliance with environmental regulations. EPA can levy penalties even for environmental damage in which a corporation had only minimal involvement or did not contribute in any way.

CERCLA is, in many ways, a harsh law. It has garnered a notorious reputation for treading on such time-honored legal concepts as the prohibition of ex post facto laws, the protection of due process, and the right of everyone to a day in court. The law defies so many precedents that an undesirable acronym has emerged—GUPI, for guilty until proven innocent. CERCLA supersedes even bankruptcy laws, imposing overriding priority on a bankrupt's debt for cleanup costs.

CERCLA applies an unprecedented form of joint, strict, and severable liability, with each PRP individually responsible for *all* the cleanup costs. Limited statutory defenses are available and, in most cases, are impossible to establish. Superfund uses the prohibition against pre-enforcement review to effectively cut off a PRP's ability to challenge the cost effectiveness of the project the PRP is forced to fund. The courts have been almost unanimous in supporting the expansion of EPA's administrative jurisdiction which, many say, comes at the cost of due process.

Despite numerous criticisms, a significant amount of remedial investigation and remedial work has been performed on Superfund and other hazardous waste sites in the United States. Many of the site characterizations, however, have been inadequate or fell short of the desired goals, as evidenced by prolonged site assessments. Often, multiple sampling efforts are required because the sampling goals were poorly defined. As a result, old data often did not contain the information required to make the necessary decisions, and new data were not compatible with the old.

Growing pains are typical of a new industry. This is especially true for an industry testing and challenging the implementation of a new legislative act. Inefficiencies have resulted in significant expenditures of time and money for all types of site remediation.

One criticism of CERCLA and other authorities is that it is process-oriented rather than goal-oriented. GEM provides a goal-oriented basis for site characterization and decision making. Many of the GEM concepts specifically target regulatory and management decision making and fit well into a regulatory framework. While the goals and objectives of the regulations may change over time, GEM is founded on scientific principles that will outlast the changing styles in legislation.

2.4 Use of Visual Tools

The need to communicate effectively escalates to a critical point when statistical or mathematical concepts are part of a courtroom litigation. It would be naive to expect lay people, such as judges and jurors, to rapidly understand and put into context the mathematical findings of a statistical expert. To transmit ideas and results successfully, a reliable method is to use visual tools. If employed appropriately, such tools can stimulate, inform, and educate without confusing those whose technical knowledge is limited. Advice to the statistician might be: When in doubt, draw a picture, and, for the geostatistician: When in doubt, make a map.

Visual representations have proved effective in overcoming difficulties in communication with the general public. The human mind generally comprehends pictures more easily than formulas. We are visual creatures and can quickly integrate significant amounts of even complicated visual information.

Unfortunately, opportunities to express data visually are routinely missed. For example, in monitoring wells, water is checked for a suite of contaminant compounds on a regular basis at many sites. These data are used to determine compliance with regulatory standards, to locate contaminant sources, and to examine trends in the levels of the specified contaminants.

Tables 2-1 and 2-2 display typical data results from groundwater monitoring

Table 2-1. Monitoring well data for a single well taken at six different times.

	Concentrations (µg/l)				
Sampling Date	**PCE**	**TCE**	**DCE**	**VC**	**TCA**
6/4/92	0.0	820.0	0.0	6395.0	0.0
6/5/92	152.0	870.0	6300.0	0.0	0.0
7/7/92	0.0	310.0	6350.0	1785.0	0.0
7/8/92	29.3	791.0	7520.0	20,500.0	0.0
8/13/92	0.0	18.3	268.0	199.0	0.0
9/1/92	25.6	601.0	8030.0	7400.0	0.0

Table 2-2. Monitoring well data for nine wells taken on the same day.

	Concentrations (µg/l)					
Well ID	**Sampling Date**	**PCE**	**DCE**	**TCA**	**DCA**	**TCE**
LC-1	12/9/86	95.0	22.0	9.0	0.0	0.0
LC-2	12/9/86	53.0	13.0	14.0	0.0	0.0
LC-3	12/9/86	32.0	6.0	3.0	0.0	0.0
LC-4	12/9/86	26.0	4.0	9.0	0.0	0.0
LC-5	12/9/86	5.8	1.6	3.3	6.4	10.0
LC-6	12/9/86	8.3	1.5	0.9	7.5	4.6
BBD-1	12/9/86	0.0	0.0	0.0	63.7	41.2
BBD-1	12/9/86	0.0	0.0	0.0	51.8	33.5
BBD-1	12/9/86	0.0	0.0	0.0	10.9	13.0

efforts. Table 2-1 displays the analyses for a single monitoring well over six quarterly time periods. Five contaminants are being tracked: TCE, PCE, TCA, DCA, and vinyl chloride (VC). Table 2-2 lists the analytical results for 9 well locations, at a different site, for similar contaminants of concern (COCs).

The visualization of these data is the pie diagram, Figure 2-1. Each pie diagram displays a quarterly sampling event. The size of each sector of the pie is proportional to the mass of the particular contaminant in comparison to the total mass of the five contaminants. Higher proportions of a particular contaminant result in larger pie segments. The radius of the pie diagram is equal to the logarithm of the total mass of the five contaminant compounds measured. Therefore, the greater the total mass, the larger the pie. From this single diagram, both the contaminant proportions (mass percent) and the total mass can be viewed. Figure 2-1 shows a generally steadily increasing total concentration over time, combined with a dramatic increase in the VC component.

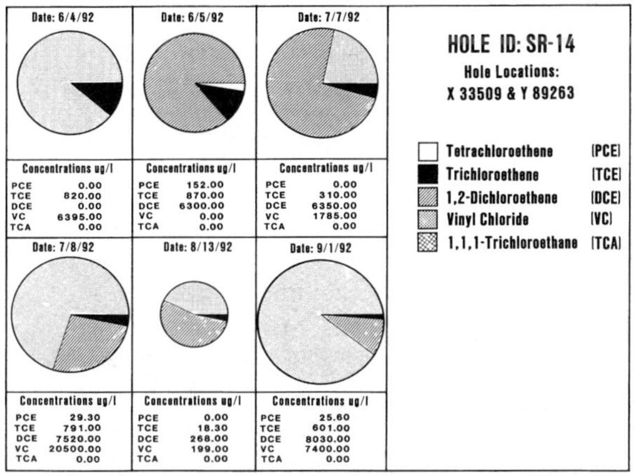

Figure 2-1. Groundwater monitoring results: single well over time.

Figure 2-2 shows multiple well locations for the second quarter of monitoring. The proportion of PCE is higher in the southwest area. The proportion of DCA is higher in the northeast portion of the map. This indicates two different potential sources. Although this conclusion can be deduced from Table 2-2, simple visual displays make the job much easier.

Concepts can often be represented visually as well as mathematically, and this book will continue to emphasize the visual aspects of the GEM process, oftentimes to the exclusion of complicated formulas. Even though the concepts of sampling theory and geostatistics cannot be discussed without formulas, the

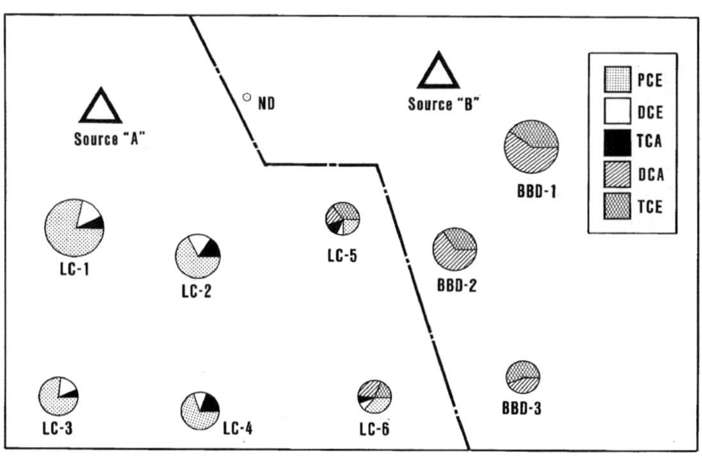

Figure 2-2. Groundwater monitoring results: multiple wells (snapshot in time).

mathematics will be kept to a minimum. An attempt will also be made to explain the components and implications of the formulas.

2.5 MAPS AND MAPMAKING

Maps have fascinated mankind for centuries. Maps can depict spatial relationships that are local, regional, or global. Maps can even chart areas beyond our own planet. Some of the earliest maps may, in fact, have charted the positions of the stars and other celestial objects. In some cultures, including that of the 20th-century, they have served as powerful tools.

The origin of the map has been lost to history. No one knows when, where, by whom, for whom, or for what purpose the first sketches were drawn to communicate a sense of place, a sense of *here* in relation to *there*. The concept seems so natural that one must assume that the first map was drawn millennia ago. Quite possibly, maps antedated written language. We do know that the first map was drawn long before the human mind was able to conceive of the worlds beyond the known shore and horizon (Wilford 1981).

Evidence indicates that separate ancient civilizations independently initiated the construction of maps. Before the first Europeans reached the Pacific, Marshall Islanders were making charts from sticks lashed together with fibers to depict prevailing winds and wave patterns; shells and coral were employed to represent islands. Tahitians communicated geographical knowledge of the South Pacific to Captain Cook, clearly demonstrating that the culture was familiar with the concept of maps. Archeological evidence unearthed in China refers to the use of maps as early as the seventh century B.C. Other evidences include pre-Columbian maps in Mexico, elaborate Incan maps, coastal maps carved by Eskimos in ivory, city maps made by the Mesopotamians, and a "world" map (perhaps the first) made by the Babylonians.

The efficiency and elegance of the visualization provided by a map have brought comfort as well as information to many over the course of time. A map is, in fact, a visual database, with information about shape, size, distance, height, as well as a host of other possible items. At a single glance, the mind can integrate many of the features incorporated in the map, making a quick assessment of a potentially complicated situation. Figure 2-3 shows how a single spatial data set may be displayed visually in a variety of ways.

Geostatistical appraisal (GA) techniques are intimately tied to mapmaking. GEM will examine the assumptions, construction techniques, and pitfalls associated with making maps from sample data. GEM also demonstrates the importance of establishing DQOs before the process so that appropriate and accurate maps result.

2.6 Can a Map Be Trusted?

As he approached the end of his flight in earth orbit in 1962, astronaut John Glenn remarked, "I can see the whole state of Florida just laid out like on a map." A number of other astronauts, and many of us who saw the photography transmitted from space, marveled at how much the Florida peninsula, the meandering Mississippi, the island of Britain, the boot of Italy, the Great Lakes, and other geographical areas matched our mental images of them.

Often, these shapes were firmly stamped in our minds in grade school. We grew up with these images and have taken it for granted that maps were faithful reflections of reality. Still, we were somehow amazed when reality turned out to be true to the maps we had trusted for so long.

Trusting the validity of a map may lead to a fascinating question: What if the map is wrong? An initial response could be: No way! The maps that we commonly use have proven to be so accurate that there is no need to doubt their validity. Even if we had doubts, until man and machine entered earth orbit, there was no visible check on the accuracy of the world map.

Interestingly, ancient mapmakers may have been more in tune with the uncertainty in mapping than we are today. Early map representations of the world often have misplaced coastlines, incorrectly shaped landforms, or missing islands. This is to be expected, given the limited tools and information available to early mapmakers.

We can also find fascinating admissions of uncertainty by mapmakers. In unexplored portions of maps, they placed fanciful sea dragons and other mythical creatures (Fig. 2-4). With little available data in these areas, they tried to indicate the uncertainties and potential dangers associated with these regions.

Today, mapmakers have more sophisticated techniques for displaying the uncertain areas of the maps. Although the ancients had little to fear from mythical creatures, modern-day man may fall victim to the "liability monster" if areas of a site are not well sampled and mapped. A monster, in the form of undiscovered contamination, becomes a liability with frightening costs.

GEM techniques offer many ways in which areas of uncertainty can be quantified and mapped so that undetected liability can be minimized. Three examples are illustrated in Figures 2-5 to 2-7. A kriging error map isolates areas of greatest uncertainty, denoting targets for additional sampling (Fig. 2-5). Probability maps show the probability that an area exceeds contaminant action levels (Fig. 2-6). Cross validation maps show the locations of misclassification errors and suggest locations for sampling or for postexcavation verification sampling (Fig. 2-7). Details of these maps will be discussed in later chapters.

The mapping capabilities furnished by modern geostatistical methods are a double-edged sword. Aided by low-cost and high-speed computers, sample data can be quickly processed into a beautiful map. Expensive and sophisticated

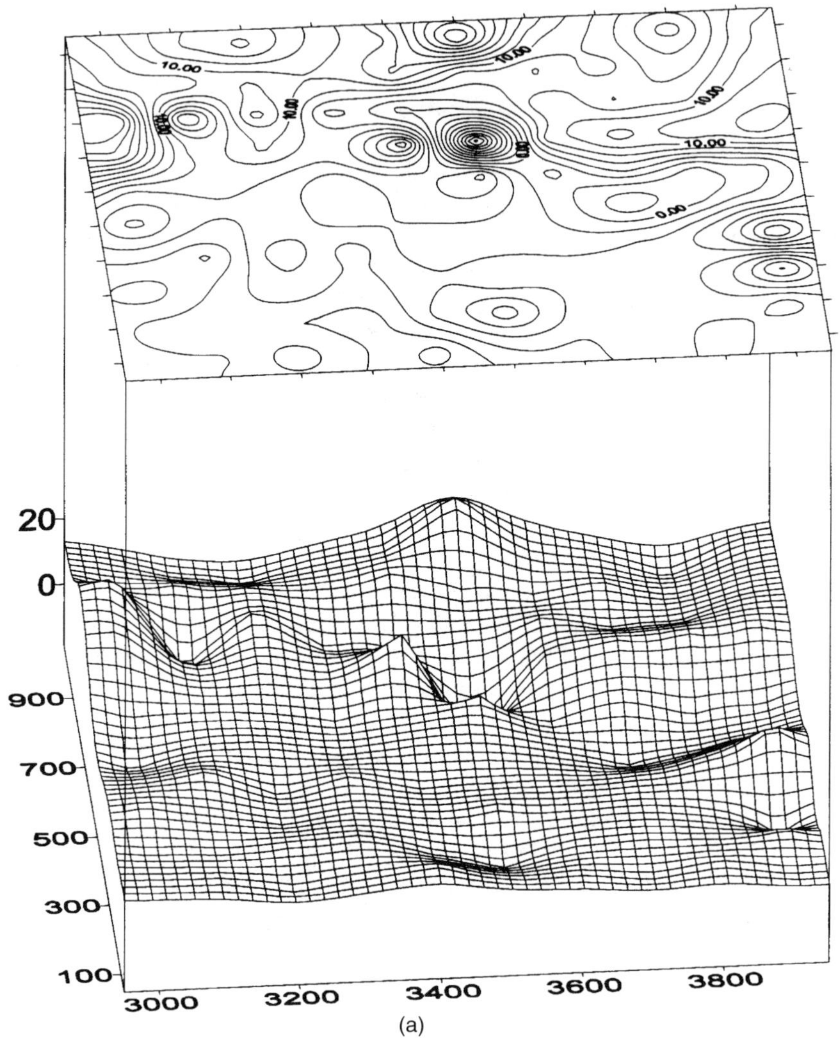

Figure 2-3. Three different visual displays of the same data: (a) contour map (top), and mesh map (bottom) and (b) block map. (Courtesy of Golden Software, Inc.)

interpolation algorithms calculate the grid values, and fine-tipped pens draw precise lines on high-quality papers or mylar; the result is a product of apparently unquestionable integrity that commands belief. As one geologist put it, "Neatness is truth."

The apparent integrity of computer-generated maps so dominates our thinking that it causes many to forget to question the computer output. The once trite

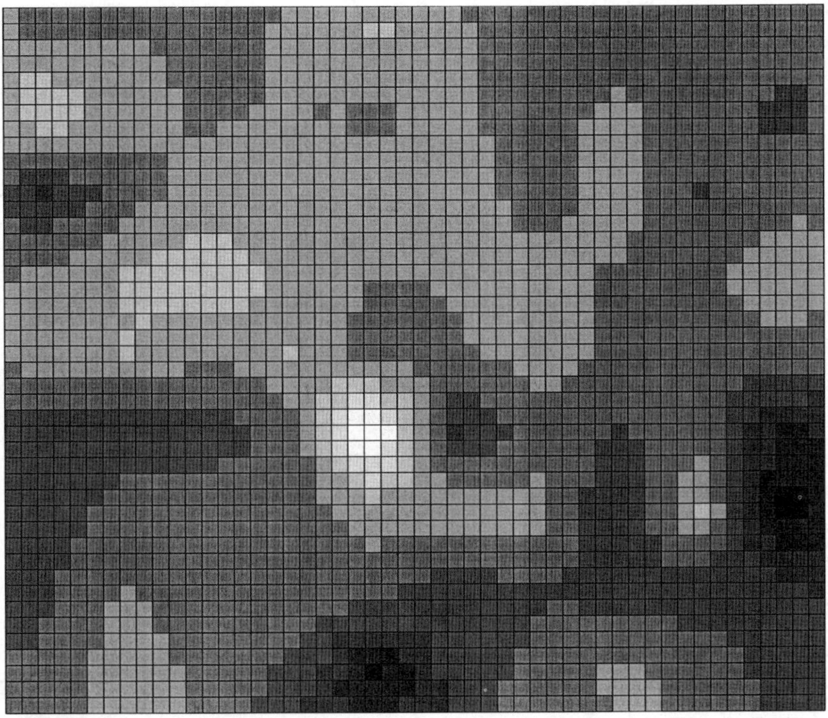

(b)

Figure 2-3. *(continued)*

phrase "garbage in, garbage out" has been corrupted to a new, more frightening aphorism, "garbage in, gospel out." Without the necessary statistical underpinnings, these maps (sometimes in true 3D with beautiful colors and shading) may be little more than "geo-art."

Often, it is impossible to know whether sample data accurately reflect the actual concentrations of the locations sampled. If the placement of a roadway on a highway map appears incorrect, it can be checked. If a chemical sample datum is suspect, it is frequently impossible to reanalyze the sample. Returning to the sample location for a "ground truth" is usually of little benefit.

Several factors limit the usefulness of return sampling. The material at the sample location may have been altered since the initial sample was taken, thus the second sample would not be representative of the first. Also, because the distribution of some contaminants is highly variable over very small distances, a sample taken very close to the original would be expected to exhibit a different concen-

Figure 2-4. Map of Iceland by Ortelius drawn in 1585. (Reprinted with permission from Richard B. Arkway, Inc., New York.)

Figure 2-5. Kriging error map.

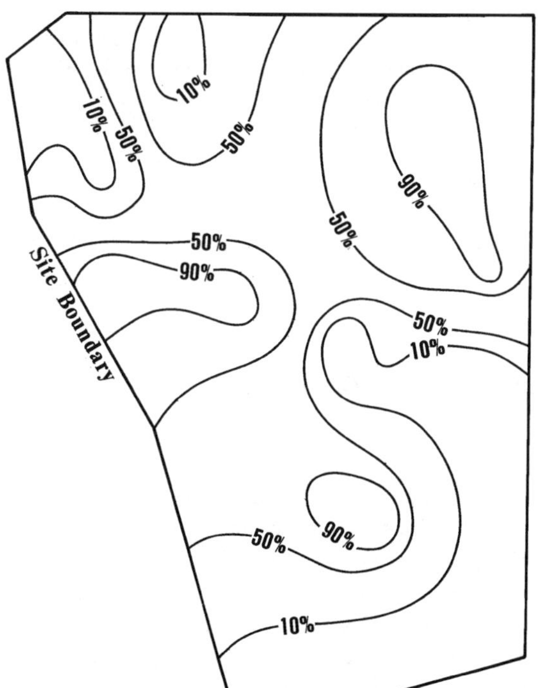

Indicator Probability Map

Figure 2-6. Probability map.

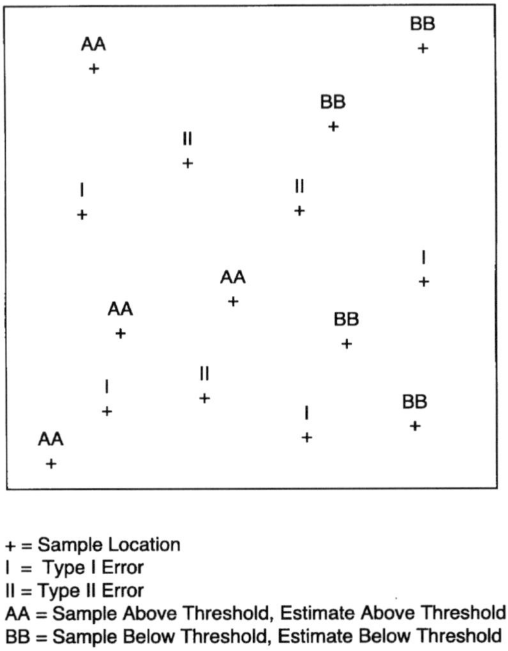

Figure 2-7. Typecast-error posting map.

tration. As a practical matter, the exact location may not be known beacuse survey information is lacking or survey markers have been destroyed.

2.7 The General Nature of Kriging

In the minds of some, kriging has become a generic checklist item to be completed in the life of an environmental project. Some commercial programs offer a menu of potential estimators, with kriging among the options. One can simply push a button to select the kriging option, and the estimation procedure is performed. While it is possible for programmers to arrange for such convenience, anyone familiar with complete geostatistical analysis will find such an approach wanting. To further illustrate why, the following analogy is presented, which demonstrates the danger in using broad, general terms when referring to spatial estimators such as kriging.

Situation A: A set of sample data has been collected. The location and PCB concentration of each sample has been posted on a base map. One objective is to produce a contour map of the PCB concentrations. A decision is made regarding the contouring method; the decision is to perform kriging.

Situation B: A patient suffering from an illness makes a visit to his family physician. After diagnosing the illness, the patient asks the physician to recommend an appropriate treatment. The physician responds he has decided to treat the patient with medicine.

In situation B, an answer such as "medicine" strikes one as laughable, with the absurdity apparent even to a person without formal medical training. One knows intuitively that many factors are considered in determining the actual treatment, including the types of effective medicine available, the amount of medicine in each dose, the number of times each dose is given, other medications interfering or producing side effects, the general health of the patient, and other factors. One may also suspect that a portion of the diagnosis and treatment procedure may not have been appropriate or, at least, should have been handled differently. One also intuits that some types of "medicine" may be totally inappropriate or ineffective, even to the point of endangering the patient's life.

"Kriging" as a response in situation A is no more considered than "medicine" in situation B. The type of kriging, the variogram parameters used, the kriging parameters used, the nature of the spatial continuity structure, bounding constraints used during kriging, data search algorithms, or other relevant details were not specified. The "diagnosis and treatment" components used in the kriging and geostatistical analysis of a data set should be subject to the same level of scrutiny that would be given to diagnosis and treatment of a sick patient.

To extend the medical analogy a bit further, kriging might be compared to the variety of "-cillins" commonly prescribed: penicillin, amoxicillin, ampicillin, and so forth. Kriging, too, comes in many varieties, including, to name a few, ordinary, indicator, lognormal, universal, probability, disjunctive, and multigaussian. Each type of kriging has strengths and weaknesses in certain situations just as the various "-cillin" type drugs are effective for different diseases. It is important to understand that a particular type of kriging is appropriate for a particular application.

It is not unexpected that a general term such as kriging is used when a more specific meaning is required. Kriging is a relatively new technique to environmental applications. As a result, many people are not aware of its breadth. Although practitioners may understand kriging well enough, managers, regulators, and others in nonpractitioner roles may not possess sufficient understanding to formulate questions that will distinguish the different forms and lead to the appropriate kind of kriging.

Kriging belongs to a family of statistical estimators known as BLUE: best linear unbiased estimators. This *best* designation implies that one can not improve on kriging and, thus, it does not generally elicit a challenge. One must remember that *best* is a relative term and does not necessarily imply *good*.

EPA inadvertently furthered simplistic misconceptions when it tacitly endorsed the kriging technique in its 1987 DQO guidance documents and by means of its geostatistical software package GeoEAS. EPA did not provide detail on the various types of kriging and a novice reader might assume that kriging is a simple, single-form process.

2.8 Variance among Geostatisticians

Despite the objectivity of geostatistical appraisal, subjective decisions are made during the GA process. The EPA Environmental Monitoring Systems Laboratory (EMSL) in Las Vegas studied the variability between different individuals performing geostatistical analysis (Englund 1990). Most practicing geostatisticians have preferences in their approach to modeling, and the impact of these differences might be cause for concern.

Starting with two large exhaustive sets of actual data (78,000 data locations), the EPA drew a sample data subset (less than 200 locations) from each. Both data subsets were sent to the 12 selected independent investigators. The investigators were asked to provide geostatistical estimates in block form. The intent was to see how closely the geostatisticians would replicate the "true" block map.

Although geostatistical estimates were requested, kriging was not specified, and investigators were allowed to employ alternative estimators. EPA granted the investigators latitude and flexibility for decision making in other areas, including outlier elimination, data transformation, data stratification, evaluation of spatial structure, and computation of interpolated estimates.

The resultant maps varied considerably as seen in Figures 2-8a to 2-8m. Figure 2-8a shows the "true" situation. Most of the remaining figures show patterns reminiscent of the true situation. The notable exception is Figure 2-8e. Each map shows a variation on the theme of the "true" map, with some more similar to it than others. Some simplify the original map, whereas others are simply reorganizations. A few could be considered acceptable representations; others would be rated inadequate.

Figures 2-9a to 2-9m show a second example of the geostatisticians' attempt to replicate the unknown true map. Again, significant variations are displayed between maps and between individual maps and the true situation, with some maps considered acceptable approximations and others poor.

The EPA concluded that site assessors need to be concerned about the methods employed to evaluate data. Variability in spatial estimation methodologies has a significant impact on the quality of the estimates, and on the quality of decisions based on the estimates.

At a hazardous waste site, maps are used to make costly remedial decisions. When cost functions were applied to determine the economic impact of the

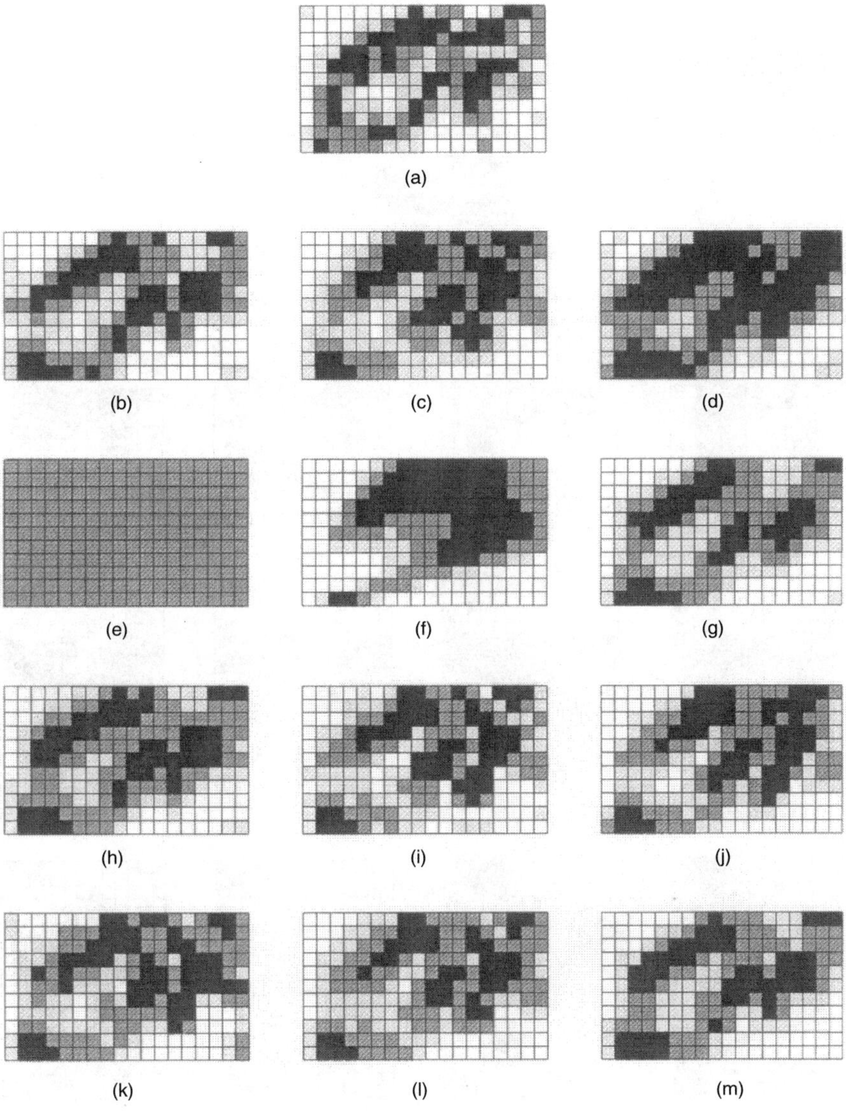

Figure 2-8. Contour maps of variable 1 in EPA's geostatistician study. (From A variance of geostatisticians, by E.J. Englund, *Journal of Mathematical Geology* 22: 436–437. Reprinted with permission from Plenum Publishing Corporation.).

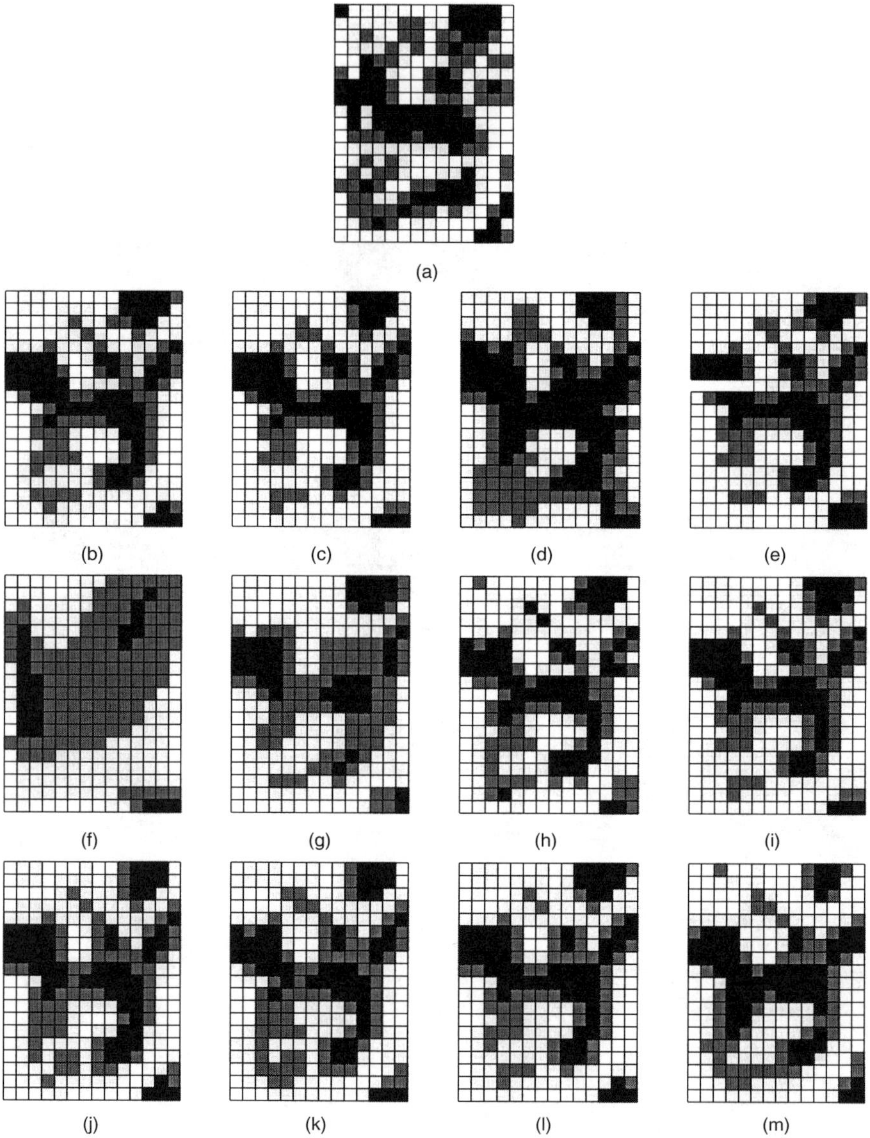

Figure 2-9. Contour maps of variable 2 in EPA's geostatistician study. (From A variance of geostatisticians, by E.J. Englund, *Journal of Mathematical Geology* 22: 438–439. Reprinted with permission from Plenum Publishing Corporation.)

remedial decisions based on the mapping provided by the assorted geostatisticians, variations in cost estimates were dramatic. The relative differences between the highest and lowest costs (measured at several different action levels) ranged from 4% to 75%, with one extreme of 98% for the first data set, and from 4% to 29% in the second data set.

2.9 Variance among Interpolators

The EPA study just discussed illustrates the variations expected among individual modelers. Similar variability can be obtained using the wide variety of interpolation functions contained in many commercially available programs.

We should ask a question similar to the one asked by EPA: Given a data set and a single geostatistician, to what extent does the choice of different interpolation functions affect the final maps?

Figures 2-10a to 2-10f show map results for a single geostatistician using six different interpolation functions: (a) inverse squared distance, (b) cubic polynomial regression, (c) ordinary kriging, (d) universal kriging, (e) multiquadrics, and (f) thin-plate splines. The differences are dramatic.

Quite similar maps were produced by ordinary kriging, universal kriging, multiquadrics, and thin-plate splines. Corresponding shapes at the high, low, and boundary areas can be seen, with relatively minor variations. In general, the inverse distance squared (ISD) map shows these same curves; at the same time, it is a strikingly different visual image. On the ISD map, the features tend to be more circular than the continuous curves of the group of four techniques that produced similar maps. The cubic polynomial regression shows extreme smoothing.

Figures 2-11a to 2-11f show more possibilities using these same interpolation methods. Three of the maps show ISD maps and three show multiquadrics. In both sets, ISD and multiquadrics, different smoothing factors were used. For ISD, factors of 0.0, 0.5, and 1.0 were used. As the smoothing factor increases, major changes in the map occur. The detailed display of factor 0.0 recedes to broad, generalized curves at factor 0.5. At factor 1.0, more detail emerges than at factor 0.5, but the map retains a look similar to the factor 0.5 map.

For the multiquadrics, factors of 0.0, 1.0, and 4.0 were used. The influence of the smoothing factor in this case is much less than that seen for ISD. The changes appear to be relatively minor, retaining the same basic forms from one map to the next. For ISD, the appearance of the factor 0.0 map differs markedly from the factor 1.0 map. For multiquadrics, the appearance of the factor 0.0 map is quite close to that of the factor 1.0 and 4.0 maps.

The map results in Figures 2-10 and 2-11 make it obvious that different interpolators produce different results for identical data sets. GEM has a number

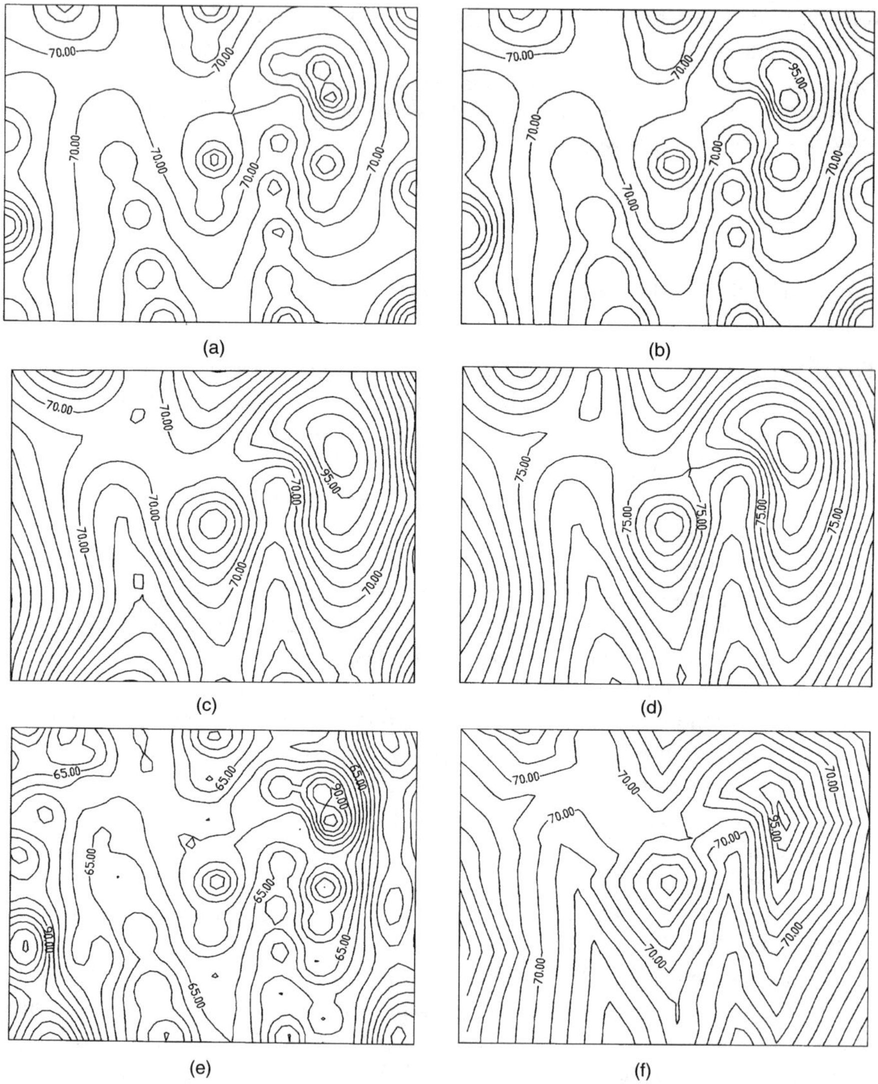

Figure 2-10. Contour maps produced by several different computerized interpolation algorithms: (a) inverse distance squared, (b) cubic polynomial regression, (c) ordinary kriging, (d) universal kriging, (e) multiquadrics, and (f) thin plate splines. (Courtesy of Golden Software, Inc.)

Figure 2-11. Contour maps showing various degrees of smoothing resulting from different interpolation algorithms: (a) inverse distance squared with smoothing factor 0.0, (b) inverse distance squared with smoothing factor 0.5, (c) inverse squared distance with smoothing factor 1.0, (d) multiquadrics with smoothing factor 0.0, (e) multiquadrics with smoothing factor 1.0, and (f) multiquadrics with smoothing factor 4.0. (Courtesy of Golden Software, Inc.)

of statistical tools that will allow the modeler to evaluate the correctness of a map.

2.10 Public Domain Geostatistical Software

The use of geostatistical tools has greatly accelerated in recent years. Microcomputer technology and geostatistical software have greatly increased the use of geostatistics. USEPA has been a driving force behind the increased use of geostatistics for environmental problems.

In the early 1980s, USEPA's Environmental Monitoring Systems Laboratory (EMSL) in Las Vegas began to investigate intensively the possibility of using geostatistical techniques for environmental problems. In 1987, USEPA released the Geostatistical Environmental Assessment Software package, commonly known as GeoEAS. GeoEAS is in the public domain and readily available. By early 1993, USEPA EMSL had distributed approximately 4000 copies of the software. It is estimated that the secondary distribution has increased the total to over 10,000 users (personal communication, Englund 1993).

In addition to GeoEAS, USEPA has provided information on the theory and application of geostatistical methods in its 1987 DQO guidance documents. Geostatistical techniques are described in both the Development Process and the Example Scenario documents. These guidance documents have boosted the acceptance of geostatistical techniques, informing many that these tools exist.

Other sources for easily accessible geostatistical software have recently emerged. While not public domain or without cost, an affordable set of geostatistical programs with documentation and source code called GSLIB for Geostatistical Software Library and User's Guide is available (Deutsch and Journel 1992). GSLIB is a full three-dimensional package geared to the professional user. It also offers more advanced types of geostatistical analysis, including cokriging, universal kriging, and conditional simulation.

A more recent public domain package is UNCERT, developed under contract to the U.S. Bureau of Reclamation. UNCERT offers a full range of geostatistical applications, as does GSLIB, and it can be obtained from the Colorado School of Mines.

With increased access to geostatistical software, a corresponding increase in the potential for misapplication exists. Such growing pains constitute an unfortunate part of the learning curve. GEM establishes some standards that minimize blatant errors and misapplications. Full coverage of geostatistical theory and techniques is not possible here. Readers are therefore strongly encouraged to consult other books on geostatistical theory, DQOs, and STP to broaden their theoretical base.

2.11 GEM and Chaos

In the last 20 years, a new and fascinating field called Chaos has emerged (Gleick 1987, Yfantis, et al. 1987). Chaos theory examines dynamical systems such as earth weather, animal populations, fluid flow, or the beating of a human heart. A key tenet of Chaos is that there is an "order in the randomness." Frequently, it is found that a phenomenon is locally unpredictable but globally stable.

There are a number of similar concepts in statistics and Chaos. Statisticians found a type of "order in the randomness" long before the development of Chaos theory. The result of the next toss of a coin or roll of a die, a local condition, cannot be predicted with certainty, but the pattern of outcomes becomes clear and stable as many experiments are performed. This global stability allows the gambling industry not only to exist but to realize a substantial profit.

In environmental sampling, the value of the next outcome is never known. A series of samples will typically produce an expected shape when the data are grouped into a histogram; an example is the normal bell-shaped curve (Fig. 2-12). As will be seen in Part 2 of this book, the stability of the normal distribution allows the statistician to predict with reliability but not certainty. Thus, by means of statistics, order comes out of chaos.

The tools discussed in GEM provide ways to maximize the amount of order that can be achieved in sampling and mapping. The Poisson distribution (Chapter 6) and STP show how to maximize the order among samples by collecting samples of appropriate weight. Geostatistics, through the variogram, evaluates data and proves that a spatial order is present.

The order found through STP and GA techniques is analogous to another concept of Chaos, that of attractors. A simple example of an attractor is the point at which a pendulum finally settles because the force of gravity "attracted" it to that point. Gravity plays an important role in STP. In a world where entropy rules, causing greater and greater disorder, gravity counteracts this tendency and causes segregation or ordering of materials. STP mitigates the effects of segregation in sampling.

Chaos also stresses a "sensitive dependence" on initial conditions, meaning that seemingly small variations in the beginning can have large effects later. For

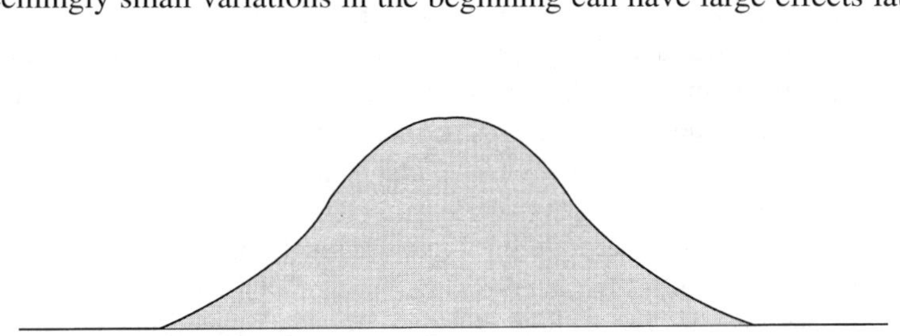

Figure 2-12. Normal bell-shaped curve.

example, if a commuter is 30 seconds late for a car pool he may miss a bus that cycles every 10 minutes. The next bus, however, will not deliver him in time to make a train that runs every hour that should take him to the plane that leaves only once a day.

If initial sample data do not represent actual conditions, clean sites may be inappropriately remediated or dirty sites may go uncleaned. Inaccurate sample results may be mapped and inappropriately classify areas of contamination. Figures 2-10 and 2-11 were mapped using identical data; different initial data would produce different contours and potentially different decisions.

Finally, the concept of fractals is an offshoot of Chaos that has significance in GEM. Fractals are items that exhibit self-similarity, typically at many different scales. The now-classic example is the coastline of England, which exhibits the same amount of roughness at many levels of view, from space down to very local levels.

GEM also uses the concept of self-similarity. In STP, self-similarity can be both good and bad, depending on sample weight. For samples composed of small volumes or weights, self-similarity often results in missed contamination and, thus, poor decision making. For larger weight samples, self-similarity promotes repeatability and, thus, a more accurate assessment of the true quantity of hazardous contaminant present. In GA, the variogram carefully examines the self-similarity between samples at various distances. Self-similar samples assist the estimation process and provide greater confidence in estimates and subsequent decisions.

Figure 2-13 shows an example of a fractal developed by applying Newton's method to solving for the cube root of minus 1. The plane has been divided into three regions shown in different shades representing different colors. All points in each region are "attracted" to the root lying in that area. An interesting feature is the almost unimaginably complex boundary, peculiar in that every point on it borders all three regions. In addition, as one zooms in further and further on the diagram, a fractal structure is disclosed, with the same pattern repeated on increasingly smaller scales.

Areas of relative stability and complex boundaries are found in environmental situations. Within a particular science, for example, relative stability exists, with accepted knowledge and conventions. The complexity arises when two or more fields attempt to interface or come together at a boundary. One such situation was intimated in Chapter 1 in the discussion concerning the scientific validity of a sample and the samples' usefulness as legal evidence. On each side of the issue, perceptions are relatively clear. When there is an attempt to interface at the boundary, complexity often arises.

Integration of fractal methods into geostatistical appraisal has begun but is not yet highly developed (Yfantis et al. 1987, Agterberg 1994, Rose and Srivastava 1994), and its long-term applicability is unclear. Detailed examination of

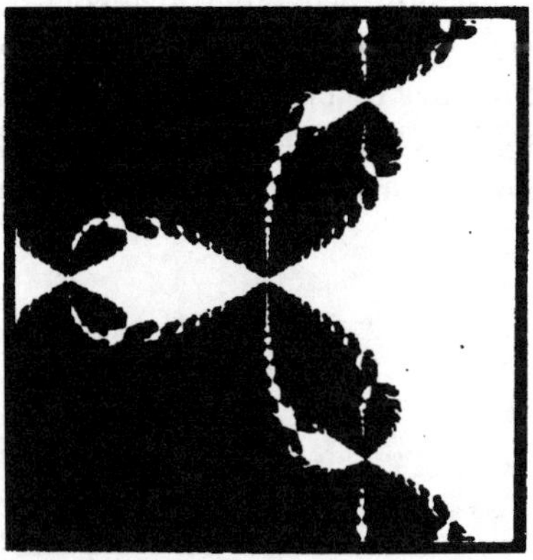

Figure 2-13. Fractal diagram of Newton's solution for the cube root of -1 showing self-similarity and complex boundary. (From *The Beauty of Fractals*, by H-O. Peitgen and P.H. Richter (Berlin: Springer-Verlag, 1986). Reprinted with permission from Springer-Verlag and Dr. Hartmut Juergens.)

fractals is beyond the scope of this book; the concepts of Chaos and fractals will be used to make qualitative comparisons to GEM concepts.

The preceding discussion has shown that Chaos has concepts that are useful in describing problems in sampling and mapping. It also reminds us that many sciences are highly interrelated, with benefits derived from viewing them together despite the boundary issues. This is fundamental to GEM: DQOs, STP, and GA are strongly interrelated and should be viewed as one. The benefits derived as a whole may be greater than the sum of the individual benefits.

Conclusion

An ever-growing awareness of environmental concerns has prompted a world-wide movement to address these issues. In the United States, this impetus has prompted the enactment of rigorous laws, process, and authority to protect and clean up the environment. Collection and mapping of sampling data play a key role in the characterization and remedial process. Chapter 2 has demonstrated that reliance on maps must be tempered with confidence that the maps produced from the sample data are of high quality. The assumptions and techniques used in creating maps must be carefully scrutinized so that appropriate and correct techniques can be implemented. With computer options growing ever more numerous, better techniques are constantly becoming available. On the other hand, this increases concerns that all techniques may be considered equally appropriate, giving rise to the possibility that quality techniques may be applied to inappropriate situations or may simply be misapplied. It is the function of the GEM process to provide tools to evaluate the available techniques so that correct methods may be implemented.

Chapter 3

Introduction to Error

SYNOPSIS: The term error is often used in a wide variety of contexts. In order to understand error management, an understanding and definition of error must be developed. Error must also be distinguished from the related concepts of confidence, probability, uncertainty, reliability, and risk. This chapter begins by discussing the "softer" nature of confidence, probability, uncertainty, reliability and risk in conceptual terms. The risks associated with statistical inference and estimation will be related to the inductive logic approach used. The error components of precision, accuracy, and bias will be introduced, as well as their relation to an error quantification model. Spatial estimation will be introduced, along with various methods for extrapolating data to surrounding regions. The final topic discussed is the important distinction between absolute error and misclassification error as it relates to environmental problems.

KEY WORDS: Error, Confidence, Uncertainty, Risk, Probability, Reliability, Estimation, Deductive Logic, Inductive Logic, Neighborhood, Extension, Precision, Accuracy, Bias, Estimation Error, Misclassification Error, Type I Error, Type II Error, False Positive, False Negative

3.1 Errors vs. Confidence

The field of statistics developed from probability theory, which had its own origins with games of chance in the seventeenth century. Activities such as throwing dice, spinning a roulette wheel, tossing a coin, or drawing a card all constitute games of chance. In each situation or game, the outcome of a particular action remains unknown until the event occurs. Nevertheless, predictable long-term outcomes exist that provide an understanding of the nature of the game and the probability that certain outcomes will result.

The notion of probability differs somewhat from the concept of error. Probability states the chance that a specific event will happen before the event takes place, a type of a priori assessment. An error can be measured only after the event has occurred or a posteriori. An error is also, in some senses, a more concrete item in that it may have a magnitude with reference to a given standard. For instance, if a true value of 25.0 was estimated to be 27.3, an error of 2.3 has occurred.

Probability is a number between zero and one that expresses the chance that a specific event will occur under stated conditions. A probability statement might be: A .33 probability exists that an estimate exceeds the true value. For the toss of a coin, one can state that a .5 probability exists that tails will result. If one predicts that tails will actually result, a .5 probability also exists that an error in the prediction will result, that is, heads.

Should the outcome of the toss happen to be heads, the incorrect prediction can be considered to be a misclassification error. The experimenter thus learns that probability is related to a statistical *confidence* before the event, whereas errors are measured after the fact in terms of magnitude and classification.

Probability-related terms such as *confidence*, *uncertainty*, and *reliability* refer to a priori conditions and thus cannot guarantee any particular result for an experiment. Thus, an estimate is only a rational guess as to the actual value or outcome. It is not possible to determine what the errors will be or when they will occur, but one may wish to quantify the potential magnitude or probability of the anticipated errors.

In contrast, error relates to a known outcome or value, with its magnitude calculated exactly or quantified in terms of a class of error. The error may also yield the sign of the discrepancy (positive or negative), thus offering insights as to any potential biases.

When driving an automobile, the driver understands that a small probability exists that an accident will occur during the trip, but is generally *confident* that all will go well. Insurance statistics can quantify the probability of an accident's occurring yet cannot predict when an actual accident will occur.

Drivers and insurance companies both use the available data provided by probabilities to make decisions based on risk, a concept that implies a type of loss function. Risk is often defined as the probability of a certain outcome multiplied by the cost (loss) resulting from that outcome. The greater the potential loss, the greater the potential risk. Similarly, the greater the probability, the greater the risk.

Environmental risk assessment examines the risks associated with exposures to hazardous chemicals that may impact human health or ecological systems. In general, the greater the concentration, the greater the risk and, thus, a greater potential loss to human health or the environment or both. Environmental loss functions are discussed in Chapter 17.

3.2 Deductive and Inductive Logic vs. Statistical Inference

Figure 1-2 showed small sample masses used to infer masses orders of magnitude larger. Such massive generalization of sample concentrations to the larger area will undoubtedly generate estimation errors, given that the process of generalizing is inductive. The task then becomes one of managing the errors produced in estimation. This section will begin the error management process by introducing inductive inference as it relates to statistical estimators.

3.2.1 Deductive Logic

Arguments have been traditionally categorized into two different types, deductive and inductive. All arguments claim that the premises or propositions provided offer some grounds for the truth of the conclusions. Whereas a conclusion is offered by both types of argument, only deductive arguments involve the claim that the premises provide conclusive grounds.

A classical example of a deductive argument is (Copi 1978):

All humans are mortal.
Socrates is human.
Therefore Socrates is mortal.

Also associated with deductive arguments are the more technical terms *valid* and *invalid.* A deductive argument is valid when its premises, if true, do provide conclusive grounds for its conclusion. The task of deductive logic is to explain the nature of the relation between the premises and conclusions in valid arguments, thus discriminating valid from invalid arguments. The preceding example contains two true premises and a valid conclusion.

All deductive arguments are either valid or invalid. To confuse the issue, an argument may be valid while containing false propositions. For example:

All spiders have six legs.
All six-legged creatures have wings.
Therefore all spiders have wings.

If the premises were true, the preceding conclusion would be valid. Thus, this is still a valid argument.

As a contrast to valid arguments, consider the following syllogism:

If you owned all the gold in Fort Knox, you would be wealthy.
You do not own all the gold in Fort Knox.
Therefore, you are not wealthy.

Although both premises of the argument are true, the argument must be considered invalid because the conclusion is false. It is impossible for the premises of a valid argument to be true while having a false conclusion.

3.2.2 Inductive Logic

It is also possible to deduce the wrong conclusions based on the available empirical data. One is reminded of the story involving the researchers and the frog. The researchers yelled at the frog and the frog jumped. The researchers cut off the frog's legs and yelled again. This time the frog did not jump. The researchers were divided as to their conclusions: One group claimed that amputating the legs of a frog makes it deaf, while the others concluded that the frog's ears were located in the legs!

While these conclusions are good for a small chuckle, history reveals that erroneous conclusion occurs frequently in the scientific world. For centuries, man observed that the sun revolved around the earth. Man therefore concluded that the earth was a stationary object and that all other heavenly bodies orbited the earth. Similarly, barely five hundred years ago, it was thought that since the earth looked flat (especially the oceans), it truly was flat and the possibility existed of falling off if one were to sail too far.

Nineteenth-century pundits, using existing population growth trends, predicted that twentieth-century man would experience extreme transportation problems in urban areas from overwhelming volumes of horse waste in the streets. Modern man puzzles over problems of the present day: As industrial societies pump large quantities of greenhouse gases into the earth's atmosphere, will problems develop related to global warming or ozone layer depletion? Based on current empirical data, some people draw conclusions that indicate problems while others reject the notion that significant problems loom.

It is even possible to be aware that some arguments are invalid yet perform serious scientific work. Two of the major breakthroughs in the twentieth century are the theory of relativity and quantum mechanics. Still, it is recognized that these two important fields have irreconcilable discrepancies between them, indicating that at least one of them is invalid. Yet, until the questions are resolved, each branch of physics will continue to advance the science and solve day-to-day problems.

The preceding examples show that inductive arguments do not prove a conclusion; they merely provide grounds to support it. For example,

Hitler was a dictator and was ruthless.
Stalin was a dictator and was ruthless.
Castro is a dictator.
Therefore Castro is probably ruthless.

Inductive arguments may be categorized as better or worse, depending on the degree to which they add strength in supporting a conclusion. In any event, they call for a speculative approach.

One of the underlying components of statistical theory is that of inductive inference. Statistics are employed to infer probable conclusions based on empirical evidence, often accompanied by a statement of confidence associated with the estimate. Techniques such as statistical confidence intervals can be used to assess the probability that an inductive argument is true.

In the world of science, conclusions derived often go beyond the results and boundaries of a particular experiment to the class of all similar experiments. Progress in science often results from the experimentation process, which provides one avenue for obtaining new knowledge. In the most simplistic terms, an experiment is performed, data are obtained, data are analyzed, and conclusions are drawn based on the resulting data. Results of the analyses are empirical data, and empirical data are always incomplete.

Statistical inference enters our lives in many ways. A familiar example is predicting political election results based on polling. Modern scientific polls use extremely sparse data (compared to the entire voting population) to predict election results. Typically, these results are accurate in assessing the winner.

Inductive inference has a reputation as a hazardous endeavor. In fact, a theorem of logic states that uncertainty is present in inductive inference. By definition, one simply cannot infer generalizations and be assured that these inferences are correct, especially not for all circumstances. While modern pollsters possess a respectable track record, history will not soon forget the election of “President Dewey.”

In the red bead experiment of Deming (see Chap. 1), empirical evidence predicted well for the short term but could not be absolutely relied on for the future. Nevertheless, inferences can be made, and the degree of risk or uncertainty associated with these "educated guesses" can be measured. One of the main functions of statistical methods is to provide ways in which the degree of uncertainty surrounding an inference may be quantified.

Geostatistical methods have been derived from existing classical statistical theory and methods. GEM recognizes the hazards and uncertainties surrounding inferences or estimation and develops tools to deal with the problems encountered. It is also important for geostatisticians to keep in mind that errors in deductive logic can impute errors.

Using sample data, the geostatistician estimates the contaminant concentration at unsampled locations during the mapping process. Estimation errors may result from an insufficient number of samples, sample data of poor quality that do not represent the actual concentrations, or poor estimation procedures. This chapter will discuss these concepts as they relate to estimation.

The red bead experiment also demonstrated that a known population can be sampled and a wrong conclusion drawn with regard to the content of red vs. white beads. If the sample collection device introduces a bias or an error, incorrect conclusions and inferences may result. STP endeavors to minimize the errors introduced by sampling protocols and equipment.

3.3 Error Components: Precision, Accuracy, and Bias

3.3.1 Bias

Many statistical estimators have the property of being BLUE, which is an acronym for best linear unbiased estimators. This distinction points out that an unbiasedness condition adds to the desirability of the estimator.

Bias is defined as any systematic or consistent over- or underestimation of the true values of the quantity being estimated. For example, if a quantity consistently receives an estimated value that is 50 ppm too low, then a negative bias of 50 ppm is present. A bias may be proportional as well, over- or underestimating by perhaps 10%; for example, a value of 100 ppm estimated as 110 ppm, a value of 200 ppm estmated as 220 ppm, and so forth.

In the GEM process, three different types of bias are commonly introduced:

Sampling bias: the bias introduced during the sampling process. This bias is a function of the error components of sampling related to heterogeneities, sampling devices, and sample sizes. It is independent of any preparation and analytical bias introduced.

Measurement bias: the bias introduced as part of the measurement or preparation and analytical process. This is introduced by measurement equipment, analytical devices, and human factors.

Estimation bias: the bias introduced by the estimation process that predicts the value of an unsampled point, block, or volume. This bias is related to the extensions performed by estimators such as kriging or inverse power distance.

Random errors are unpredictable entities that occur during a sampling, measurement, or estimation procedure, causing deviations from the true concentration of the quantity being estimated. These errors result from many factors, and the magnitude changes for each individual error.

3.3.2 Precision

Precision is a measure of the variability between individual estimates or measurements and indicates how closely individual calculations agree with one another. It can also be thought of as repeatability. Precision and accuracy should not be confused or used synonymously.

3.3.3 Accuracy

Accuracy summarizes the closeness of measurements to the true (but possibly unknown) value. The analogy of a target is often used when describing accuracy, indicating how close to the target center measurements or estimates come.

3.3.4 Visual Relationships between Error Components

Figure 3-1 shows the classic visual relationship among precision, accuracy, and bias (Jessen 1978). The center of each target represents the true value that one is attempting to measure or estimate. The dots represent numerous repeat measurements or estimations. Figures 3-1a and 3-1c show that bias is present, given that the scatter of measurements does not fall at the center of the target.

Figures 3-1a and 3-1c illustrate also the concept of precision. The tight cluster in Figure 3-1a indicates a high degree of repeatability in the measurements or estimates. The estimates in Figure 3-1c are not very repeatable, indicating a larger degree of randomness.

The pattern in Figure 3-1b indicates low bias, low precision, and low accuracy. Measurements do not often come close to the true value, nor are they

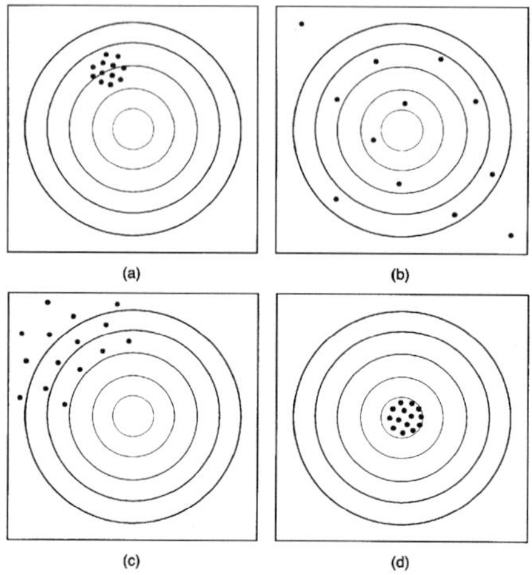

Figure 3-1. Illustration of the concepts of precision, accuracy, and bias. (From *Statistical Survey Techniques*, by R.J. Jessen. Copyright ©1978. Reprinted by permission of John Wiley & Sons, Inc.)

repeatable. The low bias offers only slight consolation in that the mean would be correct.

Of the four conditions presented, Figure 3-1d shows the most optimal situation. In this case, the clustering indicates a desirable high accuracy, combined with the other desired attributes of high precision and low bias.

3.4 Definition of the Estimation Error

In measurement or analytical situations, it is not common to have more than a very limited number (one or two) of repeat measurements. Thus, production of a target diagram generally does not help. Note that the true value is not known, and so precision but not accuracy may be determined.

For cases involving mapping, large numbers of estimates are made for individual blocks or points. If extraction of the units is being performed, some situations provide chemical analysis of the block units, against which the estimated values can be compared. This can also be achieved by means of cross validation procedures to be discussed in Chapter 14.

The true concentrations of the blocks can be defined as Z_1, Z_2,...,Z_n, and the estimated concentrations can be defined as $Z_1^*, Z_2^*,...,Z_n^*$. The estimation error can now be defined as the true value minus the estimate:

$$e = Z - Z^* \tag{3.1}$$

or individually:

$$\begin{aligned} e_1 &= Z_1 - Z_1^* \\ e_2 &= Z_2 - Z_2^* \\ &\cdot \quad \cdot \quad \cdot \\ &\cdot \quad \cdot \quad \cdot \\ &\cdot \quad \cdot \quad \cdot \\ e_n &= Z_n - Z_n^* \end{aligned}$$

The block estimate can then be viewed in terms of the following model:

$$Z^* = m + e \tag{3.2}$$

where m is the mean and e the error. The error term e is composed of individual errors and biases:

$$e = b_1 + b_2 + \ldots + e_1 + e_2 + e_3 + \ldots \tag{3.3}$$

where

b_1 = sampling bias
b_2 = analytical bias
e_1 = sampling error
e_2 = analytical error
e_3 = spatial (field) error

3.5 What Is Spatial Estimation?

To develop an appreciation for the estimation process using spatial data, it is necessary to review the actual practices used in the field of estimation and mapping. By means of this exercise, we can isolate the underlying concepts and the associated questions that will arise.

The general problem can be stated in the following manner: The actual concentrations of a few localized samples are known, and we wish to determine the concentration of an unsampled location or unsampled area. We may even wish to know the average concentration of the entire site. It goes without saying that the existing sample data should be used as part of the estimation process. We can then manipulate these few values (s_1, s_2,..., s_n) in order to produce a combination of these concentrations that will be considered the actual concentration of the unsampled entity of interest.

The process by which we manipulate the sample data can take a multitude of different forms. Many techniques have been developed to assist in this type of process, and a few will be discussed in detail. Whatever process is chosen and applied by the practitioner, three fundamental concepts are implied (David 1977):

- The concept of a neighborhood
- The concept of extension
- The concept of error of estimation

3.5.1 The Neighborhood

The idea of a neighborhood was introduced in Chapter 1 and indicated that sample data located relatively close to one another often exhibited similar concentration values. A neighborhood can be defined as a localized area within which conditions remain relatively constant. This can refer to contaminant concentration, aquifer thickness, annual rainfall, soil permeability, and so forth. An example of neighborhoods can be seen in Figure 3-2.

Data located outside this local neighborhood typically show concentrations of greater variability. The similarity of nearby samples is related to the statistical correlation present among the samples. Knowing the concentration of one or more of the sample data in the neighborhood provides an assessment of the relative concentration of other samples or even of unsampled locations.

Figure 3-3 shows an example of sample concentration data that exhibit neighborhood characteristics. Whereas individual sample concentrations vary locally, localized samples remain more similar than those found in other neighborhoods.

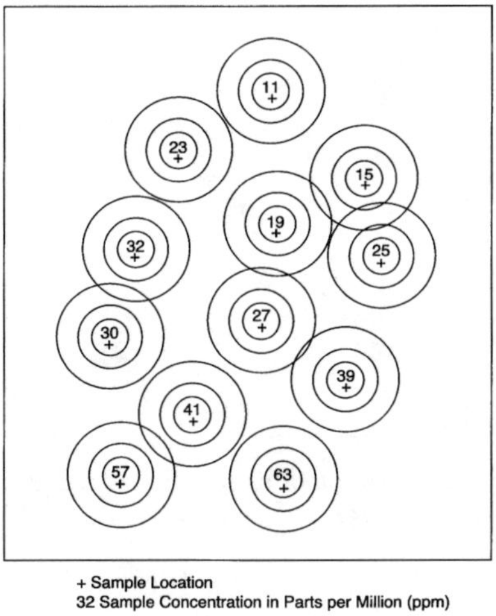

Figure 3-2. Increasingly larger neighborhoods surrounding central sample points.

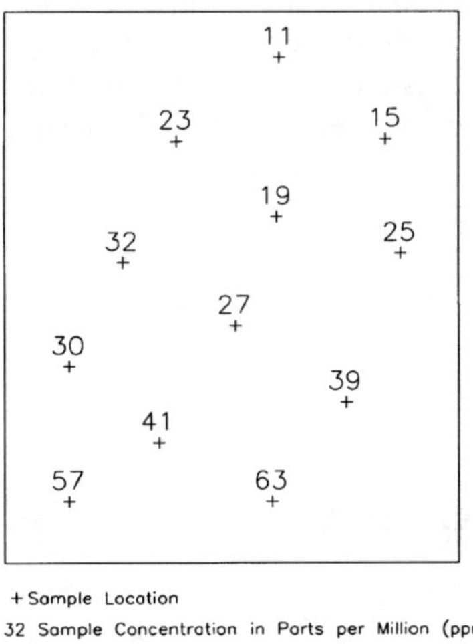

Figure 3-3. Samples showing similar concentrations in a neighborhood.

3.5.2 Extension of Sample Data

The neighborhood concept gives rise to the idea of extending sample concentrations to other localized points, areas, or volumes. Because the localized samples exhibit a spatial continuity, it is logical to assume that there is also a zone of influence associated with the sample. Figure 3-4 shows an example of a simple

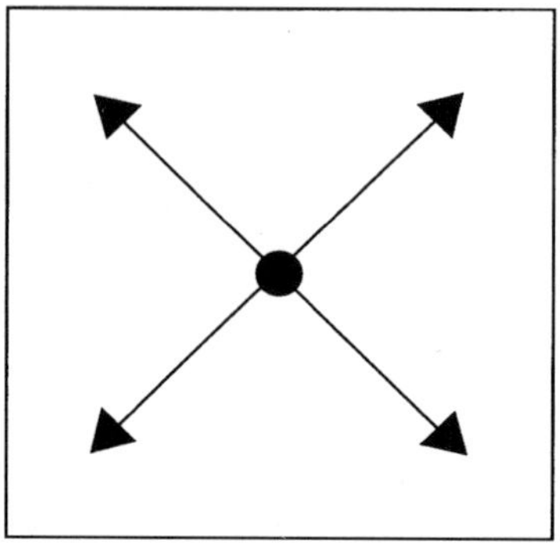

Figure 3-4. Central-point extension method applied to a square block. (From *Geostatistical Ore Reserve Estimation*, by M. David, Elsevier, 1977. Courtesy of Michel David.)

extension of a sample to an area around itself. The block being estimated is given the concentration of the centrally located sample value. Figure 3-5 illustrates another type of simple extension. In this case, a similar block retains the average concentration of the four sample values located at each corner of the block. This has sometimes been referred to as "poor man's kriging."

All mapping estimations can be separated into successive extensions of existing sample concentration values. Extension techniques can be grouped into two categories: geometric and distance-weighting methods. Both, however, are based on the same principle.

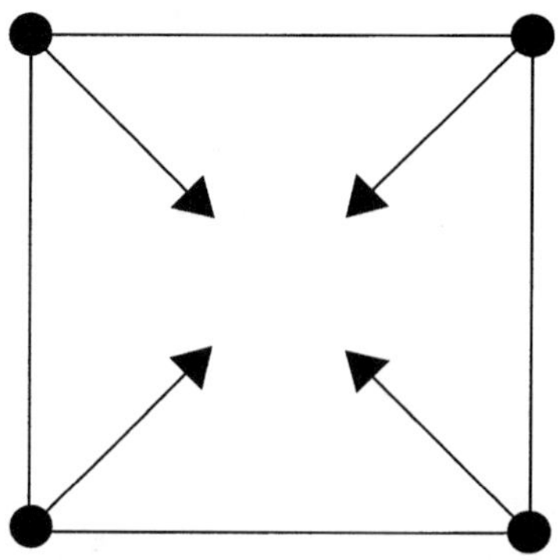

Figure 3-5. Four-corner extension method applied to a square. (From *Geostatistical Ore Reserve Estimation*, by M. David, Elsevier, 1977. Courtesy of Michel David.)

3.5.2.1 Geometric Methods

Geometric approaches have been used for many years for estimation. Before the advent of computers, they were quite necessary because more advanced techniques required excessive amounts of time to perform and to verify. Interestingly, they were also the first techniques to be computerized. Geometric methods are still used for estimation purposes. A few of the most common types of geometric estimation methods are summarized below.

Regular Blocks Samples that have been collected on a regular grid spacing can be extended in several ways. Figure 3-4 shows a square grid with the central point extended to the entire block. This is one of the simplest types of extension. Figure 3-5 illustrates a four-corner averaging approach, this time extending in rather than out. Figures 3-6 and 3-7 show that the blocks into which the sample is extended

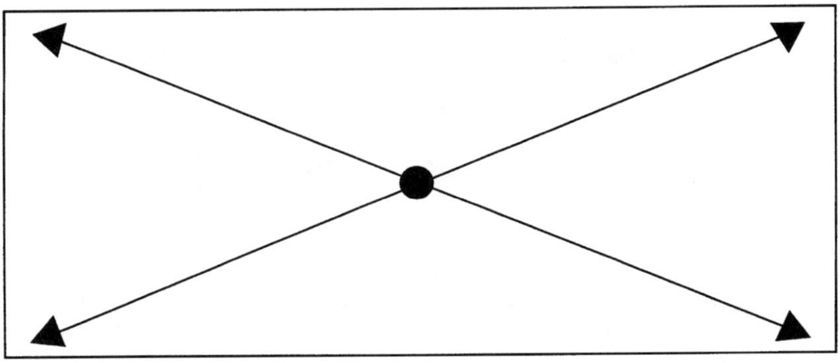

Figure 3-6. Central-point extension applied to a rectangular block. (From *Geostatistical Ore Reserve Estimation*, by M. David, Elsevier, 1977. Courtesy of Michel David.)

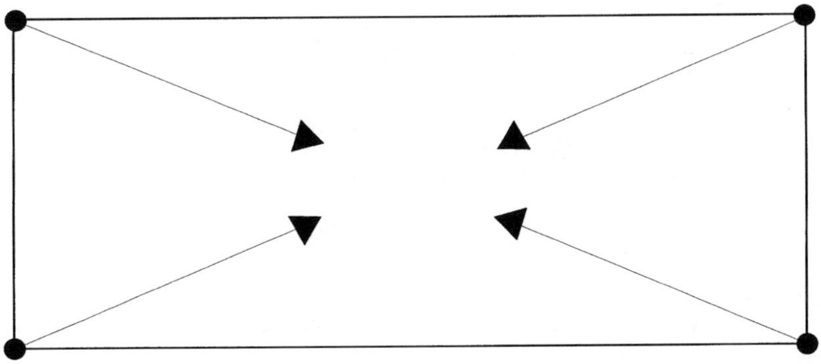

Figure 3-7. Four-corner extension applied to a rectangular block.

need not necessarily be square. In Figure 3-8, staggered, or "triangular," grids may be extended in a similar fashion.

Irregular Blocks Samples taken on an irregularly spaced grid can be extended as the central point into a four-sided block of varying shape. An example of this type of extension is diagrammed in Figure 3-9.

Polygons A more general form of the irregular block format is the polygon extension. Figure 3-10 shows a typical polygon extension. In this case, several surrounding samples create the shape of the polygon through the construction of

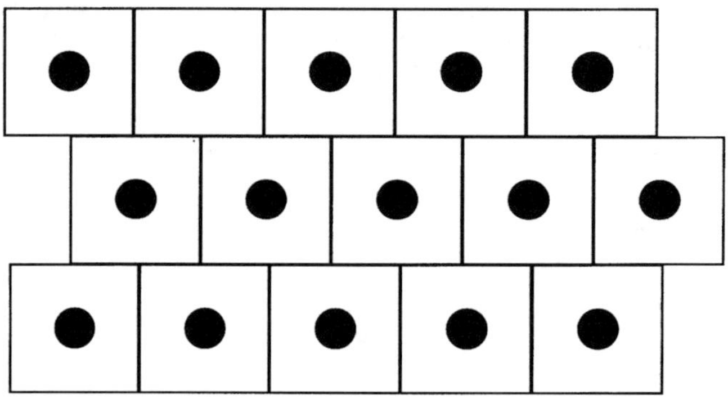

Figure 3-8. Central-point extension method applied to a staggered grid with square blocks. (From *Geostatistical Ore Reserve Estimation*, by M. David, Elsevier, 1977. Courtesy of Michel David.)

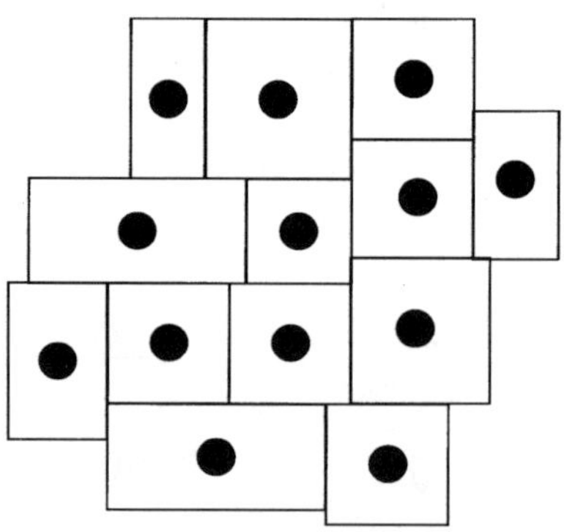

Figure 3-9. Central-point extension method applied to irregularly shaped blocks. (From *Geostatistical Ore Reserve Estimation*, by M. David, Elsevier, 1977. Courtesy of Michel David.)

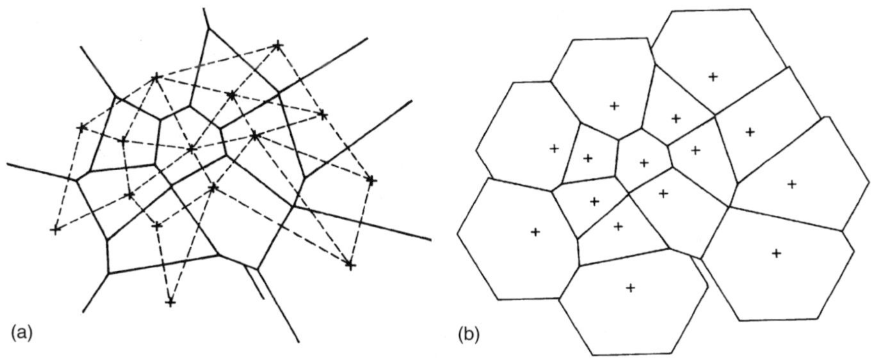

Figure 3-10. Polygonal extension using six-sided polygons. Polygon construction details are shown in (a) and finished polygons are shown in (b).

perpendicular bisectors. The polygons of varying shape again take on the value of the central sample data point. By tradition, most polygons constructed have been limited to six sides, with the polygonal shapes affected mostly at the margins of the data and isolated data points. Although six-sided polygons have some desirable properties, polygons are not restricted to six sides and may have 3, 4, 5, 7, 8, or virtually any number of sides desired. Some polygons having 3, 4, 5, 8, and 12 sides are shown in Figure 3-11a through 3-11e.

One unpleasant aspect of polygonal techniques is the discontinuous nature between points. Polygons create essentially a series of plateaus between data. Whereas data are generally continuous, polygons break away from the idea of continuity and impose an artificial breaks reminiscent of mineral crystal prisms (Fig. 3-12).

Triangular Extension Figure 3-13 shows a triangular form of extension, which is often used for rapid computer contouring. Triangles also overcome the discontinuity problem presented by polygonal approaches. Instead of vertical prisms, a plane is fit between three adjacent points. The question then becomes, which three points? While several methods of triangular construction are available, the technique of Delauney triangulation offers easy calculation, with the additional benefit of having resultant triangles as close to equilateral as possible.

Cross Sections The more three-dimensional (actually 2½ dimensions) type of extension technique of cross sections is shown in Figure 3-14. Cross-section outlines in two dimensions are extended and matched with others at either regular or varying cross-section interval distances.

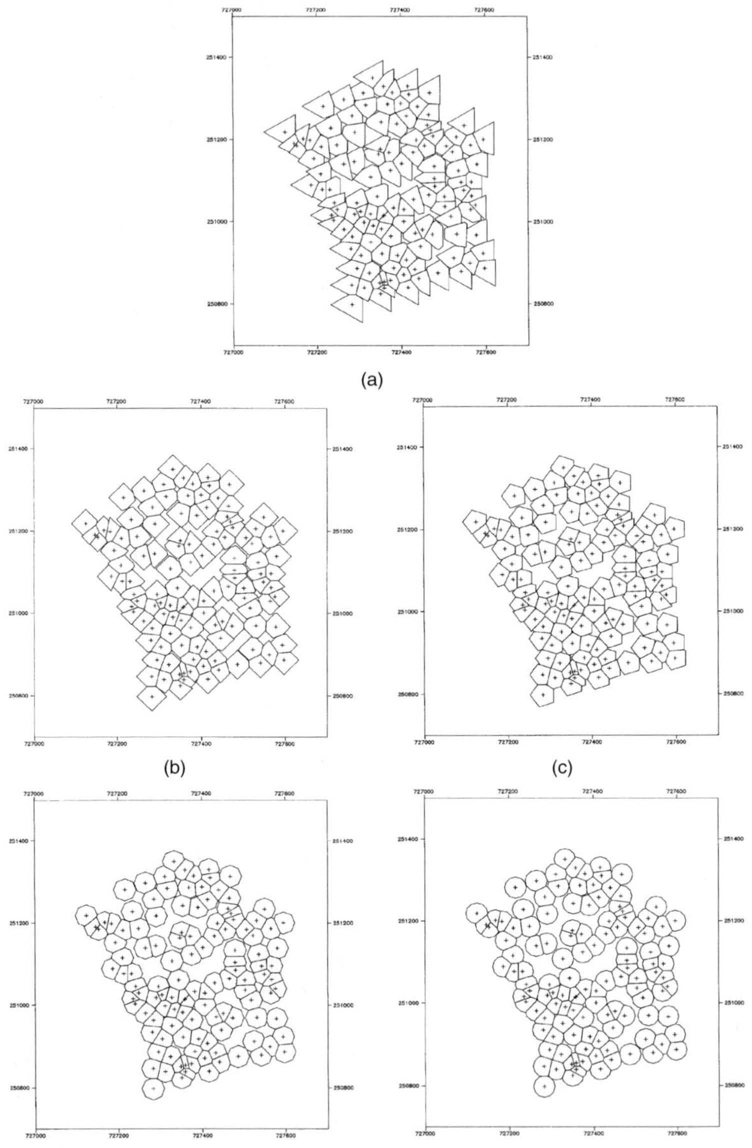

Figure 3-11. Polygonal extension using polygons with 3, 4, 5, 8, and 12 sides. (Courtesy of Geostat Systems International, Inc.)

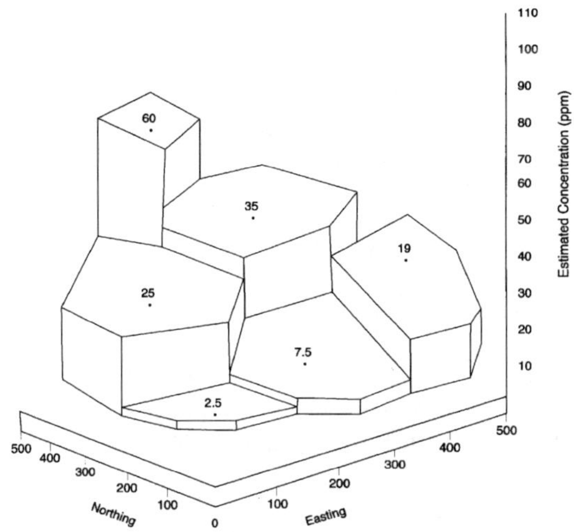

Figure 3-12. Demonstration of the abruptness of polygonal extension showing vertical discontinuities. (From *An Introduction to Applied Geostatistics*, by E.H. Isaaks and R.M. Srivastava. Copyright ©1989 by Oxford University Press, Inc. Used by permission of Oxford University Press, Inc.)

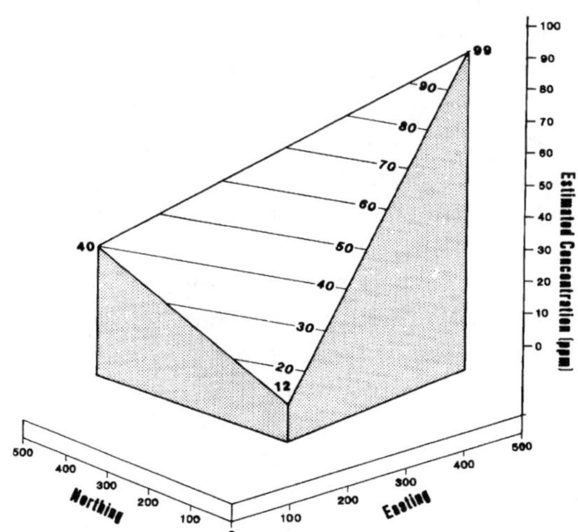

Figure 3-13. Triangular extension method. (From *An Introduction to Applied Geostatistics*, by E.H. Isaaks and R.M. Srivastava. Copyright ©1989 by Oxford University Press, Inc. Used by permission of Oxford University Press, Inc.)

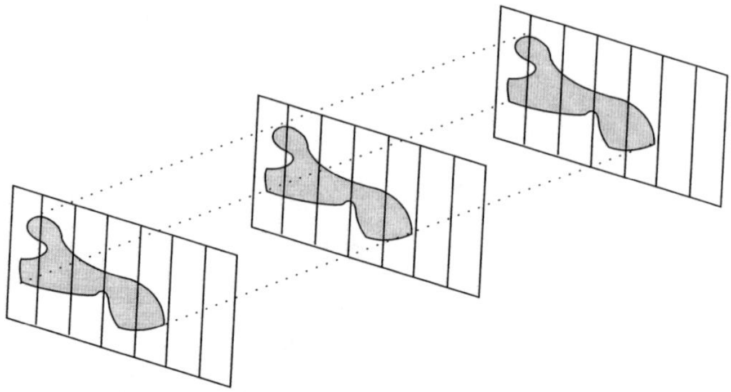

Figure 3-14. Cross-section extension method. (From *Geostatistical Ore Reserve Estimation*, by M. David, Elsevier, 1977. Courtesy of Michel David.)

Distance-Weighting Methods Many geometric extension methods use only a single or very few sample data points to estimate the value of an area or volume. The neighborhood concept suggests that perhaps it is appropriate to incorporate more than one sample in the estimation of an area.

Distance-weighting methods are known for the large number of repetitive calculations required for each location estimate. As such, these approaches gained rapidly in popularity after computers became generally available. These methods assign to a point, block, or area a linear combination of the concentrations of the surrounding samples. Whatever the weighting method, this is again the extension of the concentration of some samples to a larger volume.

Many distance-weighting functions have been developed for extension and estimation. The geostatistical kriging estimator is one such function. Kriging, like many other estimators, is a fairly complicated mathematical approach. The details of the kriging estimator will be described in Chapter 13. Other types of estimators include spline functions, minimum curvature, and trend surface analysis (Fig. 2-10).

One distance-weighting function that is easily described is known as *inverse power distance* (IPD). The IPD approach embodies two basic ideas: (1) More than a single sample should be used in the estimate; and (2) samples closer to the quantity being estimated should receive more weight than those farther from the block.

In addition, IPD allows the estimator to select a power that is deemed appropriate for the estimation. Typical powers used by practitioners include 1 and 2 (d^1 and d^2) for contouring applications. A power of 20 (d^{20}) can be used to simulate polygonal estimation. Because the power of 2 is commonly used, this technique is often called *inverse squared distance* (ISD). An example IPD calculation appears in Figure 3-15.

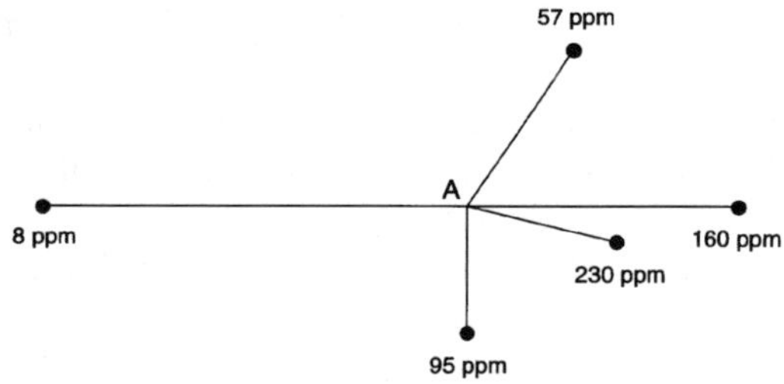

$$A = \frac{\frac{8}{70^2} + \frac{95}{30^2} + \frac{57}{33^2} + \frac{160}{50^2} + \frac{230}{21^2}}{\frac{1}{70^2} + \frac{1}{30^2} + \frac{1}{33^2} + \frac{1}{50^2} + \frac{1}{21^2}}$$

$$A = 152 \text{ ppm}$$

Inverse Power Distance With Power = 2 = Inverse Squared Distance

Figure 3-15. Estimation of an unsampled point location using the technique of inverse power distance.

3.5.3 Estimation Error

It has been shown that samples in a local neighborhood are often similar but not strictly equivalent. Even though small at times, there is a gradient of concentration between nearby samples. This suggests that, if a line were drawn halfway between two nearby samples (a perpendicular bisector perhaps), the actual concentration at the halfway point might be roughly the average of the two samples.

The construction of such a divisor is standard practice in geometric weighting methods and is implicit in distance-weighting approaches. Thus, when we extend the value of a central sample to the edges and corners of the block (the halfway points), we now expect that some degree of change will have occurred in the concentration by the time the extension reaches the edge.

If the concentration is presumed to change at the edges, then, by definition, the estimation (at least at the edges) is wrong and an error has been made. The block average is likely not to be strictly equal to the value of the central sample; instead, it will be equal to some combination of the actual concentration gradients throughout the block. Anticipating that an extension may produce an error is not just academic supposition; experience has shown that errors do indeed occur during the estimation process.

The validity of an estimation method can be assessed by the magnitude of the errors that occur during the estimation process. Unfortunately, we cannot predict the magnitude and sign of an individual error—if the error were known, we would simply fix it! In fact, only the distribution of errors can be accurately predicted in advance.

Armed with this knowledge, we could attempt to design the best possible estimator. What characteristics should this estimator have? It is first necessary to define the term *best*. This would, of course, be the estimator that, on average, yielded the smallest errors, considering all the blocks being estimated. Another desirable feature would be the absence of systematic overestimation or underestimation. Later on, it will also be shown that, when a threshold cutoff is involved, the classification should also be correct, so that correct remedial decisions can be made.

3.6 Distribution of the Estimation Error

Now that actual errors are being calculated, the concepts of precision, accuracy, and bias become increasingly important. Optimally, the average of the estimate must not deviate from the actual concentration value. Deviation indicates a lack of accuracy. Since estimates will never exactly mirror the true values, we hope that the magnitude of the errors will have a low absolute value. Statistically, this equates to a low variance. We may know that some of the errors will be larger than others, but we hope that a very small number of them will be large. Finally, it will be much more satisfying if there is no systematic error or bias in the estimates.

Once again, these concepts can be presented visually. For error distributions, it is useful to employ the histogram graph. Figure 3-16 shows three identical histogram graphs, each representing the distribution of the errors made during an estimation process. These errors are often referred to as *residuals*. The distribution in Figure 3-16a has an average value of less than zero, indicating that a bias toward underestimation exists. Conversely, the distribution of residuals in Figure 3-16b has a positive average, resulting from consistent overestimation. Figure 3-16c presents a more balanced and acceptable situation, where the average value is equal to zero, indicating a lack of biased estimates.

It should be noted that each of these diagrams shows the same precision for the estimator employed. Only Figure 3-16c can be considered unbiased (also somewhat more accurate than the others).

Figure 3-17 is the histogram visualization of precision. Each residual histogram has a mean of zero, indicating unbiasedness. However, the width or spread of the distribution (indicating the variance) is significantly smaller in Figure 3-17a than in Figure 3-17b. These distributions represent two different types of estimators, perhaps polygonal estimation vs. inverse square distance. Given that both distributions are unbiased, the estimator used to produce Figure 3-17a would be preferred.

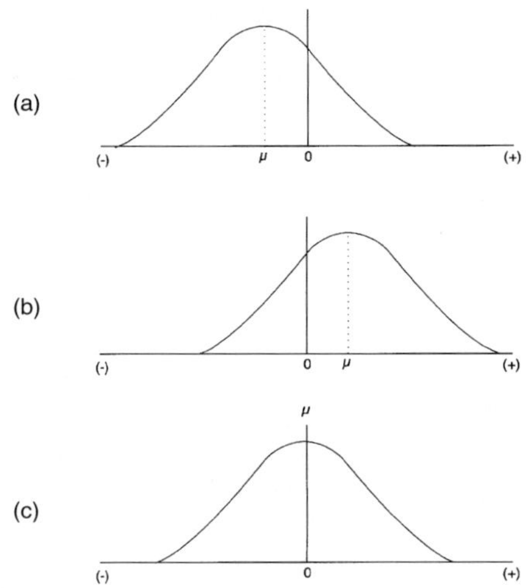

Figure 3-16. Examples of bias in sample distributions: (a) negative bias, (b) positive bias, and (c) no bias.

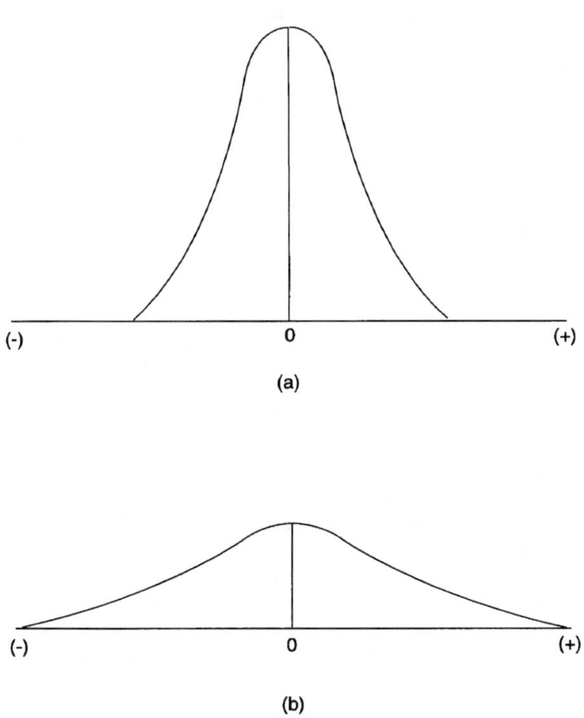

Figure 3-17. The concept of precision shown by means of distributions: (a) The narrower distribution indicates greater precision; (b) the wider distribution indicates less precision.

Table 3-1. Comparison of estimates produced by different estimators (Kriging and Inverse Squared Distance) to the real data.

	Value				
Sample Number	**Real**	**Kriged**	**ISD**	**Real–Kriged**	**Real–ISD**
1	1671	1536.0	1304.1	−135.0	−366.9
2	993	1146.2	1219.0	153.2	226.0
3	1630	1647.8	1546.8	17.8	−83.2
4	1439	1509.3	1559.0	70.3	120.0
5	746	738.5	766.1	−7.5	20.1
6	667	757.4	786.8	90.4	119.8
7	1254	1284.1	1447.2	30.1	193.2
8	1318	1267.9	1368.3	−50.1	50.3
9	1154	1143.6	991.9	−10.4	−60.2
10	1012	1012.6	1172.3	0.6	−20.3
11	827	1017.3	1206.0	190.3	379.0
12	1726	1736.9	1807.6	10.9	144.6
13	1105	933.7	913.8	−171.3	−191.2
14	1239	1209.3	1218.9	−29.7	−20.1
15	921	919.2	808.0	−1.8	−113.0
16	644	689.8	674.4	45.8	30.4
17	1452	1410.8	1512.7	−41.2	60.7
18	1220	1176.1	1169.6	−43.9	−50.4
19	1954	1803.0	1713.0	−151.0	−241.0
20	1389	1588.3	1850.6	199.3	461.6
21	976	996.7	1131.1	20.7	155.1
22	1330	1208.6	1087.5	−121.4	−242.5
23	863	893.5	1053.5	30.5	190.1
24	2034	1805.4	1813.2	−183.6	−220.8
25	1587	1717.1	1838.1	130.1	251.1
26	956	936.9	938.7	−19.1	−17.3
27	1131	1106.0	1080.7	−25.0	−50.3
28	1338	1227.5	1320.2	−110.5	−17.8
29	2267	1973.5	1868.8	−293.5	−398.2
30	1541	1365.3	1450.9	−175.7	−90.1
31	792	1071.6	1132.0	279.6	340.0
32	974	962.6	994.3	−11.4	20.3
33	1275	1244.4	1254.9	−30.6	160.3
34	1553	1302.5	1362.6	−250.5	−190.4
35	1076	1453.1	1813.3	377.1	737.3
36	588	738.0	908.0	150.0	320.0
37	964	984.1	1145.1	20.1	181.1
38	1298	1349.1	1459.1	51.1	161.3

3.6.1 Example Comparison of Estimators

Table 3-1 lists a series of real concentrations along with two series of estimated concentrations for the same blocks. The first estimated series was produced using kriging, the second using inverse squared distance (ISD). The blocks reside on a site that has mine tailings containing significant concentrations of lead (Pb).

For each estimation series, a sum of errors, an average error and standard deviation has been calculated. For kriging, the sum of errors is a very low 4.70 ppm. The average error for kriged estimates is ±98.2 ppm, with a standard deviation of 95.0 ppm. For ISD, the sum of errors is 192.3 ppm. The average error is 176.2 ppm with a standard deviation of 152.6 ppm.

Comparing the error sums for each estimator, we see that the kriging has (on average) a much lower estimation error. Whereas both estimators are biased to the high side, kriging gives a much lower bias in the estimates than ISD. Kriged estimates are also more accurate. Based on these results, kriging would be considered the estimator of choice in this situation.

Comparing the standard deviations, kriging also yields a much lower variance or spread of the errors. The range of kriging errors is −293.5 ppm to 377.1 ppm (a spread of 670.6 ppm), whereas the ISD range is −398.2 ppm to 737.3 ppm (a spread of 1135.5 ppm). This means that precision is also higher for kriging, making kriging the clear winner of the two methods for this data set.

Figures 3-18 and 3-19 use histogram graphs to the visualize the relationship of the estimation errors. Both error distributions exhibit a roughly normal shape,

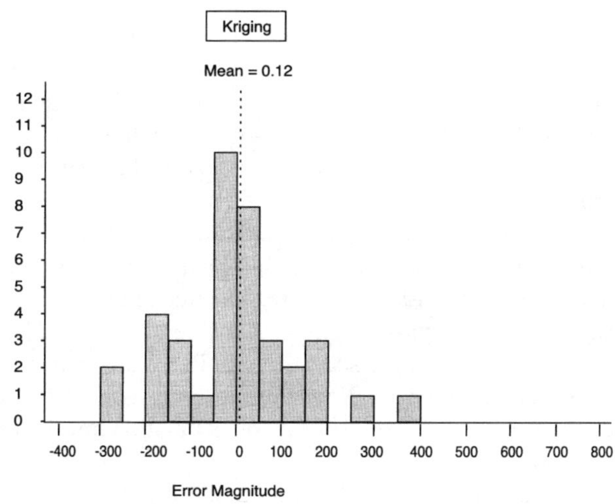

Figure 3-18. Histogram of error values resulting from estimation using kriging.

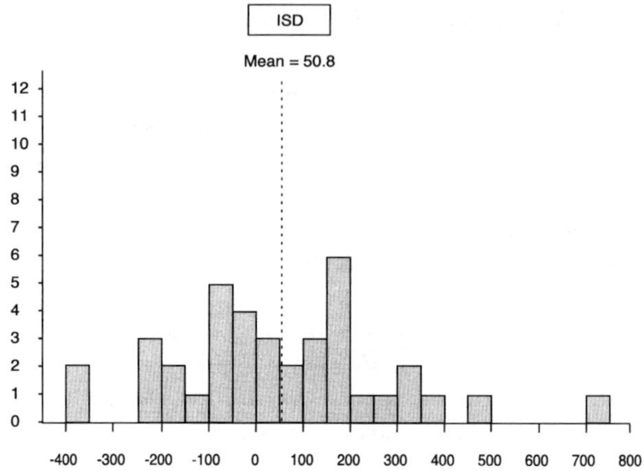

Figure 3-19. Histogram of error values resulting from estimation using inverse squared distance.

but the distribution for the kriged estimates is more tightly centered around zero; that is, it has a preferable smaller variance.

3.7 Misclassification Errors

Precision, accuracy, and unbiasedness are not the only criteria by which we may wish to select an estimator. In many cases, the statistics of the two distributions will be closer to one another than the two presented here. Equivalent performance makes the selection of an optimal estimator more difficult. In this situation and in situations in which only one threshold concentration is of concern, another procedure may be utilized.

Often, maintaining small errors and unbiasedness will be superseded by classification issues. For example, if an unsampled area has a concentration of 1000 ppm, an estimator will seldom estimate the value of the area to be 1000 ppm exactly. An estimate of 980 ppm might be considered a "good" estimate or a "good" approximation based on the accuracy. Similarly, an estimate of 1010 ppm might also be considered accurate. Note that, in each case (980 ppm and 1010 ppm), the percent error is 2% or less. For spatial estimation, this level of accuracy is considered very good.

For remedial activities, critical threshold concentrations are established for each contaminant of concern. Areas estimated to have concentrations above this threshold will receive treatment, but areas estimated to be below the threshold will not. Given a strict threshold level, "small" estimation errors or inaccuracies may be quite significant. In the following discussion, the difference between a quantitative error (real minus estimated) will be contrasted with a qualitative error (misclassification).

3.7.1 Type I Error: False Positives

Consider another situation in which the true soil PCB concentration of a block area is 22 ppm and the established threshold for remediation is equal to 25 ppm. Any soils estimated to be 25 ppm or above will be excavated and treated. If the block is estimated to have a PCB concentration greater than or equal to 25 ppm (29 perhaps), two errors have been made. The first is the *error of estimation* (29 ppm minus 22 ppm or 7 ppm). The second is the *misclassification* error, so named because the block has been incorrectly classified above the 25 ppm threshold.

Despite a relatively "small" error of estimation, the misclassification error in this case (as well as generally) will have a greater impact on the remedial consequences than will the estimation error. The reason for this is that a treatment decision is being made based on the classification of the estimate. An area whose estimated concentration is above 25 ppm that truly has a concentration less than 25 ppm will be remediated. This remediation is inappropriate and increases the remedial expense to those who are funding the cleanup. This form of misclassification is called Type I Error (Rendu 1980). It is also referred to as *overestimation* or, as discussed in Chapter 1, a *false positive*.

If the estimated concentration had been 15 ppm, the error of estimation would have been of exactly the same magnitude as before (7 ppm vs. 7 ppm). Although the magnitude of the error is equal in both cases, an estimate of 15 ppm places the block in a different material classification. More important, this time it is a correct classification. Despite the 7 ppm error of estimation, this block will not incur inappropriate remedial costs.

3.7.2 Type II Error: False Negatives

Type II Error is the counterpart of Type I Error in terms of misclassification. If the estimated concentration of the block described above does not exceed the threshold (21 ppm, for example), the block goes untreated based on the 25 ppm action threshold. If the true concentration of the block is really above 25 ppm, a decision will be made to forgo treatment. This incorrect decision is known as Type II Error, also known as *underestimation* or *false negatives*. In this case, potential future health risk liabilities (of unknown magnitude) arise as opposed to the direct, immediate cost of unnecessary remediation. Even though the long-term liability may indeed prove to have a cost, it is very difficult to estimate and is beyond the scope of this book.

If the estimation had been 62 ppm, the error of estimation would have been much greater than that for the estimate of 21 ppm. However, an estimate of 62 ppm will cause a correct classification decision to be made despite a large estimation error. Estimation errors for areas with concentrations above the threshold limit are often very large. For example, it is not uncommon for soils PCB concentrations to reach 25,000 ppm or more although the threshold limit may be closer to 25 ppm. Also, in the case of PCBs, the contaminant distribution is often highly erratic, with concentrations exhibiting steep gradients over short distances. Therefore, an area could be estimated to have a PCB concentration of 1000 to 5000 ppm when its actual concentration may be closer to 500 ppm. Nevertheless, the correct classification will be made.

Figure 3-20 shows a generalized visual representation of the correct and incorrect classifications described above. This graph is known as a *misclassification ellipse*. To develop the ellipse, the estimated concentrations are plotted on the *X* axis, and the true concentrations are plotted on the *Y* axis. If an estimation method were perfect, the estimated concentration would always equal the true concentration. Plotting true vs. estimated in this situation would result in a graph with a 45° line emanating from the origin of the graph.

Unfortunately, perfect estimation is seldom achieved and the estimated concentrations fall off the 45° line. As repeated estimation experiments are performed

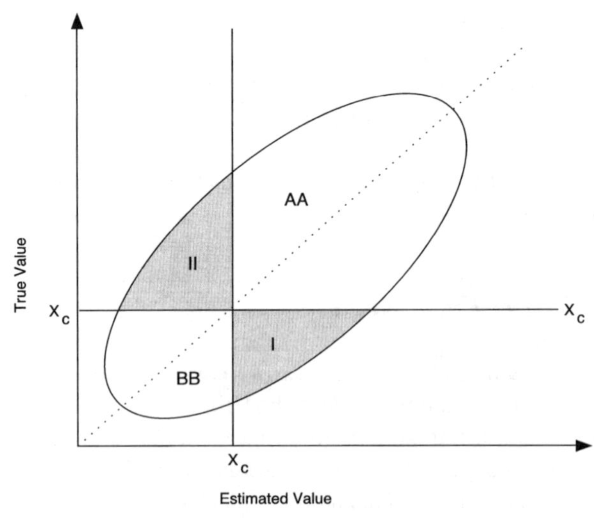

Figure 3-20. The misclassification ellipse. (From *Optimization of Sampling Policies: A Geostatistical Approach*, by J-M. Rendu, MMIJ-AIME, 1980, with permission.)

and the true vs. estimated values plotted, a scatter of points forms. This scatter plot often takes on a roughly elliptical shape; hence the name misclassification ellipse.

To ascertain the quantities and types of misclassifications present, the threshold concentration is also plotted on the graph (Fig. 3-20), perpendicular to axes for both the true and estimated values. By means of this process, four distinct quadrants of classification emerge. These correspond to the error and correct classification types, with each representing a different decision category.

The lower left quadrant of Figure 3-20 (labeled BB) contains concentration values for which both the estimated and the true values are below the threshold. In this case, correct classification has been made, so that the area will appropriately avoid treatment. Similarly, in the upper right quadrant (labeled AA), both the estimated and true values fall above the threshold. Again, correct classification decisions result, this time with the material being scheduled for appropriate treatment.

The remaining quadrants, labeled I and II, present situations of problematic classification. These correspond to Type I and Type II Errors. The lower right quadrant, labeled I, shows the estimated concentrations above the threshold limit, although the true concentrations are below the threshold. This is Type I Error. In the upper left quadrant, labeled II, the estimated values are below the threshold but the actual values fall above the threshold. This situation is Type II Error.

Table 3-2 presents the kriging and ISD data in a misclassification format. Instead of listing the actual error, the type of classification is presented. For this approach, the magnitude of the error takes a back seat to the decision made on the classification. Differences in classification between the two methods are highlighted with an asterisk.

As with the statistical comparison, the kriging estimates again emerge the winners since kriging produces fewer misclassification errors. Kriging produced three Type I Errors and one Type II Error, whereas ISD produced six Type I Errors and two Type II Errors.

Scatter plots for the kriged and ISD data can be seen in Figure 3-21. The kriged values cluster around the 45° line much more tightly than the ISD estimates do. This "slimmer" ellipse is generally an indication of better estimation and more accurate classification.

The term *Typecasting of Error* has been applied to the technique of analyzing misclassification error (Myers 1991). This approach is highly useful in conjunction with the method of cross validation. This approach can be used to determine optimal variogram and optimal kriging parameters employing performance-based goals.

The Typecasting of Error approach may not always point to kriging as the best interpolation method for a particular data set. For different data, a spline or other interpolation method may be more appropriate. This can assist greatly in problems discussed in Chapter 2. Typecasting techniques will be discussed in detail in later chapters.

Table 3-2. Comparison of number of misclassification errors produced by two estimation techniques (Kriging vs. Inverse Squared Distance).

	Value			*Class*	
Sample Number	**Real**	**Kriged**	**ISD**	**Kriged**	**ISD**
1	1671	1536.0	1304.1	AA	AA
2	993	1146.2	1219.0	I	I
3	1630	1647.8	1546.8	AA	AA
4	1439	1509.3	1559.0	AA	AA
5	746	738.5	766.1	BB	BB
6	667	757.4	787.0	BB	BB
7	1254	1284.1	1447.2	AA	AA
8	1318	1267.9	1368.3	AA	AA
9	1154	1143.6	991.9	AA	II[a]
10	1012	1012.6	1172.3	AA	AA
11	827	1017.3	1206.0	I	I
12	1726	1736.9	1807.6	AA	AA
13	1105	933.7	913.8	II	II
14	1239	1209.3	1218.9	AA	AA
15	921	919.2	808.0	BB	BB
16	644	689.8	674.4	BB	BB
17	1452	1410.8	1512.7	AA	AA
18	1220	1176.1	1169.6	AA	AA
19	1954	1803.0	1713.0	AA	AA
20	1389	1588.3	1850.6	AA	AA
21	976	996.7	1131.1	BB	I[a]
22	1330	1208.6	1087.5	AA	AA
23	863	893.5	1053.5	BB	I
24	2034	1850.4	1813.2	AA	AA
25	1587	1717.1	1838.1	AA	AA
26	956	936.9	938.7	BB	BB
27	1131	1106.0	1080.7	AA	AA
28	1338	1227.5	1320.2	AA	AA
29	2267	1973.5	1868.8	AA	AA
30	1541	1365.3	1450.9	AA	AA
31	792	1071.6	1132.0	I	I
32	974	962.6	994.3	BB	BB
33	1275	1244.4	1254.9	AA	AA
34	1553	1302.5	1362.6	AA	AA
35	1076	1453.1	1813.3	AA	AA
36	588	738.0	908.0	BB	BB
37	964	984.1	1145.1	BB	I[a]
38	1298	1349.1	1459.3	AA	AA

[a]Estimates where ISD produces a misclassification while Kriging produces the correct classification.

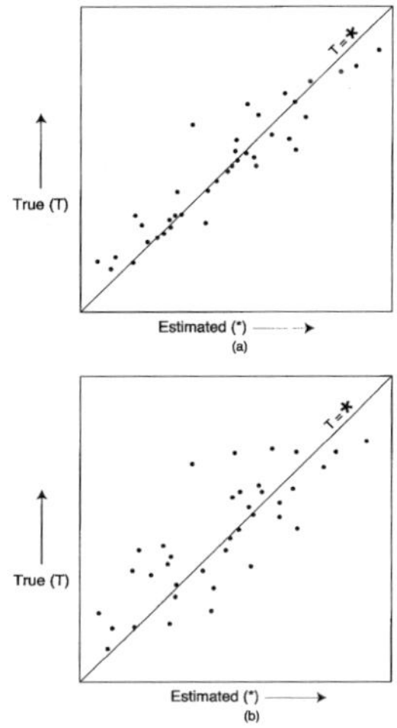

Figure 3-21. Scatter plot of estimated vs. true values: (a) Kriged estimates are used; (b) inverse squared distance estimates are used.

Conclusion

Prediction is a risky game as one tries to generalize from specific sample data to the surrounding area. Estimation is an unfortunate necessity as time and resource constraints limit other options. Risk and uncertainty are a priori entities as they describe potential problems. Error is a posteriori in that one can measure or define a specific result. A number of techniques for estimating spatial data have been presented, with varying levels of complexity. Although not comprehensive, this sample of techniques is sufficient to demonstrate that any estimation technique has a degree of associated error. It is important to realize that the difference between a deterministic algorithm and a statistical model is that the statistical algorithm provides an error of estimation. Within the realm of error, different types exist that must be managed in different ways, depending on the DQO and other needs of a project. Chapter 3 has defined some general conditions and considerations; subsequent GEM chapters will set out specific techniques for addressing, solving, and managing the errors and uncertainty surrounding estimates and maps through statistics, STP, and GA.

PART 2 Statistical Considerations

Chapter 4

Foundations of Statistics

SYNOPSIS: Statistical inference and estimation are based on information about a population derived from sampling. In order to make valid inferences, it is necessary to define and limit the population precisely. Chapter 4 introduces many of the statistical concepts that are necessary for the GEM process to attain the DQOs for a given project. This chapter begins by discussing the importance of carefully defining populations to be sampled and applicable sampling domains. Two other important GEM themes, equiprobability and support, are introduced, as well as issues related to random sampling. Support will be introduced as a concept that must be considered during all GEM phases. This chapter will also distinguish between deterministic and probabilistic (statistical) data and emphasize the selection of an appropriate model so that inferences and decisions will be well founded.

KEY WORDS: Universe, Population, Sampling Domain, Equiprobability, Sample Support, Exposure Unit, Superfund Accelerated Cleanup Model, Deterministic Data, Probabilistic Data, Model.

4.1 The Universe or Population

In order to produce successful and meaningful statistical analyses, it is important to understand the terms *population* and *sample*. These fundamental concepts often receive little or no attention in statistical texts as applications outside the earth sciences are often more straightforward. These ideas are crucial to statistical applications in geologic areas because it is often important to discriminate between populations (background vs. plume, ore vs. waste, etc.). Definition of target populations for sampling is a key part of the DQO process and must, therefore, be done with care.

A population or universe is the entire selection that is available for study. In the earth sciences, this generally relates to the total mass of material in which one is interested. Two sizes of universes exist—finite and infinite. A finite universe is measurable but may be so large that it is considered infinite for computational and practical purposes. As an example, to appraise the amount of trace dioxin contained in manufactured pesticides, the universe could be every barrel of pesticide ever produced. This value is finite but its size is vast. Generally, one is interested in only a small portion of a finite universe or population.

In the earth sciences, a universe can have one or more characteristics. Frequently it is necessary to understand whether the population is unidimensional or multidimensional. Other pertinent factors enter the decision arena in the form, for example, of geologic strata, soil units, transport mechanisms, erosional conditions, historical property use by man, and topography. It is critical for statisticians involved in the site investigation to appreciate these separate universes (or influences in a single universe) and to design the sampling and data interpretation accordingly.

Random samples or, more simply, samples are the portion of the universe that is used to calculate the population statistics. Population statistics are estimates of the true (but unknown) population parameters and provide an indication of the range and clustering of the sample data values, along with a measure of the variability. When questions of uncertainty and risk are considered, the variability around a mean value is often a key question. As such, the GEM process will frequently focus on characterizing and minimizing the variability in a system.

In the earth sciences, samples are used to map the variables of interest. The number of sample data available is always finite given that time and money resources are limited. Generally, the greater the number of samples, the more accurate the estimation of the population parameters.

4.2 Sampling Domains

The concept of sampling domains is a logical extension of the idea of a universe or population. Although a universe could be every barrel of dioxin ever produced, the population of interest may be only the PCB-contaminated groundwater at a particular hazardous waste site. This puts the population of interest in a more manageable context.

The concept of sampling domains is of critical importance to the correct design and implementation of a successful sampling program and holds true for both spatial and nonspatial statistics. Mixing domains is an invitation to both error and difficulty and should be avoided to the maximum extent possible and practicable.

If a population and domain are not clearly defined and carefully related to the objectives (DQOs) of the study, the data collected may contain little useful information for the purpose initially intended. Data may be insufficient in number or simply inappropriate to answer the questions required.

For example, if we wish to determine the average life of lightbulbs, we should first define carefully what type of lightbulbs will be appropriate for the study. Factors may be size, wattage, filament type, conditions under which the bulb may be used, constancy of use, and so on. The definition may be very specific with regard to these factors. It would probably not be appropriate, for instance, to mix fluorescent lights with incandescent.

Geologic situations pose similar and often even more complex problems of domain separation. Mining geologists and engineers, the first major users of geostatistics, have almost always needed to deal with a variety of geologic conditions at a mine. These include different rock types, different mineralization zones, different mineralization episodes, different oxidation or reduction conditions, variability in fractures in the rock, and more. An example of a geologic setting for a mining project is shown in Figure 4-1.

Sometimes, the economic mineralization is confined to a single geologic unit or layer, greatly simplifying domain analysis. Often, however, the mineralization encompasses several units and features. Some of these units may be more suited than others to mineral deposition. Thus, the quantities of minerals to be extracted may be different in the different layers. The different domains are likely to have different population and spatial correlation statistics as well.

The sampling and subsequent geostatistical analysis should take these differences into consideration. This is not simply an academic exercise; worldwide practice throughout the mining industry has confirmed that these different geologic domains are statistically different, particularly with respect to the spatial correlation parameters (Barnes 1982).

As with lightbulbs or mines, a hazardous waste site often needs to be subdivided into different areas for sampling purposes. This concept is similar to what commonly occurs in the designation of operable units (OUs) at a site by regulatory agencies. The division into sampling domains often goes beyond the OU distinction. For example, a single OU may contain two or more domains that need to be treated separately. Conversely, a sampling domain may cross OU boundaries.

One simple example of domain separation is the case of multiple aquifers (Fig. 4-2). Surface contamination may have a greater impact on the shallow

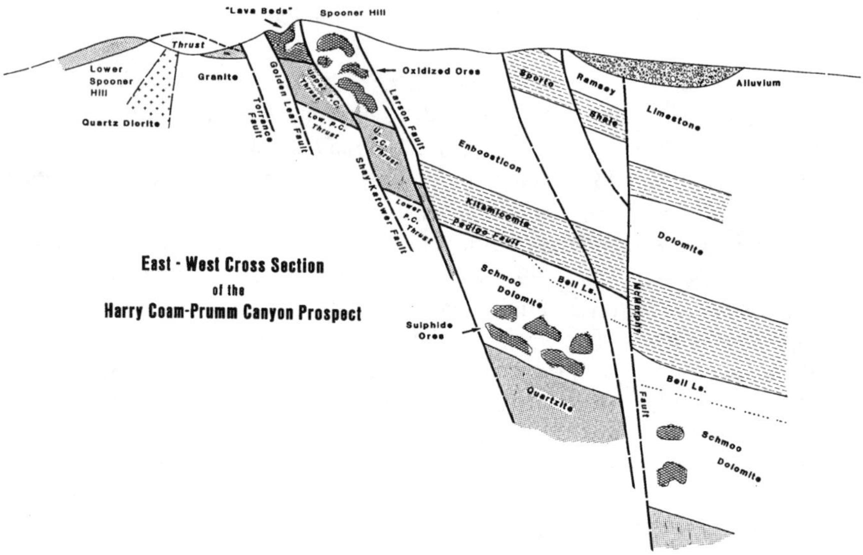

Figure 4-1. Domain illustration using a geologic setting from a mining project.

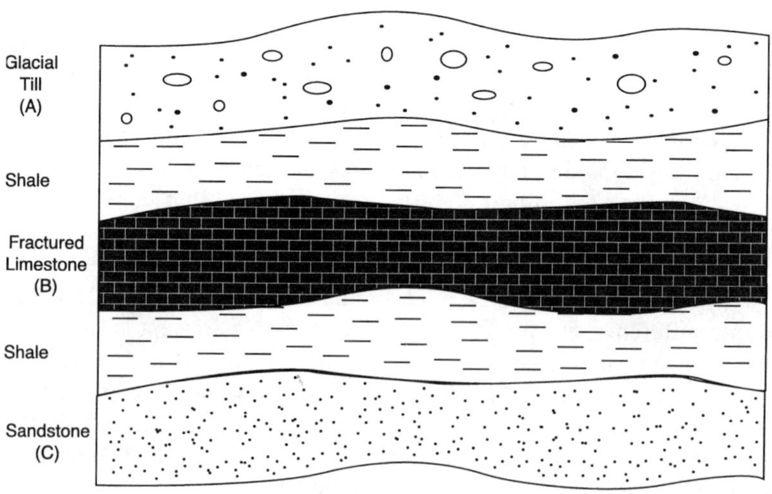

Figure 4-2. Domain illustration: multiaquifer situation.

aquifer while deeper units remain relatively unaffected. In the case in which all aquifers exhibit significant contamination, each would still need to be sampled individually for the porosity and permeability values to be used in hydrologic flow models. It would be risky to assume, without verification, similar conditions in aquifers composed of glacial till, fractured limestone, and porous sandstone.

Figure 4-3 shows an example of domain separation taken from the ReSolve site. The ReSolve site was contaminated with PCBs. For sampling and geostatistical appraisal, the site was divided into several areas. The Waste Management Area (WMA) contained the central processing facility, along with some lagoon areas. Based on operational history and geologic conditions, the Waste Management Area was sampled and modeled as a separate domain.

Surrounding the WMA is a large Wetland Area. The Wetland Area had no active processing components during the site oil reprocessing operations. Thus, it was considered a separate domain for sampling and estimation. It is physically more distant from the former source of contamination and topographically lower than the WMA. The physical composition of the soil was also different, with more clay than the sandy WMA.

Figures 4-4 and 4-5 are histograms of soil PCB concentration data for the WMA and Wetland Area, respectively. Both distributions are skewed, but they differ in shape as well as in number of samples. The WMA population

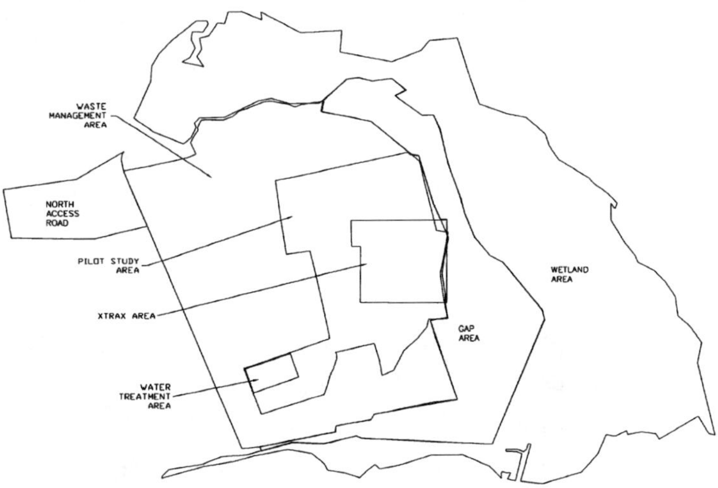

Figure 4-3. Domains for the ReSolve Site. Domains include the Waste Management Area (WMA), the Gap Area, the Wetland Area, and the North Access Road Area. Engineering subdomains of the WMA include the Pilot Study Area, the XTRAX Area, and the Water Treatment Area. (From A geostatistical analysis of PCB concentrations in soils at the ReSolve site, by J.C. Myers, 1992.)

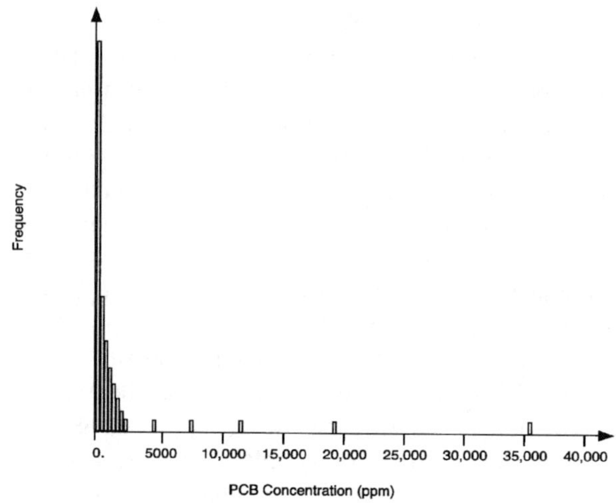

Figure 4-4. Histogram showing the PCB concentration data for the Waste Management Area of the ReSolve site. (From A geostatistical analysis of PCB concentrations in soils at the ReSolve site, by J.C. Myers, 1992.)

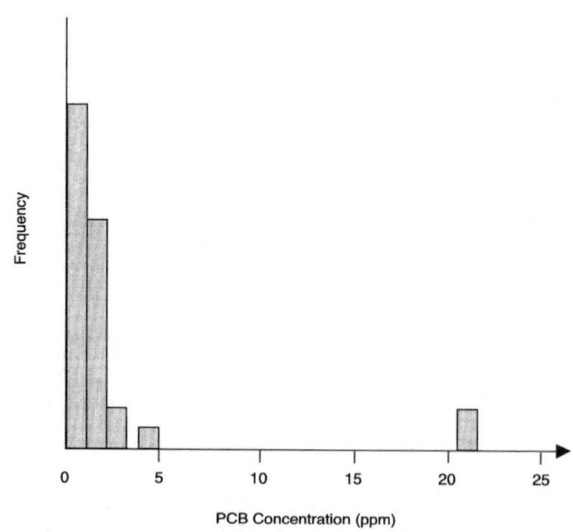

Figure 4-5. Histogram showing the PCB concentration data for the Wetland Area of the ReSolve site. (From A geostatistical analysis of PCB concentrations in soils at the ReSolve site, by J.C. Myers, 1992.)

contained more than 1300, the Wetland Area less than 200. Most noticeable, however, is the dramatic difference in population statistics.

The WMA has a mean, variance, and standard deviation of 365.2 ppm, 3,239,701 $(ppm)^2$, and 1800 ppm, respectively, whereas the Wetland Area parameters are 0.89 ppm, 9.75 $(ppm)^2$, and 3.12 ppm. It is apparent that two different statistical populations exist and that the domain separation was justified, which can be confirmed by statistical comparisons of the estimated population parameters.

Between the WMA and the Wetland Area is a Gap Area. This area was a transition zone between the previous two domains. A final area of concern was the North Access Road. This area was a main access to the site and was composed primarily of road-fill material. The Gap and North Access Road areas were sampled and modeled as separate domains.

In addition to the different physical domains that may be encountered at a site, it may also be important to consider the various temporal domains. Many environmental phenomena oscillate or are cyclical in time. Examples of this behavior include seasonal groundwater levels, tidal oscillation, daily or hourly airborne contaminant levels in a metropolitan area (ozone, carbon monoxide, sulfur dioxide, etc.), seasonal air inversions, even periodic fluctuations in pollutant discharge from industrial sources that can cause "pulses" in air or groundwater plumes. Figure 4-6 is an example of yearly oscillations of average monthly carbon monoxide concentrations in downtown Denver, Colorado, for the years 1990 to 1994.

Domains are subject to change over time. One example comes from the educational field, where Scholastic Aptitude Test (SAT) scores have been declin-

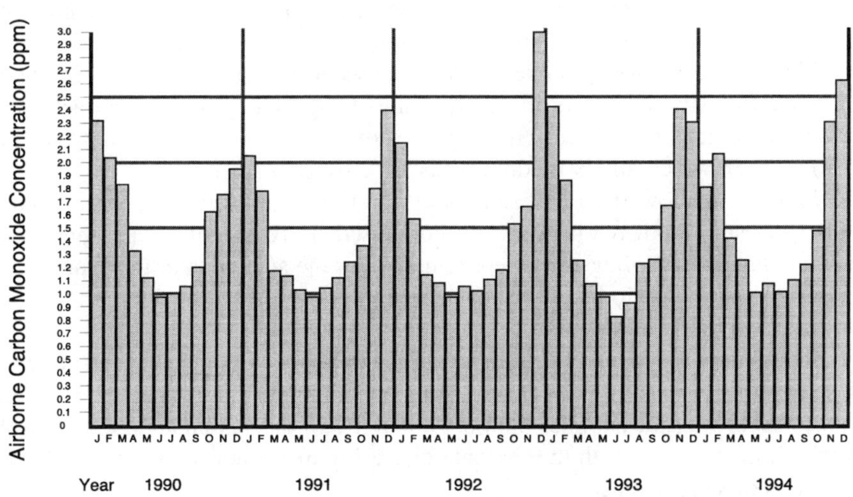

Figure 4-6. Average monthly carbon monoxide concentrations (ppm) in Denver, Colorado, for 1990 through 1994, showing yearly cycles.

ing over the past 30 years. This drop in average SAT scores has been blamed widely on a failing educational system. Whether this claim is true or not, one must note that the tested population has changed. In the past, only the best students were encouraged to attempt the SAT test. Today, a more diverse and more numerous population of students takes the test. This alone would be expected to bring changes to average test scores.

From a visual standpoint, mixing two domains or populations may produce a histogram with two peaks or modes. This is generally a clear indication that two different populations exist. Mixing two populations for use in spatial analysis may cause interference between correlation structures that typically have different spatial correlation characteristics. The individuality of the spatial structure exhibited by separate domains will be discussed in Chapter 10.

4.3 Equiprobability and Random Sampling

To achieve representative sampling, it is often said, one simply needs to take random samples. Random samples are indeed necessary, but pure randomness is not a sufficient condition and often leads to sampling that is not representative.

A major problem with strict random sampling is that clusters tend to arise. Figure 4-7 shows some simple examples using randomly selected numbers from Colorado's state lottery. In Figure 4-7a, six numbers have been highlighted at evenly spaced intervals in order to visualize unclustered selections. In Figure 4-7b, four examples of relatively unclustered random picks are presented, although some clustering occurs in each example.

Figure 4-7c shows examples of four consecutive lottery drawings in 1995. Significant clustering occurs in nearby samples at some point along each line, leaving large gaps elsewhere. Figure 4-7d shows some extreme (but not uncommon) cases in which all six random picks are closely spaced.

Figure 4-8 shows two realizations of purely random sampling, each from a spatial domain. Both realizations are composed of 10 samples each, and both indicate that the sampling is not representative of the spatial area. In Figure 4-8a, the samples show regional clustering, with more samples in the southern half of the area than in the north.

In addition, localized clustering occurs, with pairs of samples occurring very close together. The southwest corner alone contains four samples. This grouping represents 40% of the samples but only about 4% of the spatial area. Figure 4-8b shows similar results, with the regional clustering in the northern half and localized pairs of sample clusters.

Figure 4-9 shows two different sampling programs in which no spatial region appears to be either significantly favored or excluded from the sampling process.

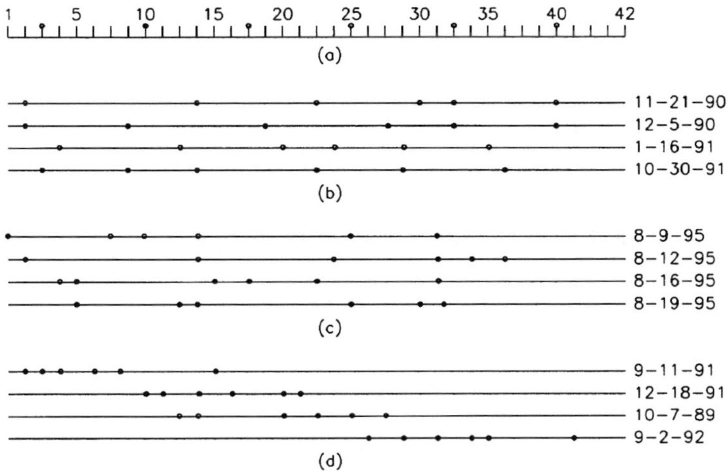

Figure 4-7. Examples of clustering in randomly selected data. (a) Selections are evenly spaced along the line; (b) examples of "evenly" spaced random picks are shown; (c) examples of localized clustering; (d) examples of extreme clustering.

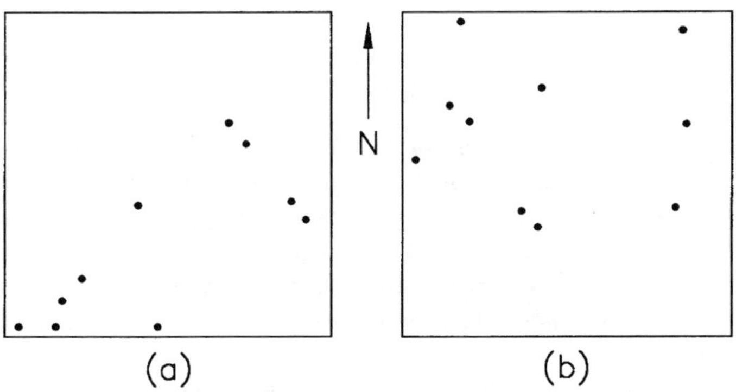

Figure 4-8. Two outcomes of a strict random sample location selection process showing clustering and spatial biases.

These results are much preferred to those obtained in Figure 4-8. Techniques for designing equiprobable spatial sampling programs will be discussed in Chapter 16.

Although random sampling does not actively promote clustering, it should not be surprising when it occurs. Using the lottery as an example, a clustered selection result of 1, 2, 3, 4, 5, 6, is just as likely as 32, 33, 34, 35, 36, 37 or as 3, 10, 17, 24, 31, 38 (evenly spaced) or any other result, assuming that the selections are truly random. By extension, spatial clusters can also be expected.

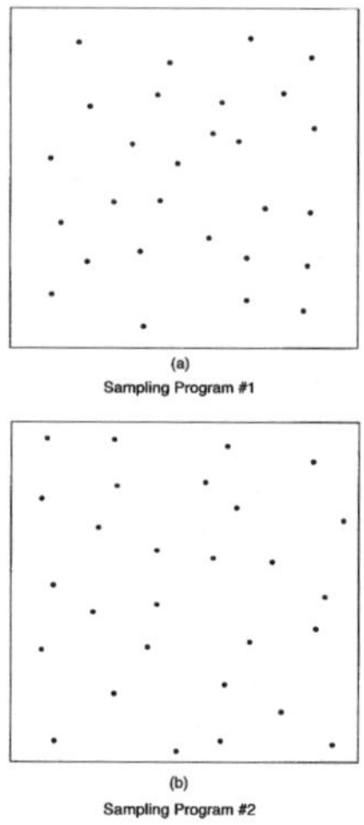

Figure 4-9. Results of equiprobable random sampling for a spatial situation.

4.4 THE CONCEPT OF SUPPORT

4.4.1 The Concept of a Sampling Unit

In Chapter 1, a simple experiment from TQM produced a complicated result. In a similar vein, a simple question will be posed regarding the nature of a sample, and this will be followed by a more complex answer than is intuitively anticipated.

The question is: What constitutes a sample? This seems like a harmless enough question, but it is one that must be precisely defined for earth science applications. To understand this, it is first necessary to examine some traditional sample types.

It is true that, in many areas, including some sciences, a "natural" sampling unit exists for the purposes of measurement. Often, this is an individual organism such as a human being, a dog, or a tree. Measurements of height, weight, lifespan, diameter, and so forth can all be related to the particular organism under study. Political polls measure approval of individual candidates or whether people agree or disagree with particular policy decisions of political leaders.

Earth science data, however, do not typically carry such fundamental, logical, and easily applied units. It is especially true for solid samples such as soil, sediment, and rock. To generalize, paraphrasing Gertrude Stein that a sample is a sample is a sample would be highly misleading. Instead, we must look more closely at the problem; specifically, we must look to a volumetric solution.

Suppose that the objective of a sampling program is to characterize the permeability in an aquifer. During the DQO stage, we discover that many different types of sampling units could be used. A sampling unit could be 5-ft lengths of 2-in.-diam. core, 1-ft lengths of 3-in.-diam. core, 10-ft lengths of 1-in.-diam. core, or some other variation on this theme (Fig. 4-10a to 4-10c).

Obviously, the entities shown in Figure 4-10 are different physical sampling units. What is not obvious is that these different units will generate different sample populations, particularly with respect to the population variances. One or more of these sampling units may be very inappropriate for characterizing the permeability parameters of interest. Variability of the sampling unit may exceed the DQO error tolerance limits.

This revelation makes it clear that the physical size of a sampling unit is very important in characterizing a geologic or environmental area. This concept is known as *support*. With a "natural" sampling unit, support is always implied and understood. Natural units are not implied for earth science sampling; therefore, units can and should be defined.

Support implies that additional information must be supplied in describing estimated population parameters. We must always refrain from making statements regarding a population of concentrations or values unless the support is specified. For example, consider the following statement: The concentrations of lead at the site follow a lognormal distribution, with a mean of 1743 ppm and a standard deviation of 587 ppm. Although the statement sounds quite scientific, it is mean-

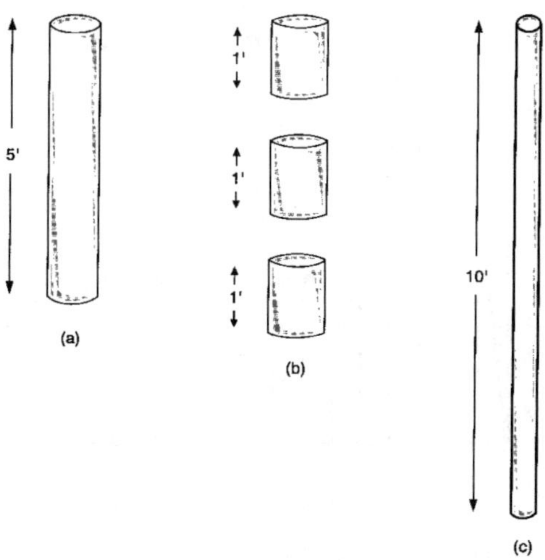

Figure 4-10. Examples of different-sized sample support.

ingless because it does not disclose the nature of the support used in the sampling or subsampling units.

In the field, soil boring cores may be analyzed with x-ray fluorescence devices every 3 to 6 in. because of the speed of the analytical technique. It is not practical, however, to attempt to remediate soil in increments of less than 6 in. or 1 ft due to equipment constraints.

As a practical matter, therefore, it may be necessary to composite mathematically two or more vertical soil samples into a single vertical sample unit. The variance of 3-in. samples would be greater than the variance of 6-in. or 1-ft samples. This compositing process has an effect on both the population statistics and the geostatistical appraisal. Larger variances typically produce estimates that contain correspondingly larger amounts of uncertainty.

4.4.2 Support Effects on Sample Distributions and Statistics

Figure 4-11 shows the effect of sample support on the sample statistics. The distribution of 3-in. samples is shown by the vertical bars, and the distribution of 1-ft samples is shown by the curve. The spread of the 3-in. samples is greater in both directions than that for the 1-ft samples. We can observe that the maximum value, having been diluted by the more prevalent lower concentration material,

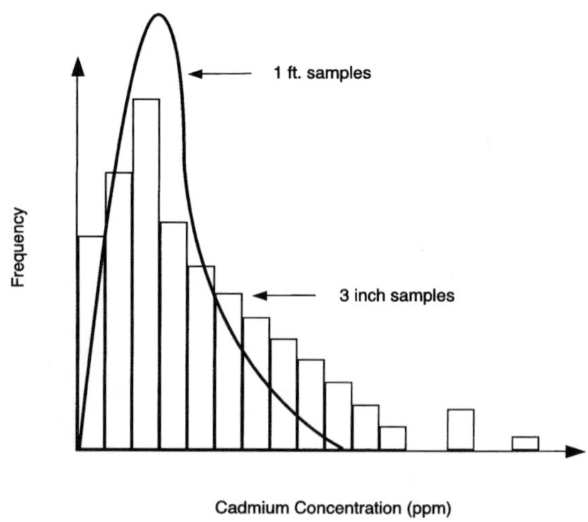

Figure 4-11. Histogram of 3-in. samples (vertical bars) vs. 12- in. samples (smooth curve).

decreases as support increases. Conversely, there is a corresponding decrease in the number of extremely low concentrations although the skewness of the original data makes this change less noticeable.

Increasing the support "squeezes" both ends of the distribution, reducing the spread, along with the variance and standard deviation. Note, however, that the mean value remains unaffected.

Although not universally intuitive, this reduction in variability due to support has a logical origin. Consider a case in which sampling is performed by means of a teaspoon. One teaspoon contains a nugget of galena (Pb) from mine tailings or a PCB-laden glob of oil. Analysis of this sample reveals extreme concentrations. The next teaspoon may be completely barren of the contaminant. Further samples are likely to reflect this dramatic fluctuation between samples.

The natural heterogeneity of materials and contaminant distribution within the materials produces wide variations in concentrations for teaspoon-sized samples. These fluctuations can be dampened by increasing the support. For example, if we were to fill a number of quart jars with teaspoon-sized samples, mix thoroughly, and analyze, the extreme highs and lows of the teaspoon units used to fill the jars would be mitigated. The variability of the quart units would be less than the variability of the teaspoon units. This is the same phenomenon as that seen in Figure 4-11.

Use of a teaspoon would be considered an extreme case for field sampling, and we might conclude that such an example has no practical relevance. On the contrary; when a field sample reaches the laboratory, just such a small amount of material may actually be used to represent the entire sample. Successively smaller volumes of material are taken as the sample approaches the laboratory analytical device. Thus, the variability of the initial sample data will increase unless appropriate subsampling methods are applied.

4.4.3 Support Effects in Geostatistical Appraisal

In GA, the concept of support is expanded into what is known as the volume-variance relationship. Just as samples with smaller support have higher variances, smaller units of estimation have higher variances. For example, the variance of $10 \times 10 \times 1$ ft blocks is higher than that for $50 \times 50 \times 1$ ft blocks. An example of this relationship is shown in Figure 4-12.

Estimating smaller and smaller blocks often appears to result in more map resolution. This is a mistake because the data are being asked to deliver more information than they contain. In reality, the sample data are merely spread thinner and thinner while the uncertainty associated with any particular block estimate increases rapidly.

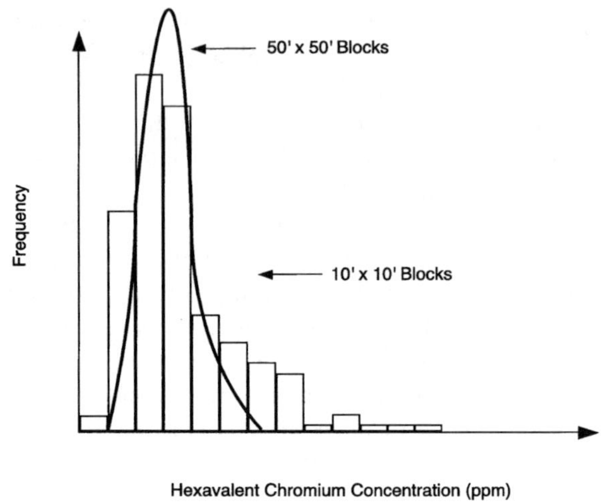

Figure 4-12. Histogram of 10 × 10 ft (vertical bars) blocks vs. 50 x 50 ft blocks (smooth curve).

4.4.4 Exposure Units: DQO and Risk-Based Supports

DQOs and the Superfund Accelerated Cleanup Model (SACM) approach of risk-based decision making also use a type of support called the *exposure unit* (EU). SACM establishes the tolerable postremedial risk that serves as a goal for cleanup (Blacker and Goodman 1994b). Figure 4-13 is an expanded version of Figure 1-3 in which the process includes the determination of an appropriate EU size. EUs are commonly on the order of 5000 ft^2 to 1/2-acre (approximately 20,000 ft^2). DQOs establish portions of the site to be designated as EUs, a spatial area over which the risk will be integrated and assessed (Neptune et al. 1990, Ryti et al. 1991).

4.4.5 Support Summary

Support is a factor that must be considered at many stages of the GEM process. These stages include selection of sample volume (STP), laboratory subsampling (STP), compositing for variogram analysis (GA), and selection of block size for estimation by kriging or other interpolator (GA), and in the DQO process where risk-based decisions are made in relation to variously sized exposure units.

An old saying in the retail business identifies the three most important criteria for success as "location, location, location." Paraphrasing, we can specify the most important considerations in the GEM process as support, support, support.

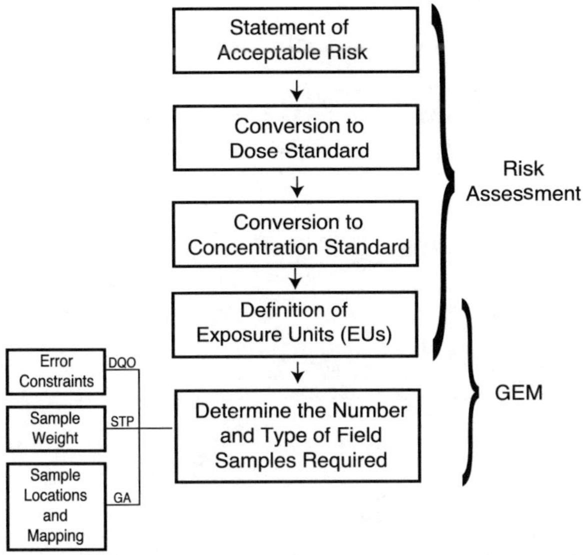

Figure 4-13. Relationship between environmental risk assessment and the GEM process showing exposure unit (EU) definition.

Figure 4-14 summarizes the relationship between support volumes for DQOs, STP, and GA.

A final note on support: Henceforth the term *sample size* will be reserved to describe the number of samples collected, whereas the physical size of the sample will be termed *sample support* or *sample volume*.

4.5 Data Types and Models: Deterministic vs. Probabilistic

4.5.1 Deterministic Data

Only two types of data are used to classify all numerical results. The first is deterministic data. These are data that are measurable to a high degree of accuracy within a specific context. Scalar quantities such as a 10-kg mass or a $20 bill represent deterministic data. Calculating the area of a circle or the volume of a sphere is deterministic. Approximations in physics are often close enough to reality to be considered deterministic. An example is the time it takes an object to

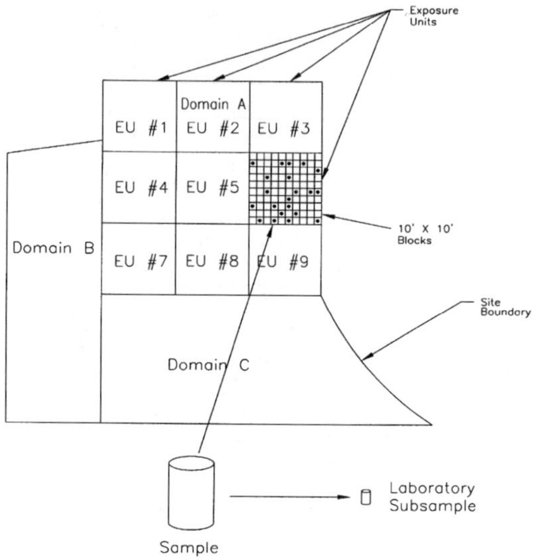

Figure 4-14. Relationship between support units for DQOs, STP, and GA.

fall from a height of 100 ft in a vacuum, which can be calculated accurately using Newton's gravitational formula.

Deterministic data can be used effectively in a variety of situations. A good example of deterministic approaches is provided by Isaaks and Srivastava (1989). Figure 4-15 shows seven sample data points collected over time. The magnitude of each datum is charted on the *Y* axis, time appears on the *X* axis. Two possible

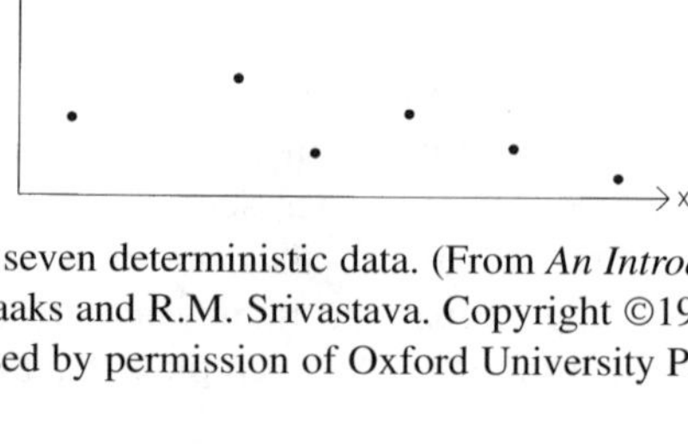

Figure 4-15. Graph of seven deterministic data. (From *An Introduction to Applied Geostatistics*, by E.H. Isaaks and R.M. Srivastava. Copyright ©1989 by Oxford University Press, Inc. Used by permission of Oxford University Press, Inc.)

deterministic solutions appear in Figures 4-16 and 4-17. In Figure 4-16, the data are shown to reflect the height of a bouncing ball that has been released from a selected height. The data do not necessarily reflect the maximum height achieved by the ball during a particular bounce cycle. Nevertheless, the data accurately reflect the height of the ball at the instant in time at which the sample was taken.

Figure 4-17 demonstrates an alternative type of deterministic data. In this example, the seven data points reveal the interest rates charged by a particular

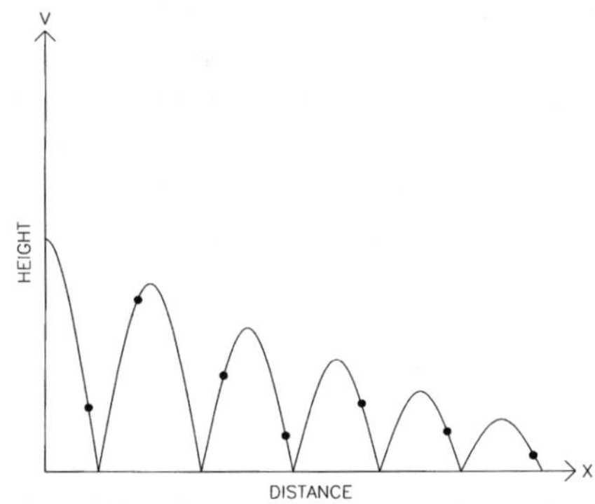

Figure 4-16. One possible deterministic model for the data presented in Figure 4-12. Here the path of a bouncing ball fits the data. (From *An Introduction to Applied Geostatistics*, by E.H. Isaaks and R.M. Srivastava. Copyright ©1989 by Oxford University Press, Inc. Used by permission of Oxford University Press, Inc.)

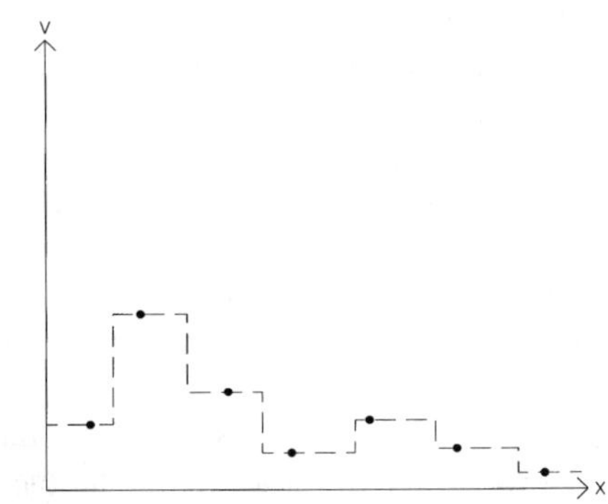

Figure 4-17. Alternative model for the deterministic data. In this case, the data model interest rates as a step function that changes at regular weekly intervals. (From *An Introduction to Applied Geostatistics*, by E.H. Isaaks and R.M. Srivastava. Copyright ©1989 by Oxford University Press, Inc. Used by permission of Oxford University Press, Inc.)

bank. Assuming that interest rates are measured on Tuesday of each week and that the bank adjusts the interest on each Thursday, the profile seen in Figure 4-17 can be developed.

For either deterministic model just described, the value can be obtained exactly for any selected point in time. The predictability and convenience of deterministic models make them easy to use. The models are sufficiently accurate to eliminate concern about the degree of error associated with the estimates. As this example demonstrates, it is mandatory to provide further information or assumptions to the raw data.

Taken by themselves, deterministic data do not supply a means of predicting values between existing data. It is only within the context of a model that they begin to make sense. One should also note that the bouncing ball model can continue to predict beyond the data available whereas the interest rate model is limited to existing data.

Unfortunately, not all data or models are so consistent and predictable. Often, in fact, the outcome is largely uncertain. This gives rise to the notion of probabilistic data and probabilistic models.

4.5.2 Probabilistic Data

The second type of data classification is probabilistic data. Probabilistic (or stochastic) data are generally referred to as random variables (RVs). A random variable is a variable that can result in one of several (or many) possible values. One fixed value is not sufficient to describe probabilistic data. In some cases, all the possible outcome values may be known; in others, an infinite range of values may occur.

For example, throwing a die produces random values from the set {1,2,3,4,5,6}. All possible outcomes are evident in advance. Similarly, tossing a coin produces either heads or tails, with each potential outcome known before the toss. These two types of random experiments also share other characteristics: The outcomes (also termed *events*) possible are equiprobable and finite. Equally likely events have the same chance of occurring under stated conditions. The die has a 1/6 possibility of resulting in the top side having a 1, 2, 3, 4, 5, or 6; both heads and tails have a 50% chance of occurring. For a die, only six outcomes are possible; for a coin, only two.

A sample collected from a hazardous waste site is a random variable. This definition applies because we do not know a priori what the resultant concentration will be. All of the possible concentrations are "known" (but infinite), however, since a concentration cannot exceed 100%. The set of outcomes and their associated probabilities are referred to as the *probability law* or *probability distribution* of a random variable.

Consider the case in which the total number of samples at the site is known, along with the number of samples that have particular characteristics (e.g., PCB concentrations of 25–50 ppm). The probability of a defined event can then be determined, assuming that the samples are representative of the actual contaminant distribution. Reflecting again on the toss of a die, we know that only six possible outcomes are possible. As each outcome is equally likely, each has a 1/6 probability of occurring. This is simply the ratio of the possible elementary outcomes to the total possible outcomes.

Sample data, such as PCB concentrations, usually have different probabilities of occurrence for the different concentration classes. For example, the probability of a PCB concentration occurring in the range of 0–25 ppm may be .3, for the 25–50 ppm range .25, and so on. On a sitewide basis, there may be a higher probability of obtaining a sample below 25 ppm than above 500 ppm. However, in certain portions of the site, it may be highly likely that samples will have concentrations above 500 ppm. Whereas, when rolling a die, the odds of particular outcomes are equal and random, spatial data are generally not equiprobable for all concentration categories.

The processes that distribute contaminants at a hazardous waste site can be extremely complicated though not necessarily random. Our level of understanding with regard to these complex processes is typically very limited, to the point at which the behavior of the contaminants may appear to be random. It is the purpose of statistics and geostatistics to demonstrate that there is generally some order to the randomness.

Despite our ignorance, it will become necessary to predict the concentrations of unsampled locations based on the data already collected. It is always desirable to be able to ascertain the quality of the estimates provided by the model. Unlike deterministic data, probabilistic data can provide insight as to the nature of an appropriate model for estimation.

Statistical models are also able to quantify uncertainty surrounding estimates, whereas deterministic models cannot. To assist in the endeavors of model selection and uncertainty quantification, Chapters 5 and 6 will introduce the concepts of data distributions and distribution models.

Conclusion

Chapter 4 has introduced several important concepts that are central to the GEM process. The importance of defining a statistical population and appropriate domains is key to satisfying project DQOs. The notions of equiprobability and randomness play an important role in obtaining representative data for the site. These concepts have been introduced with some caveats, which will be expanded on in later chapters. One of the most important GEM themes, support, was introduced. In later chapters, support issues will be taken up as they affect all three GEM areas (DQOs, STP, and GA) in terms of sample volume, laboratory subsampling, population statistics, variogram analysis, kriged block estimation, and risk characterization through DQO exposure units. Differences between deterministic and probabilistic data have been described, and the distinction and limitations between deterministic and probabilistic models discussed.

Chapter 5

Data Distributions

SYNOPSIS: Collection of sample data for random variables inevitably produces a range of sample values called the distribution. The range of values may be known a priori but generally is not. Data distributions serve as a way of organizing data and classifying it into groups for more detailed analysis. Also, estimates of distribution parameters called statistics may be calculated for comparing data sets and distributions. This chapter will develop the concept of a distribution along with the idea of population parameters and sample statistics.

KEY WORDS: Discrete Variables, Continuous Variables, Probability Density Function, Histogram, Cumulative Distribution, Cumulative Density Function, Area Weighting, Indicator Data, Central Tendency, Mean, Median, Mode, Variance, Standard Deviation.

5.1 Concept of Data Distributions

The set of outcomes from a sampling operation and their corresponding probabilities are referred to as the *probability law* or *probability distribution*. This distribution may or may not be known a priori. For example, when a perfect die is rolled, the possible outcomes are 1,2,3,4,5,6, each with an equal probability of 1/6. This is a simple type of distribution model in which the probabilities are known.

In other situations, it may not be as easy to predict the probability of an event of interest. If we wish to know the probability of PCBs above 500 ppm for a 2-kg soil sample, the true distribution will never be known. It then becomes necessary to compute an experimental probability distribution. Based on the experimental data, we can then infer which theoretical distribution model will be appropriate for estimating the desired probabilities.

Observing a few outcomes of a random variable will not allow us to calculate exactly the distribution parameters as they are parameters of a conceptual model. What can be calculated are the sample statistics based on a particular data set. Note that a different set of samples taken at different times or locations would produce different estimates of the population statistics. Nevertheless, as a practical matter, we assume that the population parameters of the random variable are the same as the estimates calculated from sample data. This is an assumption we are willing to make as the number of samples available is virtually always limited with respect to the sample universe or sample domain. It does, however, introduce a degree of uncertainty.

5.2 Types of Data Distributions: Discrete vs. Continuous

There are two kinds of distributions, discrete and continuous. Discrete variables can take on only isolated values. For example, the outcome of the roll of a die can take on only integer values. Similarly, if we assign the value of zero to the result of heads in a coin toss and a value of one to tails, these integers can represent discrete variables. Such unique, "countable" values are termed *discrete*, because values between them (0.5, 0.75, 1.3, etc.) are not possible. Examples are the number of cancer deaths per year in a community or the number of rabbits per square mile.

In contrast, continuous distributions can take on both integer and noninteger values. For instance, a value of 1.73 is not possible in a discrete distribution that measures the number of releases per month from a hazardous waste facility. This, however, is a valid datum in a continuous distribution measuring dioxin concentration in parts per billion for 1-kg soil samples. Most environmental variables are continuous variables.

A distribution attributes each possible value x a probability $p(x)$. For the roll of a die, $p(x) = 1/6$ for each possible outcome. For the toss of a coin, $p(x) = .5$

for each outcome. Each $p(x)$ is nonnegative, with the sum of all possible $p(x)$ values equal to one. Also note that the probability $p(x)$ may be zero for many values or outcomes. Such values are considered to be undefined. One example for which $p(x)$ is equal to zero is a case in which the outcome of the roll of a single die is nine.

Although the range of integer and noninteger values is infinite, it is not likely that all possible values will be encountered during a sampling program. To organize the likely values, we can apply a mathematical function. The function $f(x)$ is called the *discrete probability density function* (pdf), or *distribution function*. This is referred to more simply as the density or the distribution. It is useful to think of $f(x)$ as the relative frequency with which the separate values of x will occur. The sum of $f(x)$ is equal to one, taken over all values for which $f(x)$ is defined. For example, the sum of $f(x)$ for all defined values of the roll of a die {1,2,3,4,5,6} is one, where the individual probabilities are 1/6, 1/6, 1/6, 1/6, 1/6, and 1/6. Mathematically, this is written

$$\sum f(x) = 1 \tag{5.1}$$

Sometimes it is useful to know the probability of a particular outcome falling within a range of values. For instance, we might wish to know the chance of rolling a number between 3 and 5 (inclusive) with a die. This is defined as the probability of a result between a and b. The answer is attained by summing all the individual probabilities between a and b. In the example of the die, the sum is $p(3) + p(4) + p(5) = 1/6 + 1/6 + 1/6 = 1/2$. Written in mathematical notation, these results are

$$P(a \leq x \leq b) = \sum_{x=a}^{b} f(x) \tag{5.2}$$

or

$$P(3 \leq x \leq 5) = \sum_{x=3}^{5} f(x) = \frac{1}{2} \tag{5.3}$$

where $f(x) = 1/6$ for all individual values.

Figure 5-1 shows the discrete distribution for the values permissible when rolling a die. In each case, the probability is the same. Figure 5-2 shows another discrete distribution, this time the number of days per month during which the

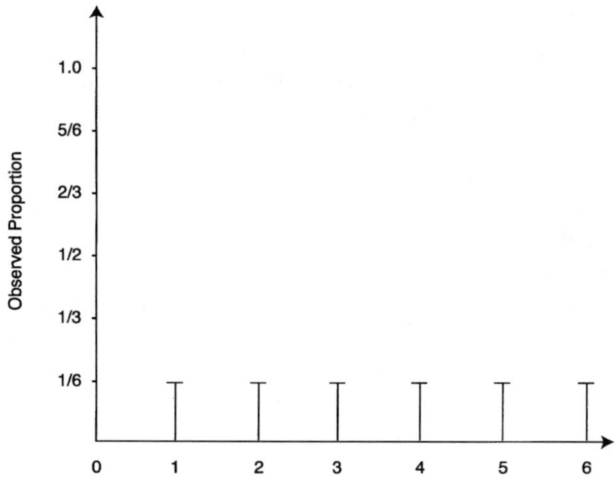

Figure 5-1. Histogram of the expected outcomes of rolling a die.

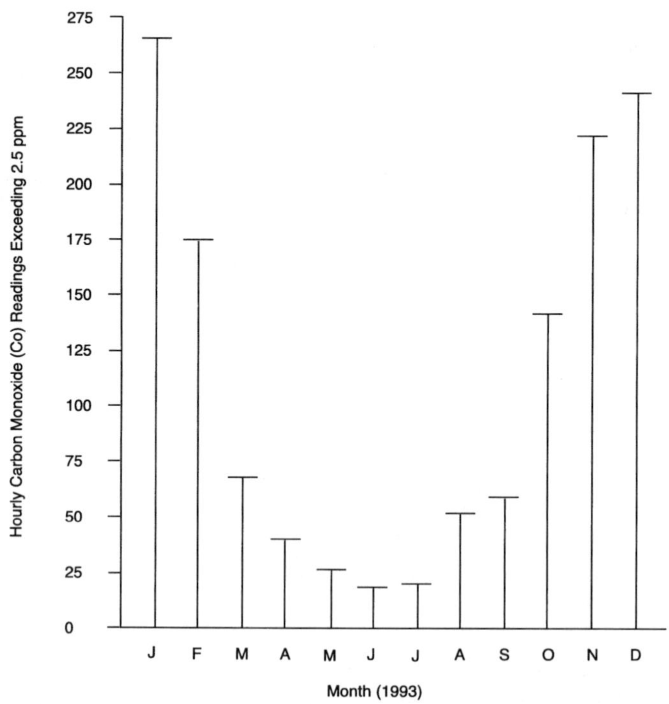

Figure 5-2. Histogram for number of hourly carbon monoxide readings exceeding 2.5 ppm.

carbon monoxide levels in Denver, Colorado, exceeded 2.5 ppm during the calendar year 1993.

Examples of discrete distribution models are the hypergeometric, the binomial, and the Poisson. Discrete distributions are less commonly used than continuous distributions for hazardous waste applications. The Poisson, however, has an extremely important relationship to sampling theory and practice (STP): Trace levels of contaminants are well modeled by the Poisson distribution techniques combined with appropriate sample support volumes.

For continuous distributions, the situation regarding probability outcomes becomes more complicated. The interval between one and two contains an infinite number of real numbers, 1.001 and 1.002, for example. The interval between 1.001 and 1.002 is divisible into increasingly smaller intervals such as 1.0011, 1.00111, 1.001111, 1.0011111, and so on.

The difference between each probability increment for a continuous distribution is, in fact, infinitesimally small. A density of probability $f(x)$ is defined so that the probability of one value lying between x and $x + dx$, where dx is infinitesimal, will be $f(x)\,dx$. The notation using the summation symbol Σ for the discrete distribution is replaced by the integral symbol $\int$ for the continuous distribution. Using this convention, the probability of x lying between a and b is:

$$P(a \leq x \leq b) = \int_a^b f(x)\,dx \tag{5.4}$$

Often, one wants to know the probability of an event occurring between two values in the distribution, where the lowest value (a) is infinitely small ($-\infty$) and the upper value is some type of threshold or cutoff value. As a practical matter, zero is usually an acceptable substitute for $-\infty$, given that few variables are measured in the negative range. A caveat for hazardous waste applications is necessary, however. The variable acid-base potential is expected to have negative values for many situations.

The probability of x being less than or equal to a given value x_o is called the cumulative probability $F(x_o)$ and is written as follows for a continuous distribution:

$$P(x \leq x_o) = \int_{-\infty}^{x_o} f(x)\,dx \tag{5.5}$$

with $F(-\infty) = 0$ and $F(+\infty) = 1$. Note that the complementary probability of the value being above the limit threshold is

$$1 - \int_{\infty}^{x_o} f(x)\,dx \tag{5.6}$$

Practical applications of the cumulative probability are shown in Chapter 6 by using examples with the normal distribution.

5.3 Visual Representations of a Distribution

A frequency distribution is a fundamental graphic format used to present the observed values from a sample set for a particular variable. The frequency distribution graph is commonly called a *histogram.* The histogram can be used for either grouped or ungrouped data, usually corresponding to discrete and continuous variables, respectively. The histogram rapidly and pictorially provides a good representation of the shape of the data's distribution.

Construction of the histogram is relatively simple. The values for the variable of interest are placed on the x axis. Histograms categorize data, starting with the lowest observed values at the origin of the graph, proceeding along the x axis through the highest values. It is customary to use a constant cell or class width to depict the interval. The frequency of occurrence for each class of data is scaled proportionally on the y axis. The frequency can be expressed in actual counts or occurrences (50, 100, 150, 200, etc.) or in relative terms (10%, 20%, 30%, etc.).

The histogram displays how often observed values fall within certain intervals or classes such as between 25 and 50 ppm. These types of categories were implied in equations 5.2 and 5.3:

$$P(a \leq x \leq b) = \sum_{x=a}^{b} f(x) \tag{5.7}$$

$$P(a \leq x \leq b) = \int_{a}^{b} f(x)\,dx \tag{5.8}$$

where a and b correspond to the values of 25 and 50 ppm, respectively.

The variables encountered in environmental data sets may assume a number of different distributional shapes and forms. Figure 5-3 shows a distribution of soluble lead (Pb) concentrations in surface soils along major highways, a result of emissions from automobiles burning leaded gasoline. This graph has a preponderance of the data occurring in the low concentration ranges. This is termed a skewed distribution. The distribution shown in Figure 5-3 has a skew to the right, also known as a positive skew. Conversely, Figure 5-4, showing heating values (BTUs/lb) in a lignite coal seam, is skewed to the left, which is a negative skew.

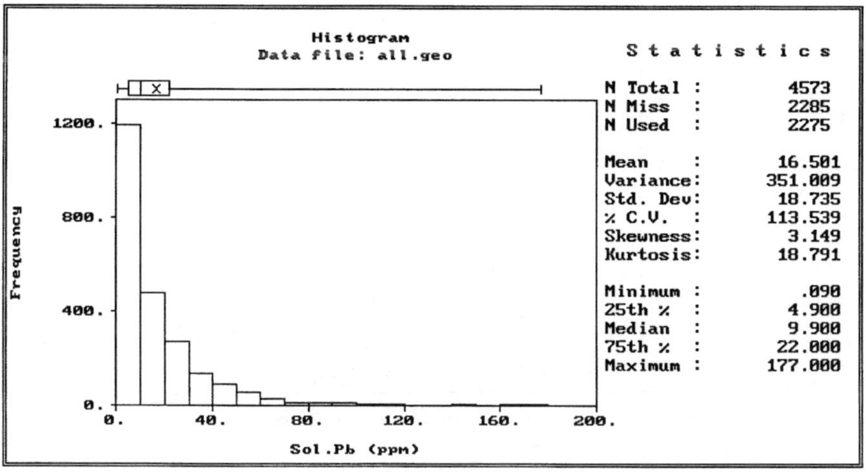

Figure 5-3. Histogram of soluble lead (Pb) concentrations showing positively skewed data.

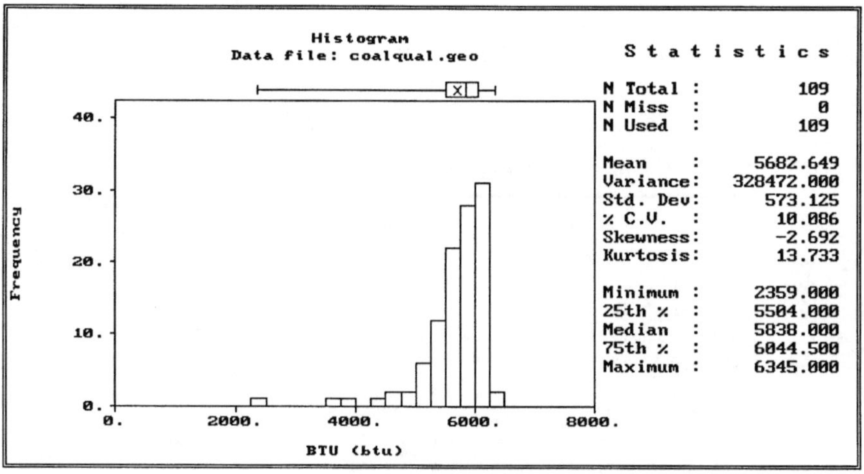

Figure 5-4. Negatively skewed distribution showing heating values (BTUs) in a lignite coal seam.

Figure 5-5 shows a distribution of aquifer thickness measurements. The shape in Figure 5-5 is a symmetric distribution known as normal or Gaussian. It is also the bell-shaped curve of classical statistics.

The histogram has a corresponding graph, the cumulative frequency graph, that is also quite useful for visual interpretation. Table 5-1 shows a frequency table

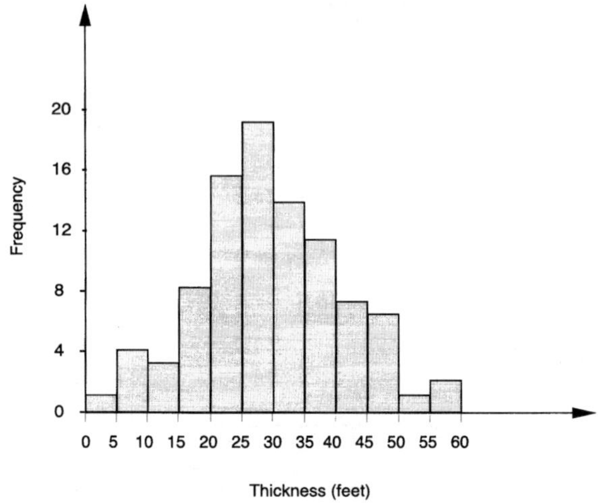

Figure 5-5. Nonskewed (symmetric) data showing measurements of aquifer thickness.

Table 5-1. Frequency table of lead (Pb) values.

Class	Number	Percentage
$500 \le x < 600$	1	2.63
$600 \le x < 700$	2	5.26
$700 \le x < 800$	2	5.26
$800 \le x < 900$	2	5.26
$900 \le x < 1000$	6	15.79
$1000 \le x < 1100$	2	5.26
$1100 \le x < 1200$	3	7.89
$1200 \le x < 1300$	5	13.16
$1300 \le x < 1400$	4	10.53
$1400 \le x < 1500$	2	5.26
$1500 \le x < 1600$	3	7.89
$1600 \le x < 1700$	2	5.26
$1700 \le x < 1800$	1	2.63
$1800 \le x < 1900$	0	0.00
$1900 \le x < 2000$	1	2.63
$2000 \le x < 2100$	1	2.63
$2100 \le x < 2200$	0	0.00
$2200 \le x < 2300$	1	2.63

for the lead (Pb) data presented in Table 3-1. Class intervals are listed, along with (1) the corresponding number of data that fall into the category and (2) the percentage of data in that class compared to the entire distribution.

Table 5-2 presents the data in a cumulative frequency format. This time, the percentages are summed from lowest to highest values and percentages for each class, providing a running total of the data. The cumulative frequency data can be shown in histogram form. Figure 5-6 shows a cumulative frequency histogram for soil lead (Pb) data from USEPA's GeoEAS example data set. The x axis plots a histogram of the data categories; the y axis plots the cumulative frequency of data in percent.

A more commonly used way to display cumulative data is the cumulative frequency plot. An example of this type of display, for aquifer thickness measurements, can be seen in Figure 5-7. This type of graph requires a special axis scaling format, generally referred to as probability paper. As with other histogram and frequency plots, the x axis shows the data categories. The y axis, however, is probability-based.

Instead of bearing a regular arithmetic scale, probability paper has the values of $F(0)$, $F(1)$, $F(2)$, and so forth, F being the cumulative normal distribution as read from tables. This means that, if a data set is normally distributed, the resulting line gained from plotting $F(x)$ vs. x will be straight. How straight is straight? There is no single answer nor objective measure. Experience has shown that the lines will

Table 5-2. Cumulative frequency table of lead (Pb) values.

Class	Number	Percentage
$x < 600$	1	2.63
$x < 700$	3	7.89
$x < 800$	5	13.15
$x < 900$	7	18.41
$x < 1000$	13	34.20
$x < 1100$	15	39.46
$x < 1200$	18	47.35
$x < 1300$	23	60.51
$x < 1400$	27	71.04
$x < 1500$	29	76.30
$x < 1600$	32	84.19
$x < 1700$	34	89.45
$x < 1800$	35	92.08
$x < 1900$	35	92.08
$x < 2000$	36	94.63
$x < 2100$	37	97.34
$x < 2200$	37	97.34
$x <$	38	100.00

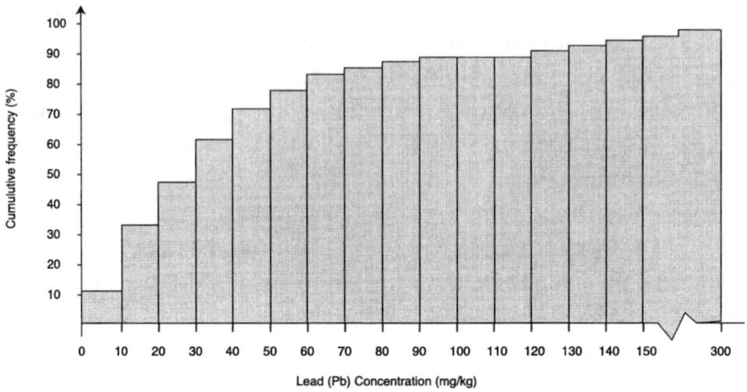

Figure 5-6. Cumulative frequency histogram for soil lead (Pb) data.

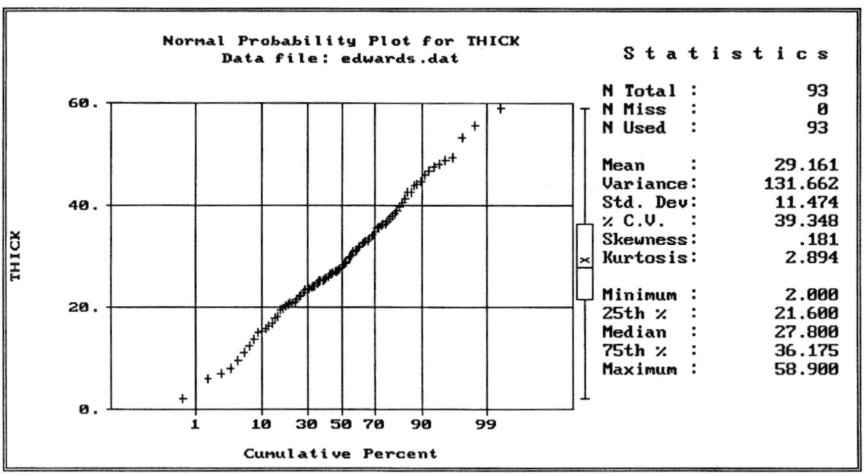

Figure 5-7. Cumulative frequency plot for aquifer thickness data.

either be straight, have a marked curvature, or zigzag. The line in Figure 5-8 shows a tendency to zigzag; a "better" line was shown in Figure 5-7. Perhaps all that can be said is that the line of the cumulative frequency graph furnishes a qualitative understanding of how close a particular data set comes to being normal.

Some other features of the cumulative frequency graph should be noted. The line intercepts the mean value at the 50% probability line. Also, the standard deviation may be directly determined by using the concentration axis between 50% and 84%.

The normal distribution provides many tools that are both useful and easy to implement. Unfortunately, most geologic data do not follow a normal distribution

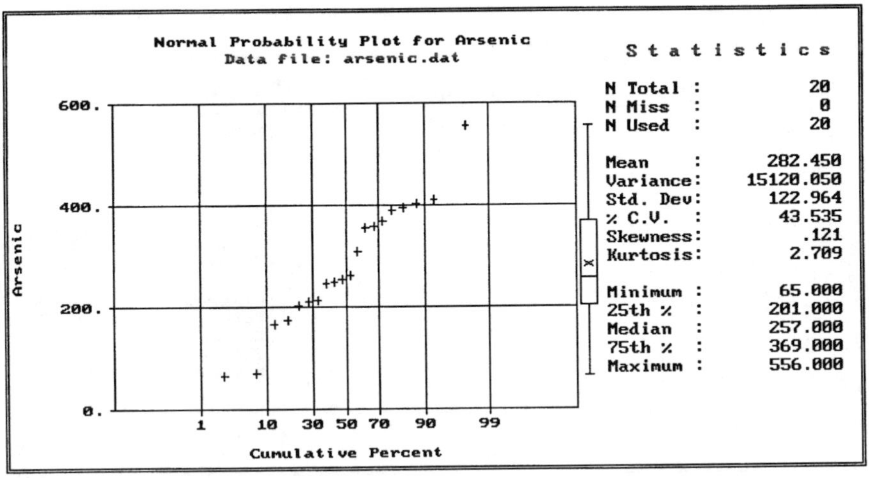

Figure 5-8. Example of a "zigzag" line for a cumulative frequency plot of soil arsenic data.

or one even remotely close to normal. The most common situation produces data sets containing an abundance of rather small values along with a few very large ones.

This plethora of smaller concentrations is perhaps a manifestation of the second law of thermodynamics, which states that nature seeks a state of maximum randomness or entropy, along with minimum energy conditions. Areas or samples of high concentration, left to the processes of weathering, tend to dissipate and become diluted in the less concentrated background areas. The particles naturally disseminate themselves, seeking a more stable state of homogenized, low-level values. The energy component can be thought of in terms of exposure to humans: A lower concentration delivers a lower energy dose.

Figure 5-9 shows a histogram of the lead (Pb) data in the GeoEAS example data set, the same data as in Figure 5-6. This distribution of raw data is markedly skewed to the right. Histograms such as this (showing a departure from the symmetrical shape of the normal distribution) are typical of environmental contaminant concentrations in both soil and groundwater.

Cumulative frequencies will plot as a straight line if the distributions are normal. Figure 5-10 shows the cumulative frequency plot for the soluble lead data in Figure 5-3. Rather than a straight line, the plot shows a markedly convex shape, highly characteristic of positively skewed data.

Many times, data do not follow a normal distribution but, rather, their logarithms tend to be normally distributed. The cumulative frequency diagram can be modified to accommodate this common condition. By using a logarithmic scale

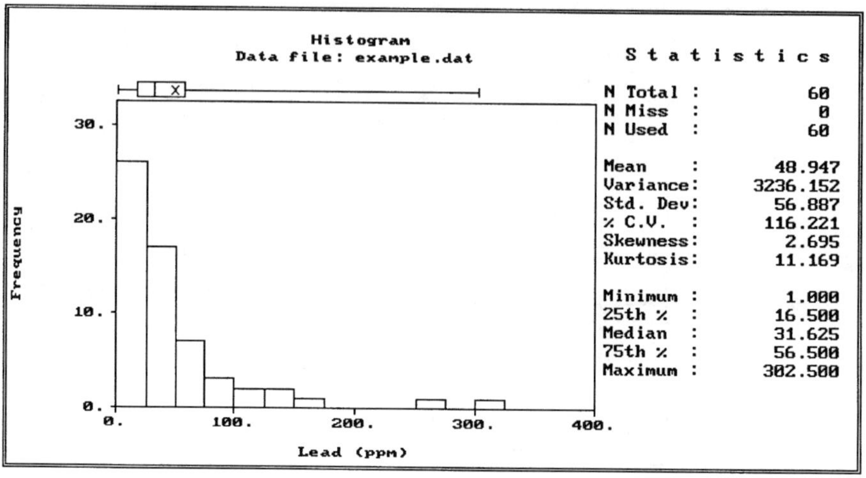

Figure 5-9. Histogram of lead (Pb) data showing the skewed shape of the lognormal distribution.

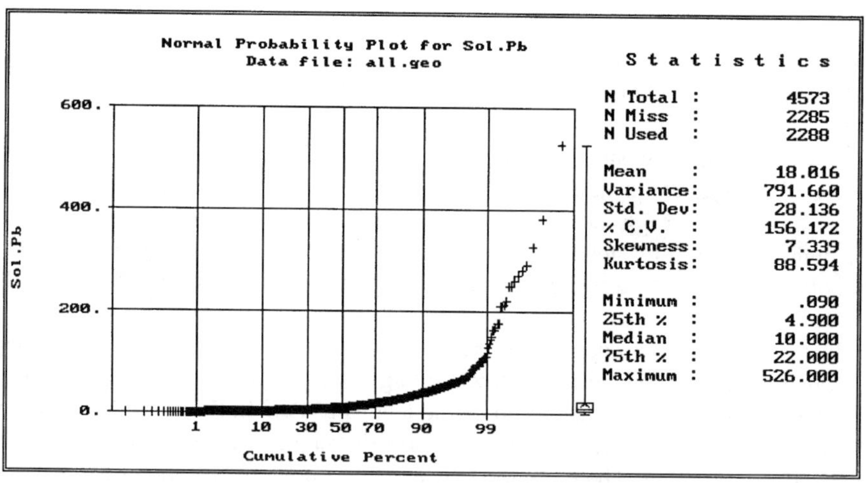

Figure 5-10. Cumulative frequency plot of the lead (Pb) data shown in Figure 5-3.

along the x axis instead of an arithmetic scale, we can check for the condition of lognormality. If the data are lognormal, a straight line will result.

An alternative to log-probability paper is to take the logarithm of each sample before plotting the data on regular probability paper. An example of log-transformed data for cumulative frequency plotting is shown in Figure 5-11, using the same soluble lead data as in Figures 5-3 and 5-10.

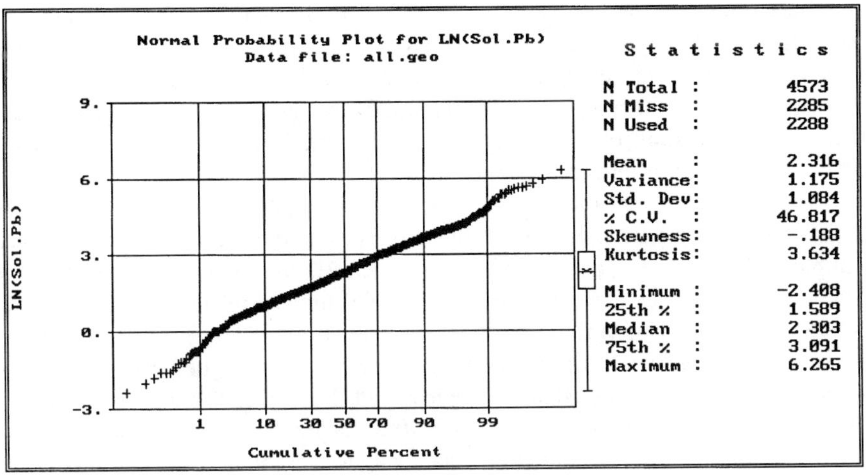

Figure 5-11. Cumulative frequency plot of the Ln-transformed data from Figure 5-3.

5.4 Distributions of Spatial Data

5.4.1 Area Weighting

Techniques such as histograms, cumulative histograms, and cumulative frequency plots work well for data that are not spatial, such as the height of a person or the diameter of a machine part. Each sample is generally independent of the next one and, as such, exerts no influence on the expected outcome for subsequent sample values.

This condition of independence means that each sample should contribute an equal weight toward the histogram, cumulative histogram, or cumulative frequency plot. By default, this weight is usually equal to one. Distribution graphs shown to this point have all used this weighting scheme.

The idea that all samples have an equal weight in their contribution toward the distribution is so natural that many people never give it a second thought. The problem is that spatial data are most frequently correlated with nearby samples. Therefore, information is being shared between two or more samples, especially in clustered data. This means that each of the samples has some of the same information and that this information is being accounted for more than once in the data distribution. This redundancy necessarily introduces a bias.

In environmental site characterization, this can become a severe problem. Clustered data can occur for numerous reasons. For instance, an initial site sample may be taken on a pseudoregular grid providing approximately equal coverage, and therefore weight, to each sample. Based on preliminary results, "hotspots" or their interpreted margins may be targeted to better define their size. Areas thought to be below the action level may be sampled sparsely or not at all.

By targeting hotspots, the spatial nature of the data introduces a bias toward high values. The hotspots are overrepresented spatially, and a clear picture of the true sample distribution cannot be obtained.

To account for this bias, it is necessary to weight each sample value in an appropriate manner. For spatial data, this means using the amount of area that each samples influences. Figure 5-12 shows a typical (though hypothetical) situation. These data represent the spatial locations of samples surrounding a lead (Pb) smelter. The density of sampling is highest around the smelter. As one moves away from the smelter, the sampling density decreases. As might be expected, the sample values also indicate that the highest lead (Pb) concentrations are closest to the smelter. These high values will distort the shape of the distribution.

Figure 5-13 shows a more appropriate way to weight the contribution of each sample to the distribution. Each sample is enclosed by a polygonal area defined by its position with respect to surrounding data. The concentration of each sample can be weighted by the area of the polygon:

$$F(x) = \frac{\sum_{i=1}^{n} w_i \cdot x_i}{\sum_{i=1}^{n} w_i} \tag{5.9}$$

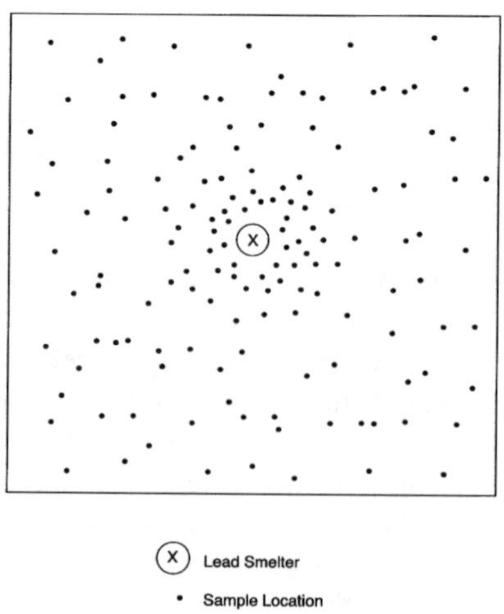

Figure 5-12. Location map of soil samples around a lead (Pb) smelter.

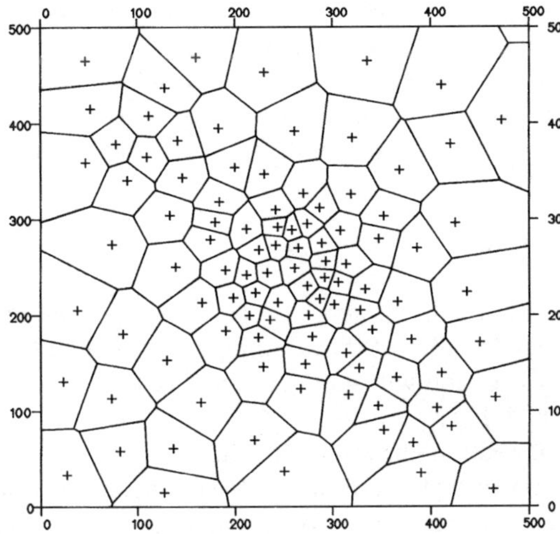

Figure 5-13. Use of polygons to decluster data by area weighting.

Thus, the high-concentration data near the smelter are given smaller and more representative weights by virtue of their smaller polygonal areas. Conversely, the values further from the smelter now receive more weight because they represent larger areas of the project.

The use of weighted data has a profound effect on the cumulative distribution. Figure 5-14 shows a comparison between weighted and unweighted data. The heavy line shows the weighted data, and the thin line shows the unweighted data. The figure shows a significant difference between the two graphs. The weighted data indicate greater numbers of low-concentration samples and cause the graph

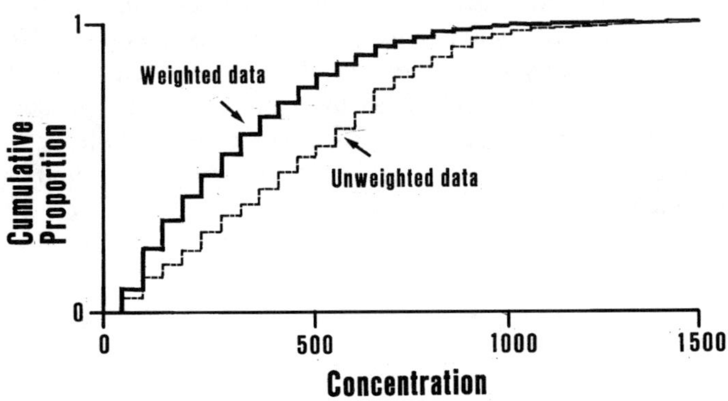

Figure 5-14. Area-weighted and non-area-weighted cumulative distributions.

to rise more quickly, pushing it to the left of that for the unweighted data, where higher-valued data appear in greater abundance.

Figure 5-14 shows the typical situation in which high-concentration zones are oversampled. It is possible, although less common, to oversample low concentration areas. In that case, the locations of the thick and thin lines would be reversed.

5.4.2 Indicator Data

To this point, most of the discussion has addressed sample data in a raw form. Earlier in this chapter, the idea of transforming raw data by the use of logarithms was introduced. This section will introduce another type of transformation commonly used for spatial data.

In the preceding section, the effort was focused on estimating the cumulative frequency distribution. From this estimation, the frequency distribution may also be calculated, and vice versa. There are two basic approaches to estimating cumulative frequency or frequency distributions. The approach described in the preceding section is referred to as a nonparametric approach in which estimates of F(x) values are calculated at multiple, and often arbitrary, values of x using the existing data.

The second approach is the parametric approach. This method applies a function that completely describes the distribution for any value of x. Although this offers a certain convenience, parametric approaches rely heavily on a random function model in which the distribution is assumed to be known, typically a strong assumption. The application of distributional models will be covered in Chapter 6.

The indicator approach is a nonparametric approach and as such does not rely as heavily on a random function model. Because it does not make a distribution hypothesis, it relies exclusively on the sample data. It is highly useful in applications in which the proportion of the distribution or the probability of a value exceeding an action limit needs to be assessed.

The indicator transformation begins with the selection of an action level or cutoff value. For environmental situations, this is generally supplied by the regulatory agency and, in most situations, only a single cutoff is appropriate. Note that, for various reasons, regulatory agencies often eschew commitment to an action cutoff. It is a bit ironic that an agency's reluctance to commit may be a larger obstacle to the use of the indicator approach than any statistical problem or challenge.

Once the cutoff is established, the data can be transformed into an indicator in the following manner:

$$i_j = \begin{pmatrix} 1 \text{ if } x_j \leq x_c \\ 0 \text{ if } x_j > x_c \end{pmatrix} \tag{5.10}$$

where x_c is the action level cutoff.

Equation 5.10 may be used at any cutoff. To determine the proportion of the distribution below any cutoff, the equation is

$$F(x_c) = \frac{1}{n} \sum_{j=1}^{n} i_j(x_c) \tag{5.11}$$

For environmental situations, it is generally desired to calculate the proportion of the distribution above the regulatory action level. In this case, equation 5.9 may be rewritten to produce

$$i_j = \begin{pmatrix} 0 \text{ if } x_j \leq x_c \\ 1 \text{ if } x_j > x_c \end{pmatrix} \tag{5.12}$$

5.5 Distribution Parameters

The foregoing discussion has shown that many types of distributions exist, both in form (discrete vs. continuous) and shape (normal, lognormal, etc.). Even within a single category of distribution—say, normal distributions—many possible variations exist. Distribution parameters assist in standardizing types of distributions. For example, a normal distribution is completely specified by its mean and variance.

In order to compare distributions, estimates of population parameters are calculated. First, however, a word of caution. For many real populations, the complete distribution cannot be inferred from just a few parameters. Thus, the true population parameters always remain unknown. Population statistics are only summaries of the features of the distribution and do not provide complete information. Second, the parameters cannot be deduced exactly from only a few observed outcomes of the population. Instead, they are parameters of a theoretical model. However, in practice, we must assume that the sample statistical parameters accurately represent the actual distribution parameters. Differences between the two can generally be mitigated by taking a greater number of samples.

5.5.1 Central Tendency

5.5.1.1 The Mean

Calculation of the Mean The mean or average value of the population provides information as to the "center of gravity" around which other values in the random variable population are distributed. The mean is the best known and most used of the central tendency parameters. Note that the mean is not necessarily the most commonly observed sample result nor the center of the distribution. Figure 5-15 shows a hypothetical data distribution in which the mean is far from any of the frequently observed values.

Figure 5-15 shows but one example of an unusually shaped distribution, perhaps monthly home heating costs over the course of a year. Many oddly shaped histograms could be developed to represent a variety of situations. The mean is simply a type of balancing point for the data values. In other words, if a histogram shape were cut from a piece of wood, the mean would be the balancing point, regardless of the shape of the distribution.

The mean is calculated by summing the value of each discrete sample observed in the data experimental set and then dividing by the number of samples used. This provides a simple formula:

$$\overline{x} = \frac{1}{n} \sum_{i=1}^{n} x_i \tag{5.13}$$

where the individual values (x_i) are summed from 1 to n and then divided by n, the number of samples.

For a theoretical probability distribution, formula 5.1 can be written

$$m = \int_{-\infty}^{\infty} xf(x)\, dx \tag{5.14}$$

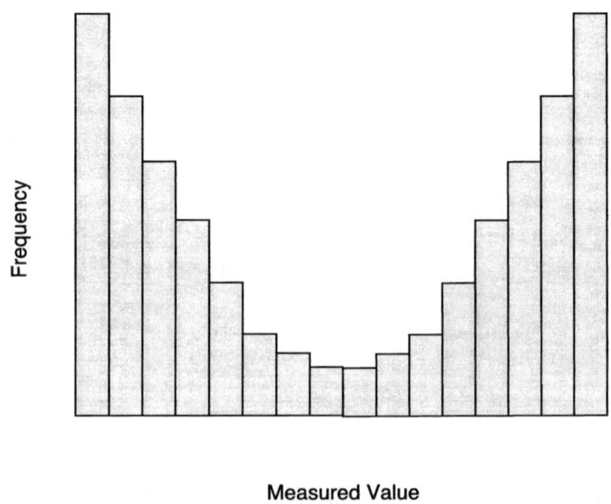

Figure 5-15. Histogram of a hypothetical distribution.

The mean value m is commonly called the *expected value* of x and is written

$$m = E(x) \tag{5.15}$$

Other types of averages can be calculated for sample data sets. These include:

Quadratic average

$$m_Q = \sqrt{\frac{1}{n}\sum_{i=1}^{n} x_i^2} \tag{5.16}$$

Geometric average

$$\log m_G = \frac{1}{n}\sum_{i=1}^{n} \log x_i \tag{5.17}$$

Weighted average

$$m_w = \frac{1}{\sum_{i=1}^{n} w_i}\sum_{i=1}^{n} w_i x_i \tag{5.18}$$

Harmonic average

$$m_H = \frac{n}{\sum_{i=1}^{n} 1/x_i} \tag{5.19}$$

The reader is again reminded about the difference between population parameters and sample statistics. The quantities m and σ are population parameters; $\bar{x}$ and s are sample statistics. The true population parameters will remain unknown; all that is available in practice is the suite of estimated population statistics based on sample data.

Sensitivity to Outliers The mean is a highly sensitive sample statistic. It is sensitive to *outliers* or other very high values. Extreme values often occur in environmental data; sample data commonly vary by several orders of magnitude. For example, an initial sampling program may be performed at a site and an average value calculated. A second round of sampling may introduce one or more extreme values that change the average value dramatically.

By definition, the mean is equal to the expected value *E(x)*. The mean of the first round of sampling is expected to be equal to the mean of the second round. This is because the same population is being sampled both times and the expected value *E(x)* is always the same for each sampling event. The population parameter *m* remains constant (but unknown); only the sample statistic changes.

This brings up some practical questions. First, which mean value is closer to the true population mean? Was there a bias in one of the sampling programs? Would a third round of sampling produce a mean lower than the first, in between the two observed means, or higher than either observed mean? What effect has the sample support had on the mean? The answers to these questions are examined in the chapters on STP and GA.

The extreme values encountered in environmental data make it necessary to exercise caution when calculating an average value. On a lighter note, we are reminded of the statistician who had one hand in the freezer and one hand on the stove. On the average, he was comfortable!

Declustering Previously, it was shown that the data distribution can be biased because of spatial clustering of sample data. The mean is also sensitive to such clustering. Spatial clustering introduces a bias, usually upward. A biased mean will also bias risk assessment model results that rely on average concentrations over the site. Therefore, it is crucial to obtain a mean value that is representative of the site. Much care is often used to assure representative sampling; it seems fair to continue the effort by calculating representative mean values.

The same principle applies to the mean as that presented for the cumulative distribution. To calculate a simple average of the smelter data (Fig. 5-12), using equations 5.13, 5.16, 5.17, 5.18, or 5.19 for these sample data would be inappropriate. The average would be biased upward because a disproportionate number of samples represent areas close to the smelter. Instead, we should use equation 5.9, taking the area of the polygon surrounding the sample to weight the data appropriately in calculating the average.

Another method for obtaining a better estimate of the mean value is shown in Figure 5-16. A project area has been divided into equal area cells on a square grid. Using random selection techniques, only one sample from each cell is selected to represent the contaminant concentration in the cell. The selected samples are then averaged (eq. 5.13). This technique is called *declustering*. The weighted polygonal technique described above is also a form of declustering.

Everyone's Favorite: Lying With Statistics The calculation of the mean value provides a convenient way to lie with statistics. One April Fool's Day, it was reported by National Public Radio that Canada had purchased Arizona. Pundits were concerned because the average temperature of Canada had suddenly been raised several degrees as a result of the purchase. The conclusion was that the wheat-producing area of Canada would be shifted many miles to the north. This would greatly expand wheat production capacity, flooding international markets

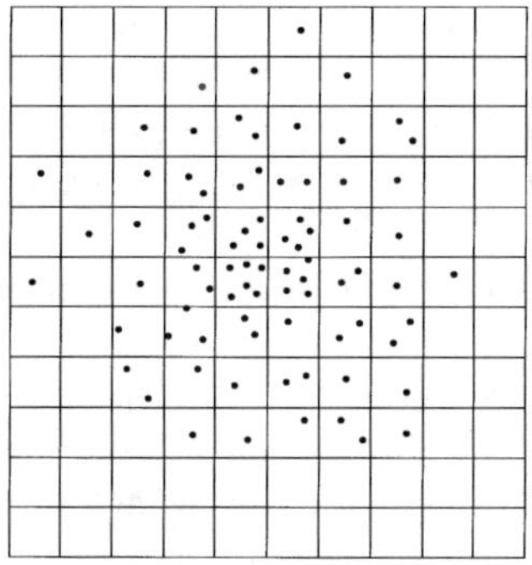

Figure 5-16. Example of a declustering grid.

and driving prices down. This, in turn, would cause extreme hardship to smaller wheat-producing countries. The report might have added that frogs had spontaneously gone deaf or had sprouted ears on their legs.

Obviously, these market predictions are not correct and do not respect the rules of domain analysis. The point is that the mean is easily misrepresented or lacks meaning. One is cautioned to examine carefully the data and assumptions of anyone purporting to calculate a representative mean value.

5.5.1.2 The Median

The median is another measure of central tendency. The median represents the midpoint of the observed sample values arranged in increasing order. This places half of the data values below the median and half of the values above the median. For practical computation, a computer program can easily sort the data so that each individual value is greater than or equal to the preceding value. In mathematical terms, this is

$$x_1 \le x_2 \le, \ldots, \le x_n \tag{5.20}$$

The median can easily be calculated from sorted data by means of one of the following equations:

$$M = x_{(n+1)/2} \quad \text{if } n \text{ is odd} \tag{5.21}$$

or

$$M = \frac{x_{n/2} + x_{n/2+1}}{2} \quad \text{if } n \text{ is even} \tag{5.22}$$

A nice feature of the median is that it is less sensitive to outliers, or extreme values, than the mean. Figure 5-17a is a histogram of a skewed distribution typical of 200 environmental data. The mean value is plotted by a dashed vertical line and the median by a solid vertical line. Note that the median and the mean values are at some distance apart.

Figure 5-17b shows the same distribution after 50 additional samples have been acquired. In this new round of sampling, several samples exhibiting extremely high concentrations were obtained. The mean and median values have again been plotted. The extreme values have had a significant effect on the mean value of the data. The line indicating the mean value has shifted further to the right, reflecting a higher average concentration.

The median, however, has made only a slight shift. This shift reflects the addition of the new samples. These graphs demonstrate that the mean value is highly sensitive to additional data, especially high-valued data or *outliers*. The median is shown to be much more stable than the mean. We might also check the

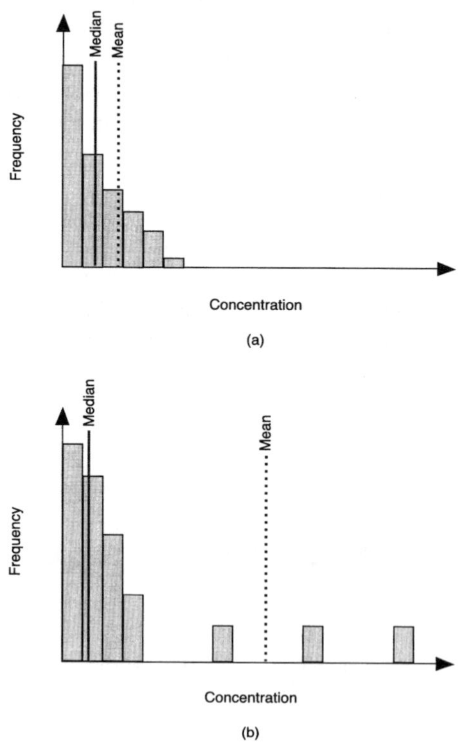

Figure 5-17. Examples of the sensitivities of the mean and median statistics: (a) The mean and median are relatively close; (b) additional high-concentration samples push the mean to the right while the median remains almost stationary.

location of the 50 new samples because they may correlate to a hotspot. If so, declustering may be appropriate.

The median can even be described as a more "democratic" statistic. Each American is given one vote for electoral purposes, regardless of whether the voter has ambivalent or extremely vehement feelings. Likewise, all sample data are given but one count, and the median is calculated with minimal regard to any extreme sample "opinion" from the outliers. Simply, half the data are placed above the median and half are placed below it.

5.5.1.3 The Mode

The final important measure of central tendency is the mode. The mode is the value that occurs most often in the distribution. For discrete data, this is the individual value that occurs most often. For continuous (grouped) data, the mode is the midpoint of the *modal cell*, which is the most frequently observed cell. The mode of a set of grouped data is shown in Figure 5-18.

It should be noted that the value of the mode can change slightly depending on the size of the cells. Figure 5-19 shows the same distribution as the previous figure but with smaller cells. The smaller cell size has shifted the mode of the distribution to the left.

Another situation that can arise is two modes in one distribution. Figure 5-20 is a histogram in which two distinct modes exist. The first mode contains a greater number of values; the second has fewer values. This indicates the possibility that the sampling program has encountered two different types of data populations.

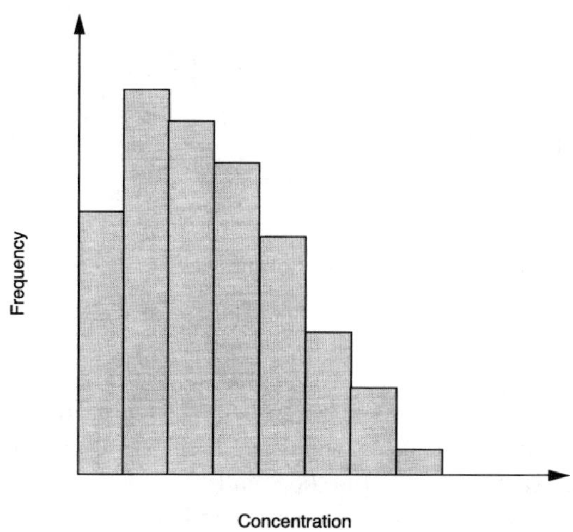

Figure 5-18. Occurrence of the mode where the histogram cell width is large.

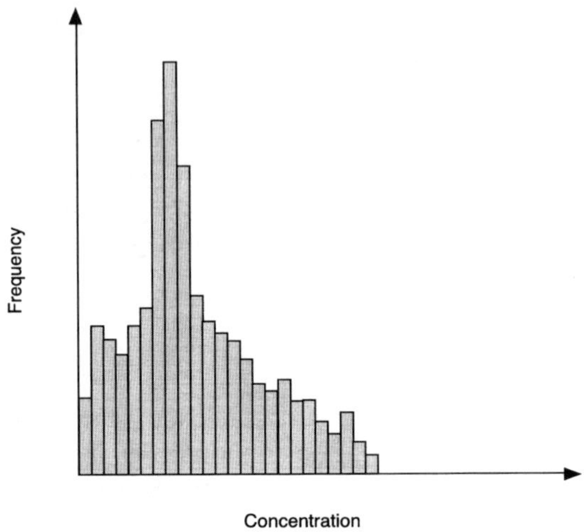

Figure 5-19. Apparent shift in the mode where the histogram cell width for the same data is smaller.

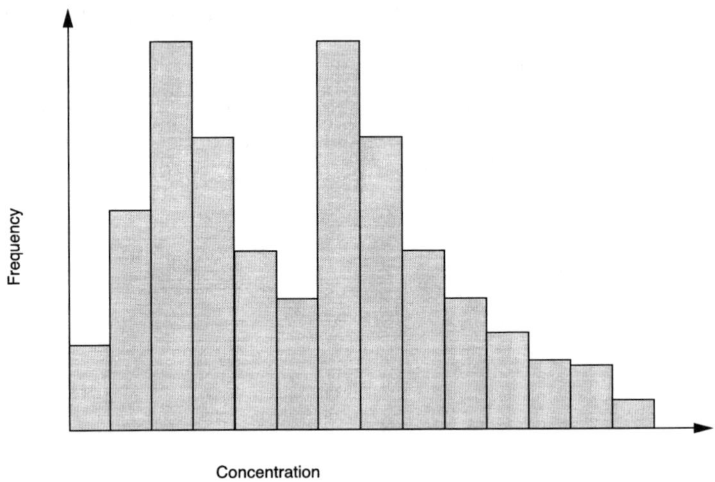

Figure 5-20. Example of a bimodal distribution.

The mode is not commonly used to a large degree in environmental data analysis. It is, however, an important concept in STP, sometimes providing an indication that the sample support is too small.

A final note on the measures of central tendency: For a normal distribution, the mean = the median = the mode, a useful property in decision theory.

5.5.2 Measures of Dispersion

Two closely related distribution parameters are used to measure the spread of values around the mean. These are the variance (σ^2) and the standard deviation (σ). For a limited set of samples, the variance can be represented by

$$\sigma^2 = \sum_{i=1}^{n} \frac{(x_i - m)^2}{n} \tag{5.23}$$

The standard deviation of the population is simply the square root of the variance. It can be expressed as

$$\sqrt{\sigma^2} = \sigma = \sqrt{\sum_{i=1}^{n} \frac{(x - m^2)}{n}} \tag{5.24}$$

A convenient computational formula for calculating the variance is

$$\sigma^2 = \left(\frac{1}{n} \sum_{i=1}^{n} x_i^2 \right) - \mu^2 \tag{5.25}$$

The variance is the parameter used most commonly by statisticians and is easier to handle than the standard deviation. Geostatistical appraisal also makes extensive use of variances because of their additive properties.

Note that the variance is expressed in percent squared if the original units are percentages. Similarly, the variance is parts per million squared if the original units are parts per million. The standard deviation is expressed to the same power as the original units, not to a power of two.

The dispersion parameters play a very important role in error management. They represent the levels of uncertainty associated with estimates of many types. The larger the dispersion, the greater the uncertainties. Figure 5-21 shows two distributions with the same mean value but with different variances.

Many types of variances will be introduced in both STP and GA that will be used to manage the uncertainties at hazardous waste sites. Much of the focus of GEM will be on minimizing the variances, thus reducing the uncertainty. This implies that other sample statistics, especially the mean, are interesting but are of limited value for practical applications in error and uncertainty management.

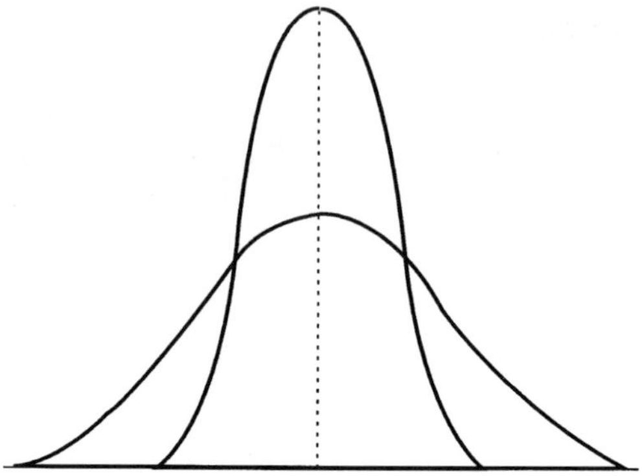

Figure 5-21. Distributions with identical means but different variances.

Conclusion

A distribution represents the probability density function (pdf) for the data being analyzed. The pdf is the probability that a datum of a given value or value range will occur when a random sample is taken. Visually, the pdf takes the form of a histogram, whose shape may vary markedly depending on the type of data in the sample population. In Chapter 4, the effect of sample support volume on the spread or variance of the distribution was shown. This chapter presented another unique aspect of spatial data where area weighting for clustered samples must be factored into the cumulative distribution. Distribution parameters, including the mean, variance, median, and mode statistics, were presented, along with ways to estimate these quantities using sample statistics. These sample statistics offer ways in which to compare distributions and their parameters quantitatively as well as visually.

Chapter 6

Distributional Models

SYNOPSIS: As seen in Chapter 5, data distributions can take a variety of shapes, including symmetric and nonsymmetric. Chapter 4 showed that it is advantageous to apply a model to deterministic data; the same is often true for data involving random variables. This chapter will introduce a few of the commonly used distributional models for statistical data, a number of which are highly useful in the GEM process.

KEY WORDS: Normal Distribution, Z Statistic, Normal Tables, Level of Confidence, Level of Significance, Confidence Interval, Central Limit Theorem, Hypothesis Testing, Type I Error (α), Type II Error (β), Power of a Test, Power Curve, Lognormal Distribution, t-Test, Binomial Distribution, Bernoulli Trial, Poisson Distribution, Fundamental Error

6.1 THE NORMAL DISTRIBUTION

The most familiar distribution is the *normal*, or *Gaussian*, distribution. This distribution is the familiar bell-shaped curve used in many statistical applications. The normal distribution should occur when the value of a phenomenon is the resultant sum of several independent causes. The normal distribution can be concisely defined, and it provides symmetry and other properties that make it convenient for theoretical estimation.

6.1.1 Normal Distribution Function

The equation for the normal distribution is a function of two parameters m and σ:

$$f(x)=\frac{1}{\sigma\sqrt{2\pi}}\exp\left[-\frac{1}{2}\left(\frac{x-m}{\sigma}\right)^2\right] \tag{6.1}$$

A typical normal curve is pictured in Figure 6-1. All normal distributions are centered at the mean and are symmetrical around this statistic as well. This implies that, for a normal distribution, the mean, median, and mode are all coincident.

Figure 6-2 shows the relationship of the standard deviation to the distribution and the mean. This figure shows that, at a distance of $\pm 1\sigma$ from the mean, 68% of the distribution is encompassed. It also shows that, at $\pm 1\sigma$, the normal curve has a point of inflection at which the curve changes from convex to concave. At a distance of $\pm 2\sigma$ from the mean, more than 95% of the distribution is accounted for. At $\pm 3\sigma$, more than 99.7% of the distribution is covered.

It is interesting to note that $\pm 1\sigma$, or 68% is almost $\frac{2}{3}$. Given this, one can say that, if a sample is picked at random from a normal population, there is a one out of three chance that it will be more than one standard deviation away from the

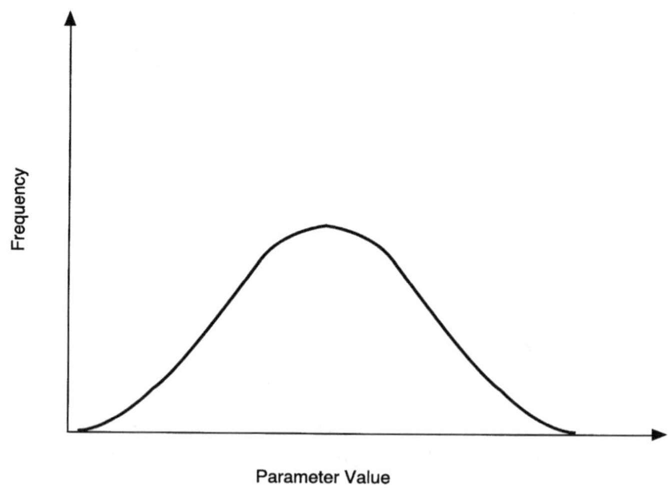

Figure 6-1. The normal curve

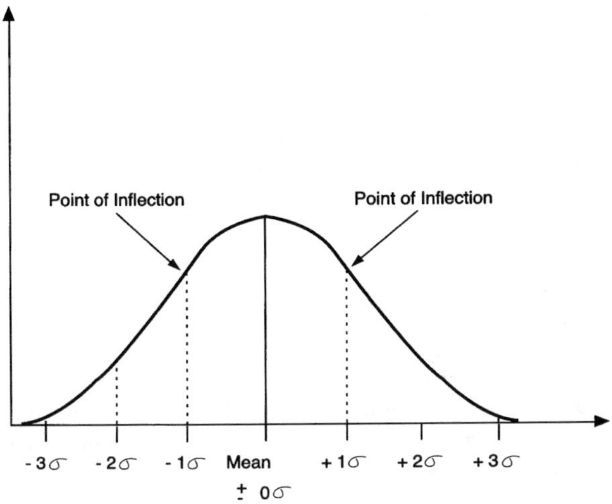

Figure 6-2. The normal distribution curve with standard deviation intervals and inflection points. (From *Pierre Gy's Sampling Theory and Sampling Practice: Heterogeneity, Sampling Correctness, and Statistical Process Control,* 2nd ed., by F.F. Pitard. Copyright © 1993 CRC Press, Boca Raton, FL. Courtesy of Francis Pitard.)

mean. Conversely, there is a two out of three chance that the sample will be within one standard deviation on either side of the mean. Thus, for a random sample, there is a better than average chance that a random sample from a normal distribution will be within one standard deviation of the mean.

To compare distributions, we look to the cumulative distribution function. The cumulative distribution, expressed as

$$\int_{-\infty}^{x_0} f(x)\,dx = F(x_o) \tag{6.2}$$

does not lend itself to simple calculation, however. It is, nonetheless, a widely used entity, simplified by the use of tables. In order to accommodate the myriad of μ and σ values encountered in practice, the normal distribution is transformed into a more universal format. This is called the standard normal distribution (SND) and is computed as a function of $(x_o - m)/\sigma$. The variable quantity Z is defined as

$$z = \frac{\text{deviation around the mean}}{\text{standard deviation}} = \frac{x_o - m}{\sigma} \tag{6.3}$$

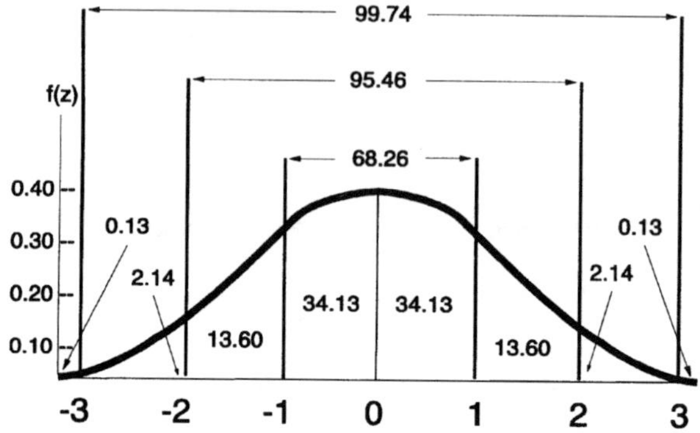

Figure 6-3. Visual representation of the SND standardization with percent areas under the SND curve. (From *Statistical Procedures for Engineering, Management, and Science,* by L. Blank, McGraw-Hill, 1980, with permission.)

Figure 6-3 is a visual interpretation of this standardization process. If x is at the mean, the value of z is equal to zero. If x_o is $\pm 1\sigma$, z is correspondingly ± 1.

As an example, consider a normal distribution with a mean of 10 and a standard deviation of 6. If x_o is equal to 16, we notice that this value is also equal to the sum of the mean and one standard deviation (10 + 6). Using equation 6.3, the z value for the SND is equal to $(16 - 10)/6 = +1$ or one standard deviation. If x_o is 10 (equal to the mean), the z value for the SND is $(10 - 10)/6 = 0$.

In this way, a normally distributed variable can be transformed into a standard normal distribution (SND), no matter what the value of the mean and standard deviation. Once normalized, the z statistic always has a mean of zero and a standard deviation of one. This is commonly written as

$$z \sim N(0,1) = z \sim \text{SND} \tag{6.4}$$

to be read: z is distributed normally with a mean of zero and a standard deviation of one; or z is distributed as a standard normal distribution.

Note that the sign of the z statistic is very important. If the z statistic is positive, x_o is larger than the mean value. If z is negative, x_o is smaller than the mean.

6.1.2 Use of Normal Tables

Once the z value is obtained, standard tables can be used to derive the values and solve problems. Table 6-1 provides values for the cumulative normal distribution.

Table 6–1. Cumulative normal distribution.

$$F(z) = \int_{-\infty}^{z} \frac{1}{\sqrt{2\pi}} e^{-t^2/2}\, dt; \left(z = \frac{x_0 - m}{\sigma}\right)$$

z	.00	.01	.02	.03	.04	.05	.06	.07	.08	.09
−3.5	.00023	.00022	.00022	.00021	.00020	.00019	.00019	.00018	.00017	.00017
−3.4	.00034	.00033	.00031	.00030	.00029	.00028	.00027	.00026	.00025	.00024
−3.3	.00048	.00047	.00045	.00043	.00042	.00040	.00039	.00038	.00036	.00035
−3.2	.00069	.00066	.00064	.00062	.00060	.00058	.00056	.00054	.00052	.00050
−3.1	.00097	.00094	.00090	.00087	.00085	.00082	.00079	.00076	.00074	.00071
−3.0	.00135	.00131	.00126	.00122	.00118	.00114	.00111	.00107	.00104	.00100
−2.9	.0019	.0018	.0017	.0017	.0016	.0016	.0015	.0015	.0014	.0014
−2.8	.0026	.0025	.0024	.0023	.0023	.0022	.0021	.0021	.0020	.0019
−2.7	.0035	.0034	.0033	.0032	.0031	.0030	.0029	.0028	.0027	.0026
−2.6	.0047	.0045	.0044	.0043	.0041	.0040	.0039	.0038	.0037	.0036
−2.5	.0062	.0060	.0059	.0057	.0055	.0054	.0052	.0051	.0049	.0048
−2.4	.0082	.0080	.0078	.0075	.0073	.0071	.0069	.0068	.0066	.0064
−2.3	.0107	.0104	.0102	.0099	.0096	.0094	.0091	.0089	.0087	.0084
−2.2	.0139	.0136	.0132	.0129	.0125	.0122	.0119	.0116	.0113	.0113
−2.1	.0179	.0174	.0170	.0166	.0162	.0158	.0154	.0150	.0146	.0143
−2.0	.0228	.0222	.0217	.0212	.0207	.0202	.0197	.0192	.0188	.0183
−1.9	.0287	.0281	.0274	.0268	.0262	.0256	.0250	.0244	.0239	.0233
−1.8	.0359	.0351	.0344	.0336	.0329	.0322	.0314	.0307	.0301	.0294
−1.7	.0446	.0436	.0427	.0418	.0409	.0401	.0392	.0384	.0375	.0367
−1.6	.0548	.0537	.0526	.0516	.0505	.0495	.0485	.0475	.0465	.0455

Table 6-1. Cumulative normal distribution. *(continued)*

z	.00	.01	.02	.03	.04	.05	.06	.07	.08	.09
−1.5	.0668	.0655	.0643	.0630	.0618	.0606	.0594	.0582	.0571	.0559
−1.4	.0808	.0793	.0778	.0764	.0749	.0735	.0721	.0708	.0694	.0581
−1.3	.0968	.0951	.0934	.0918	.0901	.0885	.0869	.0853	.0838	.0823
−1.2	.1151	.1131	.1112	.1093	.1075	.1057	.1038	.1020	.1003	.0985
−1.1	.1357	.1335	.1314	.1292	.1271	.1251	.1230	.1210	.1190	.1170
−1.0	.1587	.1562	.1539	.1515	.1492	.1469	.1446	.1423	.1401	.1379
−0.9	.1841	.1814	.1788	.1762	.1736	.1711	.1685	.1660	.1635	.1611
−0.8	.2119	.2090	.2061	.2033	.2005	.1977	.1949	.1922	.1894	.1867
−0.7	.2420	.2389	.2358	.2327	.2297	.2266	.2236	.2207	.2177	.2148
−0.6	.2743	.2709	.2676	.2643	.2611	.2578	.2546	.2514	.2483	.2451
−0.5	.3085	.3050	.3015	.2981	.2946	.2912	.2877	.2843	.2810	.2776
−0.4	.3446	.3409	.3372	.3336	.3300	.3264	.3228	.3192	.3156	.3121
−0.3	.3821	.3783	.3745	.3707	.3669	.3632	.3594	.3557	.3520	.3483
−0.2	.4702	.4168	.4129	.4090	.4052	.4013	.3974	.3936	.3897	.3859
−0.1	.4602	.4562	.4522	.4483	.4443	.4404	.4364	.4325	.4286	.4247
−0.0	.5000	.4960	.4920	.4880	.4840	.4801	.4761	.4721	.4681	.4641
0.0	.5000	.5040	.5080	.5120	.5160	.5199	.5239	.5279	.5319	.5359
0.1	.5398	.5438	.5478	.5517	.5557	.5596	.5636	.5675	.5414	.5753
0.2	.5793	.5832	.5871	.5910	.5948	.5987	.6026	.6064	.6103	.6141
0.3	.6179	.6217	.6255	.6239	.6331	.6368	.6406	.6443	.6480	.6517
0.4	.6554	.6591	.6628	.6664	.6700	.6736	.6772	.6808	.6844	.6879
0.5	.6915	.6950	.6985	.7019	.7054	.7088	.7123	.7157	.7190	.7224
0.6	.7257	.7291	.7324	.7357	.7389	.7422	.7454	.7486	.7517	.7549
0.7	.7580	.7611	.7642	.7673	.7704	.7734	.7764	.7794	.7823	.7852
0.8	.7881	.7910	.7939	.7967	.7995	.8023	.8051	.8078	.8106	.8133
0.9	.8159	.8186	.8212	.8238	.8264	.8289	.8315	.8340	.8365	.8389
1.0	.8413	.8438	.8461	.8485	.8508	.8531	.8554	.8577	.8599	.8621

1.1	.8643	.8665	.8686	.8708	.8729	.8749	.8770	.8790	.8810	.8830
1.2	.8849	.8869	.8888	.8907	.8925	.8944	.8962	.8980	.8997	.9015
1.3	.9032	.9049	.9066	.9082	.9099	.9115	.9131	.9147	.9162	.9177
1.4	.9192	.9207	.9222	.9236	.9251	.9265	.9279	.9292	.9306	.9319
1.5	.9332	.9345	.9357	.9370	.9382	.9394	.9406	.9418	.9429	.9441
1.6	.9452	.9463	.9474	.9484	.9495	.9505	.9515	.9525	.9535	.9545
1.7	.9554	.9563	.9574	.9582	.9591	.9599	.9608	.9616	.9625	.9633
1.8	.9641	.9649	.9656	.9664	.9671	.9678	.9686	.9693	.9699	.9706
1.9	.9713	.9719	.9726	.9732	.9738	.9744	.9750	.9756	.9761	.9767
2.0	.9773	.9778	.9783	.9788	.9793	.9798	.9803	.9808	.9812	.9817
2.1	.9821	.9826	.9830	.9834	.9838	.9842	.9846	.9850	.9854	.9857
2.2	.9861	.9864	.9868	.9871	.9875	.9878	.9881	.9884	.9887	.9890
2.3	.9893	.9896	.9898	.9901	.9904	.9906	.9909	.9911	.9913	.9916
2.4	.9918	.9920	.9922	.9925	.9927	.9929	.9931	.9932	.9934	.9936
2.5	.9938	.9940	.9941	.9943	.9945	.9946	.9948	.9949	.9951	.9952
2.6	.9953	.9955	.9956	.9957	.9959	.9960	.9961	.9962	.9963	.9964
2.7	.9965	.9966	.9967	.9968	.9969	.9970	.9971	.9972	.9973	.9974
2.8	.9974	.9975	.9976	.9977	.9977	.9978	.9979	.9979	.9980	.9981
2.9	.9981	.9982	.9983	.9983	.9984	.9984	.9985	.9985	.9986	.9986
3.0	.99865	.99869	.99874	.99878	.99882	.99886	.99889	.99893	.99896	.99900
3.1	.99903	.99906	.99910	.99913	.99915	.99918	.99921	.99924	.99926	.99929
3.2	.99931	.99934	.99936	.99938	.99940	.99942	.99944	.99946	.99948	.99950
3.3	.99952	.99953	.99955	.99957	.99958	.99960	.99961	.99962	.99964	.99965
3.4	.99966	.99967	.99969	.99970	.99971	.99972	.99973	.99974	.99975	.99976
3.5	.99977	.99978	.99978	.99979	.99980	.99981	.99981	.99982	.99983	.99983

This table provides the values for the right or upper half of the normal distribution. Note that a z value of zero (the mean) has an F value of 0.5. This confirms that the mean of the population is centered, with half the population values above and half below.

At $z_o = 1.0$, the table provides a value of $F(z_o) = 0.8413$. This indicates that approximately 84% of the values in the population of interest are below this point (cumulative distribution), with approximately 16% above. $F(1.0)$ is also equal to the mean plus one standard deviation ($m + 1\sigma$); therefore, 84% of a normal population is less than the mean plus 1σ. This can be confirmed visually from Figure 6-3.

If we wish to know the value below which 90% of the population fell, Table 6-1 can again be used. A z value of 1.282, interpolated to the third decimal place, shows an $F(2.0)$ value of 0.90. This indicates that 90% of the distribution is below this point. To obtain the actual value in terms of data concentration, we simply multiply 1.282 by the standard deviation and add this result to the mean. Using the previously described normal distribution that has a mean of 10 and a standard deviation of 6, the calculation would be

$$[(1.282)(6)] + (10) = 17.692 \approx 17.7 \tag{6.5}$$

Thus, 90% of the distribution is below 17.7. Note that this same value could be obtained from a cumulative frequency plot by examining where the 90% line intersects the cumulative data line.

One - sided — outside			Two - sided — inside		
z	Percentage area%	Either graph	z	Percentage area (%)	Graph
3.0	0.135	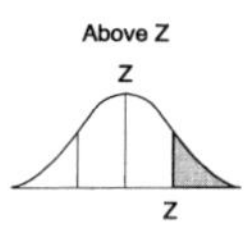	3.0	99.73	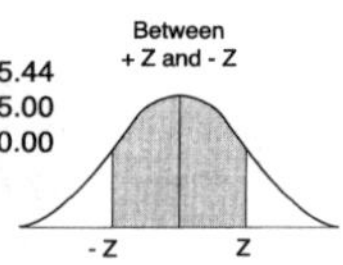
2.5	0.620		2.5	98.76	
2.326	1.0		2.326	98.00	
2.0	2.28	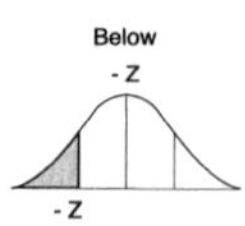	2.0	95.44	
1.96	2.5		1.96	95.00	
1.645	5.0		1.645	90.00	
1.5	6.68		1.5	86.64	
1.282	10.0		1.282	80.00	
1.0	15.87		1.0	68.26	

Figure 6-4. Commonly used z values and percentage areas for the standard normal distribution. One-tailed percentages are outside and two-tailed percentages are inside. (From *Statistical Procedures for Engineering, Management, and Science,* by L. Blank, McGraw-Hill, 1980, with permission.)

An understanding of the symmetry embodied in the normal distribution can be noted further. $F(1.0)$ is roughly 0.84. If one subtracts $F(0)$, the mean, the result is about 0.34. Because of the symmetry around the mean, $\pm 1\sigma$ or, equivalently, $F(1.0 \leq z_0 \leq (1.0)$ results in $(0.34 + 0.34) = 0.68$, or 68% of the population. This confirms the result shown in Figure 6-3, that the area between one standard deviation on either side of the mean encompasses 68% of the population.

Some statistical texts provide tables that have only positive z values. To derive the values below the mean, one must subtract. For example, $F(1.0)$ is 0.84. To get the value for $F(-1.0)$, it is necessary to subtract $F(1.0)$ from 1.00. This is equal to $1.00 - 0.84 = 0.16$. This indicates, as we have seen, that only 16% of the distribution lies between $-\infty$ and one standard deviation below the mean. Commonly used z values and percentage areas for the standard normal distribution can be seen in Figure 6-4.

Example 6.1

The concentration of a pollutant in wastewater being discharged is distributed $N(8,1.5)$. What are the chances, that on any given day, the pollutant concentration in the discharge water will exceed the regulatory limit of 10 ppm?

The solution to the problem lies in determining the proportion of the distribution that lies above 10 ppm. Using the z statistic:

$$z_0 = \frac{10-8}{1.5} = +1.33 \tag{6.6}$$

A value of $z_o = 1.33$ corresponds to an $F(z_o)$ value of 0.9082 in the cumulative distribution tables for the normal distribution. This means that 90.82% (or approximately 91%) of the distribution lies below the regulatory limit of 10 ppm. As such, the discharge water can be expected to exceed regulatory limits about 9% of the time.

Example 6.2

The cadmium concentration in fly ash waste is distributed $N(1.0,0.72)$. What are the chances that a random sample of the ash will have a cadmium concentration between 0.5 and 1.75 ppm?

The lower limit can be defined using the z statistic:

$$z_1 = \frac{0.5 - 1.0}{0.72} = -0.69 \tag{6.7}$$

corresponding to an $F(z_1) = 0.2451$ in the normal table. The upper limit can be defined as

$$z_2 = \frac{1.75 - 1.0}{0.72} = +1.04 \tag{6.8}$$

corresponding to an $F(z_2)$ value of 0.8508. Thus the area under the normal curve is $0.8508 - 0.2451 = 0.6057$, or 60.57%. One would expect that a random sample from the ash would have a concentration between 0.5 and 1.75 ppm approximately 6 out of 10 times.

6.1.3 Confidence and Significance

The mean value is a true but unknown population parameter, estimated by $\bar{x}$ derived from sample data. It is a common desire to determine the confidence with which the mean of the population is known. *Confidence*, and the related entity *significance*, are commonly used words in statistics and it is necessary to distinguish between the two.

Confidence refers to the amount of faith one has in the statistical conclusion reached. *Significance* is a measure of the amount of mistrust in the conclusion. Formal definitions are:

Confidence level: The degree of assurance that a particular statistical statement is correct under specified conditions.

Significance level: The degree of uncertainty about the statistical statement under the same conditions used to determine the confidence level.

Significance levels are symbolized by α, where α is between zero and one ($0 \leq \alpha \leq 1$). Confidence levels are represented mathematically by $1 - \alpha$. The sum of the confidence level and the significance level is equal to 1.0. This can be shown as

$$\text{Confidence level} + \text{significance level} = 1.0 \tag{6.9}$$

or

$$(1 - \alpha) + \alpha = 1.0 \tag{6.10}$$

For example, if the significance level is $\alpha = .10$, the corresponding confidence level is $1 - .10 = .90$, or 90%. For this situation, the conclusion drawn would be expected to be incorrect about 10 times in 100 or about 1 time in 10. The smaller the significance level, the less chance of error in decision making. A significance level of .02 would be expected to produce an erroneous conclusion only 1 time in 50. This assumes equal sample sizes as confidence is n-dependent.

6.1.4 Confidence Intervals

In Chapter 5, formulas were presented for the calculation of population statistics, most notably for the mean and variance. These formulas yield a particular estimated value for the population parameter known as a *point estimate*. Because these estimates are based on a limited number of sample data, we would not expect that the true value of the population parameter would exactly equal the calculated estimate.

As such, a degree of uncertainty exists surrounding the true value of the population parameter such as the mean. One weakness of the point estimation technique is that it offers no way to quantify the probability of error associated with the estimate. An *interval estimation* provides a range of values within which the population parameter is likely to be.

A *confidence interval* describes an interval in which the true population parameter—say, the mean—will fall with a specified level of certainty, perhaps 90%. This means that, on average, 90 such intervals out of every 100 calculated on the basis of means of sample size n will include the population mean μ.

A note of caution on a subtle point: It is not correct to state that the value of μ has a probability of .90 of being within the interval. The value μ is a constant, not a random variable. If μ is in the interval, the probability of it being in the interval is 1.0; if not, the probability is 0.0. The population parameter does not have a probability distribution.

In contrast, the intervals are random variables. We can appropriately make probability statements about the *proportion of intervals* that would include the value μ, whereas it is not appropriate to make probability statements regarding μ.

To construct a confidence interval, the general practice is to specify n and α in advance. These parameters may be considered as unknowns, however, and one can solve for the value of this unknown. In addition, it is necessary to decide whether the interval will be one-sided or two-sided. For a one-sided test, α is assigned to only one tail of the distribution. This is usually the upper tail for environmental applications as we most commonly wish not to exceed a regulatory action level. For a two-sided test, α is divided equally ($\alpha/2$) between the upper and lower tails.

Figure 6-4 is a graphic representation of significance and confidence. Both one- and two-tail significance areas exist for a significance level, but only a single area exists for confidence. A one-tail significance (Fig. 6-4a) has the complete α area in one tail of the distribution, either upper or lower. The remainder of the distribution is the confidence zone $1 - \alpha$. A two-tail significance places one-half α ($\alpha/2$) in each tail of the distribution (Fig. 6-4b). The confidence $1 - \alpha$ falls symmetrically between these two significance areas.

For a two-sided confidence interval with $\alpha = .05$, the probability that z falls between the $\alpha/2$ limits is

$$P(-1.96 \leq z \leq +1.96) \tag{6.11}$$

or, more generally,

$$P(z_{\alpha/2} \leq z \leq z_{\alpha/2}) = 1 - \alpha \tag{6.12}$$

It should be noted that this is the probability interval for the particular standardized normal variable:

$$z = \frac{(\bar{x} - \alpha)}{\sigma/n} \tag{6.13}$$

6.1.4.1 *Known* σ

For a two-sided confidence interval on the mean μ, the expression is

$$\bar{x} - z_{\alpha/2}\frac{\sigma}{\sqrt{n}} \leq \mu \leq \bar{x} - z_{\alpha/2}\frac{\sigma}{\sqrt{n}} \tag{6.14}$$

or

$$\mu \leq \bar{x} + z_{\alpha}\frac{\sigma}{\sqrt{n}} \tag{6.15}$$

for a one-sided test.

Example 6.3

A sample of size $n = 16$ is taken from a normal population with a known variance. The sample average was calculated to be 137, with a standard deviation of 21. Determine a two-sided 80% confidence interval for the mean.

Using equation 6.9,

$$\bar{x} - z_{\alpha/2}\frac{\sigma}{\sqrt{n}} \leq \mu \leq \bar{x} - z_{\alpha/2}\frac{\sigma}{\sqrt{n}} \tag{6.16}$$

$$137 - (1.28)\frac{21}{\sqrt{16}} \leq \mu \leq 137 + (1.28)\frac{21}{\sqrt{16}} \tag{6.17}$$

$$137 - 7.04 \leq \mu \leq 137 + 7.04 \tag{6.18}$$

$$129.96 \leq \mu \leq 144.04 \tag{6.19}$$

6.1.4.2 Unknown σ

A major assumption for the equations set forth in Section 6.1.4.1 is that the population standard deviation σ is known. This is commonly not the case in practical situations. In fact, it is usually the case that, when the population mean is unknown, the population standard deviation σ is unknown as well. In such cases, the population standard deviation is estimated by using the sample standard deviations.

Under these circumstances, it is appropriate to use the t-distributed random variable:

$$\frac{\bar{x} - \mu}{s/\sqrt{n}} \tag{6.20}$$

with $n - 1$ degrees of freedom. Again the underlying population is assumed to be normal. Determining the confidence interval using a t distribution is very similar to that for a normal distribution. The limits of the interval are found by using t values (Table 6-2) as opposed to z values. Also, an additional parameter ν that specifies the degrees of freedom must be used. The parameter ν is calculated:

$$\nu = n - 1 \tag{6.21}$$

Table 6-2. The *t* distribution.

	One-tail α					
	0.10	**0.05**	**0.025**	**0.01**	**0.005**	**0.001**
	Two-tail α					
ν	**0.20**	**0.10**	**0.05**	**0.02**	**0.01**	**0.002**
1	3.078	6.314	12.706	31.821	63.657	318.300
2	1.886	2.920	4.303	6.965	9.925	22.327
3	1.638	2.353	3.182	4.541	5.841	10.214
4	1.533	2.132	2.776	3.747	4.604	7.173
5	1.476	2.015	2.571	3.305	4.032	5.893
6	1.440	1.943	2.447	3.143	3.707	5.208
7	1.415	1.895	2.365	2.998	3.499	4.785
8	1.397	1.860	2.306	2.896	3.355	4.501
9	1.383	1.833	2.262	2.821	3.250	4.297
10	1.372	1.812	2.228	2.764	3.169	4.144
11	1.363	1.796	2.201	2.718	3.106	4.025
12	1.356	1.782	2.179	2.681	3.055	3.930
13	1.350	1.771	2.160	2.650	3.012	3.852
14	1.345	1.761	2.145	2.624	2.977	3.787
15	1.341	1.753	2.131	2.602	2.947	3.733
16	1.337	1.746	2.120	2.583	2.921	3.686
17	1.333	1.740	2.110	2.567	2.898	3.646
18	1.330	1.734	2.101	2.552	2.878	3.611
19	1.328	1.729	2.093	2.539	2.861	3.579
20	1.325	1.725	2.086	2.528	2.845	3.552
21	1.323	1.721	2.080	2.518	2.831	3.527
22	1.321	1.717	2.074	2.508	2.819	3.505
23	1.319	1.714	2.069	2.500	2.807	3.485
24	1.318	1.711	2.064	2.492	2.797	3.467
25	1.316	1.708	2.060	2.485	2.787	3.450
26	1.315	1.706	2.056	2.479	2.779	3.435
27	1.314	1.703	2.052	2.473	2.771	3.421
28	1.313	1.701	2.048	2.467	2.763	3.408
29	1.311	1.699	2.045	2.462	2.756	3.396
30	1.310	1.697	2.042	2.457	2.750	3.385
40	1.303	1.684	2.021	2.423	2.704	3.307
60	1.296	1.671	2.000	2.390	2.660	3.232
80	1.292	1.664	1.990	2.374	2.639	3.195
100	1.290	1.660	1.984	2.365	2.626	3.174
∞	1.282	1.645	1.960	2.326	2.576	3.090

A 100(1 − α) confidence interval for μ can be calculated:

$$\frac{\bar{x} - t_{\alpha/2,\nu}}{s/\sqrt{n}} \leq \mu \leq \frac{\bar{x} + t_{\alpha/2,\nu}}{s/\sqrt{n}} \tag{6.22}$$

For the data given in Example 6.3 using an unknown standard deviation estimated by s with 15 degrees of freedom, the 95% confidence interval using the t distribution is

$$\bar{x} - \frac{t_{\alpha/2,\nu}}{s/\sqrt{n}} \leq \mu \leq \bar{x} + \frac{t_{\alpha/2,\nu}}{s/\sqrt{n}} \tag{6.23}$$

$$137 - (1.341)\frac{21}{\sqrt{16}} \leq \mu \leq 137 + (1.341)\frac{21}{\sqrt{16}} \tag{6.24}$$

$$137 - 7.04 \leq \mu \leq 137 + 7.04 \tag{6.25}$$

$$129.96 \leq \mu \leq 144.04 \tag{6.26}$$

6.1.5 The Central Limit Theorem

Although sample data are seldom so accommodating as to follow a normal distribution, the normal distribution still plays an important role in the GEM process. Estimation errors from interpolators such as kriging often follow a normal distribution, allowing some convenient applications. Also, a normal distribution of replicate samples is approached for appropriate sample volumes and corresponding acceptable levels of error in STP.

For both normal and non-normal data, the central limit theorem (CLT) allows one to use the normal distribution to infer about a population that is not normally distributed. The central limit theorem states:

The population may have any unknown distribution with a mean μ *and finite variance* σ^2. *Take samples of size* n *from the population. As the size of* n *increases, the distribution of sample means will approach a normal distribution with mean* μ *and variance* σ^2/n.

The CLT implies that, if a distribution of any shape is sampled repeatedly, the distribution of the means will be normal. Figure 6-5 demonstrates the principle

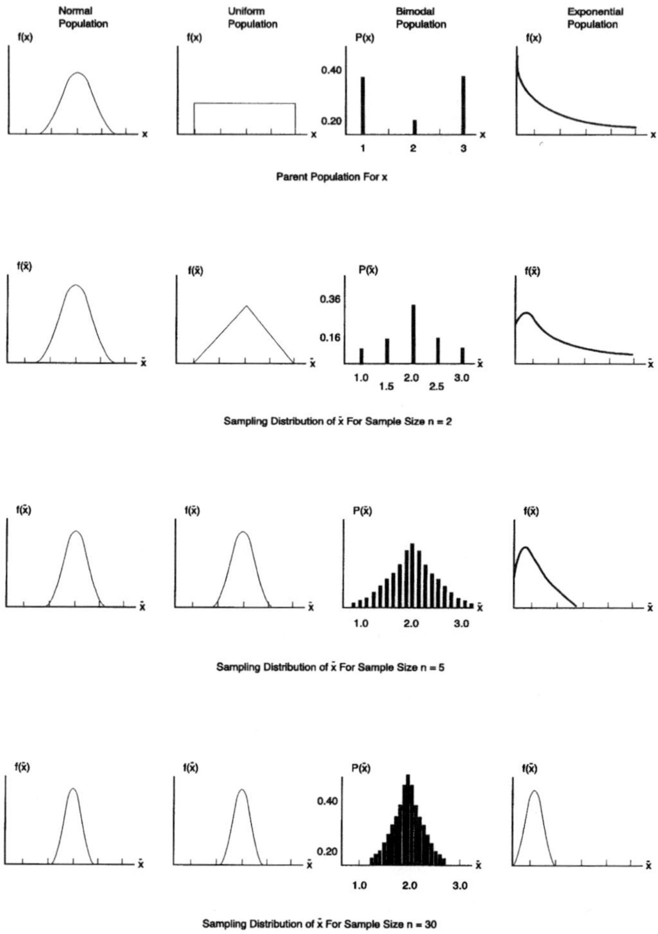

Figure 6-5. Sample distributions of x values for different initial population distributions with $n = 2, 5, 30$. (From *Statistical Analysis for Business and Economics*, 3rd ed., by D.L. Harnett and J.L. Murphy. Copyright © 1985 Addison-Wesley Publishers Limited. All rights reserved. Reproduced by permission.)

involved. The first row of Figure 6-5 shows four different distributions: a normal, a uniform, a bimodal, and a skewed population. Row 2 shows the distribution of sample means when each of the populations is sampled with a sample size of $n = 2$. The distribution of means from the normal population remains normal, and each of the other three distributions of sample means more closely resembles a normal population than the distribution of sample data.

Row 3 in Figure 6-5 shows the results of sampling each population with a sample size of $n = 5$. The distribution of sample means now closely follow a normal distribution in most cases. Finally, when the sample size is increased to $n = 20$, all distributions of sample means are normally distributed. Note also that the variances of the distributions of sample means are less than the variances of the starting populations. This is consistent with the CLT, which states that the variance decreases as the square root of n increases.

The sample mean $\bar{x}$ is itself a random variable, having a real mean μ and a standard deviation σ_m:

$$\sigma_m = \frac{\sigma}{\sqrt{n}} \tag{6.27}$$

As n increases, more of the population is included in the distribution of $\bar{x}$. If n is infinity, the sample is the same as the population, and the mean is known with absolute certainty, driving the standard error of the mean to zero. Figure 6-6 shows an illustration of a 95% upper confidence limit (UCL) for a given number of samples. The diagram indicates that the 95% UCL is far below the 95th percentile of the data.

The standard error of the mean may assist one who questions how well the average concentration of a hazardous waste site is known for purposes of risk assessment after n samples have been collected. The formula can also be used in

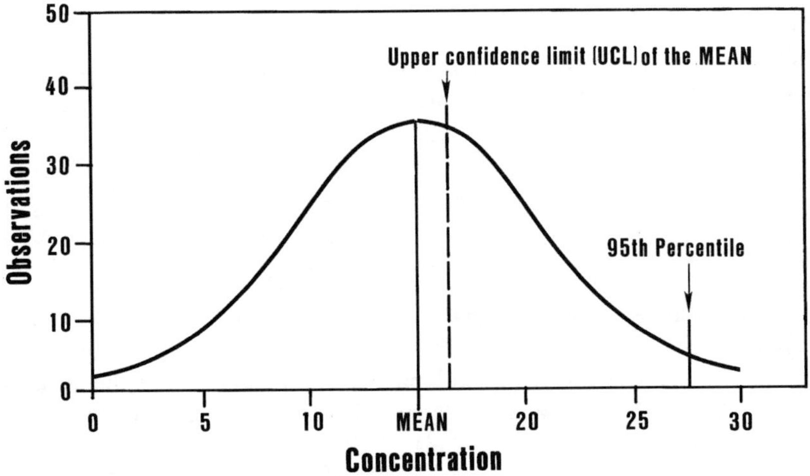

Figure 6-6. Comparison of the UCL and the 95th percentile. As sample size increases, the UCL of the mean moves closer to the true mean, while the 95th percentile of the distribution remains at the upper end of the distribution.

reverse to calculate how many samples will be necessary to achieve a desired level of accuracy.

The standard normal distribution, along with the CLT given in equation 6.27, is used to calculate the necessary sample size, provided that two questions can be answered: (1) What level of confidence is desired? and (2) What is the maximum difference (*D*) permitted between the estimate of the mean and the true value of the mean? *D* is defined as the largest allowable error between the estimated value and the true value:

$$D \geq |\text{ estimated value} - \text{true value }| \tag{6.28}$$

or

$$D \geq \text{UCL} - \text{DCL} \tag{6.29}$$

For a starting population following a normal distribution with a known standard deviation, the variable

$$Z = \frac{(x_0 - m)}{\sigma/\sqrt{n}} \tag{6.30}$$

is distributed *N*(0,1). For a desired confidence level of 1 − α, equation 6.8 becomes the following 100(1 − α) interval for x_o − m:

$$-z_{\alpha/2}\frac{\sigma}{\sqrt{n}} \leq (x_0 - m) \leq z_{\alpha/2}\frac{\sigma}{\sqrt{n}} \tag{6.31}$$

For environmental applications, one is generally concerned only if the true population mean exceeds an action limit. Thus, the left side of equation 6.9 is not relevant. This changes $z_{\alpha/2}$ to z_α and implies that the largest value the quantity $x_o - m$ can take is $z_\alpha(\sigma/\sqrt{n})$. With the largest allowable error *D* equal to $x_o - m$, it can be stated as

$$D = x_0 - m = z_\alpha \frac{\sigma}{\sqrt{n}} \tag{6.32}$$

By solving this relationship for n, the minimum sample size required to estimate the mean at the specified level of confidence is

$$n = \frac{z_\alpha^2 \sigma^2}{D^2} \tag{6.33}$$

EXAMPLE 6.4

Sampling for PCBs has been performed at a site with a sample size of 5. The mean PCB concentration has been determined to be 471 ppm, with a standard deviation of 72 (ppm)^2. Regulations dictate that PCB concentrations above 500 ppm must undergo incineration. The regulatory agency wishes to know: (1) How well is the mean value known with the current level of sampling? and (2) How many samples would be needed to assure that the error on the mean is less than 29 ppm, with a confidence of 95% that the true mean is less than 500 ppm?

Whereas the distribution of the PCB values is likely to be skewed, the sample means should approach a normal population. For $n = 5$, the distribution of the PCB means is expected to be normal, with a mean of 471 ppm and a variance of 5184/5 = 1036.8 (ppm)^2 and a corresponding standard deviation of 32.2 ppm.

To determine sample size, the substitution becomes

$$n = \frac{(1.645)^2(72)^2}{(500 - 471)^2} = 16.68 \simeq 17 \tag{6.34}$$

Note that the value of n is always rounded up to the nearest integer to ensure that a sufficient number of samples will be taken. In this case, the mean value is poorly known, and a large number of additional samples will be required to assure that it is below the established limit of 500 ppm.

6.1.6 Caveats for a Trusted Formula

The central limit theorem is one of the most widely used statistical formulas for determining precision of the mean and for determining sample sizes. The convenient feature of the CLT is that it allows us to use a normal distribution to infer about a population of interest that is not itself normally distributed. This offers a potential convenience since most environmental data have distributions far from normal.

Now, the caveat. David (1977) states that the standard error of the mean "... has been responsible for the largest number of mistakes made on the account of statistics." Although there are no facts or figures to back up this claim, there is no doubt that this formula has been widely misapplied.

Use of this formula assumes that the sample data are independent of each other. Violation of this assumption makes the formula invalid. In most environmental applications, spatial data are not independent. In fact, we generally hope that they are not independent so that spatial correlation methods can be applied. Even if a first round of sample data proves uncorrelated, it is highly likely that subsequent samples will no longer be independent.

For spatially correlated data, equation 6.33 produces an error value that is higher than reality. In other words, through the assumption of independence, a lot of information has been ignored that could produce a smaller error value using the same number of sample data. When calculating the number of samples to reach a specified level of confidence, assuming uncorrelated data will produce a result demanding more samples than necessary. This has potential to cause unnecessary expenditures in sampling budgets.

Note also the implied assumption that the correct sample support has been used for the sample data. In order to produce normally distributed mean values, samples of sufficient volume must be taken to overcome the effects of the Poisson distribution. If sample support is too small, we may have a statistic more representative of the standard error of the mode.

6.1.7 Hypothesis Testing

The calculation of point estimates allows a unique and clear representation of the unknown population parameters. Such calculations were shown in Chapter 5. In this chapter, interval estimates have been introduced as a way of expressing the uncertainty associated with a given population parameter. Confidence intervals allow us to test how valid the assumed values of the population parameter are.

The process of hypothesis testing presents us with an objective method for determining whether certain statistical assumptions made are valid. The purpose of hypothesis testing is to select the appropriate assumption. The hypothesis test is structured so that two mutually exclusive and jointly exhaustive competing hypotheses concerning a population parameter are evaluated, with only one selected.

Two different categories of hypotheses exist: simple and composite. Examples of simple hypotheses include simple tests of means, such as $\mu = 25.0$ and $\mu_1 - \mu_2 = 0$. Alternatively, composite hypotheses specify a range of values that a population parameter might assume, such as $\mu \leq 0.25$, $\mu \neq 1000$, or $\mu_1 - \mu_2 \neq 50$.

The hypothesis test is characterized by the statement of two conflicting conditions termed the *null hypothesis* (H_0) and the *alternative hypothesis* (H_a). The null hypothesis can take either the condition presumed to be true or that presumed to be untrue. EPA recommends in the 1994 DQO document that the null hypothesis H_0 take the value presumed to be true or as EPA states, the "true state of nature." An example of a hypothesis statement set for a two-sided or simple test is

$$H_0: \mu = 500$$
$$H_a: \mu \neq 500$$

An hypothesis statement for a one-sided or composite test is

$$H_0: \mu = 500$$
$$H_a: \mu > 500$$

When the statistical test is performed to compare the validity of the competing hypothesis statements, the result will cause the null hypothesis H_0 to be either *rejected* or *not rejected*. This phraseology is similar to the pleas that may be used by defendants and verdicts rendered by juries in the American legal system, *guilty* or *not guilty*. In hypothesis testing, a decision not to reject may still carry some lingering doubt about the truth, just as a verdict of not guilty does not necessarily imply total innocence. Nevertheless, for the sake of simplicity, the phrase "accept the null hypothesis H_0" will be used synonymously with "not reject the null hypothesis H_0."

In the courtroom, the defendant is initially presumed innocent whereas, in hypothesis testing, H_0 is initially presumed true. In both cases, the burden lies in proving the assumption wrong beyond a reasonable doubt. Unfortunately, in both the courtroom and in statistics, an incorrect decision may be made. In court, this means falsely convicting an innocent person or inappropriately exonerating a criminal. In statistics, these decision errors are called Type I and Type II errors.

A Type I error occurs when the null hypothesis is rejected based on the available sample information when, in fact, it is true. If one sets up the hypothesis conditions to be H_0: the person is innocent; and H_a: the person is guilty, a Type I error is equivalent to wrongly convicting the person.

A Type II error results when the null hypothesis H_0 is retained when in fact it should be rejected, that is, the alternative hypothesis H_a is true. This is the same as letting a guilty person go free.

Of course, correct decisions may also be made by accepting a true H_0 and rejecting a false H_0. The four possible decision outcomes are illustrated in Figure 6-7.

It is important to take into account the probabilities of both Type I and Type II errors before performing a hypothesis test. The probability of a Type I error is denoted by alpha (α), the level of significance or probability of an error described in Section 6.1.3. Equivalently,

$$\begin{aligned}\alpha &= \text{level of significance} = P(\text{type I error})\\ &= P(\text{reject } H_0 | H_0 \text{ is true})\end{aligned}$$

The complement of this situation is the confidence level of the test where no mistake is made. This can be represented as

$$\begin{aligned}(1-\alpha) &= \text{confidence level} = 1 - P(\text{type I error})\\ &= P(\text{accept } H_0 | H_0 \text{ is true})\end{aligned}$$

In statistical testing, it is desirable to keep the probability of a Type I error as low as possible. This can be done easily by using small values of α. However, there is a tensional relationship between Type I and Type II errors: As Type I errors decrease, Type II errors increase. The probability of a Type II error is referred to as beta (β) and is shown by the following relationships:

$$\beta = P(\text{type II error}) = P(\text{accept } H_0 | H_0 \text{ is false})$$

		If the true situation is	
		H_0 is true	H_0 is false
Action	Accept H_0	Correct Decision	Incorrect Decision (Type II error)
	Reject H_0	Incorrect Decision (Type I error)	Correct Decision

Figure 6-7. The four possible decision outcomes of statistical hypothesis testing. (From *Statistical Analysis for Business and Economics*, 3rd ed., by D.L. Harnett and J.L. Murphy. Copyright © 1985 Addison-Wesley Publishers Limited. All rights reserved. Reproduced by permission.)

		If the true situation is H_o is true	H_o is false
Action	Accept H_o	$1-\alpha$ (Confidence level)	β (beta)
	Reject H_o	α (alpha) (Significance level)	$1-\beta$ (Power of the test)
	Sum	1.00	1.00

Figure 6-8. The probability of the decision outcomes in hypothesis testing. (From *Statistical Analysis for Business and Economics*, 3rd ed., by D.L. Harnett and J.L. Murphy. Copyright © 1985 Addison-Wesley Publishers Limited. All rights reserved. Reproduced by permission.)

Just as with α, β has a complement that is known as the power of the test or

$$1-\beta = P(\text{type II error}) = P(\text{accept } H_0 \mid H_0 \text{ is false})$$

Ideally, the test should be structured so that the power is large (close to 1), providing a small β when H_O is false. Figure 6-8 shows an alternative version of Figure 6-7 that includes significance, confidence, and power information.

To perform a statistical test of hypotheses, a six-step process may be used. The components are listed below:

1. State the null and alternative hypotheses.
2. Select the significance level.
3. Determine the appropriate test statistic.
4. Determine the critical region.
5. Compute the value of the test statistic.
6. Make the statistical decision and interpretation.

An example calculation of the β (Type II) error will demonstrate the listed steps.

Example 6.5

Lead (Pb)-contaminated soil at a site under remediation follows a normal distribution with a standard deviation of 175 ppm. Each 100 y^3 unit volume is tested with a series of nine samples to determine the average lead concentration. Soil with excessive Pb concentrations must be buried and capped. The latest 100 y^3 unit volume showed a mean Pb concentration of 425 ppm. What is the probability of a Type II error if, in reality, the true concentration is 500 ppm? Use a .05 level of significance.

The hypotheses are

$$H_0: Pb = 425 \text{ ppm}$$

$$H_a: Pb \neq 425 \text{ ppm}$$

The test statistic used to test the mean value is

$$Z = \frac{w - \mu_o}{\sigma/\sqrt{n}} \tag{6.35}$$

The critical region (Fig. 6-9) is defined by

$$P(z \leq z_1) = P\left(\frac{w_1 - 425}{120/\sqrt{9}}\right) = .025 \tag{6.36}$$

$$\left(\frac{w_1 - 425}{120/\sqrt{9}}\right) = -1.96 \tag{6.37}$$

$$w_1 = 346.6 \text{ ppm} \tag{6.38}$$

and

$$P(z \geq z_2) = P\left(\frac{w_2 - 425}{120/\sqrt{9}}\right) = .025 \tag{6.39}$$

$$\frac{w_2 - 425}{40} = +1.96 \tag{6.40}$$

$$w_1 = 503.4 \text{ ppm} \tag{6.41}$$

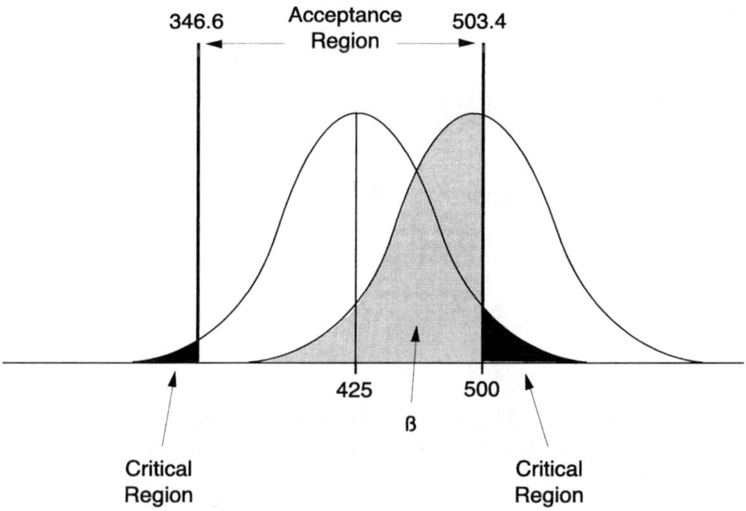

Figure 6-9. Acceptance region, critical region, α errors, and β error for the statistical test of hypothesis in Example 6.5.

The critical region is defined by w_1 = 346.6 and w_2 = 503.4. Any Pb concentration that falls between these two values will cause the null hypothesis H_0 to be accepted.

The probability of a Type II (β) error is

$$P(w_1 \leq 346.6) = P\left(z \leq \frac{346.6 - 500}{40}\right) = P(z \leq -3.835) = .0 \tag{6.42}$$

combined with

$$P(w_2 \leq 503.4) = P\left(z \leq \frac{503.4 - 500}{40}\right) = P(z \leq .085) = .5339 \tag{6.43}$$

The likelihood of a Type II error (β), the chance of accepting H_0 when H_a is true, is equal to the area between −3.835 and .5339, or

$$P(-3.835 \leq z \leq .085) \tag{6.44}$$

and, equivalently,

$$P \text{ (type II error)} = F(0.085) - F(-3.835) = .5339 - .0 = .5339 \quad (6.45)$$

This yields a probability of .5339, or a 53.4% chance of occurrence.

EXAMPLE 6.6

The Remedial Project Manager (RPM) has stated that errors on the low side are acceptable at any level. Using a one-sided test, what is the new probability of a Type II error?

The hypotheses are

$$H_0: \text{Pb} = 425 \text{ ppm}$$

$$H_a: \text{Pb} > 425 \text{ ppm}$$

The test statistic used to test the mean value is

$$Z = \frac{w - \mu_o}{\sigma/\sqrt{n}} \quad (6.46)$$

The critical region (Fig. 6-10) is defined by:

$$P(z \leq z_1) = P\left(\frac{w - 425}{120/\sqrt{9}}\right) = .05 \quad (6.47)$$

$$\frac{w - 425}{120/\sqrt{9}} = +1.645 \quad (6.48)$$

$$w = 490.8 \text{ ppm} \quad (6.49)$$

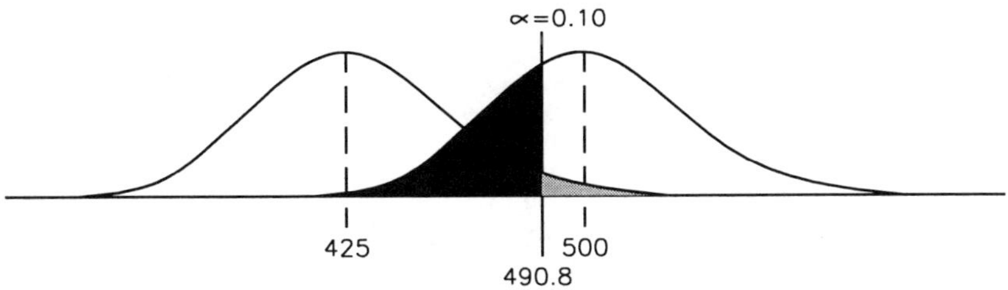

Figure 6-10. Critical region for the hypothesis test in Example 6.6.

The probability of a Type II (β) error is

$$P(w \leq 490.8) = P\left(z \leq \frac{490.8 - 500}{40}\right) = P(z \leq -.23) = .4090 \qquad (6.50)$$

or a 40.9% chance.

Examples 6.5 and 6.6 demonstrate that, depending on whether the test is one- or two-sided, the chance of a Type II error is between approximately 41% and 53% for the situation described. Example 6.6 has also proven a previous statement: As the level of significance α increases, the probability of a Type II error decreases.

It should be intuitive that the likelihood of a Type II error will decrease as the means being compared get further apart in value. For instance, it is harder to distinguish between means of 425 and 450 ppm than between 425 and 525 ppm. This relationship of increasing difference between means can be presented graphically in what is called a *power curve*.

Tables 6-3 and 6-4 show the calculated values for β and power (1 − β) for several sample mean values using sample sizes of $n = 9$ and $n = 25$ respectively. As predicted, the probability of a Type II error decreases as the means get further apart. Similarly, the power of the test increases as the difference between means increases. A visual representation of these results is presented in Figure 6-11.

It should also be intuitive that, with more sample information, the power of the test will increase. The examples shown so far have held n constant at 9. Figure 6-11 also shows a second power curve representing a sample size of 25. This power curve is steeper than that for $n = 9$, indicating better resolving power. This is substantiated by the graph where for each point on the $n = 25$ curve the value of β is smaller and the power of the test is higher.

For environmental situations, one-sided hypotheses are often preferred because the sample mean must be compared to the action limit. Figure 6-12 shows a one-sided test for a sample mean below the action limit. Figure 6-13 shows a

Table 6-3. Calculated values for β and power (1 − β) for several mean values. The Table includes values for one-sided and two-sided tests with sample size $n = 9$.

$\mu = 500, \sigma = 120$, n = 9				
	β = P (*Accept* H_o)		*Power* = P (*Reject* H_o) *Power* = $(1 - \beta)$	
Mean	**One-Sided**	**Two-Sided**	**One-Sided**	**Two-Sided**
325	.0032	.0079	.9968	.9921
350	.0177	.0367	.9823	.9633
375	.0694	.1220	.9306	.8780
400	.1963	.2946	.8037	.7054
425	.4090	.5339	.5910	.4661
450	.6535	.7604	.3465	.2396
475	.8461	.9042	.1539	.0958
500	1 − α = .95	1 − α = .95	α = .05	α = .05
525	.8461	.9042	.1539	.0958
550	.6535	.7604	.3465	.2396
575	.4090	.5339	.5910	.4661
600	.1963	.2946	.8037	.7054
625	.0694	.1220	.9306	.8780
650	.0177	.0367	.9823	.9633
675	.0032	.0079	.9968	.9921

Table 6-4. Calculated values for β and power (1 = β) for several mean values. The Table includes values for one-sided and two-sided tests with sample size $n = 25$.

$\mu = 500, \sigma = 120$, n = 25				
	β = P (*Accept* H_o)		*Power* = P (*Reject* H_o) *Power* = $(1 - \beta)$	
Mean	**One-Sided**	**Two-Sided**	**One-Sided**	**Two-Sided**
325	.0000	.0000	1.0000	1.0000
350	.0000	.0000	1.0000	1.0000
375	.0002	.0006	.9998	.9994
400	.0059	.0136	.9941	.9864
425	.0694	.1220	.9306	.8780
450	.3309	.4503	.6691	.5497
475	.7277	.8190	.2723	.1810
500	1 − α = .95	1 − α = .95	α = .05	α = .05
525	.7277	.8190	.2723	.1810
550	.3309	.4503	.6691	.5497
575	.0694	.1220	.9306	.8780
600	.0059	.0136	.9941	.9864
625	.0002	.0006	.9998	.9994
650	.0000	.0000	1.0000	1.0000
675	.0000	.0000	1.0000	1.0000

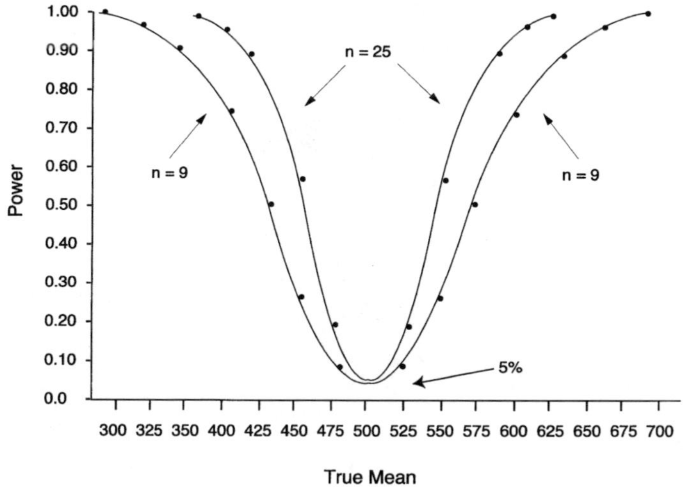

Figure 6-11. Power curves for a two-sided hypothesis test with population parameters: $\mu = 500$, $\sigma = 120$. (From *Statistical Analysis for Business and Economics*, 3rd ed., by D.L. Harnett and J.L. Murphy. Copyright © 1985 Addison-Wesley Publishers Limited. All rights reserved. Reproduced by permission.)

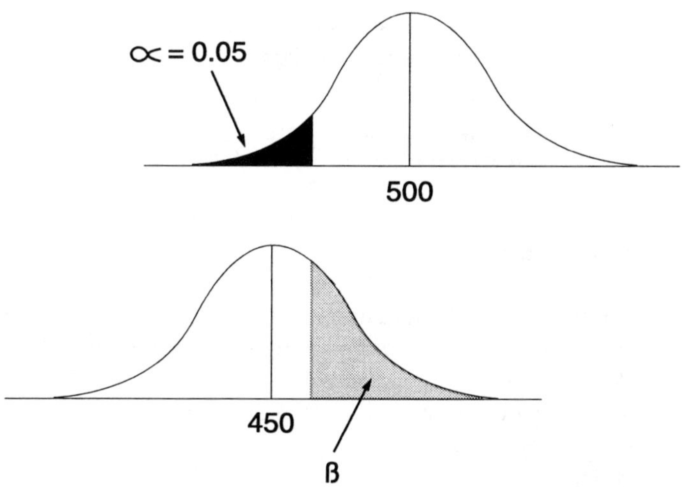

Figure 6-12. Critical region and β region for a one-sided hypothesis test where the sample mean is below the action level.

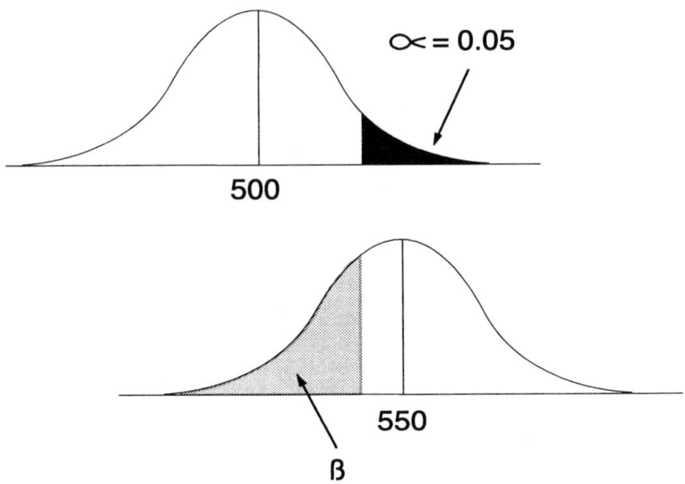

Figure 6-13. Critical region and β region for a one-sided hypothesis test where the sample mean is above the action level.

one-sided test for a sample mean above the action limit. Power curves for one-sided tests are shown in Figure 6-14 for $n = 9$ and $n = 25$. As with the two-sided test, the power for $n = 25$ is better able to resolve between the two hypotheses. Figure 6-15 shows that one-sided tests are more powerful than two-sided tests for equal numbers of samples.

For the DQO process, it is important to define the acceptable levels of both α and β. By specifying these parameters, it is implied that a certain number of samples will be necessary to satisfy both α and β constraints. This can be accomplished by using the following formula:

$$n \geq \frac{\sigma^2 (z_{1-\beta} + z_{1-\alpha})^2}{\Delta^2} + (0.5) z_{1-\alpha}^2 \tag{6.51}$$

for a one-sided test, where

n = number of samples required
$z_{1-\alpha}$ = percentile of the standard normal distribution such that $P(z \geq z_{1-\alpha}) = \alpha$
$z_{1-\beta}$ = percentile of the standard normal distribution such that $P(z \geq z_{1-\beta}) = \beta$
α = probability of a Type I error
β = probability of a Type II error
Δ = minimum relative detectable difference

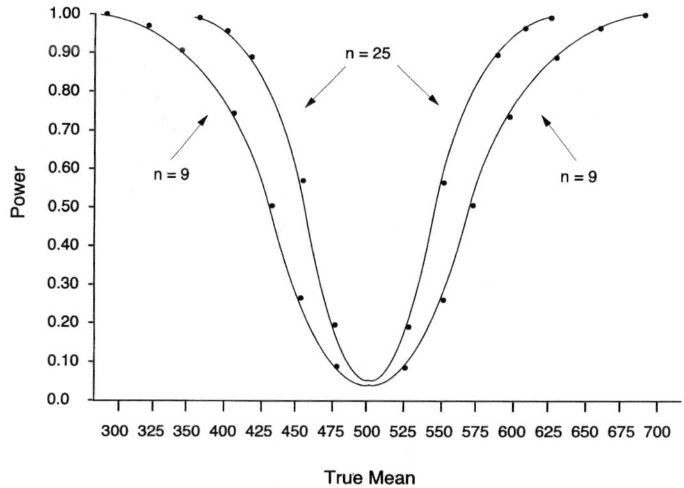

Figure 6-14. Power curves for a one-sided hypothesis test with population parameters: $\mu = 500$, $\sigma = 120$. (From *Statistical Analysis for Business and Economics*, 3rd ed., by D.L. Harnett and J.L. Murphy. Copyright © 1985 Addison-Wesley Publishers Limited. All rights reserved. Reproduced by permission.)

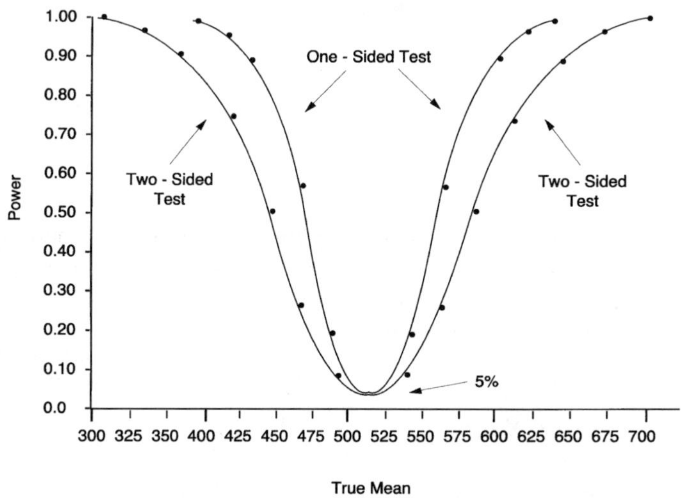

Figure 6-15. Comparison of power curves for one-sided vs. two- sided tests of hypothesis for identical population parameters: $\mu = 500$, $\sigma = 120$, $n = 9$. (From *Statistical Analysis for Business and Economics*, 3rd ed., by D.L. Harnett and J.L. Murphy. Copyright © 1985 Addison-Wesley Publishers Limited. All rights reserved. Reproduced by permission.)

Example 6.7

Lead (Pb)-contaminated soil has been excavated and placed in windrows for classification before disposal. A pilot study has indicated that the variance of Pb concentrations is 100 ppm. The RPM has set DQO goals of a 95% confidence level on the action level of 500 ppm, with a tolerable difference of 50 ppm. The power of the test should be 90%. How many samples will be necessary to achieve these error constraints?

Using equation 6.51,

$$n \geq \frac{\sigma^2 (z_{1-\beta} + z_{1-\alpha})^2}{\Delta^2} + (0.5) z_{1-\alpha}^2 \tag{6.52}$$

$$n \geq \frac{(100)^2 (1.282 + 1.645)^2}{(50)^2} + (0.5)(1.645)^2 \tag{6.53}$$

$$n \geq 34.3 + 1.35 = 35.6 = 36 \text{ samples} \tag{6.54}$$

rounded to the next highest integer.

6.2 The Lognormal Distribution

As described in Chapter 5, data often do not follow a normal distribution but their logarithms do. This is the most common case for geologic or earth science problems. The lognormal distribution will be presented here as one alternative to the normal curve in distribution modeling.

6.2.1 Equation and Shape

The equation for the lognormal distribution is as follows:

$$f(x) = \frac{1}{x\beta\sqrt{2\pi}} \exp\left[-\frac{1}{2}\left(\frac{\ln \gamma - \ln x}{\beta} \right)^2 \right] \tag{6.55}$$

Typically, the lognormal distribution is described by the median of the distribution, equal to $\gamma = e^{\alpha}$ where α is the average of the logarithms, and β the standard deviation of the logarithms.

One should note that, on a lognormal cumulative frequency plot, the 50% line provides an average of logarithmic values, α. This is actually the antilog of the median ($\gamma = e^{\alpha}$). Note that this is not the mean value. The mean value can be obtained by

$$m + \gamma e^{\beta^2/2} = e^{\alpha + \beta^2/2} \tag{6.56}$$

where β is the standard deviation in log space. This helps to correct a common problem in skewed distributions: overestimation of the mean due to the presence of extreme values.

It is a common error to use the median value of the logarithmic distribution for the mean. This mistake arises because it is counterintuitive to add one-half the variance to the average of the logarithms before exponentiating.

The standard deviation can be obtained from the cumulative frequency plot by subtracting the logarithms of the grades associated with the interval 16% to 50% or 50% to 84%.

Despite the computational quirk, the mean value of a population of lognormal data is often desired. Because a few extreme values are present, the true mean can be hard to pin down. These extreme values can exert significant, if not drastic, influence on the calculated mean value. This problem is even more pronounced for small data sets. The following example will take advantage of the known median value ($m = e^{\alpha}$) and assist in deducing a more appropriate mean.

6.2.2 Example: Use of the Sichel t Estimator

Occasionally, data plotted on lognormal probability paper continue to show a convex shape, especially at the origin. This indicates that the data are even more highly skewed, perhaps "on-beyond" lognormal. An example of this behavior from Sichel is given in Figure 6-16.

Sichel (1966) has produced tables to assist in this situation, especially when available data are scarce. Often, the variable that is lognormal is not the data themselves but, rather, the data values plus a constant, say, θ. In this case, the variable $\ln(x+\theta)$ is normally distributed. This is referred to as a three-parameter lognormal distribution. Figure 6-16 is an example of how the cumulative frequency plot of a three-parameter lognormal distribution taken from Krige (1962) and David (1977) may be adjusted to produce a straight line.

A group of five soil samples has been collected on a site where PCBs were

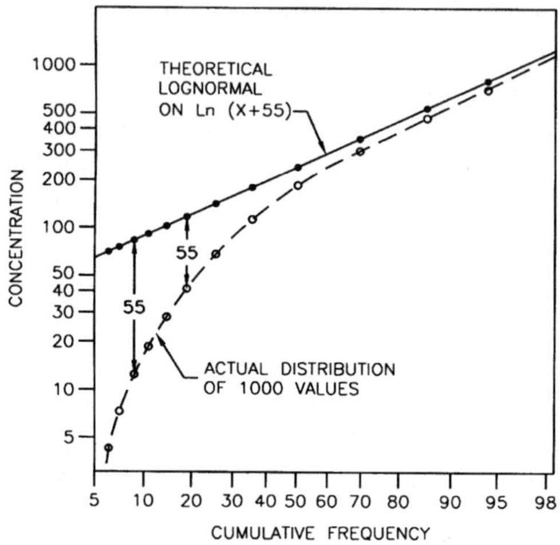

Figure 6-16. Cumulative frequency plot for a three-parameter lognormal distribution. (From *The Estimation of Means and Associated Confidence Limits for Small Samples from Lognormal Populations*, by H.S. Sichel, South African Institute of Mining and Metallurgy, 1966.)

known to have been in use. The individual concentrations are 12, 18, 23, 51, and 2250 ppm. The natural logarithm of (concentration + 50) follows a lognormal distribution. We will now paraphrase Sichel (1966): Table 6-5 shows the layout of the computations and definition of notations. It is easy to show that

$$e^{\bar{x}} = \text{antilog}_{10}(x') \tag{6.57}$$

and

$$V = 5.3019[m'_2 - (x')^2] \tag{6.58}$$

where 5.3019 is the square of ln10 to transform decimal logs into natural ones, m'_2 and $\bar{x}'$ are defined in Table 6-5.

$$e^{\bar{x}} = \text{antilog}_{10}(2.1708) = 148.2 = \text{median} \tag{6.59}$$

$$V = 5.3019\,[5.0722 - (2.1708)^2] = 1.9078 \tag{6.60}$$

Table 6-5. Notational definitions and computation summary for the example using Sichel's estimator.

Sample	z (ppm)	$z + \theta$ (ppm + 50)	$x' = \log_{10}(z + \theta)$	$(x')^2$
1	12	62	1.7924	3.2127
2	18	68	1.8325	3.3581
3	23	73	1.8633	3.4719
4	51	101	2.0043	4.0172
5	2250	2300	3.3617	11.3010
Total	2354	2604	10.8542	25.3609
Means	$\bar{z} = 470.8$	$z + \alpha =$ 520.8	$\bar{x}' = 2.1708$	$m_2' = 5.0722$

Referring to Table 6-6, with $V = 1.9$ and $V = 2.0$, for a sample size of $n = 5$, it is seen that

$$\gamma_5(1.9) = 2.303 \tag{6.61}$$

$$\gamma_5(2.0) = 2.395 \tag{6.62}$$

Linear interpolation between these tabulated figures yields

$$\gamma_5(1.91) = 2.308 \tag{6.63}$$

The t estimator is then

$$t = e^{\bar{x}}\gamma_5(V) = (148.2)(2.308) = 342.0 \tag{6.64}$$

To obtain an estimate for the mean, it is necessary to subtract the constant $\gamma = 50$ from t; thus,

$$\text{Estimate of mean concentration} = t - \gamma = 342.0 - 50 = 292.0 \tag{6.65}$$

Experience has shown that the estimate of the mean using the Sichel t estimator is much closer to the actual mean than a simple arithmetic average or geometric average. This, of course, implies that sufficient additional samples are

Table 6-6. Sichel's lognormal tables. Factor $\gamma_n(V)$ for estimation of mean of lognormal population

	n										
V	2	3	4	5	6	7	8	9	10	11	12
0.00	1.000	1.000	1.000	1.000	1.000	1.000	1.000	1.000	1.000	1.000	1.000
0.02	1.010	1.010	1.010	1.010	1.010	1.010	1.010	1.010	1.010	1.010	1.010
0.04	1.020	1.020	1.020	1.020	1.020	1.020	1.020	1.020	1.020	1.020	1.020
0.06	1.030	1.030	1.030	1.030	1.030	1.030	1.030	1.030	1.030	1.030	1.030
0.08	1.040	1.040	1.040	1.040	1.040	1.041	1.041	1.041	1.041	1.041	1.041
0.10	1.050	1.051	1.051	1.051	1.051	1.051	1.051	1.051	1.051	1.051	1.051
0.12	1.061	1.061	1.061	1.061	1.061	1.061	1.061	1.061	1.061	1.062	1.062
0.14	1.071	1.071	1.071	1.072	1.072	1.072	1.072	1.072	1.072	1.072	1.072
0.16	1.081	1.082	1.082	1.082	1.082	1.082	1.082	1.083	1.083	1.083	1.083
0.18	1.091	1.092	1.092	1.093	1.093	1.093	1.093	1.093	1.093	1.093	1.094
0.20	1.102	1.102	1.103	1.103	1.104	1.104	1.104	1.104	1.104	1.104	1.104
0.3	1.154	1.156	1.157	1.158	1.158	1.159	1.159	1.159	1.160	1.160	1.160
0.4	1.207	1.210	1.212	1.214	1.215	1.216	1.216	1.217	1.217	1.217	1.218
0.5	1.260	1.266	1.269	1.272	1.273	1.275	1.276	1.276	1.277	1.278	1.278
0.6	1.315	1.323	1.328	1.332	1.334	1.336	1.337	1.338	1.339	1.340	1.341
0.7	1.371	1.382	1.389	1.393	1.397	1.399	1.401	1.403	1.404	1.406	1.406
0.8	1.427	1.442	1.451	1.457	1.462	1.465	1.468	1.470	1.472	1.473	1.475
0.9	1.485	1.503	1.515	1.523	1.529	1.533	1.537	1.540	1.542	1.544	1.546
1.0	1.543	1.566	1.580	1.591	1.598	1.604	1.608	1.612	1.615	1.618	1.620
1.1	1.602	1.630	1.648	1.661	1.670	1.677	1.682	1.687	1.691	1.694	1.697
1.2	1.662	1.696	1.718	1.733	1.744	1.752	1.759	1.765	1.770	1.774	1.777
1.3	1.724	1.764	1.789	1.807	1.820	1.831	1.839	1.846	1.851	1.856	1.860
1.4	1.786	1.832	1.862	1.884	1.900	1.912	1.922	1.930	1.936	1.942	1.947
1.5	1.848	1.903	1.938	1.963	1.981	1.996	2.007	2.017	2.025	2.032	2.037
1.6	1.912	1.975	2.015	2.044	2.066	2.082	2.096	2.107	2.116	2.124	2.131
1.7	1.977	2.049	2.095	2.128	2.153	2.172	2.188	2.201	2.212	2.221	2.229
1.8	2.043	2.124	2.177	2.214	2.243	2.265	2.283	2.298	2.310	2.321	2.330
1.9	2.110	2.201	2.260	2.303	2.336	2.361	2.382	2.399	2.413	2.425	2.436
2.0	2.178	2.280	2.347	2.395	2.431	2.460	2.484	2.503	2.519	2.533	2.545
2.1	2.247	2.360	2.435	2.489	2.530	2.563	2.589	2.611	2.630	2.645	2.659
2.2	2.317	2.442	2.526	2.586	2.632	2.669	2.698	2.723	2.744	2.762	2.778
2.3	2.388	2.526	2.618	2.686	2.737	2.778	2.811	2.839	2.863	2.883	2.900
2.4	2.460	2.612	2.714	2.788	2.846	2.891	2.928	2.959	2.986	3.008	3.028
2.5	2.533	2.699	2.812	2.894	2.957	3.008	3.049	3.084	3.113	3.138	3.160
2.6	2.607	2.789	2.912	3.003	3.073	3.128	3.174	3.213	3.245	3.274	3.298
2.7	2.682	2.880	3.015	3.114	3.191	3.253	3.304	3.346	3.382	3.414	3.441
2.8	2.759	2.973	3.120	3.229	3.314	3.382	3.437	3.484	3.524	3.559	3.589
2.9	2.836	3.068	3.228	3.347	3.440	3.514	3.576	3.627	3.671	3.710	3.743
3.0	2.914	3.166	3.339	3.469	3.570	3.651	3.718	3.775	3.824	3.866	3.902
3.1	2.994	3.265	3.453	3.593	3.703	3.792	3.866	3.928	3.981	4.028	4.068
3.2	3.075	3.366	3.569	3.721	3.841	3.938	4.018	4.086	4.145	4.195	4.240
3.3	3.157	3.469	3.688	3.853	3.983	4.088	4.176	4.250	4.314	4.369	4.418
3.4	3.240	3.574	3.810	3.988	4.129	4.243	4.338	4.419	4.489	4.549	4.603
3.5	3.324	3.682	3.935	4.127	4.279	4.403	4.506	4.594	4.670	4.736	4.794
3.6	3.409	3.792	4.063	4.270	4.434	4.568	4.680	4.775	4.858	4.929	4.993
3.7	3.496	3.903	4.194	4.416	4.593	4.738	4.859	4.962	5.052	5.130	5.198
3.8	3.583	4.017	4.329	4.567	4.757	4.913	5.044	5.156	5.252	5.337	5.412
3.9	3.672	4.134	4.466	4.721	4.925	5.093	5.234	5.355	5.460	5.552	5.633
4.0	3.762	4.252	4.607	4.880	5.099	5.279	5.431	5.562	5.675	5.774	5.862
4.1	3.853	4.373	4.751	5.042	5.277	5.471	5.634	5.775	5.897	6.004	6.099
4.2	3.946	4.496	4.898	5.209	5.460	5.668	5.844	5.995	6.127	6.242	6.345
4.3	4.040	4.622	5.049	5.380	5.649	5.872	6.060	6.223	6.364	6.489	6.599
4.4	4.135	4.750	5.203	5.556	5.843	6.081	6.283	6.458	6.610	6.744	6.863
4.5	4.231	4.881	5.361	5.736	6.042	6.297	6.513	6.700	6.863	7.008	7.136
4.6	4.328	5.014	5.522	5.921	6.247	6.519	6.750	6.950	7.126	7.281	7.419
4.7	4.427	5.149	5.687	6.111	6.457	6.747	6.995	7.209	7.397	7.563	7.711
4.8	4.527	5.288	5.856	6.305	6.674	6.983	7.247	7.476	7.677	7.855	8.014
4.9	4.629	5.428	6.029	6.505	6.896	7.225	7.507	7.751	7.966	8.157	8.328
5.0	4.732	5.572	6.205	6.709	7.124	7.474	7.774	8.036	8.265	8.470	8.652
5.1	4.836	5.718	6.386	6.919	7.359	7.731	8.050	8.329	8.574	8.792	8.988
5.2	4.941	5.866	6.570	7.134	7.600	7.995	8.335	8.631	8.893	9.126	9.335
5.3	5.048	6.018	6.759	7.354	7.847	8.266	8.628	8.944	9.222	9.471	9.695
5.4	5.156	6.172	6.951	7.579	8.102	8.546	8.930	9.265	9.563	9.828	10.07
5.5	5.266	6.329	7.148	7.811	8.363	8.833	9.240	9.598	9.914	10.20	10.45
5.6	5.376	6.489	7.350	8.048	8.631	9.129	9.561	9.940	10.28	10.58	10.85
5.7	5.489	6.652	7.555	8.290	8.906	9.433	9.890	10.29	10.65	10.97	11.26
5.8	5.603	6.818	7.766	8.539	9.188	9.745	10.23	10.66	11.04	11.38	11.68
5.9	5.718	6.987	7.980	8.794	9.478	10.07	10.58	11.03	11.44	11.80	
6.0	5.834	7.159	8.200	9.054	9.776	10.40	10.94	11.42	11.85		

(Sichel's Table); n = number of samples.

n										
13	**14**	**15**	**16**	**17**	**18**	**19**	**20**	**50**	**100**	**1000**
1.000	1.000	1.000	1.000	1.000	1.000	1.000	1.000	1.000	1.000	1.000
1.010	1.101	1.010	1.010	1.010	1.010	1.010	1.010	1.010	1.010	1.010
1.020	1.020	1.020	1.020	1.020	1.020	1.020	1.020	1.020	1.020	1.020
1.030	1.030	1.030	1.030	1.030	1.030	1.030	1.030	1.030	1.030	1.030
1.041	1.041	1.041	1.041	1.041	1.041	1.041	1.041	1.041	1.041	1.041
1.051	1.051	1.051	1.051	1.051	1.051	1.051	1.051	1.051	1.051	1.051
1.062	1.062	1.062	1.062	1.062	1.062	1.062	1.062	1.062	1.062	1.062
1.072	1.072	1.072	1.072	1.072	1.072	1.072	1.072	1.072	1.072	1.072
1.083	1.083	1.083	1.083	1.083	1.083	1.083	1.083	1.083	1.083	1.083
1.094	1.094	1.094	1.094	1.094	1.094	1.094	1.094	1.094	1.094	1.094
1.104	1.104	1.104	1.104	1.104	1.104	1.105	1.105	1.105	1.105	1.105
1.160	1.160	1.160	1.160	1.160	1.160	1.160	1.161	1.161	1.162	1.162
1.218	1.218	1.218	1.219	1.219	1.219	1.219	1.219	1.220	1.221	1.221
1.278	1.279	1.279	1.279	1.280	1.280	1.280	1.280	1.282	1.283	1.284
1.342	1.342	1.343	1.343	1.343	1.344	1.344	1.344	1.348	1.349	1.350
1.407	1.408	1.409	1.409	1.410	1.410	1.411	1.411	1.416	1.417	1.419
1.476	1.477	1.478	1.478	1.479	1.480	1.480	1.481	1.487	1.490	1.492
1.547	1.549	1.550	1.551	1.552	1.552	1.553	1.554	1.562	1.565	1.568
1.622	1.623	1.625	1.626	1.627	1.628	1.629	1.630	1.641	1.645	1.649
1.699	1.701	1.703	1.705	1.706	1.708	1.709	1.710	1.723	1.728	1.733
1.780	1.782	1.785	1.787	1.789	1.790	1.792	1.793	1.810	1.816	1.822
1.864	1.867	1.870	1.872	1.874	1.876	1.878	1.880	1.900	1.908	1.916
1.951	1.955	1.958	1.961	1.964	1.966	1.969	1.971	1.995	2.004	2.014
2.042	2.047	2.051	2.054	2.058	2.060	2.063	2.065	2.095	2.106	2.117
2.137	2.142	2.147	2.151	2.155	2.158	2.161	2.164	2.199	2.212	2.226
2.236	2.242	2.247	2.252	2.256	2.260	2.264	2.267	2.308	2.323	2.340
2.338	2.345	2.352	2.357	2.362	2.367	2.371	2.375	2.422	2.440	2.460
2.445	2.453	2.460	2.467	2.473	2.478	2.483	2.487	2.542	2.563	2.586
2.556	2.565	2.574	2.581	2.588	2.594	2.599	2.604	2.668	2.692	2.718
2.671	2.682	2.691	2.700	2.707	2.714	2.721	2.726	2.800	2.827	2.858
2.791	2.803	2.814	2.824	2.832	2.840	2.847	2.854	2.937	2.969	3.004
2.916	2.929	2.942	2.952	2.962	2.971	2.979	2.987	3.082	3.118	3.158
3.045	3.060	3.074	3.086	3.098	3.108	3.117	3.125	3.233	3.274	3.320
3.180	3.197	3.212	3.226	3.238	3.250	3.260	3.270	3.391	3.438	3.490
3.320	3.339	3.356	3.371	3.385	3.398	3.410	3.420	3.557	3.610	3.669
3.465	3.486	3.505	3.522	3.538	3.552	3.565	3.577	3.730	3.791	3.857
3.616	3.639	3.661	3.680	3.697	3.713	3.727	3.740	3.912	3.980	4.055
3.772	3.799	3.822	3.843	3.862	3.880	3.986	3.911	4.102	4.176	4.263
3.935	3.964	3.990	4.013	4.035	4.054	4.072	4.088	4.301	4.387	4.482
4.104	4.136	4.164	4.190	4.214	4.235	4.255	4.273	4.510		
4.279	4.314	4.346	4.374	4.400	4.424	4.446	4.465	4.728		
4.461	4.500	4.534	4.566	4.594	4.620	4.644	4.666	4.956		
4.650	4.692	4.730	4.764	4.796	4.824	4.850	4.875	5.195		
4.846	4.892	4.933	4.971	5.005	5.037	5.065	5.092	5.445		
5.049	5.099	5.145	5.186	5.223	5.258	5.289	5.318	5.706		
5.260	5.315	5.364	5.409	5.450	5.488	5.522	5.554	5.980		
5.478	5.538	5.592	5.641	5.686	5.726	5.764	5.799	6.266		
5.705	5.770	5.829	5.882	5.930	5.975	6.016	6.054	6.566		
5.940	6.011	6.074	6.132	6.185	6.234	6.278	6.319	6.789		
6.184	6.260	6.329	6.392	6.450	6.502	6.551	6.596			
6.437	6.519	6.594	6.662	6.724	6.781	6.834	6.883			
6.698	6.788	6.869	6.942	7.010	7.072	7.129	7.182			
6.970	7.066	7.154	7.233	7.306	7.373	7.435	7.493			
7.251	7.355	7.449	7.536	7.614	7.687	7.754	7.816			
7.543	7.655	7.756	7.849	7.934	8.013	8.085	8.152			
7.845	7.965	8.075	8.175	8.266	8.351	8.429	8.502			
8.157	8.287	8.405	8.512	8.611	8.703	8.787	8.865			
8.481	8.620	8.747	8.863	8.970	9.068	9.159	9.243			
8.817	8.966	9.102	9.227	9.341	9.447	9.545	9.636			
9.164	9.324	9.470	9.604	9.727	9.841	9.946				
9.524	9.696	9.852	9.996	10.13	10.25					
9.897	10.08	10.25	10.40	10.54						
10.28	10.48	10.66	10.82							
10.68	10.89	11.08								
11.09	11.32									
11.52										

eventually taken and a new mean is calculated. The above example shows how one very high value can inappropriately skew the mean value.

As a practical matter, we would be taking a serious and unjustified risk in assuming that the arithmetic mean is valid based on only five samples. This is true for normal distributions but especially true for the lognormal case.

In addition, the practitioner should always calculate upper and lower confidence limits in this situation. The confidence limits will provide an indication of the amount of uncertainty present in the estimation of the mean. It will also show, to the specified level of confidence, the concentration interval in which the unknown true mean is expected to lie.

6.2.3 Calculation of the UCL for Lognormal Distributions

Experience has shown that most large or "complete" data sets resulting from soil sampling follow a lognormal distribution (USEPA 1992). Lognormality can be seen in histograms of the data (Figs. 4-4, 5-3, and 5-9); statistical tests such as the Shapiro-Wilk *W* test (Gilbert 1987) can be applied to determine if a data set is consistent with a normal or lognormal distribution.

For data sets following a lognormal distribution, the UCL may be calculated by the following equation:

$$\text{UCL} = e^{\bar{x} + \frac{\beta^2}{2} + \beta H/\sqrt{n-1}} \tag{6.66}$$

where

H = H statistic (Table 6-7)
n = number of samples

6.3 The Binomial Distribution

6.3.1 Bernoulli Trials

The binomial distribution is a discrete, or integer-based, distribution. Each integer in the distribution has an associated probability of occurrence. This sort of distribution can be of use in STP when two types of completely liberated fragments are mixed. It can also be used for concentration data by grouping classes of data and then assigning the values 0, 1, 2,...,n to each class.

Table 6-7. Quantiles of the Shapiro-Wilk W test for normality (values of W such that $100p\%$ of the distribution of W is less than W_p).

n	$W_{0.01}$	$W_{0.02}$	$W_{0.05}$	$W_{0.10}$	$W_{0.50}$
3	0.753	0.756	0.767	0.789	0.959
4	0.687	0.707	0.748	0.792	0.935
5	0.686	0.715	0.762	0.806	0.927
6	0.713	0.743	0.788	0.826	0.927
7	0.730	0.760	0.803	0.838	0.928
8	0.749	0.778	0.818	0.851	0.932
9	0.764	0.791	0.829	0.859	0.935
10	0.781	0.806	0.842	0.869	0.938
11	0.792	0.817	0.850	0.876	0.940
12	0.805	0.828	0.859	0.883	0.943
13	0.814	0.837	0.866	0.889	0.945
14	0.825	0.846	0.874	0.895	0.947
15	0.835	0.855	0.881	0.901	0.950
16	0.844	0.863	0.887	0.906	0.952
17	0.851	0.869	0.892	0.910	0.954
18	0.858	0.874	0.897	0.914	0.956
19	0.863	0.879	0.901	0.917	0.957
20	0.868	0.884	0.905	0.920	0.959
21	0.873	0.888	0.908	0.923	0.960
22	0.878	0.892	0.911	0.926	0.961
23	0.881	0.895	0.914	0.928	0.962
24	0.884	0.898	0.916	0.930	0.963
25	0.886	0.901	0.918	0.931	0.964
26	0.891	0.904	0.920	0.933	0.965
27	0.894	0.906	0.923	0.935	0.965
28	0.896	0.908	0.924	0.936	0.966
29	0.898	0.910	0.926	0.937	0.966
30	0.900	0.912	0.927	0.939	0.967
31	0.902	0.914	0.929	0.940	0.967
32	0.904	0.915	0.930	0.941	0.968
33	0.906	0.917	0.931	0.942	0.968
34	0.908	0.919	0.933	0.943	0.969
35	0.910	0.920	0.934	0.944	0.969
36	0.912	0.922	0.935	0.945	0.970
37	0.914	0.924	0.936	0.946	0.970
38	0.916	0.925	0.938	0.947	0.971
39	0.917	0.927	0.939	0.948	0.971
40	0.919	0.928	0.940	0.949	0.972
41	0.920	0.929	0.941	0.950	0.972
42	0.922	0.930	0.942	0.951	0.972
43	0.923	0.932	0.943	0.951	0.973
44	0.924	0.933	0.944	0.952	0.973
45	0.926	0.934	0.945	0.953	0.973
46	0.927	0.935	0.945	0.953	0.974
47	0.928	0.936	0.946	0.954	0.974
48	0.929	0.937	0.947	0.954	0.974
49	0.929	0.937	0.947	0.955	0.974
50	0.930	0.938	0.947	0.955	0.974

Source: After Shapiro and Wilk, 1965.

The null hypothesis of a normal distribution is rejected at the α significance level if the calculated W is less than W_α.

The toss of a coin can be used to describe and develop the idea of a binomial distribution. Tossing a coin is a type of experiment or trial with only two outcomes: heads and tails. For development purposes, one of these trials will be called a success and the other a failure. It does not matter which is the success or the failure. The result of a single experiment—in this case, the toss of a coin—is called a Bernoulli trial.

A more practical demonstration of a Bernoulli trial could be found in a remedial action. For example, consider a site where contaminated soil is being excavated. A field portable screening device is sampling soil material in advance of a front-end loader. The result of each sample test, a Bernoulli trial, is compared to the site threshold value. If the result of the trial is higher than the threshold, the soil is subjected to remedial treatment. If the result is lower, the soil is not treated.

Each sample test is either a success or a failure. Whether a sample reading above the threshold is considered a success or failure depends on your point of view. If you are the remedial contractor and are paid for each cubic yard of material excavated and treated, high concentrations are successes. If you are the PRP and must pay for the cleanup, success would probably be defined as a trial below the threshold. For purposes of example, a sample above the action level will be defined as a success. After a large number of tests, the probability of success may be determined based on the ratio of the total number of successes to the total number of trials. A probability value of, say, .46 might be obtained.

For any type of trial, the probability of a particular outcome is between zero and one, or

$$P(x=r)=(0\leq x\leq 1) \tag{6.67}$$

If only two outcomes are possible, the probability of other outcomes is $q = 1 - p$. The probability for each outcome remains constant for each trial.

6.3.2 Equation and Shape

A discrete random variable follows a binomial distribution law if the probability density function can be written

$$P(x=k)=\binom{n}{k}(1-p)^{n-k}p^{k} \tag{6.68}$$

which can be interpreted as the probability of selecting k balls in n draws (with replacement) from a population in which black balls occur with proportion p.

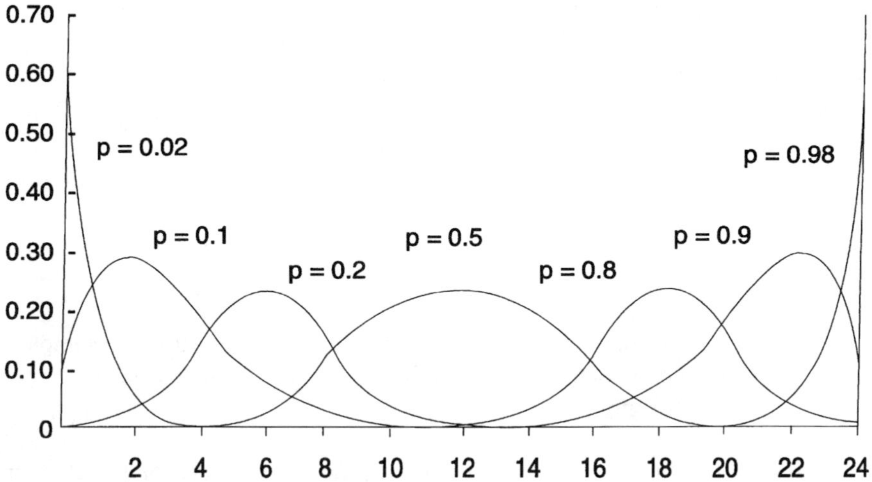

Figure 6-17. Example shapes for the binomial distribution. (From *Geostatistical Ore Reserve Estimation*, by M. David, Elsevier, 1977. Courtesy of Michel David.)

The mean of the binomial distribution is np, and the variance is $np(1 - p)$. Depending on the value of p, the binomial distribution can take on a variety of shapes. A few such shapes are presented in Figure 6-17.

6.4 Poisson Distribution

The Poisson distribution is a limit case of the binomial distribution. It seldom occurs in elevated contaminant concentration distributions as it implies that no spatial structuring is present (Pitard 1993). This type of distribution often occurs, however, when a component is present at trace levels. Trace levels occur frequently in environmental sampling, where a contaminant of concern may be present in concentrations of a few parts per million, billion, or trillion.

The Poisson is thus highly important for applications in STP. Commonly, the component of interest occurs as rare, small grains in a vast sea of inert material. The detection and clustering of these grains pose significant challenges relating to sampling errors. Because some contaminants have regulatory threshold limits at the detection limit, the challenge is often simply determining the presence or absence of contamination with some degree of reliability.

6.4.1 Equation and Shape

The equation for the Poisson distribution is

$$P(x=r)=\frac{\lambda^r}{r!}e^{-\lambda} \tag{6.69}$$

where

$P(x = r)$ = probability that r low-frequency particles appear in the sample
λ = hypothetical average number of low-frequency particles per sample

Figure 6-18 shows a number of different possible shapes of the Poisson distribution for various values of λ.

6.4.2 Implications of the Mode and Sample Support

Some important features of the Poisson distribution should be noted. The Poisson has a variance equal to the mean value, the mean being the average number of low-frequency particles per sample. Of special importance is the mode of the distribution. The highest probability $P(x = r)$ results when r is less than λ, except when λ is an integer. When λ is an integer, we can see that

$$P(x=\lambda)=P(x=\lambda-1) \tag{6.70}$$

The shape of the Poisson distribution changes rapidly between the λ values of zero and three, being highly skewed at zero and becoming less so as λ increases. By the time λ reaches four or five, the mode of the distribution closely approaches the mean and the distribution is nearing symmetry.

The behavior of the Poisson distribution at low values of λ has meaningful practical implications. With smaller and smaller λ values, the most probable result is that a single sample will drift farther and farther from the true average. This can be an insidious result as even a triplicate analysis may produce exactly the same result. We might, quite naturally, be lulled into a false sense of security that no contaminant is present, when, in fact, we are simply experiencing the most probable result, that is, the mode.

In order to solve this problem, it is necessary to increase the number of contaminant particles collected in each sample. We increase the sample support

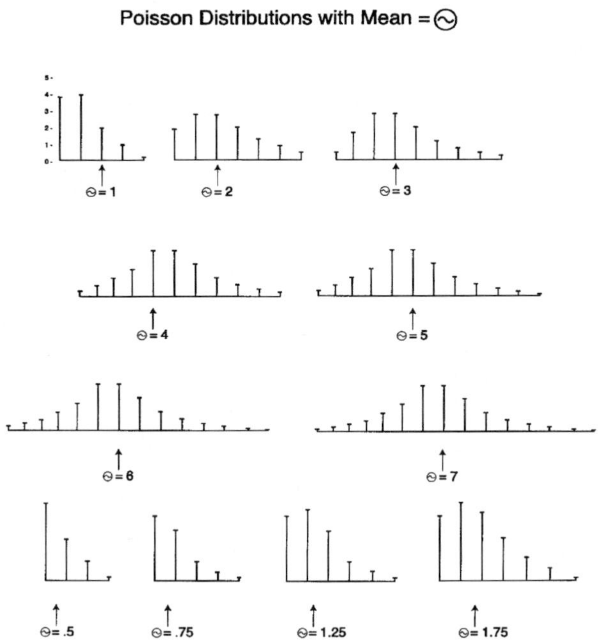

Figure 6-18. Example shapes for the Poisson distribution. (From *Pierre Gy's Sampling Theory and Sampling Practice: Heterogeneity, Sampling Correctness, and Statistical Process Control,* 2nd ed., by F.F. Pitard. Copyright © 1993 CRC Press, Boca Raton, FL. Courtesy of Francis Pitard.)

so that at least four or five particles abide in each sample. At this level of particles, the Poisson approaches a normal distribution, where the mean and the mode can be expected to coincide. This helps to assure that replicate samples do not deviate significantly from the mean.

This concept of particle capture and sample support is visualized in Figures 6-19 and 6-20. Figure 6-19a shows a contaminant distributed somewhat uniformly as trace particles in a nine-cell matrix, along with a small square representing sample support size. In this case, the sample support volume is too small. Of the nine samples taken, four result in a complete failure to detect the presence of the contaminant, a false negative or Type II error. This demonstrates that, with such a sample support, the presence of the contaminant will be detected only about half the time. Not coincidentally, zero is the Poisson mode or most frequently observed value for this sample support.

In other situations, a false positive (Type I error) may result. Consider the situation in which two "hits" are allowed but three are not. The result, in the upper right corner of Figure 6-19a (three hits), would be interpreted as being above the tolerable level. The irony is that the distribution of the contamination in this area

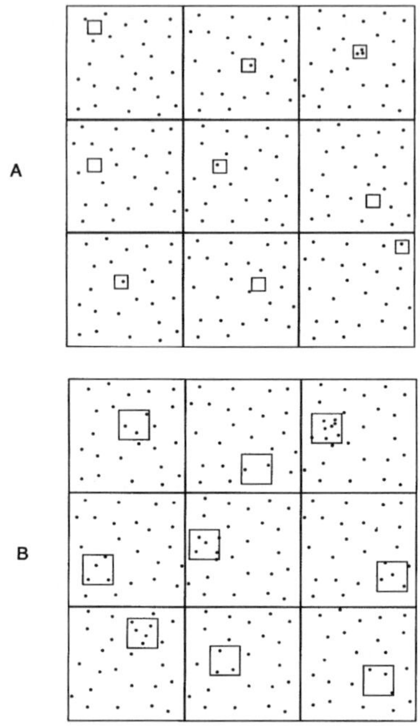

Figure 6-19. Differences in small vs. large sample support for the Poisson distribution where the contaminant particles are small. (From *Pierre Gy's Sampling Theory and Sampling Practice: Heterogeneity, Sampling Correctness, and Statistical Process Control,* 2nd ed., by F.F. Pitard. Copyright © 1993 CRC Press, Boca Raton, FL. Courtesy of Francis Pitard.)

is, for practical purposes, no greater and no less than in any other area yet, because of an inappropriate sample support, it is perceived as greater.

In order to correct this problem, the sampling unit must be increased in size. Figure 6-19b shows the effect of a larger support volume for sampling. Using this approach, no sample contains less than two hits and averages just over four hits per sample. Note that samples with only one or even zero hits are possible, suggesting that even this larger volume may not be completely sufficient, depending on the data quality needs.

Figure 6-20 shows yet another situation that may be encountered during the sampling of Poisson-distributed contaminants. Here, the average contaminant concentration may be roughly equal to that shown in Figure 6-19 yet, because of the size of the contaminant particle or contaminant cluster, both proposed volumes shown are too small. Even a support size equal to one of the nine cells is insufficient. A more appropriate size would be the square that outlines the nine cells, resulting, in this case, in the collection of three contaminant particles.

As we saw in Figure 6-18, Poisson distributions with $\lambda > 4$ approach a normal distribution. This implies that an appropriate sample volume that is collecting four or more "hits" will produce a normal distribution for replicate sample analyses from the same original sample material at a location. This does not mean, however, that the sample data across the entire site will be normally distributed. This concept is shown in Figure 6-21.

6.4.3 A Caveat for Resampling or Reanalyzing

Figures 6-19 and 6-20 provide a graphic explanation of why a common procedure, resampling or reanalyzing, should be performed with caution. Frequently, a series of samples will exhibit a range of values deemed acceptable or typical, with the exception of one *outlier* value that is outside the expected range or above an action threshold. In order to determine the validity of the sample, the sample is resubmitted to the lab for a second analysis or a colocated sample is taken.

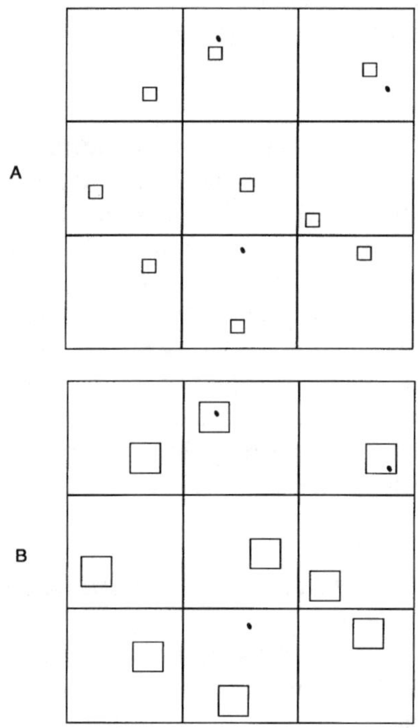

Figure 6-20. Differences in small vs. large sample support for the Poisson distribution where the contaminant particles are large. (From *Pierre Gy's Sampling Theory and Sampling Practice: Heterogeneity, Sampling Correctness, and Statistical Process Control,* 2nd ed., by F.F. Pitard. Copyright © 1993 CRC Press, Boca Raton, FL. Courtesy of Francis Pitard.)

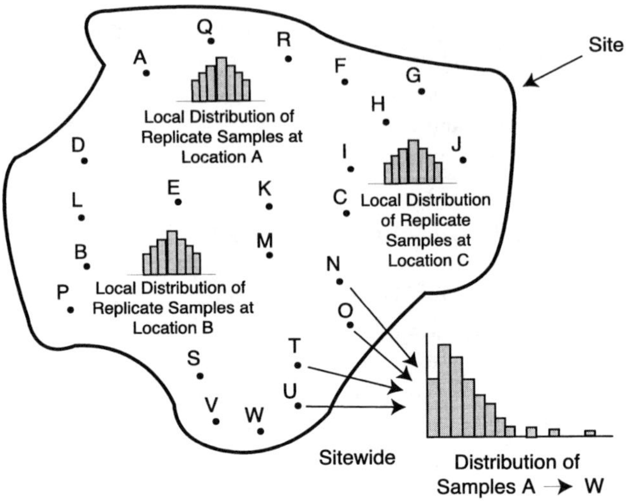

Figure 6-21. Distinction between normally distributed replicate samples on a local level vs. the skewed distribution of sitewide data.

If the contaminant distribution is Poisson and the sample weight is small (especially the laboratory subsample), the odds of the new sample coming back "clean" are very high (Figs. 6-18 to 6-20). Assuming no laboratory error, the original sample was probably a good estimate of the material analyzed. If a high-concentration sample is resubmitted and the new value is much lower, a Poisson distribution should be suspected. Excluding lab error, this also means that the suite of clean values should also be suspect as they may have missed the contamination. If outliers exist, *inliers* probably also exist. Inliers—samples in the main body of the distribution that contain a few but insufficient number of particles to be representative—generally do not come under close scrutiny but, like outliers, they indicate a sampling problem.

6.4.4 Example of Sample Support and Variability

Table 6-8 presents an abbreviated listing of two series of sample results from Appendix B-3 of the 1994 EPA manual "Preparation of Soil Sampling Protocols: Sampling Techniques and Strategies." Both series contain nine samples. The first series, called small samples, has a support volume of 2 g each. The second series has a support volume of 20 g each.

The mean concentrations for the small samples (183 ppm) is fairly close to that for the large samples (202 ppm). The variances, however, show a greater difference, with the variance of the small samples almost 10 times that for the

Table 6-8. Sample concentrations for 2-g and 20-g samples, demonstrating the volume-variance relationship for a Poisson distribution.

Small Sample			*Large Sample*		
Location	**Weight (g)**	**Conc.**	**Location**	**Weight (g)**	**Conc.**
153, 173	2	235.7	181, 114	20	285.9
166, 297	2	250.0	170, 271	20	245.0
130, 318	2	350.0	138, 353	20	225.0
219, 193	2	100.0	275, 140	20	210.6
231, 276	2	125.9	248, 203	20	225.0
261, 348	2	125.8	283, 311	20	235.0
305, 132	2	225.9	390, 196	20	275.8
354, 242	2	128.6	333, 256	20	267.5
327, 349	2	100.9	358, 340	20	276.8

large samples (6679 vs. 670), and demonstrate the existence of the volume-variance relationship.

The data also suggest that a Poisson distribution would model the process. The data for small samples have a much greater range, with samples from 100 to 350 ppm. With a mass of only 2 g, sample values may be affected in the same manner we saw in Figure 6-19, with mostly low levels of contaminants found and a low frequency of high concentrations. In fact, five of the samples show low levels in the 100- to 125-ppm range, not unlike the results in Figure 6-19. Also, a single high value was encountered in the sample data, corresponding to the single high value in Figure 6-19.

The range of the concentrations for the large samples is much narrower than the range for the small samples. All values are between approximately 210 and 285 ppm. This also results in a reduced variance. As might be expected for Poisson distributed data, the mean of the large samples has increased and, presumably, the mode has increased as well.

Although no sophisticated analysis offers physical proof that the contaminant in Table 6-8 is following a Poisson distribution, the circumstantial evidence strongly suggests that the 2- and 20-g samples follow a model similar to that illustrated in Figure 6-19.

6.4.5 Sample Support and the Fundamental Error

Sample support directly influences the fundamental error of sampling. The larger the support, the smaller the fundamental error. Figure 6-22 shows the generalized relationship between sample support and fundamental error. This shows the drift

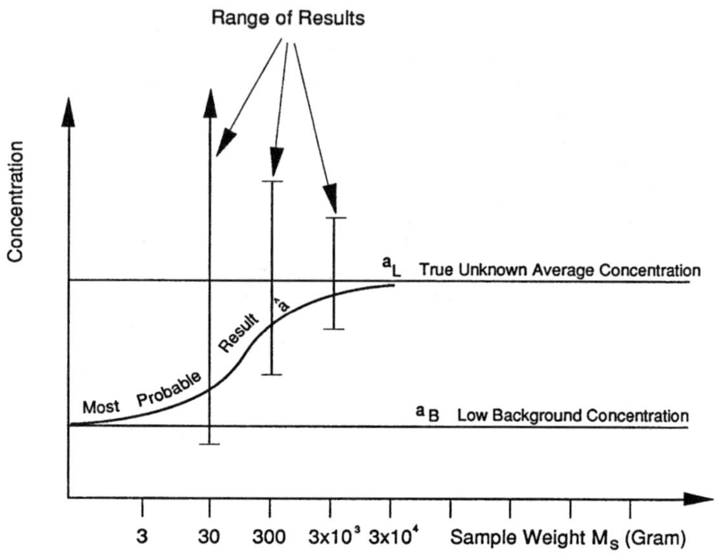

Figure 6-22. Relationship between sample support and the fundamental error *FE* for the Poisson distribution. (From *Pierre Gy's Sampling Theory and Sampling Practice: Heterogeneity, Sampling Correctness, and Statistical Process Control,* 2nd ed., by F.F. Pitard. Copyright © 1993 CRC Press, Boca Raton, FL. Courtesy of Francis Pitard.)

of the most probable result from the true average when a single sample is too small and is thus afflicted by a large fundamental error.

Pitard (1993) recommends that the fundamental error be kept below ±16%. Experience has shown that this level provides a reasonable assurance that a single sample will contain sufficient particles to produce a result close to the true average. A more complete discussion of the fundamental error will be presented in Chapters 7 and 8.

Conclusion

A good model for data is a valuable asset, whether the data are modeled using deterministic or random approaches. The aleatory nature of environmental spatial data forces us to use a probabilistic model. This implies the responsibility of selecting an appropriate model and applying it correctly. All models have assumptions that may or may not be valid for given situations. Care should be used when applying classical statistical models to spatial data as they are typically different types of data (uncorrelated vs. correlated).

PART 3 Sampling Theory & Practice

Chapter 7

Heterogeneity and Sampling

SYNOPSIS: Sampling Theory & Practice (STP) is a technically based approach to the endeavors of sampling and data aquisition. STP assumes that the medium to be sampled is not uniform, and that heterogeneities within the material will cause sampling errors. This chapter begins with a brief introduction to the seven basic sampling errors, followed by an introduction to sample selection models and heterogeneity. The concepts of precision and accuracy will be revisited and amplified, and a definition of representativeness is introduced. Heterogeneity will be discussed in detail, covering structural vs. circumstantial properties, the two types of heterogeneity, and dimensional modules of observation. STP brings many new terms and concepts to the GEM process; the reader is strongly advised to review the definitions provided in Appendix A as the discussion assumes a basic familiarity with these terms.

KEY WORDS: Fundamental Error, Grouping and Segregation Error, Long-Range Heterogeneity Fluctuation Error, Periodic Heterogeneity Fluctuation Error, Increment Delimitation Error, Increment Extraction Error, Preparation Error, Continuous Selection Error, Materialization Error, Total Sampling Error, Overall Sampling Error, Constitution Heterogeneity, Distribution Heterogeneity, Fragment, Particle, Discrete Model, Continuous Model, Precision, Accuracy, Bias, Representativeness, Equitable

Sample, Structural Property, Circumstantial Property, Comminution, Equiprobability, Support, Dimensional Module of Observation, Statistical Population, Correlated Population.

7.1 Introduction to the Seven Sampling Errors

Sampling Theory & Practice recognizes the existence of seven basic sampling errors. They are:

1. Fundamental Error *FE*
2. Grouping and Segregation Error *GE*
3. Long-range Heterogeneity Fluctuation Error CE_2
4. Periodic Heterogeneity Fluctuation Error CE_3
5. Increment Delimitation Error *DE*
6. Increment Extraction Error *EE*
7. Preparation Error *PE*

Figure 7-1 shows all the components of the overall estimation error OE of sampling. A definition of each error can be found in Appendix A. In addition, Figure 7-1 also shows that STP combines certain errors for convenience to shorten mathematical formulas. These combinations include the short-range heterogeneity fluctuation error CE_1 ($FE+GE$), the continuous selection error CE ($CE_1+CE_2+CE_3$), and the materialization error ME ($DE+EE$).

All sampling errors arise from heterogeneities within the lot to be quantified. A lot can be represented by either a discrete or continuous model, depending on the distance from which one chooses to view it. The scale of the heterogeneity dictates the appropriate model, either discrete or continuous.

The diagram in Figure 7-1 incorporates the concepts of discrete and continuous selection models for sampling. These models are analogous to, but not the same as, discrete and continuous distribution models in statistics. In general, the discrete model applies to small-scale or local situations, the continuous model to larger-scale or regional conditions.

The discrete model is based on the idea of discrete units, such as particles or fragments. Particles or fragments critically impact heterogeneities, which themselves are the sources of all sampling errors. The discrete model suggests that constituent components can be discretely counted. Interest in using a discrete model stems from the desire to quantify heterogeneity fluctuations introduced by these various fragments. This makes the discrete model applicable to chemicals, wastes, ores, vegetables, fruits, cereals, and any of their constituents, such as

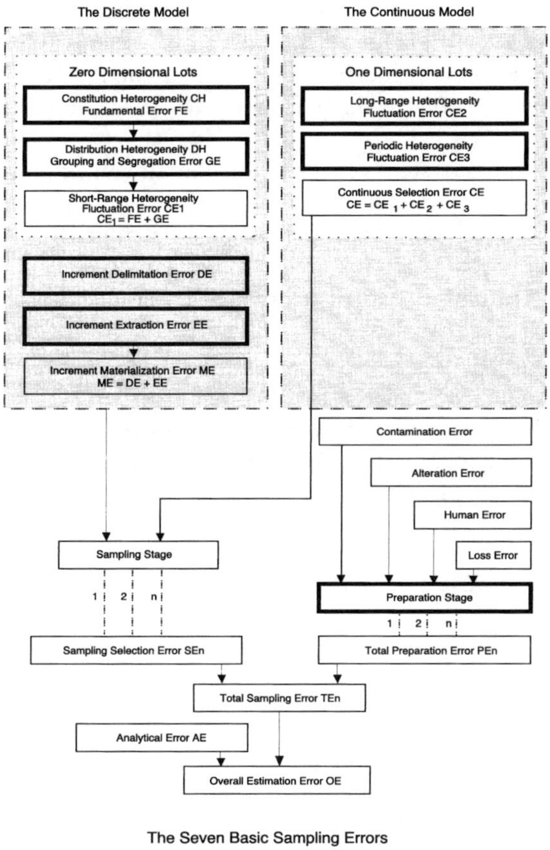

Figure 7-1. Components of the overall estimation error *OE* of sampling. (From *Pierre Gy's Sampling Theory and Sampling Practice: Heterogeneity, Sampling Correctness, and Statistical Process Control,* 2nd ed., by F.F. Pitard. Copyright © 1993 CRC Press, Boca Raton, FL. Courtesy of Francis Pitard.)

solids in slurries or pulps, particle size distributions, moisture content, minerals, and foreign material in cereals.

The continuous model is applicable to material that is moving in a stream. Examples are rivers, conveyor belts, ducts, and pipes. It also applies to stationary elongated piles or channels, provided that the section remains relatively constant. The continuous model may be described as looking at the lots with a wide-angle lens, that is, not paying attention to small details. This contrasts with the discrete model, where the lots are examined with a magnifying lens.

The choice of the model used is dictated by the scale of the heterogeneity of interest. A lot L can be considered in the following two different manners:

1. As a discrete and discontinuous set, if one is interested in fluctuations in the amount of heterogeneity introduced by various fragments. This is a small-scale and very localized characterization. It is related to the geostatistical nugget effect, which accounts for variation surrounding a point location.
2. As a continuous set, such as a flowing stream of material where the continuity is essentially a function of time or distance, if one is mainly interested in its long-range and periodic heterogeneity fluctuations. The continuous model addresses regional and large-scale characterizations. It is related to the geostatistical variogram that quantifies spatial variation and temporal fluctuations.

The use of both models is necessary in order to arrive at the overall estimation error *OE*. Figure 7-1 provides a flowchart of the process. In addition, Figure 7-1 implies the following relationships for error sum components:

Sampling Selection Error:

$$SE = FE + GE + CE_2 + CE_3 + DE + EE \quad (7.1)$$

or

$$SE = CE + ME \quad (7.2)$$

Total Sampling Error:

$$TE = SE + PE \quad (7.3)$$

Overall Estimation Error *OE*:

$$OE = TE + AE \quad (7.4)$$

The seven-error STP model summarizes the errors that result from incorrect sampling. It also provides a bridge to incorporate the analytical error introduced by the laboratory. Appropriate DQOs must take both sampling and analytical errors into account as part of the GEM process, given that both affect data reliability to a large degree.

7.2 Precision, Accuracy, and Representativeness

In Chapter 3, the concepts of precision, accuracy, and bias were introduced. Along with these ideas, visual analogues were presented. The error relationships set forth in Section 7.1 provide opportunities to develop even more objective and quantifiable definitions. In addition, sampling errors provide a mechanism to define the concept of representativeness as a combination of precision and accuracy. Defining representativeness in quantitative terms, which is not a trivial accomplishment, represents a major step forward in uncertainty analysis.

A fundamental notion to remember is that precision and accuracy are independent. Precision is concerned with measuring the variability of a sample around the average value. If little dispersion exists, the sample is said to be precise. STP defines samples as precise when the variation in the sampling error (*SE*) around the average is small. Precision is defined as the variance of the sampling error $\sigma^2(SE)$ and relates to the degree of clustering seen in Figure 3-1.

Accuracy is defined as a situation in which the mean of the sampling error $m(SE)$ approaches zero. This implies that $m(SE)$ represents the value of the sampling bias. Computationally, it is necessary to use the square, $m^2(SE)$, in order to compare $m(SE)$ with $\sigma^2(SE)$. It relates to the amount of offset from the center of the target in Figure 3-1.

A common objective is to collect "representative" samples. The problem is that the term *representative* is always loosely defined in qualitative terms. Often, one is instructed to try to make the sampling representative but is not told how to go about reaching this goal. Consequently, the goal is seldom if ever achieved.

Representativeness is one of the precision, accuracy, representativeness, completeness and comparability (PARCC) parameters described in "Data Quality Objectives for Remedial Response Activities: Development Process," the 1987 Agency guidance document. This document provides the following definition:

> Representativeness expresses the degree to which sample data accurately and precisely represent a characteristic of a population, parameter variations at a sampling point, or an environmental condition. Representativeness is a *qualitative* parameter which is most concerned with the proper design of the sampling program. The representativeness criterion is best satisfied by making certain that sampling locations are selected properly and a sufficient number of samples are collected. Representativeness is addressed by describing sampling techniques and the rationale used to select sampling locations.

This qualitative definition may serve general purposes but does not quantitate the term for useful development of DQOs. The explicit definitions of precision and accuracy of STP can be combined to produce an exact definition of representativeness. When the mean square sampling error $r^2(SE)$, equal to the sum of $m^2(SE)$ and $\sigma^2(SE)$, is smaller than a specified standard, the sample selection is said to be representative. This relationship can be stated as

$$r^2(SE) = m^2 + \sigma^2(SE) \leq r_o^2 \tag{7.5}$$

where r_o^2(SE) is the allowable limit or standard. Note the advantage for error management purposes: A standard is set, and sampling either attains the goal or fails to do so. The DQO decision rule will determine the next course of action for either result.

Building on the concept of representative samples, one can develop the definition of an equitable sample. An equitable sample is sufficiently representative to satisfy all parties, such as the laboratory, the PRP, and the regulatory agency.

The visual representations of precision, accuracy, and bias in Chapter 3 used the traditional "target" approach. This represents the process of repetitive or duplicate sampling in which a quantity of material with a specific concentration is analyzed many times, each time with the hope of producing the same analytical result. This type of model is appropriate for repetitive sampling at the laboratory or repeated tests on standard reference materials.

In practice, only one sample is usually analyzed for a given location or time. Yet many individual samples are usually analyzed throughout the site, resulting in a wide distribution of sample concentrations. Graphing sitewide data allows use of a format similar to the misclassification ellipse presented in Chapter 3. Note that the target approach is equivalent to plotting repeated tests on a standard reference material (single value). The approach presented below covers a wide range of concentration values. Also, the reader should remember that these diagrams are for sampling and not for analytical results.

Figures 7-2 through 7-9 illustrate several types of qualities and defects that can be encountered in a sampling process. These hypothetical data show the true unknown values of the content a_L of the lot on the x axis, along with the true unknown values of the content a_s of each hypothetical sample.

Figure 7-2 shows the unreachable limit case of exact sampling. In this case, the sampling is both exact and reproducible. Figure 7-3 shows the situation to which one can aspire in actual practice. This situation portrays sampling that is both accurate and reproducible. This implies that the sampling error mean $m(SE)$ and the sampling error variance $\sigma^2(SE)$ are approaching zero. Figure 7-4 shows this same situation but, this time, showing that the sampling falls within equitable limits.

Figures 7-5 through 7-9 show problem situations for sampling. In Figure 7-5, sampling is neither reproducible nor accurate. The temptation to conclude that this

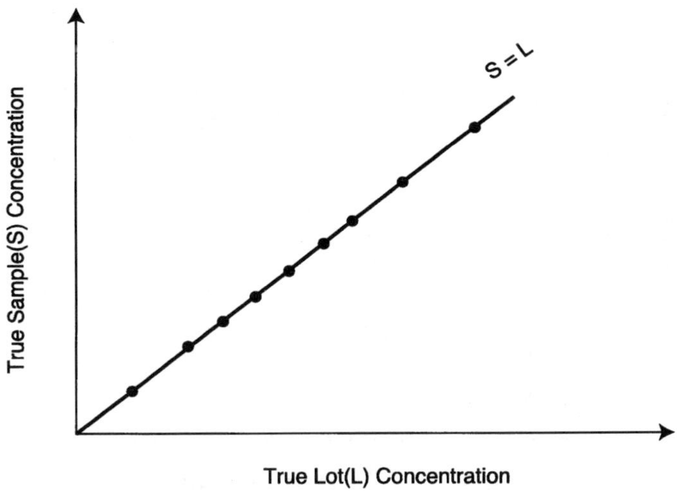

Figure 7-2. Limit case for sampling where results are unbiased and perfectly reproducible. (From *Pierre Gy's Sampling Theory and Sampling Practice: Heterogeneity, Sampling Correctness, and Statistical Process Control,* 2nd ed., by F.F. Pitard. Copyright © 1993 CRC Press, Boca Raton, FL. Courtesy of Francis Pitard.)

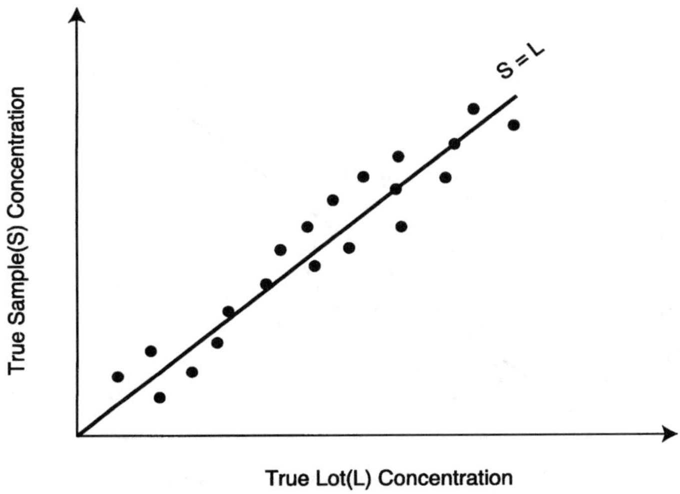

Figure 7-3. Accurate and reproducible sampling. (From *Pierre Gy's Sampling Theory and Sampling Practice: Heterogeneity, Sampling Correctness, and Statistical Process Control,* 2nd ed., by F.F. Pitard. Copyright © 1993 CRC Press, Boca Raton, FL. Courtesy of Francis Pitard.)

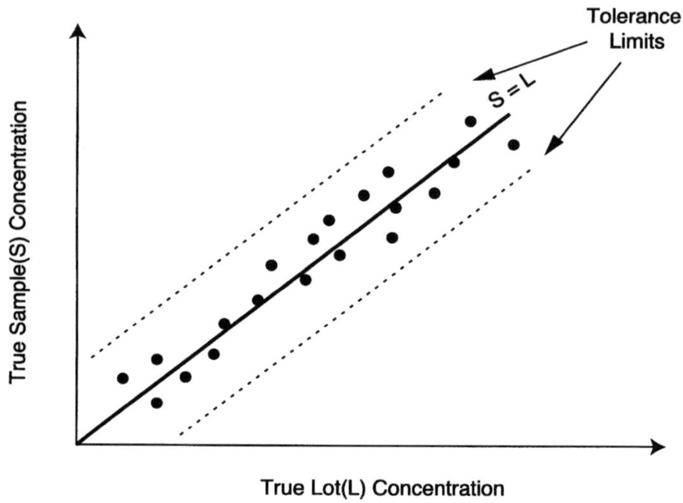

Figure 7-4. Accurate and reproducible sampling within equitable limits. (From *Pierre Gy's Sampling Theory and Sampling Practice: Heterogeneity, Sampling Correctness, and Statistical Process Control,* 2nd ed., by F.F. Pitard. Copyright © 1993 CRC Press, Boca Raton, FL. Courtesy of Francis Pitard.)

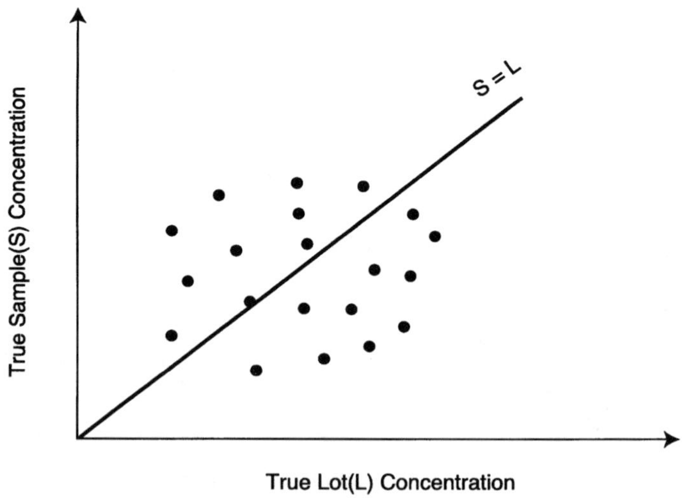

Figure 7-5. Sampling that is neither reproducible nor accurate. (From *Pierre Gy's Sampling Theory and Sampling Practice: Heterogeneity, Sampling Correctness, and Statistical Process Control,* 2nd ed., by F.F. Pitard. Copyright © 1993 CRC Press, Boca Raton, FL. Courtesy of Francis Pitard.)

type of sampling is accurate should be suppressed as there is no correlation between the true sample content and the true lot content. The notion of accuracy is further disqualified by the large sampling error mean $m^2(SE)$.

Figure 7-6 shows a pure form of bias called an *absolute bias*. In this situation, sampling is highly reproducible but is shifted from the proper mean. It is important to recognize that this is not a realistic case for sampling. It could, however, be an analytical bias.

Figure 7-7 shows the case of a relative bias. Here, samples are reproducible but biased, with the bias increasing as a function of increasing content. As with pure bias, this case is not realistic for sampling situations. This situation may occur as an analytical bias.

Figure 7-8 shows the case in which sampling is biased and reproducible, following an unknown law. In areas of low content, a low bias provides the opportunity for false negatives. At a given threshold, the situation shifts to a high bias, with its potential for false positives.

Figure 7-9 shows what is unfortunately a common occurrence in actual sampling situations. In this case, sampling is biased and not reproducible—the worst of both worlds. One of the goals of STP is to provide, in the words of Pitard (1989), an "insurance policy" to prevent situations such as the one shown in Figure 7-9.

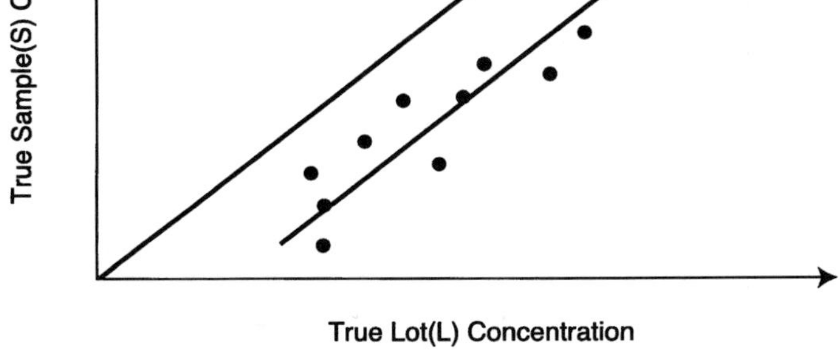

Figure 7-6. Absolute bias in sampling; not a realistic situation in sampling. (From *Pierre Gy's Sampling Theory and Sampling Practice: Heterogeneity, Sampling Correctness, and Statistical Process Control,* 2nd ed., by F.F. Pitard. Copyright © 1993 CRC Press, Boca Raton, FL. Courtesy of Francis Pitard.)

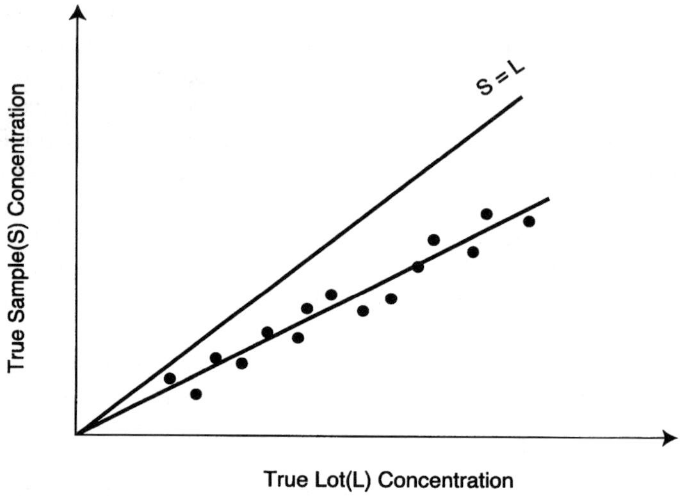

Figure 7-7. Sampling that exhibits a relative bias that is also reproducible; not a realistic sampling situation. (From *Pierre Gy's Sampling Theory and Sampling Practice: Heterogeneity, Sampling Correctness, and Statistical Process Control,* 2nd ed., by F.F. Pitard. Copyright © 1993 CRC Press, Boca Raton, FL. Courtesy of Francis Pitard.)

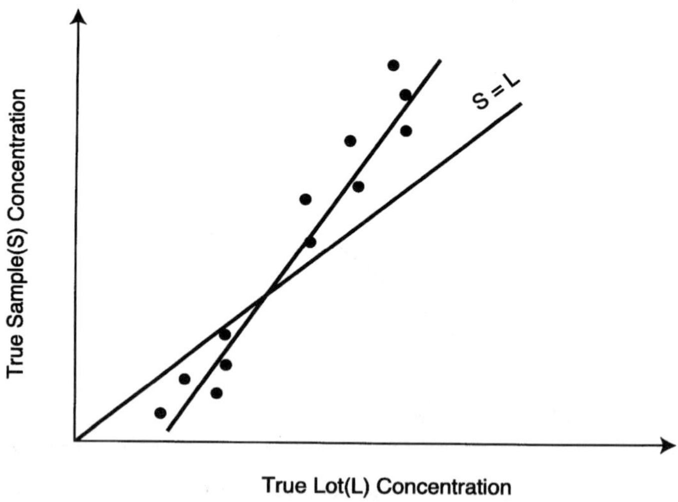

Figure 7-8. Biased and reproducible sampling following an unknown law. (From *Pierre Gy's Sampling Theory and Sampling Practice: Heterogeneity, Sampling Correctness, and Statistical Process Control,* 2nd ed., by F.F. Pitard. Copyright © 1993 CRC Press, Boca Raton, FL. Courtesy of Francis Pitard.)

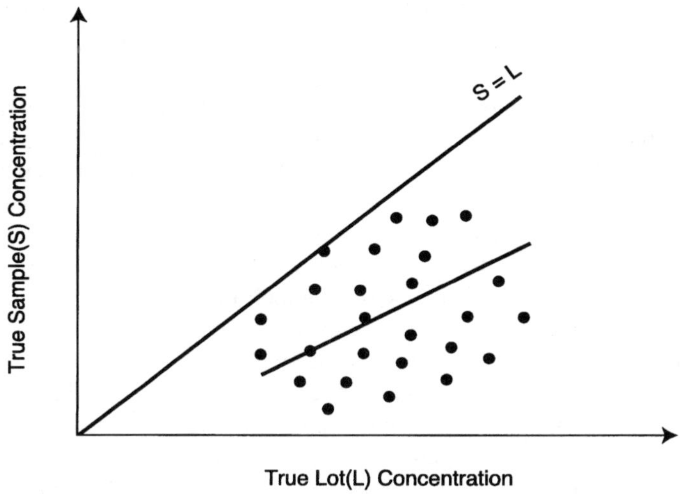

Figure 7-9. Biased sampling that is not reproducible; a common situation in sampling. (From *Pierre Gy's Sampling Theory and Sampling Practice: Heterogeneity, Sampling Correctness, and Statistical Process Control,* 2nd ed., by F.F. Pitard. Copyright © 1993 CRC Press, Boca Raton, FL. Courtesy of Francis Pitard.)

7.3 Structural and Circumstantial Properties

The primary goal of any sampling effort is to produce representative samples, defined as being both accurate and precise. To achieve this goal, sampling correctness is paramount. In the words of Dr. Pierre Gy, "Sampling must be correct. If money is to be spent on sampling, the first end to be achieved is to eliminate all sources of sampling incorrectness. This is the only way of eradicating sampling biases."

To further the understanding of sampling correctness, it is important to define and differentiate primary structural properties from secondary circumstantial properties.

A structural property is a primary property in a sampling operation. It is an intrinsic property of two components of sampling: (1) the material to be sampled and (2) the sampling equipment. A structural property is primary because it is independent of the sampling problem itself.

A structural property is one over which control can be exerted. Examples include the characteristics of a sampler, a sampling protocol, and a standard. Both heterogeneity and sampling correctness have primary structural properties. A sample is correct when all particles in the lot have an equal probability of being selected for inclusion in the sample. This is the same concept as the one discussed in Chapter 4: All locations at a site must have an equiprobable chance of being selected. Chapter 8 provides examples of both correct and incorrect sampling devices; Chapter 16 addresses correct sampling strategies.

By contrast, a property is deemed to be circumstantial if it depends on conditions that one is unable to control. Circumstantial properties depend on the sampling problem itself; therefore, circumstantial properties often depend on the primary properties. Whereas a structural property is always true and can be relied on, a circumstantial property cannot be relied on.

Circumstantial properties may depend on the primary properties; however, the reverse is never true. Perhaps the most important example of a circumstantial property is accuracy. Accuracy is a secondary property when compared to correctness. Correctness can be directly controlled but accuracy cannot.

An incorrect sampling process may become accurate under favorable circumstances but cannot be relied on to provide consistently accurate results because accuracy is a transient phenomenon. The reader should note that accuracy (and formulas or methods to prove or disprove it) is often a part of DQO plans, QAPPs, and SAPs. These good intentions do not compensate for the fact that accuracy is a circumstantial result and not under anyone's direct control. Sample replicates can determine laboratory precision but cannot quantify accuracy because the true value remains unknown.

7.4 Introduction to Heterogeneity

In one way or another, heterogeneity is responsible for all sampling errors. It is therefore essential to gain an understanding of the types of heterogeneity and their influences. First, however, it is necessary to contrast heterogeneity with homogeneity so that, for practical purposes, homogeneity can be removed from the vocabulary. Subsequently, heterogeneity will be analyzed in the context of the different varieties and the corresponding implications.

The term *homogeneous* is widely used in a variety of situations in our daily lives. Milk is homogenized, societies are referred to as homogeneous, and even mathematics speaks of homogeneous properties. Within the context of STP (and most likely other contexts as well), homogeneity is an illusion. It simply does not exist. As such, it is necessary to abandon the delusional notion that it does.

Belief in, or reliance on, the notion of homogeneity will result in many serious sampling errors. Rejecting homogeneity forces one to embrace the only alternative, heterogeneity. This chapter introduces constitution and distribution heterogeneity and their impacts on correct sampling.

Perhaps nothing illustrates the essence of heterogeneity as well as snow. When you look out on a fresh blanket of deep snow, it seems hard to imagine anything more perfect and uniform: pure, smooth, and white as far as the eye can see. Yet, as you approach the surface of the snow, you can make out individual snowflakes.

Scientists long ago discovered that every snowflake is different. While obeying the law of hexagonal crystallization, each snowflake differs from all others in size, shape, and type or family of formation. In fact, while giving the initial impression of homogeneity, snow may be the dramatic proof that homogeneity is an illusion!

An example more pertinent to environmental sampling is a pile of sand. From a distance of only a few feet, the sand looks perfectly uniform and homogeneous. It is only when you select a few individual grains and examine them closely under magnification do you see the multitude of irregularities and dissimilarities. Figure 7-10 is a visual representation of this situation. The reality: different sizes, colors, compositions, shapes, densities, opacities, hardnesses, porosities, and so on. Such an examination supports the conclusion that the world is indeed heterogeneous and that homogeneity is simply a limit case of heterogeneity.

Abandonment of the notion of homogeneity represents a paradigm shift in thinking necessary to develop the concepts and mathematics of STP. The methods of STP provide a type of insurance to achieve correct results. STP, like most types

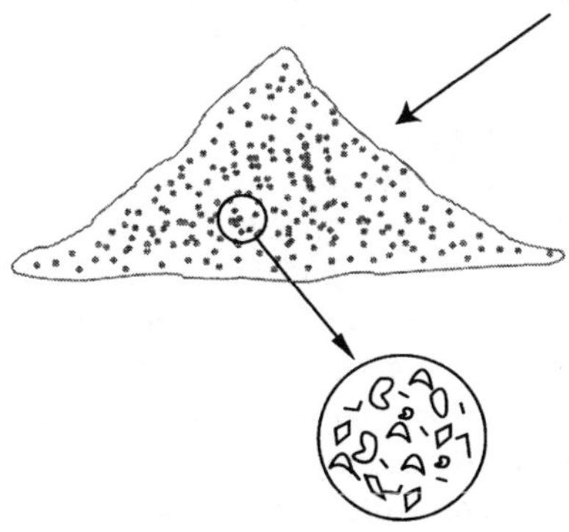

Figure 7-10. The illusion of homogeneity as demonstrated by a pile of sand, which appears homogeneous but is revealed at close range to be heterogeneous. (From *Pierre Gy's Sampling Theory and Sampling Practice: Heterogeneity, Sampling Correctness, and Statistical Process Control,* 2nd ed., by F.F. Pitard. Copyright © 1993 CRC Press, Boca Raton, FL. Courtesy of Francis Pitard.)

of traditional insurance, is based on statistics and "safe" hypotheses and assumptions. For STP, heterogeneity is the safe hypothesis.

STP clearly distinguishes two different types of heterogeneity: (1) constitution heterogeneity and (2) distribution heterogeneity. Both are primary structural properties of all materials and cannot be ignored without incurring significant risk. It is detrimental to all operations in which the composition of a specific constituent needs to be measured.

7.4.1 Constitution Heterogeneity

Constitution heterogeneity (CH_L) refers to the fundamental properties of the fragments of a lot where the fragments are examined one by one. This is equivalent to the inherent variability in the composition of each component particle of the lot L. The constitution heterogeneity of a given group of fragments subject to a given state of comminution is fixed. CH_L cannot vary unless a comminution is initiated.

As a primary structural property of the lot, constitution heterogeneity cannot be altered by mixing and homogenization processes. Only physically altering the material in a comminution process can change constitution heterogeneity. It is a common misconception that homogenization methods can affect constitution heterogeneity. Ironically, much time and effort are inappropriately spent on such activities.

Constitution heterogeneity (CH_L) is the cause of the fundamental error (FE), the first of the seven sampling errors. The fundamental error occurs when a sample S is selected from a lot L. The magnitude of this error is specifically related to the constitution heterogeneity.

The fundamental error FE is critically important because it never cancels out. The fundamental error FE is a state of minimum error that cannot be eliminated and that represents the underlying variability of the soil or waste. A minimum FE is achieved only by selecting, one by one and strictly at random, fragments that make up a sample. This, of course, is not possible in practice.

The fundamental error is a function of factors such as the mineral composition, shape, liberation, and particle size distribution of a given lot. If materials were homogeneous, this error could be eliminated. In practice, the variance of the fundamental error can only be decreased by comminution.

7.4.2 Distribution Heterogeneity

The distribution heterogeneity (DH_L) is both a complement to and a function of the constitution heterogeneity. Without constitution heterogeneity, distribution heterogeneity cannot exist. Distribution heterogeneity also has a strong spatial component that is not germane to constitution heterogeneity. Both CH_L and DH_L are influenced by support volumes.

In constitution heterogeneity, the individual fragments or particles of the lot L were of concern. The next step is to visualize the lot as separate groups of fragments. These groups are composed of a certain number of individual neighboring fragments. The distribution heterogeneity is related to how the individual fragments are represented within the groups.

If all the groups selected contain the same average composition of fragments, then, by definition, the lot can be said to have a homogeneous distribution. This being a limit case, it will never be achieved in practice.

The distribution heterogeneity (DH_L) depends for each critical constituent on three factors: (1) the constitution heterogeneity CH_L, (2) the spatial distribution of the constituent particles, and (3) the shape of the lot. The shape of the lot is not trivial as gravity plays an important and never-ending role in the distribution of particles.

The distribution heterogeneity is responsible for the grouping and segregation error *GE*. Most of the time, this error has a nonzero mean; that is, it biases the sample. For major constituents, the mean is commonly insignificant. For minor or trace constituents, such as hazardous contaminants of concern (COCs), the mean may become large.

Two factors, the grouping factor γ_s and the segregation factor ξ, influence the *GE*. The grouping factor of the *GE* is an indication of the probability that a particular fragment will be selected in a sample increment; that is, it is a measure of random selectivity. As the number of fragments in the increment increases, the grouping factor also increases. The grouping factor becomes zero when each increment is composed of only a single fragment. Note that the grouping factor γ_s is different from the variogram $\gamma(h)$.

The segregation factor ξ describes the natural range of distribution heterogeneity between a minimum DH_L and a maximum DH_L. It is a correction factor that localizes the amount of distribution heterogeneity of a critical constituent; that is, it is a measure of spatial rearrangements. The segregation factor is always between zero and one. It is zero when the material is homogeneous and a maximum of one when equal to the constitution heterogeneity.

7.4.3 Visual Representations of Heterogeneities

In order to better understand the concept of heterogeneity and the specific types CH_L and DH_L, several visual examples will be provided. Figure 7-11 shows the relationship between constitution heterogeneity CH_L and distribution heterogeneity DH_L. Constitution heterogeneity is concerned with the amount of variability among the individual fragments in the lot. Distribution heterogeneity refers to

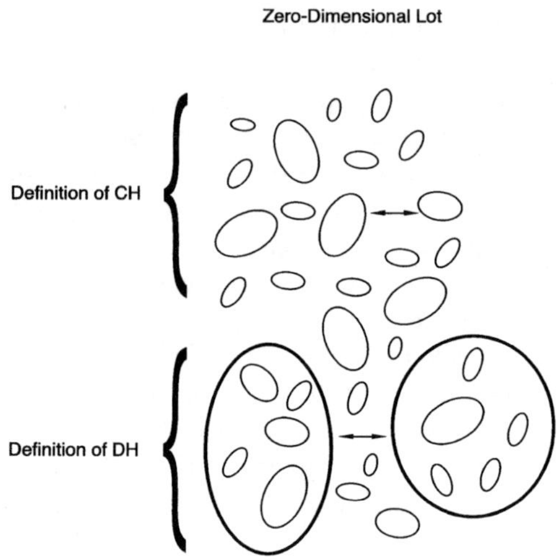

Figure 7-11. The difference between constitution and distribution heterogeneity. (From *Pierre Gy's Sampling Theory and Sampling Practice: Heterogeneity, Sampling Correctness, and Statistical Process Control,* 2nd ed., by F.F. Pitard. Copyright © 1993 CRC Press, Boca Raton, FL. Courtesy of Francis Pitard.)

the manner in which the component particles or fragments separate themselves into groups.

Figures 7-12 through 7-15 show other relationships between CH_L and DH_L in a block shading format. Figure 7-12 is a uniform grid of white blocks. The individual blocks are exactly identical and serve as a demonstration of a homoge-

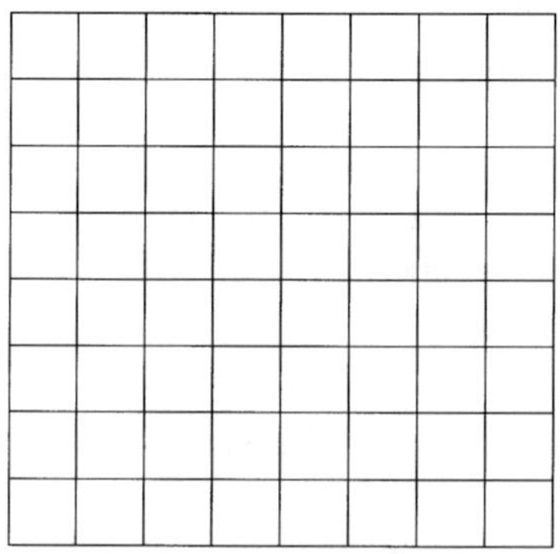

Figure 7-12. Homogeneous constitution and distribution. (From *Pierre Gy's Sampling Theory and Sampling Practice: Heterogeneity, Sampling Correctness, and Statistical Process Control,* 2nd ed., by F.F. Pitard. Copyright © 1993 CRC Press, Boca Raton, FL. Courtesy of Francis Pitard.)

neous constitution. In addition, the distribution is also homogeneous, with any imaginable grouping containing a strictly identical composition. This illustration also serves to confirm the statement made earlier that, without constitution heterogeneity, distribution heterogeneity cannot exist.

Figure 7-13 contains a similar-sized grid but with four different types of shading or patterns. Because four different components are present, the constitution is necessarily heterogeneous. However, if we consider a 4 × 1 grouping of squares either vertically or horizontally, a form of homogeneity can be observed. This can be defined as a *homogeneous modular distribution* (Pitard 1993).

Figure 7-14 shows a dangerous type of distribution heterogeneity. The component blocks are identical to those shown in Figure 7-13 and, therefore, this situation exhibits constitution heterogeneity. However, the four types of component squares have been segregated into four horizontal layers. In this case, distribution heterogeneity is at the maximum, with the constituents completely separated from each other. This is the limit case of segregation that gravity constantly attempts to impose. This demonstrates why grab samples are simply specimens and should be viewed with great caution.

In Figure 7-15, the same component blocks are distributed randomly throughout the matrix. To construct this diagram, the pattern of each square was selected at random, one by one, from the available patterns, providing equiprobabilistic opportunities. Randomizing approximates homogenizing, which tends to remove lingering correlation between the position and pattern of the individual blocks.

Note that, even with the use of a randomizing technique, small clusters can still be observed; this clustering is consistent with the discussion on random selection of sampling locations in Chapter 4. Although the probability of any two

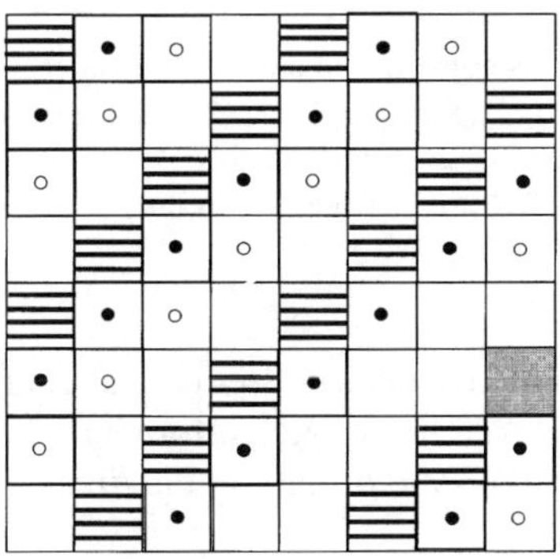

Figure 7-13. Heterogeneous constitution with homogeneous modular distribution (4 × 1 module). (From *Pierre Gy's Sampling Theory and Sampling Practice: Heterogeneity, Sampling Correctness, and Statistical Process Control,* 2nd ed., by F.F. Pitard. Copyright © 1993 CRC Press, Boca Raton, FL. Courtesy of Francis Pitard.)

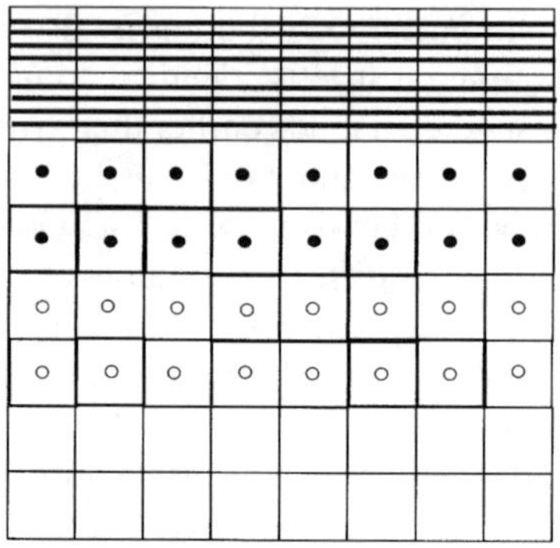

Figure 7-14. Heterogeneous constitution combined with maximum distribution heterogeneity (perfect segregation). (From *Pierre Gy's Sampling Theory and Sampling Practice: Heterogeneity, Sampling Correctness, and Statistical Process Control,* 2nd ed., by F.F. Pitard. Copyright © 1993 CRC Press, Boca Raton, FL. Courtesy of Francis Pitard.)

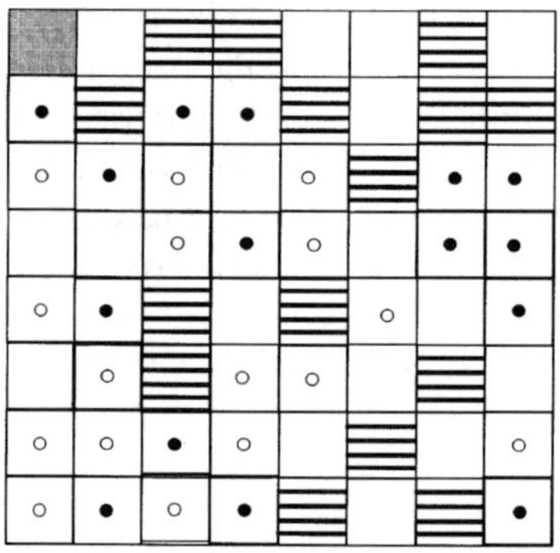

Figure 7-15. Heterogeneous constitution and natural (uncorrelated) distribution homogeneity. (From *Pierre Gy's Sampling Theory and Sampling Practice: Heterogeneity, Sampling Correctness, and Statistical Process Control,* 2nd ed., by F.F. Pitard. Copyright © 1993 CRC Press, Boca Raton, FL. Courtesy of Francis Pitard.)

specific particles combining to form a cluster is small, clusters occur with a certain frequency when sufficient trial events are performed. The implication is that even random selection produces a state of natural distribution homogeneity that lacks perfection.

These diagrams serve as icons to the general conditions of heterogeneity encountered during sampling. The reader is reminded that the magnitude of heterogeneity may not be known prior to sampling. The value of STP is to enable us to understand the possibilities and plan for any situation.

7.4.4 Distribution Heterogeneity and Support

Because the distribution heterogeneity DH_L incorporates a spatial component, it follows logically that the concept of support will be of importance as well. An example of the support relationship will now be given for distribution heterogeneity DH_L. More sophisticated types of support relationships for DH_L will be presented in the next section.

Figure 7-16 sketches the spatial relationship between groups of fragments. In Figure 7-16a, the various types of fragments are well mixed. Thus, segregation is minimized and DH_L is small. In Figure 7-16b, the material is highly segregated and DH_L is large. Figure 7-17 shows the same material with a larger grouping unit or support. The distribution heterogeneity is smaller for the groups with larger support.

An important point to note here is that the DH_L support influences not only particles or fragments but also the interstitial volume. This interstitial porosity or permeability or both may be the parameters of concern for core samples intended to measure groundwater flow parameters. Figures 7-16 and 7-17 also show the influence of support for interstitial parameters and why small-diameter cores can produce misleading information.

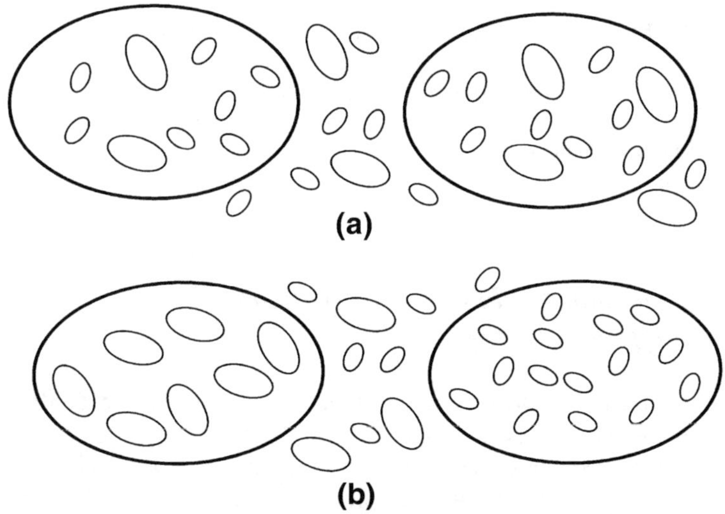

Figure 7-16. The effect of segregation on distribution heterogeneity *DH*: (a) for unsegregated material, *DH* is small, and (b) for segregated material, *DH* is large. (From *Pierre Gy's Sampling Theory and Sampling Practice: Heterogeneity, Sampling Correctness, and Statistical Process Control,* 2nd ed., by F.F. Pitard. Copyright © 1993 CRC Press, Boca Raton, FL. Courtesy of Francis Pitard.)

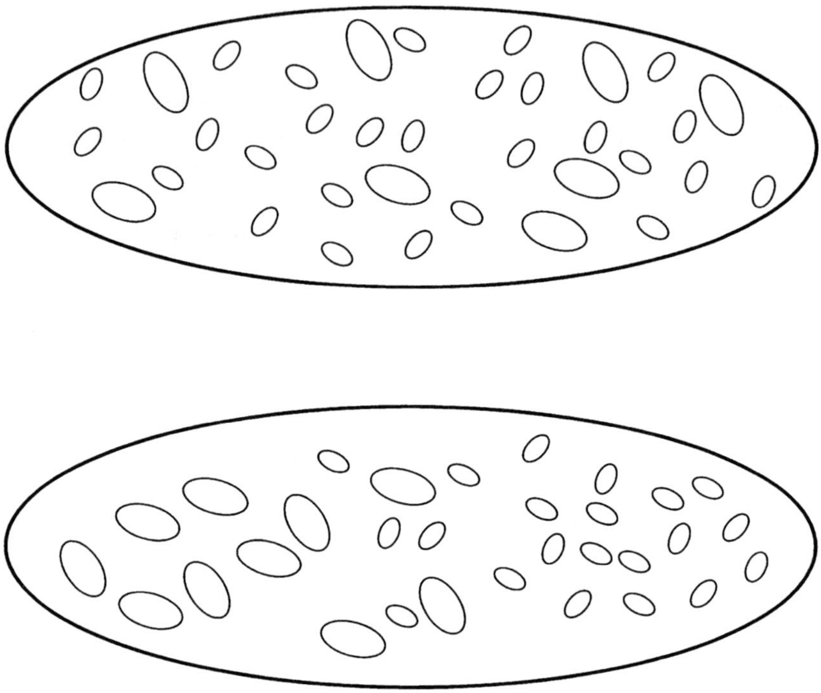

Figure 7-17. Mitigating the effects of the distribution heterogeneity *DH* by using larger support volumes; in each case, *DH* is the same. (From *Pierre Gy's Sampling Theory and Sampling Practice: Heterogeneity, Sampling Correctness, and Statistical Process Control,* 2nd ed., by F.F. Pitard. Copyright © 1993 CRC Press, Boca Raton, FL. Courtesy of Francis Pitard.)

7.4.5 Dimensional Module of Observation and Classification of Lots

The characterization of lots as dimensional entities is often a confusing step because it combines many different concepts. The relationship among the dimensional module of observation, continuity model, order/disorder model, and heterogeneity borders on a chicken and egg situation. Understanding the dimensional module of observation is vital, however, to representing the lot effectively and integrating the concepts of STP. These various concepts are all related to heterogeneities h_1, h_2, and h_3. These relationships will be discussed here and in Sections 7.4.6.

There are three criteria for classifying a lot:

1. The number of dimensions of the easiest model to represent the lot
2. The continuity or discontinuity of the selected model
3. The order or disorder of the constituents of the lot

Criterion 1 is important because the smaller the dimension, the easier it is to solve the sampling problem associated with the lot. In fact, experience has demonstrated that only zero- and one-dimensional lots are economically solvable (Pitard 1993). The dimension of a lot will also provide a link between STP and geostatistical techniques.

Criterion 2 reintroduces the discrete and continuous models for observation. These models serve as additional links between the point or punctual samples of STP and the longer-range view of geostatistics.

Criterion 3 addresses the issue of correlated and uncorrelated data. Classical statistical models can be used for random (uncorrelated) data, whereas multivariate statistics (which includes space or time series models) must be used for correlated data.

7.4.5.1 Number of Dimensions Characterizing a Lot

One can argue that a lot always has three dimensions. From a practical perspective, however, it is often possible to ignore one or even two of these dimensions. The following are descriptions of the four types of dimensional lots based on the definitions of Pitard (1989):

- Three-dimensional lot: the contents of a jar, bag, truck, ship, railroad car, and so forth, subject to the constraint that one of these three-dimensional objects is considered as the whole lot. It can also be a compact solid such as a block inside a hazardous waste site.
- Two-dimensional lot: a three-dimensional lot in which the thickness becomes negligible because it is very small compared to the two other dimensions (e.g., a three-dimensional lot that has been flattened, an aquifer, surface soil, etc.).
- One-dimensional lots: continuous or elongated piles, streams or rivers, material on conveyor belts, and so forth, or a series of bags, jars, trucks, railroad cars, and so forth, as long as these objects are considered as a set of nonrandom, discontinuous units making up the lot, the order of which is highly relevant.
- Zero-dimensional lots: the contents of a series of bags, jars, trucks, railroad cars, and so forth, as long as these objects are considered as a set of random, discontinuous units making up the lot. A zero-dimensional lot can be regarded as a suitable convention to describe a set of unarranged units. It can also be a one-dimensional lot for which the chronological order of the various units has been lost.

7.4.5.2 Continuous and Discontinuous Models for Lots

Both the continuous and discontinuous model described earlier consider spatial components. The difference is the distance of observation that relates to the scale of heterogeneity. The discrete or discontinuous set, short-range in nature, is described as looking through a magnifying lens at local fluctuations.

The discrete approach investigates the fluctuation between fragments and the heterogeneity introduced by individual fragments. Whereas from a theoretical point of view, three dimensions still exist, this scale is sufficiently small to be considered a point observation. The discrete model is associated with heterogeneity h_1.

The continuous model incorporates a spatial or space-time view that is much larger. Flowing streams of material (water, material on a conveyor belt, etc.) are examined here. The interest in the continuous model is mainly long-range heterogeneity (h_2) and periodic heterogeneity (h_3) fluctuations. This is also the type of question addressed by geostatistical appraisal.

7.4.5.3 Order and Disorder in the Lot

In Chapter 3, the concept of random vs. correlated data was introduced for spatial data. This type of perspective is also appropriate for viewing the constituents of a lot. All components of a lot may be considered in one of the following ways:

1. A statistical population of units. This implies that adjacent units are statistically independent and that the selection of any unit is the result of a Bernoulli trial, that is, it is either selected (with probability p) or not selected (with probability $1 - p$). Statistical (disordered) populations are related to heterogeneity h_1.
2. A time or space series of units. This implies that adjacent units are statistically correlated. Ordered populations are related to heterogeneities h_2 and h_3.

It is important in STP to determine whether a population is statistical or correlated. This will prevent the inappropriate application of results valid for statistical populations to spatial data.

Earlier, homogeneity was defined as the zero of heterogeneity. Similarly, a statistical population can be defined as the zero of a correlated population. Also, perfect disorder is a limit case. Forces such as gravity and time are constantly at work as chaotic attractors and often introduce a degree of correlation or self-similarity. In fact, it is generally much easier to define the conditions promoting segregation than to define the conditions that ensure homogenization.

What must be concluded is that there is a strong chaotic component applied to the lot constituents. There is a strong degree of natural disorder coupled with forces attempting to introduce order. This hybrid effect produces a lot in which each component of the lot has at least some degree of correlation with its position within the lot.

The reader should note that the same is true for spatial data in the geostatistical sense. There is a degree both of order and disorder between nearby samples. The amount of correlation between samples can be determined by the variogram, but there still remains a randomizing component that makes prediction imperfect.

7.4.6 Heterogeneity of Lots: Dimensional Classification

7.4.6.1 Zero-Dimensional Lots

Zero-dimensional lots deal with random fluctuations between neighboring fragments. This analysis occurs on a very small spatial scale, generally one that will be submitted in its entirety to the analytical device. The populations examined are largely statistical and are influenced by the constitution heterogeneity CH_L and the distribution heterogeneity DH_L. Heterogeneity h_1 applies to this situation.

The fluctuation in fragments generates the fundamental error *FE*. There are also opportunities, however, for temporal and gravitational forces to interject a degree of order in the arrangement of the fragments, causing segregation. In fact, the larger the difference in constitution, the greater the potential segregation.

Because it is not possible in a practical sense to select individual fragments by a stochastic process, increments containing groups of fragments are collected. This leads to the grouping and segregation error *GE* introduced by distribution heterogeneity. The larger the group, the larger the error.

Zero-dimensional lots are desirable because it is always possible to perform correct sampling on them. The means are available to solve the problems associated with this type of sampling and to assure that sampling is correct.

7.4.6.2 One-Dimensional Lots

One-dimensional lots are essentially linear features. Examples given previously include streams and conveyor belts or other long piles or a series of bags, jars, and so forth. One dimensional lots are subject to constitution and distribution heterogeneity (h_1) just as zero-dimensional lots are. A unique feature, however, is that one-dimensional lots are almost always the result of chronological activity.

The introduction of a chronological component often implies human activities. These can include daily, weekly, or longer cycles at industrial installations, often combined with natural cycles such as seasonal rotation. This provides a perspective beyond the simple examination of neighboring fragments in an increment.

Chronological fluctuations in one-dimensional lots often reflect man-made trends and are characterized by:

1. The heterogeneity h_2 introduced by long-range trends; also termed *large-scale segregation*. Note that this can be quantified by the variogram of geostatistics.
2. The heterogeneity h_3 introduced by cyclic phenomena.

Heterogeneities h_2 and h_3 give rise to two new errors: the long-range heterogeneity fluctuation error CE_2 and the periodic heterogeneity fluctuation error CE_3. Both CE_2 and CE_3 are errors of the continuous model and are components of the overall estimation error OE.

CE_1 is defined as the short-range heterogeneity fluctuation error based on the small-scale heterogeneity h_1. CE_1 is based on the relationship

$$CE_1 = FE + GE \tag{7.6}$$

indicating that CE_1 is the sum of the fundamental error FE and the grouping and segregation error GE.

The total heterogeneity supported by a lot of any kind may be written

$$h = h_1 + h_2 + h_3 \tag{7.7}$$

Also, an error CE, defined as the continuous selection error or the integration error (related to the heterogeneities in eq. 7.7) can be written

$$CE = CE_1 + CE_2 + CE_3 \tag{7.8}$$

with CE_1 a link between the continuous and discrete models.

As with zero-dimensional lots, it is always possible to perform correct sampling on one-dimensional lots.

7.4.6.3 Two- and Three-Dimensional Lots

Examination of two- and three-dimensional lots takes us into the realm of geostatistics, with heterogeneities h_1, h_2, and h_3 contributing to error components. Such lots could be the contaminated soil at a hazardous waste site or an aquifer. Two- and three-dimensional lots cause problems for sampling in that the correct materialization of increments is almost always insurmountable from a sampling perspective and infeasible from an economic perspective.

7.4.7 Mathematical Expressions of Heterogeneity

Figures 7-12 to 7-18 were visual expressions of heterogeneity. Corresponding mathematical expressions of constitution and distribution heterogeneity are developed in Appendix B. Of particular interest is the constant factor of constitution heterogeneity IH_L, which is expressed as

$$IH_L = \frac{CH_L M_L}{N_F} \tag{7.9}$$

where:

CH_L = constitution heterogeneity of the lot L
M_L = the mass of the lot L
N_F = the number of fragments in lot L

Equation 7.9 will be used in Chapter 8 to mitigate the impact of the fundamental error FE on sampling and subsampling operations.

Conclusion

This chapter introduced two specific types of heterogeneity—constitution CH_L and distribution DH_L—that profoundly affect the sampling process and the resultant sampling errors. Whereas distribution heterogeneity is dependent on the presence of constitution heterogeneity, this relationship does not prevent the two types from being distinct and separate entities. The fundamental error of sampling is a direct result of the constitution heterogeneity and, once incurred, never cancels out. It can be reduced only by comminution. Structural properties of sampling, such as the constitution heterogeneity and correctness of a sampling device, have been distinguished from circumstantial properties such as accuracy. In fact, accuracy has been shown to be a random event whose likelihood increases with correct sampling. A mathematical definition of representativeness has been developed, defining the concept in terms of precision and accuracy. Accompanying mathematical derivations (Appendix B) precisely define the concepts of constitution and distribution heterogeneity in objective terms. This elevates the term *heterogeneity* beyond the level of a frequently used buzzword and frees it to be used in practical error management for the GEM process.

Chapter 8

Sampling Errors

SYNOPSIS: Gaining an understanding of the types of heterogeneities that lead to sampling errors is a critical first step toward the implementation of correct sampling on which the GEM process relies. Heterogeneities are not errors in themselves; rather, they are structural properties that enable sampling errors to arise. This chapter demonstrates how heterogeneities are related to sampling error. In addition, practical ways to manage and minimize the various sampling errors will be discussed. This chapter is accompanied by a mathematical derivation of the fundamental error FE in Appendix C, which readers should review before continuing with Chapter 8.

KEY WORDS: Fundamental Error, Constant Factor of Constitution Heterogeneity, Sampling Nomograph, Grouping and Segregation Error, Homogenization Techniques, Sampling Device, Fractional Shoveling, Cone & Quarter, Riffle Splitter, Sectorial Splitter, Vibratory Spatula, V-Blender, Materialization Error, Core Sample, Japanese Slab Cake Method, Scoop, Shovel, Spatula, Trowel, Auger, Thief, Trier, Coliwasa Tube, Kemmerer Depth Sampler, Preparation Error.

8.1 The Fundamental Sampling Error *FE*

8.1.1 Definition

The fundamental error *FE*, first of the seven sampling errors, is arguably the most important. The reason: It stems solely from the constitution heterogeneity CH_L of the material being sampled and is the error that remains when a sampling operation is perfect. The fundamental error *FE* is the only error that never cancels out. The definition of the fundamental error *FE* depends on limit case conditions: The increment selection must be correct, and the sample increments must be single random fragments.

Minimizing *FE* is a critical DQO issue because it affects the estimated sample concentration and later GA mapping and decision making. Fortunately, it provides a measure of convenience in that it is also the only sampling that can be estimated a priori. Thus, it can be effectively managed.

The fundamental error is a "good news/bad news" situation: Gy (1982) has shown that the mean *m(FE)* of the fundamental error is negligible for most cases; with the exception of trace elements, which are the primary concern of much environmental sampling.

The equation for the variance of the fundamental error *FE* is

$$s_{FE}^2 = \left(\frac{1}{M_S} - \frac{1}{M_L}\right) IH_L \tag{8.1}$$

where:

M_S = sample weight
M_L = weight of the lot L
IH_L = constant factor of constitution heterogeneity

If M_L is much greater than M_S, equation 8.1 can be simplified to

$$s_{FE}^2 = \frac{IH_L}{M_S} \tag{8.2}$$

This formula gives rise to the possibility that $\sigma^2{}_{FE} = 1$. This implies that $IH_L = M_S$. In this case, the concentration a_S is distributed around a_L, with a relative standard deviation $\sigma = \pm 100\%$.

8.1.2 Minimizing and Managing the Fundamental Error FE

The importance of minimizing the fundamental error *FE* stems from the fact that, when every other part of the sampling operation is perfect, the fundamental error *FE* still remains. Unfortunately, for trace constituents, the standard deviation of *FE* is not trivial and can easily reach relative proportions of 100%, 1000%, or more.

With some assumptions, equation 8.1 can be used to establish boundaries on the fundamental error *FE*. If the lot *L* has a very large weight M_L in comparison to the sample M_S, a maximum variance for *FE* can be calculated:

$$\sigma^2(FE)_{\text{MAX}} = \left(\frac{1}{M_s} - 0\right) IH_L \tag{8.3}$$

If the difference in sample weights M_L and M_S is not large, a minimum *FE* can be calculated:

$$\sigma^2(FE)_{\text{MIN}} = \left(\frac{1}{M_s} - \frac{1}{2M_s}\right) IH_L \tag{8.4}$$

Equation 8.4 might find practical application when a sample is known to be the result of a splitting operation using a riffle splitter. However, the details of sample collection and splitting may not always be known. Thus, equations 8.3 and 8.4 provide a means by which the range of *FE* errors may be determined.

From the time a sample is collected in the field to the time a portion of that sample is submitted to the analytical device, the sample may undergo one or more splitting or reduction operations. A fundamental error *FE* is generated at *each* stage of subsampling. The magnitude of *FE* continues to grow at each subsampling stage unless proper measures are employed.

One such measure is crushing and grinding, or *comminution*. The fundamental error can only be reduced by comminution. Comminution reduces the size of the largest particles present, thus increasing the liberation factor ℓ. Fortunately, the factor d^3 will generally be decreasing at a much faster rate than the rate at which ℓ increases. Thus, comminutions offer the opportunity to reduce and minimize the fundamental error *FE*.

Because the comminution stage is generally followed by a subsampling stage, there is a chance that the *FE* may exceed acceptable limits. Therefore, the *FE* must be tracked at each stage of subsampling. This is performed by means of a sampling nomograph.

Equation C.11 defines the constant factor of constitution heterogeneity as

$$IH_L = XY = c\ell fgd^3 \tag{8.5}$$

In this equation, all terms remain constant except d^3, which decreases each time a comminution is performed. The terms $fgc\ell$ can be combined for convenience into a single constant term:

$$C = fgc\ell \tag{8.6}$$

with

$$IH_L = Cd^3 \tag{8.7}$$

Substituting equation 8.5 into equation 8.2,

$$\sigma^2(FE) = \frac{Cd^3}{M_s} \tag{8.8}$$

By using the logarithm of equation 8.8,

$$\log \sigma^2(FE) = \log C + 3 \log d - \log M_s \tag{8.9}$$

a maximum amount of information can be placed on the sampling nomograph.

In equation 8.9, two constants exist: particle size d and sampling constant C. For a given particle size d, the value of log $\sigma^2(FE)$ is directly proportional to $-$log M_S, and the derivative of the function log $\sigma^2(FE)$ is equal to -1. The implication is that the line depicting the variations of log $\sigma^2(FE)$ as a function of $-$log M_S for a given particle size d and specified sampling constant C has a slope on the nomograph of -1.

An example of a sampling nomograph is shown in Figure 8-1. The X axis plots the weight of the sample M_S in grams. The Y axis plots the variance of the fundamental error $\sigma^2(FE)$. Particle sizes d_1 to d_n are plotted with slopes of -1.

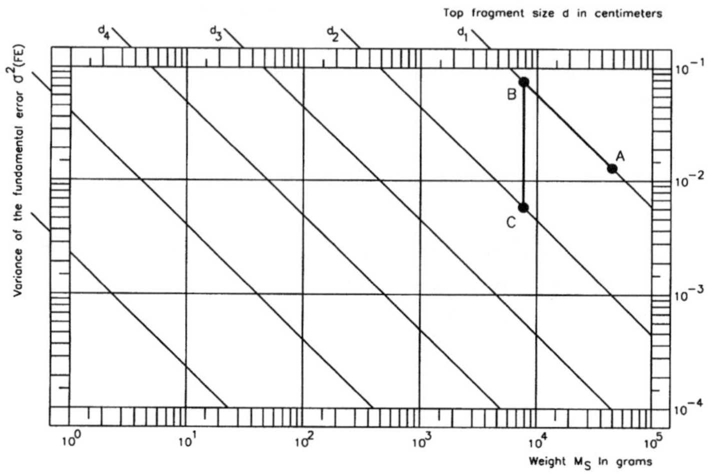

Figure 8-1. Sampling nomograph. (From *Pierre Gy's Sampling Theory and Sampling Practice: Heterogeneity, Sampling Correctness, and Statistical Process Control,* 2nd ed., by F.F. Pitard. Copyright © 1993 CRC Press, Boca Raton, FL. Courtesy of Francis Pitard.)

The "rules" of the nomograph are given in Pitard (1993):

- Every lot L in a particular stage of comminution of weight M_L is represented on the graph by one and only one point (e.g., point A in Fig. 8-1.)
- Every sampling stage that does not involve a comminution (d remains constant) is represented by a change of place of point A along the straight line of slope -1, until a new point corresponding to the weight M_S of the selected sample is reached (e.g. point B in Fig. 8-1).
- Every comminution or change of particle size d (M_S remains constant) tends to lower point B until it intersects a new straight line of slope -1 corresponding to the new particle size d (e.g., point C in Figure 8-1).

The line *ABC* represents a subsampling and comminution stage. At point *A*, a given mass of material exists with a given maximum particle size. A subsample is then taken. Since the mass of this subsample is smaller than the mass at point *A*, the variance of *FE* increases as it follows the line of slope -1 to point *B*, the new sample weight. At point B, a comminution is performed. Because comminution is the only means by which the fundamental error *FE* may be reduced, the comminution at point *B* lowers the fundamental error *FE* to point *C*.

If the material of a lot *L* contains large fragments, several successive sampling and comminution stages may be required. Figure 8-2 shows an example of a situation in which several stages were required. Each time a subsample is taken (points *A*, *C*, *E*, and *G*) the variance of the *FE* increases along the line correspond-

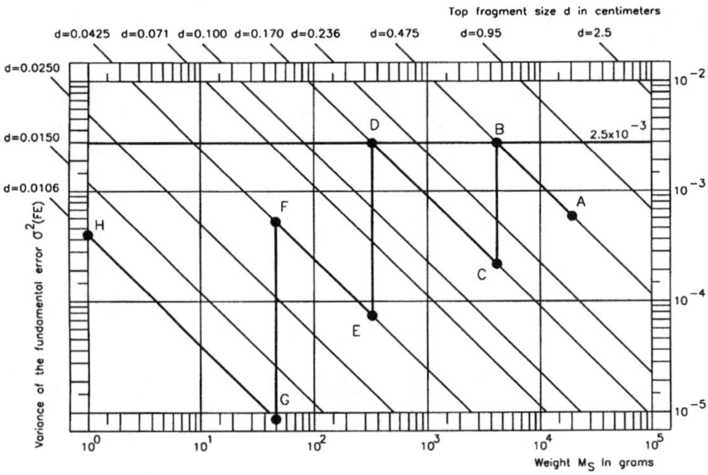

Figure 8-2. Sampling nomograph for management of the fundamental error *FE* during sampling and subsampling operations. (From *Pierre Gy's Sampling Theory and Sampling Practice: Heterogeneity, Sampling Correctness, and Statistical Process Control,* 2nd ed., by F.F. Pitard. Copyright © 1993 CRC Press, Boca Raton, FL. Courtesy of Francis Pitard.)

ing to the largest fragment diameter with slope −1. Each time a comminution takes place (points *B*, *D*, and *F*) the error is reduced to the level corresponding to the diameter of the largest particles.

Figure 8-2 also contains additional important information related to the management of the fundamental error *FE*. A horizontal line exists at a level equal to 2.5×10^{-3}. This is equal to a standard deviation $\sigma(FE)$ of ±5%. In order to keep the standard deviation of the fundamental error below ±5%, the weight of subsamples must not be less than the weight indicated by the intersection of the diameter line (slope −1) and the line equal to 2.5×10^{-3}. Note that points *B* and *D* are at this intersection; all other points lie below it.

Each subsampling stage introduces an error. Although this sampling operation was designed so that no single error exceeds ±5%, the contribution of all errors from the various subsampling stages can exceed ±5%. Pitard (1989) provides the following calculations:

1. Primary stage *AB*:

$$\sigma^2(FE)_{0.475\text{cm}} = \left(\frac{1}{4800} - \frac{1}{20{,}000}\right)12 = 1.9 \times 10^{-3} \tag{8.10}$$

$$\sigma^2(FE)_{0.475\text{cm}} = \pm 4.4\% \tag{8.11}$$

2. Secondary stage *CD*:

$$\sigma^2(FE)_{0.170\,\mathrm{cm}} = 2.3\times10^{-3} \tag{8.12}$$

$$\sigma^2(FE)_{0.170\mathrm{cm}} = \pm4.8\% \tag{8.13}$$

3. Tertiary stage *EF*:

$$\sigma^2(FE)_{0.425\,\mathrm{cm}} = 4.7\times10^{-4} \tag{8.14}$$

$$\sigma^2(FE)_{0.425\mathrm{cm}} = \pm2.2\% \tag{8.15}$$

4. Final stage *GH*:

$$\sigma^2(FE)_{0.0106\,\mathrm{cm}} = 4.1\times10^{-4} \tag{8.16}$$

$$\sigma^2(FE)_{0.0106\mathrm{cm}} = \pm2.0\% \tag{8.17}$$

This produces a total variance $\sigma^2(FE)$ of the fundamental error *FE* equal to

$$\sigma^2(FE)= (1.9\times10^{-3})+(2.3\times10^{-3})+(4.7\times10^{-4})+(4.1\times10^{-4}) \tag{8.18}$$

$$\sigma^2(FE)= 5.1\times10^{-3} \tag{8.19}$$

with

$$\sigma^2(FE)= \pm7.1\% \tag{8.20}$$

These error summaries support previous statements that the fundamental error *FE* never cancels out and only accumulates at each stage of subsampling.

Without proper planning and management, errors can quickly exceed 50%, 100%, 200%, and so forth. The overall sampling design must also leave room for any errors introduced by grouping and segregation. If not, unpredictable fluctuations are likely to result. The geostatistician will experience this random error in the form of the nugget effect on the variogram. Managing the variance of the fundamental error *FE* will also help assure that duplicate or split samples produce little deviation in laboratory analyses.

8.2 Grouping and Segregation Error *GE*

8.2.1 Minimizing the Grouping and Segregation Factors

The short-range heterogeneity fluctuation error CE_1 generated by h_1 is composed of the fundamental error *FE* and the grouping and segregation error *GE*. This can be seen in Figure 7-1 as part of the discrete model. Mathematically, it has been defined as

$$CE_1 = FE + GE \tag{8.21}$$

The variance of the grouping and segregation error *GE* is the product of three factors and can be written

$$s^2(GE) = s^2(CE_1) - s^2(FE) = \gamma_s \xi s^2(FE) \tag{8.22}$$

If one of these factors could cancel, $s^2(GE)$ could also cancel. However, this is not possible because:

- It is not possible to cancel the fundamental error variance $s^2(FE)$.
- To cancel the grouping factor γ_s, the fragments must be selected individually and at random. This is a practical impossibility.
- Cancellation of the segregation factor ξ requires a limit case and an unstable situation: The lot must reach a stage of natural distribution homogeneity in three dimensions.

It is still possible to minimize CE_1 in practice, however. The fundamental error variance $s^2(FE)$ requires that one calculate the sampling constants and then follow through by optimizing the sampling protocols. This was demonstrated in Section 8.1.2 with the sampling nomograph.

To minimize the grouping factor γ_s, the solution is to take as many small increments as possible. This approach assumes that the delimitation, extraction, and preparation of the increments will be performed correctly so as not to introduce additional error.

A tensional relationship exists between the size of the increment and the limits of delimitation, extraction, and preparation. The increment size cannot be reduced below a minimum weight in order to ensure correctness, thus restricting the ability to reduce the grouping factor.

A balance must be struck between trying to achieve a maximum number of the smallest increments possible while simultaneously respecting delimitation, extraction, and preparation. In general, a sample should be composed of at least 30 increments. Practical experience has also demonstrated that the grouping factor is by far the easiest factor to minimize in any endeavor to minimize the variance $s^2(GE)$.

Gravity imposes a severe limitation in any attempt to minimize the segregation factor ξ. It is always present and always working against efforts toward correct sampling. Because of this unidirectional force (and others), it is always easier to describe the conditions that promote segregation than those that minimize it. Among the latter are gravity, magnetic properties, electrostatic properties, differing moisture content, different angles of repose, adherence to collection or sampling devices, and adherence to specific chemicals (e.g., PCBs to oil).

Homogenization techniques are the only way to reduce the segregation factor ξ. These techniques are often difficult, expensive, and, in the end, ineffective. Ironically, some recommended homogenization procedures actually promote segregation. If achieved, three-dimensional homogenization is unstable and rapidly becomes two-dimensional homogeneity.

In summary, reduction of the variance $s^2(CE_1)$ can be achieved by:

1. Minimizing the variance of the fundamental error $s^2(FE)$
2. Minimizing the grouping factor γ_s by taking many small increments
3. Minimizing the segregation factor ξ by homogenization techniques

These steps have been listed in the order in which they should be performed. The priority is 1, then 2, then 3. Taking many increments is always possible and relatively easy to do, yet many fail to do it. People always attempt to homogenize, even though it provides the least success. It is ironic that easy-to-implement techniques that work are typically overlooked while difficult and ineffective methods are implemented.

8.2.2 Practical Aspects for Grouping and Segregation

There are many ways to implement efforts to minimize grouping and segregation. Based on the work of Pitard (1993), a few of these techniques, along with the pros and cons, will be discussed below:

Fractional Shoveling: Figure 8-3 shows true fractional shoveling with n = 5. Here the lot L is shoveled into five separate piles, with one randomly retained as the potential sample and the others going to reject. Fractional shoveling is highly effective in reducing the grouping factor as it permits a large number of increments to be collected.

Figure 8-4 shows a degenerate fractional shoveling method in which only two piles are created: one small pile for the potential sample and one large pile for the reject. The problem with this method is that it is not possible to maintain equiprobability conditions. The person performing the shoveling operation knows which scoops are reject and which are sample increment material. The shoveler may voluntarily or involuntarily alter the composition by selecting scoops richer in certain types of fragments (coarse or fine, dark or light, etc.). The method shown in Figure 8-3 is preferred, with the caution that the random selection of the pile must take place *after* the shoveling has been completed.

Coning and Quartering: Figure 8-5 shows the coning and quartering technique. The first step involves mixing and finally coning the material in the lot. Next, the cone is flattened to produce a circular pile and is then divided into four

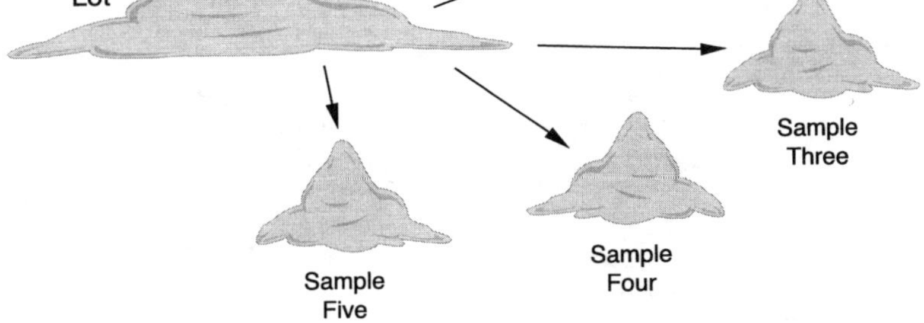

Figure 8-3. True fractional shoveling (equitable). (From *Pierre Gy's Sampling Theory and Sampling Practice: Heterogeneity, Sampling Correctness, and Statistical Process Control,* 2nd ed., by F.F. Pitard. Copyright © 1993 CRC Press, Boca Raton, FL. Courtesy of Francis Pitard.)

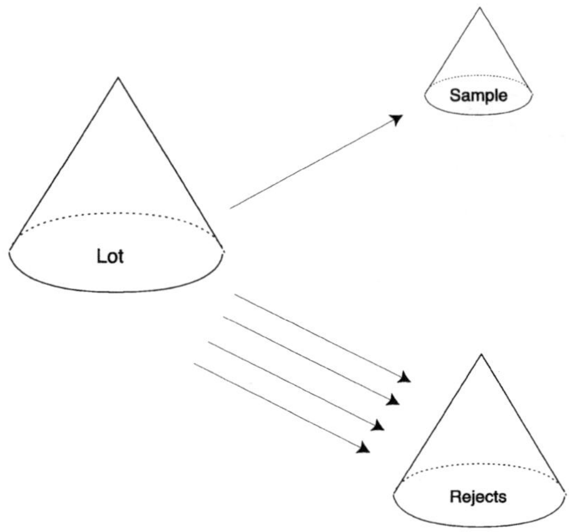

Figure 8-4. Degenerate fractional shoveling (not equitable). (From *Pierre Gy's Sampling Theory and Sampling Practice: Heterogeneity, Sampling Correctness, and Statistical Process Control,* 2nd ed., by F.F. Pitard. Copyright © 1993 CRC Press, Boca Raton, FL. Courtesy of Francis Pitard.)

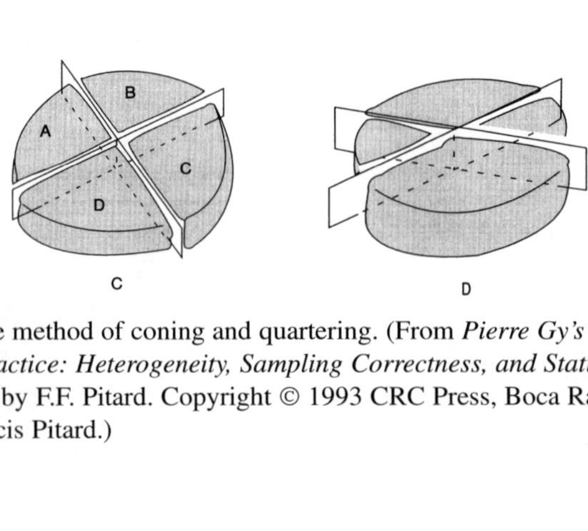

Figure 8-5. The method of coning and quartering. (From *Pierre Gy's Sampling Theory and Sampling Practice: Heterogeneity, Sampling Correctness, and Statistical Process Control,* 2nd ed., by F.F. Pitard. Copyright © 1993 CRC Press, Boca Raton, FL. Courtesy of Francis Pitard.)

quarters. Finally, two quarters are selected as a potential sample, with the other two going to reject. Coning and quartering does not reduce the grouping factor significantly as only two potential increments are available.

Alternate Shoveling: A preferred alternative to the coning and quartering approach is the alternative shoveling method shown in Figure 8-6. This approach is more effective in reducing the variance $s^2(GE)$ as it produces many more increments (which should be at least nine). Here, alternate shovel scoops are placed in a sample pile and a reject pile. The sample and reject piles must be randomly selected after the shoveling is completed.

Riffling: A riffle splitter is another way to achieve large numbers of increments. Figure 8-7 shows how a riffle splitter works. The lot material is poured from a pan into the top of the splitter. A series of chutes alternately directs the material into one of two collecting buckets. One bucket is selected at random to be part of the sample.

Sectorial Splitter: Figure 8-8 shows an example of a sectorial splitter. Material is fed into the top. The top is a revolving feeder that sends material to different jars. This method has an advantage in that the grouping factor is negligible because of the large number of increments. However, the material may be stratified, causing a potentially large segregation.

Vibratory Spatula: Sometimes, a device known as a vibratory spatula is used at the laboratory to feed material (Fig. 8-9). The vibratory spatula is a perfect

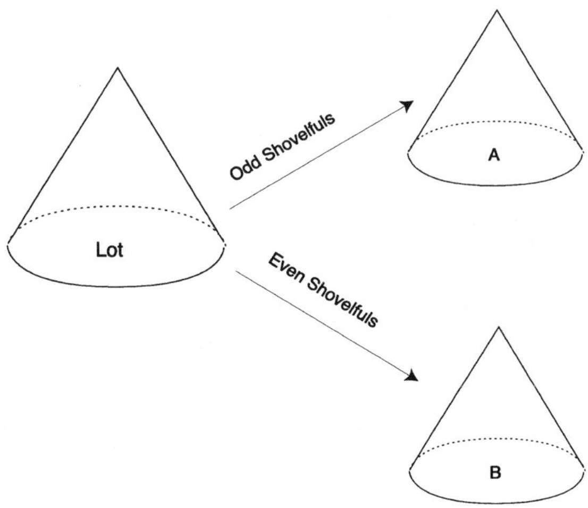

Figure 8-6. Alternative shoveling method. (From *Pierre Gy's Sampling Theory and Sampling Practice: Heterogeneity, Sampling Correctness, and Statistical Process Control,* 2nd ed., by F.F. Pitard. Copyright © 1993 CRC Press, Boca Raton, FL. Courtesy of Francis Pitard.)

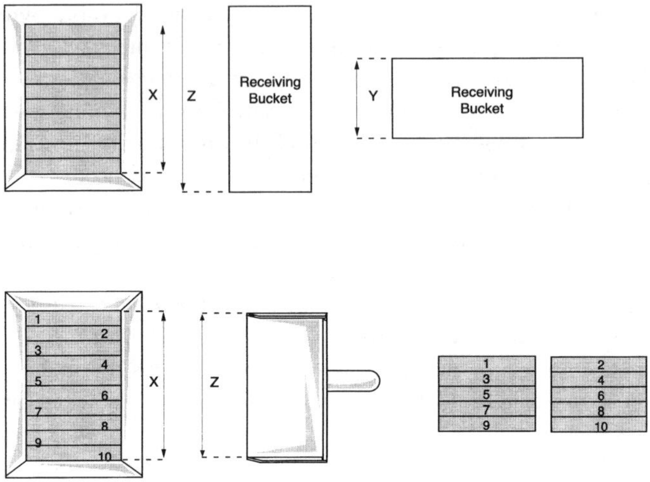

Figure 8-7. Riffle splitter for subsampling operations. (From *Pierre Gy's Sampling Theory and Sampling Practice: Heterogeneity, Sampling Correctness, and Statistical Process Control,* 2nd ed., by F.F. Pitard. Copyright © 1993 CRC Press, Boca Raton, FL. Courtesy of Francis Pitard.)

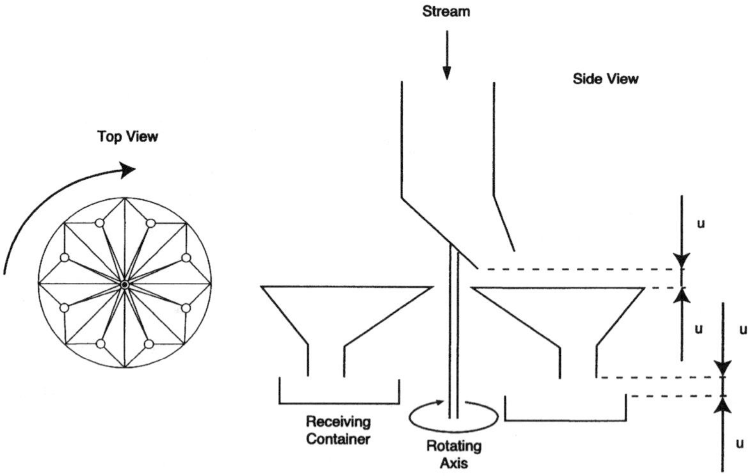

Figure 8-8. Sectorial splitter. (From *Pierre Gy's Sampling Theory and Sampling Practice: Heterogeneity, Sampling Correctness, and Statistical Process Control,* 2nd ed., by F.F. Pitard. Copyright © 1993 CRC Press, Boca Raton, FL. Courtesy of Francis Pitard.)

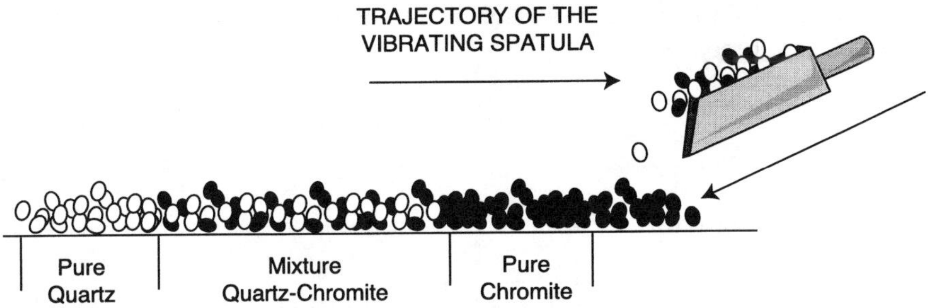

Figure 8-9. Vibratory spatula. (From *Pierre Gy's Sampling Theory and Sampling Practice: Heterogeneity, Sampling Correctness, and Statistical Process Control,* 2nd ed., by F.F. Pitard. Copyright © 1993 CRC Press, Boca Raton, FL. Courtesy of Francis Pitard.)

segregating machine and results can be very bad. Use of this type of device is strongly discouraged.

V-blenders: These devices (Fig. 8-10) are often used in an attempt to homogenize material. If the material is left in the blender for a sufficient length of time (say, 10 minutes), it is possible to achieve a near-homogeneous state. However, it segregates in 1 sec when the material is discharged. A sample should therefore be taken before discharging. Note that the blender should not be filled

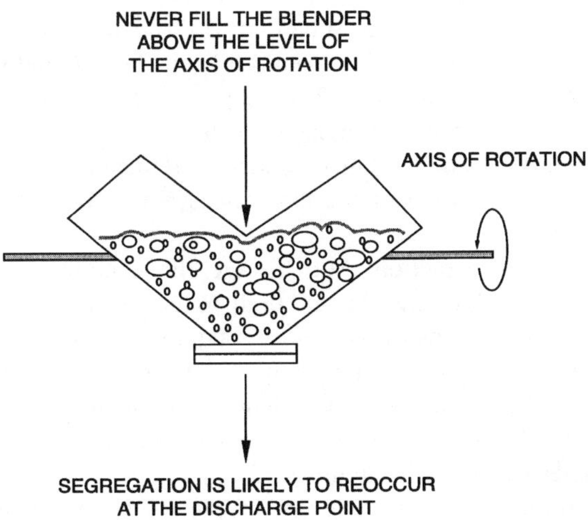

Figure 8-10. V-blender. (From *Pierre Gy's Sampling Theory and Sampling Practice: Heterogeneity, Sampling Correctness, and Statistical Process Control,* 2nd ed., by F.F. Pitard. Copyright © 1993 CRC Press, Boca Raton, FL. Courtesy of Francis Pitard.)

above the axis of rotation because homogenization time increases exponentially as the blender becomes fuller.

8.3 Materialization Error *ME*

While often very large, the errors introduced by the heterogeneity of materials are typically random errors. In contrast, materialization errors are notoriously biased and are probably the major cause of all biases in sampling. The concept of equiprobability is crucial to correct sampling for, without it, biases are always introduced. Management of the materialization errors (increment delimitation error *DE* and increment extraction error *EE*) must be achieved to minimize or prevent biases. This is possible because sampling correctness of the sampling device is a primary structural property. This puts the responsibility on the sampler to exercise control over the *ME* and the associated biases.

The increment sampling process consists of elementary and independent sequential steps:

1. The sampling point selection: All points in the lot must be available for sampling under equiprobability conditions. If this condition is not met, sampling is always biased to an unknown extent. Equiprobability conditions can be met by using a selection scheme that is (a) systematic with a random starting point (e.g., square grid), (b) a random stratified approach (e.g., random location within each grid cell), or (c) random.
2. The increment delimitation: The shape and size of the sampling device delimits the geometric boundaries of the increment called the *extended increment*. An example of an extended increment is shown in Figure 8-11. The particulate nature of the material is not a concern of the extended increment.
3. The increment extraction: At this stage, the sampling tool extracts material—soil, liquid, or gas. The particulate nature of the material is now an issue and must follow the center of gravity rule. For any type of sampling device, the fragment must be taken if the center of gravity of the fragment lies within the boundary of the extended increment. An example is shown in Figure 8-12. The fragmental increment must be consistent with the set of particles as shown in Figure 8-13.
4. The increment reunion: The set of punctual increments is called the *punctual sample*. The set of extended increments is called the *extended sample*. The set of fragmental increments is called the *fragmental sample*. This sequential relationship is shown in Figure 8-14.

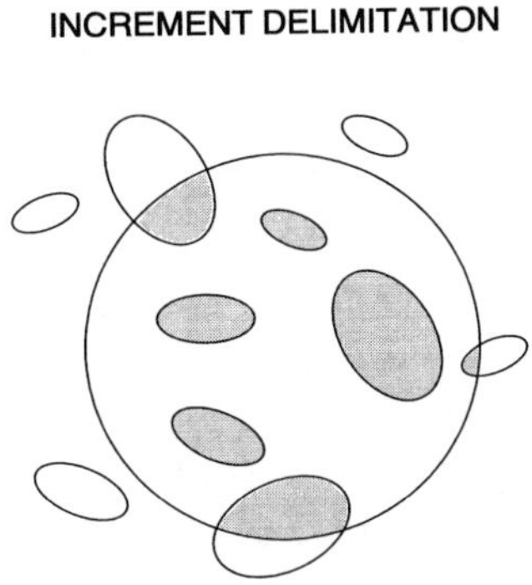

Figure 8-11. Depiction of the extended increment. (From *Pierre Gy's Sampling Theory and Sampling Practice: Heterogeneity, Sampling Correctness, and Statistical Process Control,* 2nd ed., by F.F. Pitard. Copyright © 1993 CRC Press, Boca Raton, FL. Courtesy of Francis Pitard.)

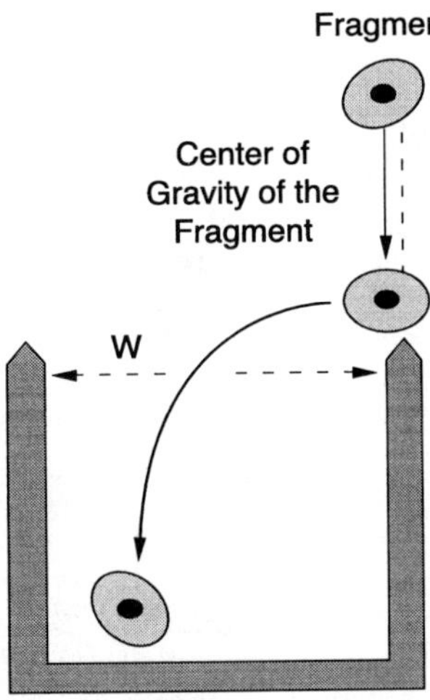

Figure 8-12. Depiction of the center of gravity rule. (From *Pierre Gy's Sampling Theory and Sampling Practice: Heterogeneity, Sampling Correctness, and Statistical Process Control,* 2nd ed., by F.F. Pitard. Copyright © 1993 CRC Press, Boca Raton, FL. Courtesy of Francis Pitard.)

Correct materialization depends on two conditions, both related to equiprobability. First, the increment must coincide with the defined dimensional module of observation. This can be stated as equiprobability in all directions. Second, the fragments that have their centers of gravity within the correct theoretical boundaries must be part of the extracted increment. There can be nothing more and nothing less. This is termed *equiprobability on impact.*

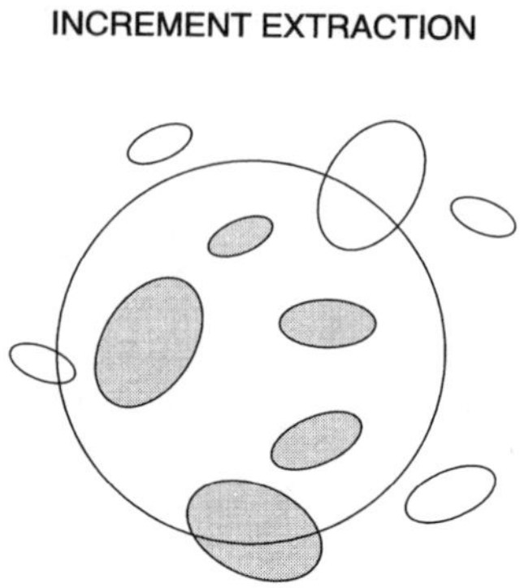

Figure 8-13. Illustration of the fragmental increment. (From *Pierre Gy's Sampling Theory and Sampling Practice: Heterogeneity, Sampling Correctness, and Statistical Process Control,* 2nd ed., by F.F. Pitard. Copyright © 1993 CRC Press, Boca Raton, FL. Courtesy of Francis Pitard.)

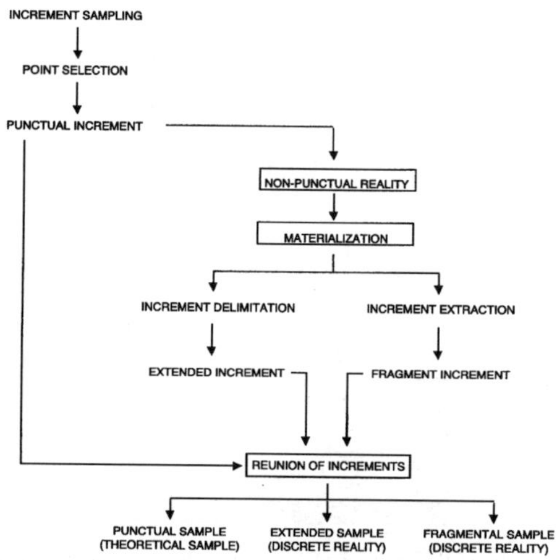

Figure 8-14. Flowchart of increment reunion (materialization process). (From *Pierre Gy's Sampling Theory and Sampling Practice: Heterogeneity, Sampling Correctness, and Statistical Process Control,* 2nd ed., by F.F. Pitard. Copyright © 1993 CRC Press, Boca Raton, FL. Courtesy of Francis Pitard.)

For a three-dimensional lot, the isotropic module is a sphere represented by its two-dimensional analogue in Figures 8-11 and 8-13. For practical purposes, the collection of correct samples from three-dimensional lots is not possible. This is because access to the selected point is difficult and the sampling device will destroy or alter the material. Acceptable three-dimensional sampling is possible in liquids since the distribution heterogeneity may not be significant at the scale normally considered.

For a two-dimensional lot, the isotropic module is a circle whose three-dimensional approximation is a cylinder with a constant cross section (Fig. 8-15). To maintain equiprobability conditions, it is essential to take the increment all the way through the thickness of the lot.

Two-dimensional lots are consolidated flat soil units or three-dimensional soil units that can be subdivided into an imaginary series of two-dimensional strata. This subdivision is a standard approach in three-dimensional geostatistical appraisal.

Two-dimensional lots are frequently sampled by means of coring or augering devices. Figure 8-16 shows the correct delimitation of a solid core sample (highly fractured solids, gravels, sand, etc.). The appropriate material for extraction falls within the lines X_1 and X_2. This is consistent with the general center of gravity rule with an example for fragments to be taken by cores shown in Figure 8-17.

Figure 8-18 illustrates a situation that may introduce bias even when the delimitation is correct. A fragment that is large in comparison to the diameter of the drill core can prevent appropriate large fragments from entering the core. Instead, finer fragments are preferentially selected, causing a bias.

Figure 8-19 shows the correct delimitation for a soft core sample (thick muds, sediments, clays, etc.). The appropriate material for extraction falls this time

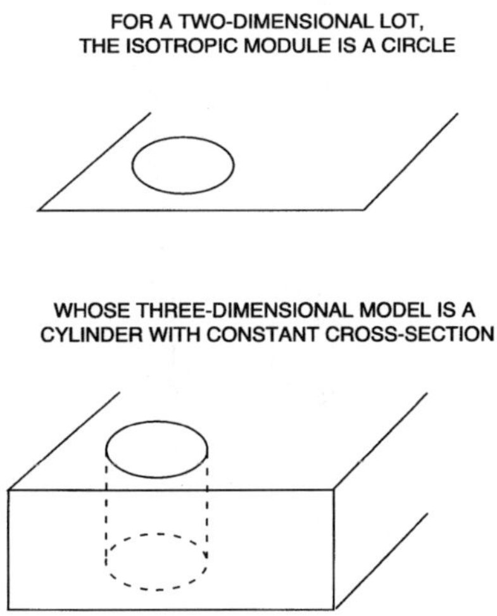

Figure 8-15. Two-dimensional isotropic module of observation. (From *Pierre Gy's Sampling Theory and Sampling Practice: Heterogeneity, Sampling Correctness, and Statistical Process Control,* 2nd ed., by F.F. Pitard. Copyright © 1993 CRC Press, Boca Raton, FL. Courtesy of Francis Pitard.)

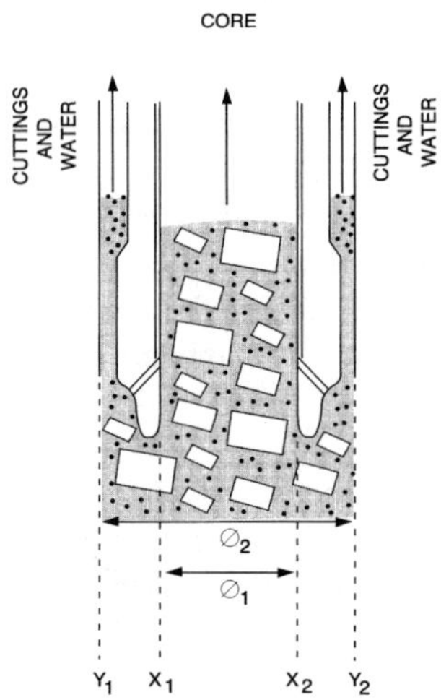

Figure 8-16. Delimitation of a solid-core sample. (From *Pierre Gy's Sampling Theory and Sampling Practice: Heterogeneity, Sampling Correctness, and Statistical Process Control,* 2nd ed., by F.F. Pitard. Copyright © 1993 CRC Press, Boca Raton, FL. Courtesy of Francis Pitard.)

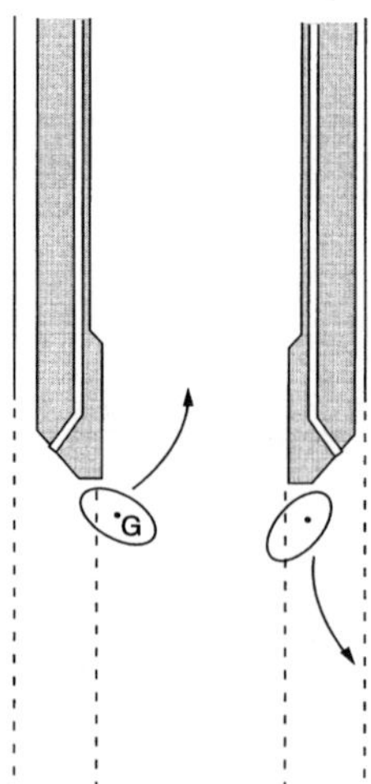

Figure 8-17. Center of gravity rule for fragments taken by coring where extraction is correct. (From *Pierre Gy's Sampling Theory and Sampling Practice: Heterogeneity, Sampling Correctness, and Statistical Process Control,* 2nd ed., by F.F. Pitard. Copyright © 1993 CRC Press, Boca Raton, FL. Courtesy of Francis Pitard.)

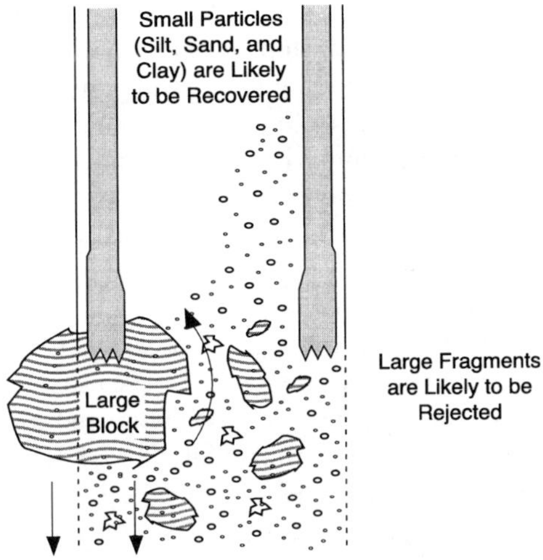

Figure 8-18. Extraction error *EE* resulting from a partial obstruction of the coring bit, causing loss of material that should be included in the increment. (From *Pierre Gy's Sampling Theory and Sampling Practice: Heterogeneity, Sampling Correctness, and Statistical Process Control,* 2nd ed., by F.F. Pitard. Copyright © 1993 CRC Press, Boca Raton, FL. Courtesy of Francis Pitard.)

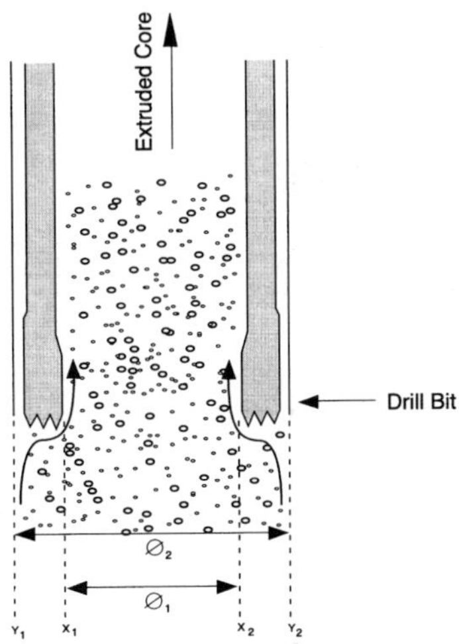

Figure 8-19. Delimitation of a soft-core sample. (From *Pierre Gy's Sampling Theory and Sampling Practice: Heterogeneity, Sampling Correctness, and Statistical Process Control,* 2nd ed., by F.F. Pitard. Copyright © 1993 CRC Press, Boca Raton, FL. Courtesy of Francis Pitard.)

within the lines Y_1 and Y_1. While appearing to obtain more than 100% recovery, techniques such as drilling with tungsten carbide bits produce a strong and tight contact between the wall and the tubing. The result is that the only place for the material between Y_1 and X_1, and between Y_2 and X_2 to go is into the core.

It is also important to use appropriate delimitation techniques when logging core samples. A core sample may penetrate and contain various materials such as fractured material, alternating gravels, silt, clay, sand, and so forth. As such, full core recovery is not often achieved. Figure 8-20 shows the correct and incorrect delimitation of core samples with partial recovery.

Although drilling techniques offer good opportunities to ensure sampling correctness, some other commonly employed devices for two-dimensional sampling present problems. These include augers, thiefs, and triers (Fig. 8-21). Like coring devices, the cross section is uniform circular. However, the conical head prevents correct delimitation and extraction at the bottom of the lot, either failing to extract the entire length or taking material that is outside the domain of the lot. As such, these devices cannot be relied on for correct sampling.

Sampling of a three-dimensional pile presents a great problem and has been termed "a perfect exercise in futility" (Pitard 1989). Correct delimitation is not possible because of constantly changing variables. Even if the pile is flattened, the problems in extraction at the bottom of the pile, as just discussed, still exist. Thus, one must expect imperfect results when sampling piles of mine tailings, fly ash, and so forth. Examples of increment delimitation and increment extraction for three-dimensional piles are shown in Figures 8-22 and 8-23.

Two-dimensional piles are also problematic. At the laboratory level, a three-dimensional pile may be reduced to a two-dimensional pile of constant thickness.

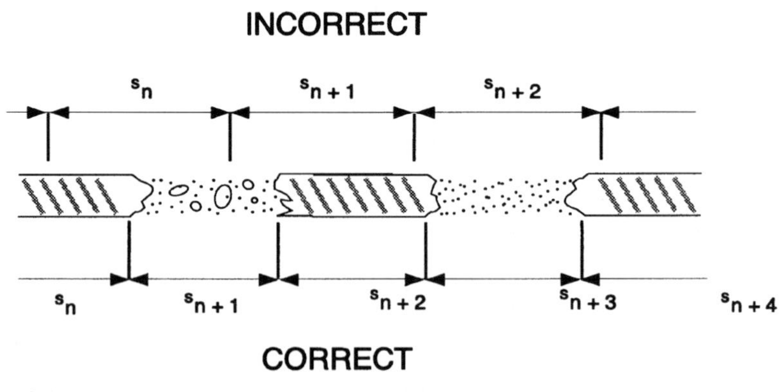

Figure 8-20. Correct and incorrect methods for delimitation of core samples. (From *Pierre Gy's Sampling Theory and Sampling Practice: Heterogeneity, Sampling Correctness, and Statistical Process Control,* 2nd ed., by F.F. Pitard. Copyright © 1993 CRC Press, Boca Raton, FL. Courtesy of Francis Pitard.)

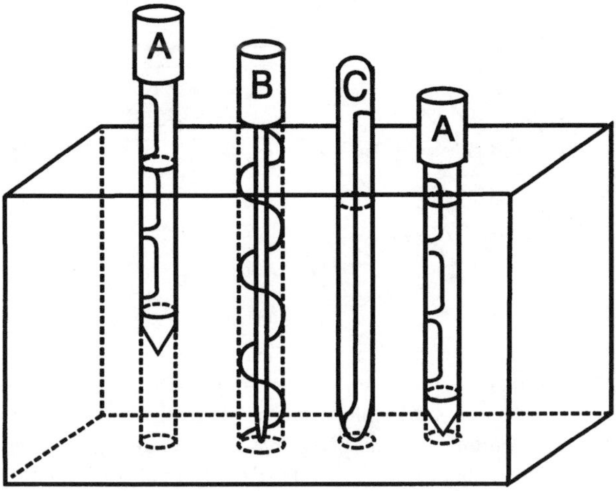

Figure 8-21. Augers, thiefs, and triers. (From *Pierre Gy's Sampling Theory and Sampling Practice: Heterogeneity, Sampling Correctness, and Statistical Process Control,* 2nd ed., by F.F. Pitard. Copyright © 1993 CRC Press, Boca Raton, FL. Courtesy of Francis Pitard.)

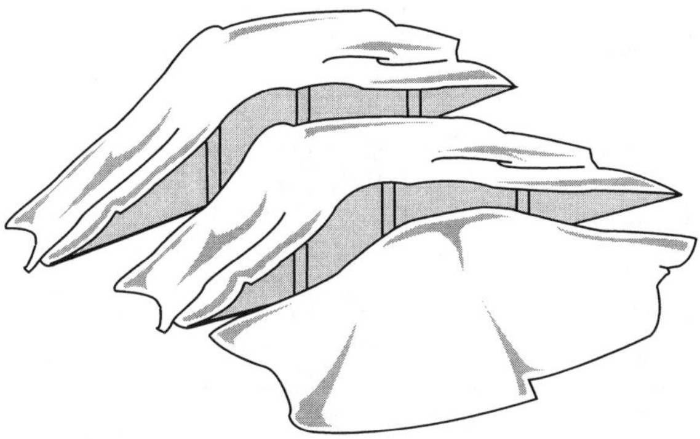

Figure 8-22. Direct sampling of a coned or ridged three-dimensional pile is virtually impossible because the increment delimitation is almost never correct. (From *Pierre Gy's Sampling Theory and Sampling Practice: Heterogeneity, Sampling Correctness, and Statistical Process Control,* 2nd ed., by F.F. Pitard. Copyright © 1993 CRC Press, Boca Raton, FL. Courtesy of Francis Pitard.)

This is called the Japanese slab-cake method. It is used primarily for large bulk materials but may be employed for small lots.

Figure 8-24 shows an example of a Japanese slab-cake operation. The lot is spread into a rectangular pile and divided into a matrix of squares. A scoop of material is taken from the center of each square or alternate squares to make the

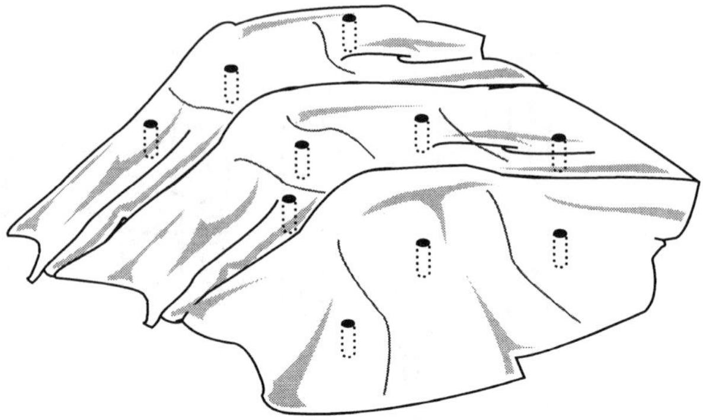

Figure 8-23. Correct increment extraction of a three-dimensional pile is nearly unachievable because of incorrect sampling devices (Fig. 8-21). (From *Pierre Gy's Sampling Theory and Sampling Practice: Heterogeneity, Sampling Correctness, and Statistical Process Control,* 2nd ed., by F.F. Pitard. Copyright © 1993 CRC Press, Boca Raton, FL. Courtesy of Francis Pitard.)

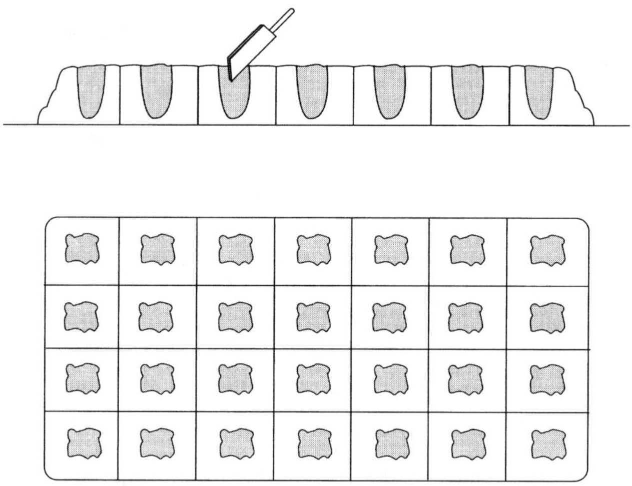

Figure 8-24. Japanese slab cakes. This two-dimensional method is almost never correct. (From *Pierre Gy's Sampling Theory and Sampling Practice: Heterogeneity, Sampling Correctness, and Statistical Process Control,* 2nd ed., by F.F. Pitard. Copyright © 1993 CRC Press, Boca Raton, FL. Courtesy of Francis Pitard.)

increments, with the increments composited to make the sample. The problem can be seen in the top portion of Figure 8-24. The sampling tool will preferentially collect material from the top of the lot while failing to extract material from the bottom. This introduces a delimitation error and bias.

As can be seen, it is very difficult or impossible to perform correct materialization on two- and three-dimensional lots. Whenever possible, it is recommended that the problem be reduced to one dimension, which is essentially an elongated pile of material or stream.

A one-dimensional lot is a cross section defined by two parallel planes (Fig. 8-25). Note again the shading of the fragments for delimitation and extraction according to the center of gravity rule.

Elongated one-dimensional piles can be created effectively at the laboratory. The pile should be composed of several layers, the more layers the better. A few random increments may then be taken across the pile. The width and height of the pile should remain constant and may be achieved by the use of a receiving frame.

It is important to maintain correct delimitation of the sampling tool. The shovel, scoop, or spatula should have vertical walls so that the entire increment is available for extraction. Figure 8-26 shows the incorrect delimitation produced by a scoop with a round section vs. a scoop with a square section (vertical walls). The shaded area for the round section represents the material that belongs in the increment but is systematically rejected as a result of incorrect delimitation.

Figure 8-27 shows the situation produced by using a flat spatula for increment extraction. This time, the reject is on the top of the pile and simply falls back into the pile as the spatula is lifted. Figure 8-28 provides several examples of both correct and incorrect sampling devices.

The two-dimensional Japanese slab-cake method may be transformed into a one-dimensional approach. The lot may simply be shaped into an elongated pile

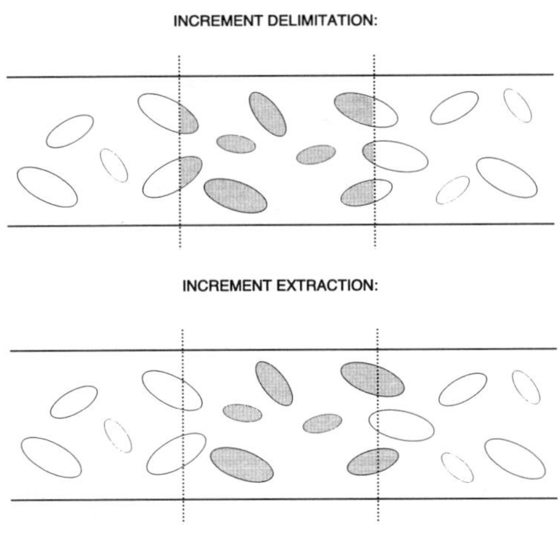

Figure 8-25. Extended increment and fragmental increment for one-dimensional lots. (From *Pierre Gy's Sampling Theory and Sampling Practice: Heterogeneity, Sampling Correctness, and Statistical Process Control,* 2nd ed., by F.F. Pitard. Copyright © 1993 CRC Press, Boca Raton, FL. Courtesy of Francis Pitard.)

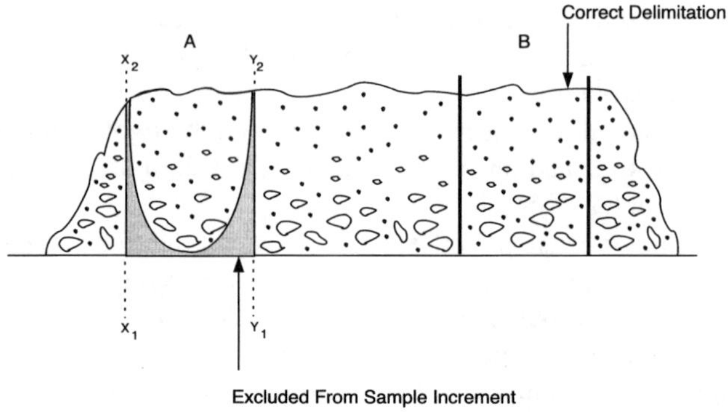

Figure 8-26. Incorrect delimitation produced by a rounded scoop vs. correct delimitation by a scoop with a square cross section. (From *Pierre Gy's Sampling Theory and Sampling Practice: Heterogeneity, Sampling Correctness, and Statistical Process Control,* 2nd ed., by F.F. Pitard. Copyright © 1993 CRC Press, Boca Raton, FL. Courtesy of Francis Pitard.)

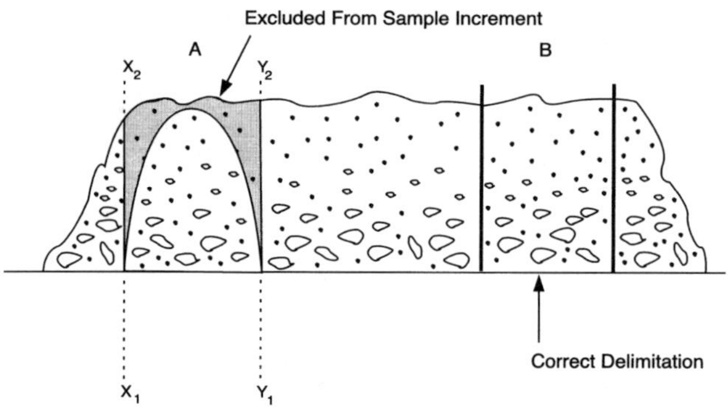

Figure 8-27. Incorrect delimitation produced by a spatula with no side walls vs. correct delimitation produced by a scoop with vertical side walls. (From *Pierre Gy's Sampling Theory and Sampling Practice: Heterogeneity, Sampling Correctness, and Statistical Process Control,* 2nd ed., by F.F. Pitard. Copyright © 1993 CRC Press, Boca Raton, FL. Courtesy of Francis Pitard.)

instead of a two-dimensional rectangular pile. Use of a correctly designed sampling tool for transversal cuts will produce appropriate increments.

Pitard (1993) recommends the following procedure for the one-dimensional Japanese slab-cake technique:

- The bag or vial containing the sample material should be emptied above a smooth and clean surface or appropriate material.

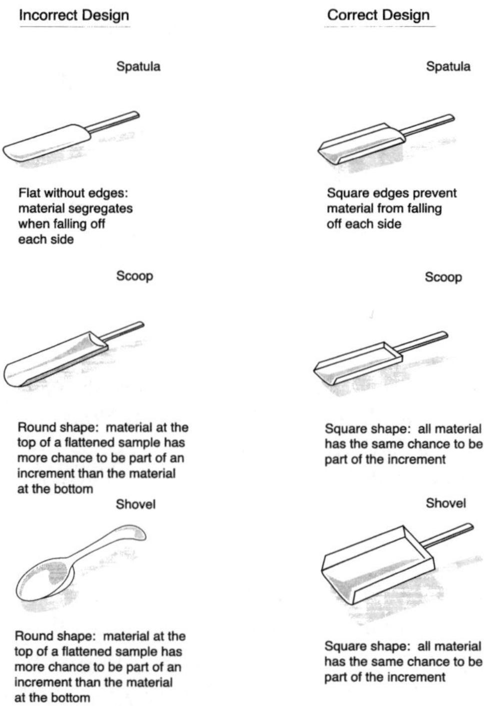

Figure 8-28. Correct and incorrect sampling devices. (From *Pierre Gy's Sampling Theory and Sampling Practice: Heterogeneity, Sampling Correctness, and Statistical Process Control,* 2nd ed., by F.F. Pitard. Copyright © 1993 CRC Press, Boca Raton, FL. Courtesy of Francis Pitard.)

- Do not try to homogenize the sample; the homogenization of heterogeneous material is often wishful thinking and may promote segregation.
- Reduce the sample using a fractional shoveling method until a sample 5 to 10 times larger than the appropriate analytical subsample is obtained.
- Shape the remaining material into a uniform elongate pile with uniform section.
- Take increments all across the one-dimensional pile and from the entire thickness as described in Figure 8-29.
- If necessary, reshape the pile, this time perpendicular to the X-X' axis shown in Figure 8-29.
- Continue to take increments all across the pile and from the entire thickness until the appropriate sample weight is reached.
- Ideally, the sample should be composed of at least 25 increments but may vary with the purpose of the sampling operation.

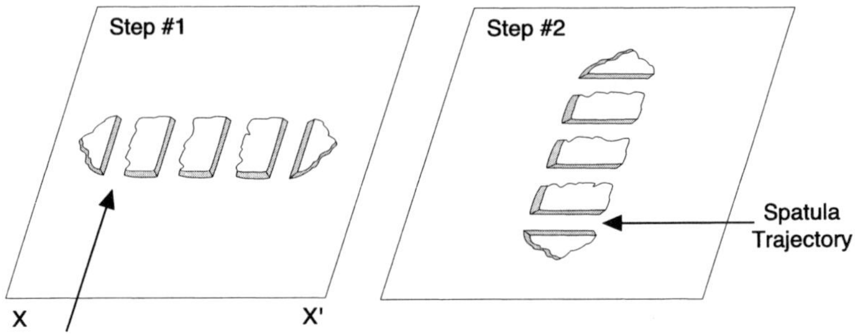

Figure 8-29. Correct delimitation through the use of one-dimensional Japanese slab cakes. (From *Pierre Gy's Sampling Theory and Sampling Practice: Heterogeneity, Sampling Correctness, and Statistical Process Control,* 2nd ed., by F.F. Pitard. Copyright © 1993 CRC Press, Boca Raton, FL. Courtesy of Francis Pitard.)

While not a perfect solution, this procedure is highly superior to that often encountered in practice: An analyst scoops material directly out of the bag or vial, taking material only from the top and placing it on the balance pan until the appropriate weight is achieved and, if too much has been added, removing additional material from only the top of the material in the balance. This procedure is guaranteed to produce a sample bias.

Extraction correctness must also be maintained in the splitting process. One reliable way of splitting sample material is to use a riffle splitter, as already discussed. However, a few precautions will be noted and reviewed.

Although a riffle splitter is a correct device, improper use can introduce an extraction bias. First, the material must be evenly distributed in the scoop. Next, it is important to feed the material slowly in the middle of the splitter, where all chutes have the same depth. Careless feeding can produce unequal distributions of material in the collecting bucket (Fig. 8-30). Feeding too fast can cause material to overflow into a different set of chutes, causing a bias, especially toward large-sized fractions.

Also, a scoop that is wider than the riffle splitter is occasionally used to feed the splitter (Fig. 8-31). This is inappropriate because unequal amounts of material will be distributed to the end chutes.

Another guideline pertains to the correct width of the riffles. Riffles that are too narrow can cause bridging by large particles. The following guideline should be used to determine the correct riffle width:

$$W \geq W_o = 2d + 5\,\text{mm} \tag{8.23}$$

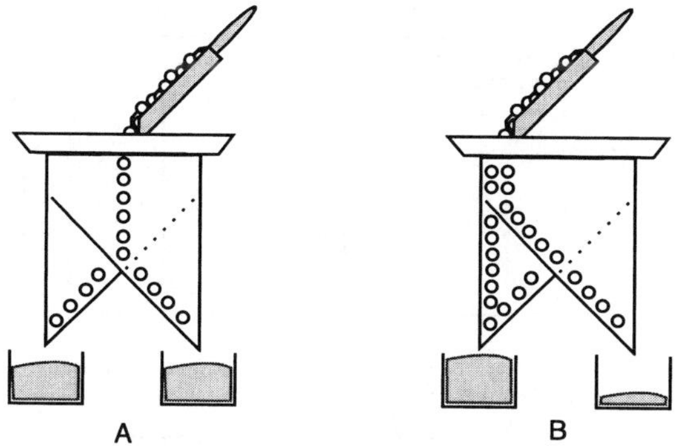

Figure 8-30. (a) Correct use of a riffle splitter, where the device is fed slowly in the middle; (b) incorrect usage, where it is quickly fed on one side. (From *Pierre Gy's Sampling Theory and Sampling Practice: Heterogeneity, Sampling Correctness, and Statistical Process Control,* 2nd ed., by F.F. Pitard. Copyright © 1993 CRC Press, Boca Raton, FL. Courtesy of Francis Pitard.)

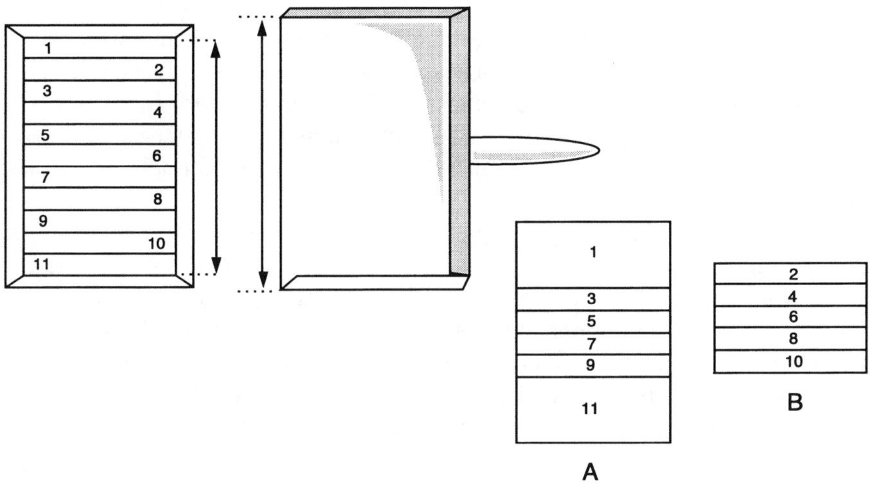

Figure 8-31. Illustration of the incorrect use of a riffle splitter where the feeding scoop is wider than the riffle chutes. (From *Pierre Gy's Sampling Theory and Sampling Practice: Heterogeneity, Sampling Correctness, and Statistical Process Control,* 2nd ed., by F.F. Pitard. Copyright © 1993 CRC Press, Boca Raton, FL. Courtesy of Francis Pitard.)

One problem that occurs is feeding the riffle splitter in the wrong direction (Fig. 8-32). Generally this is not a problem as the scoop is significantly larger than the splitter width. A more common occurrence is that an operator pours a bag directly into the riffle splitter. This practice will introduce a bias and should always be avoided.

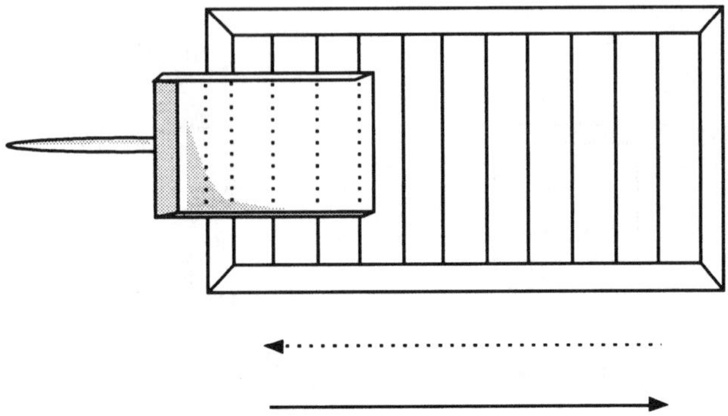

Figure 8-32. Illustration of the incorrect use of a riffle splitter, where the device is fed in the wrong direction. (From *Pierre Gy's Sampling Theory and Sampling Practice: Heterogeneity, Sampling Correctness, and Statistical Process Control,* 2nd ed., by F.F. Pitard. Copyright © 1993 CRC Press, Boca Raton, FL. Courtesy of Francis Pitard.)

Unfortunately, most of the sampling devices currently available for environmental sampling are incorrect. Examples already discussed are augers, thiefs, and triers. Others include devices to sample sediments that disturb the material or selectively suction more fine material. Depth sampling of water-using devices composed of a bottle or container also produces both incorrect delimitation and incorrect extraction.

It is known that incorrect sampling tools cannot ensure accurate sampling. Often, sampling accuracy is claimed because some form of randomness has been incorporated into the sampling design. While this is a necessary condition, it is not sufficient. Sampling must also be correct. An incorrect device may produce results that are accurate one day, biased high the next, and biased low on the third. What is both random and unknown is the direction and the bias from one day to the next. As such, these devices are unreliable, inaccurate, and inequitable.

It is often claimed that correct sampling implies unjustifiable added time and expense. This is a myth that needs to be dispelled. A good example is provided by the Coliwasa and Kemmerer depth samplers. These devices are neither complex nor expensive.

Figure 8-33 shows an illustration of a Coliwasa sampler and Figure 8-34 shows a Kemmerer depth sampler. In each case, a glass, plastic, or metal tube outfitted with an end closure device characterizes the sampler. The end closure can be opened or closed at a preselected depth under water. A continuous column of liquid may be incrementally sampled with the liquid representing the entire thickness of the two-dimensional liquid.

Provided that these devices are allowed to enter the liquid slowly, they are

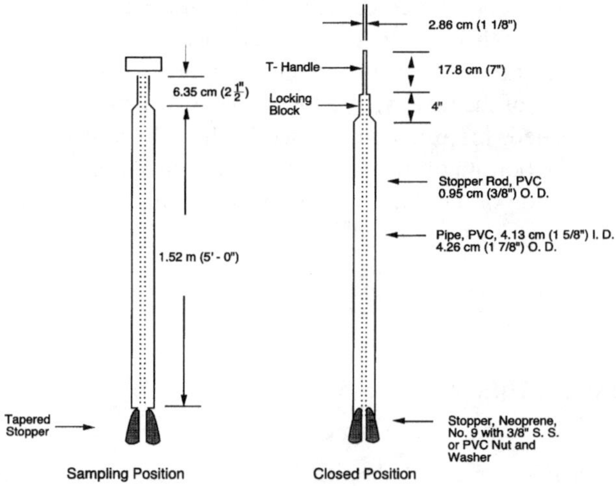

Figure 8-33. Composite liquid waste sampler (Coliwasa). (From *Pierre Gy's Sampling Theory and Sampling Practice: Heterogeneity, Sampling Correctness, and Statistical Process Control,* 2nd ed., by F.F. Pitard. Copyright © 1993 CRC Press, Boca Raton, FL. Courtesy of Francis Pitard.)

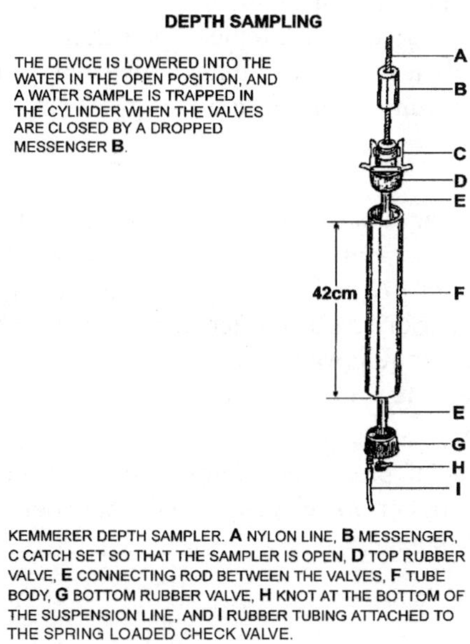

Figure 8-34. Kemmerer depth sampler. (From *Pierre Gy's Sampling Theory and Sampling Practice: Heterogeneity, Sampling Correctness, and Statistical Process Control,* 2nd ed., by F.F. Pitard. Copyright © 1993 CRC Press, Boca Raton, FL. Courtesy of Francis Pitard.)

correct. As with most correct devices (the riffle splitter just discussed provides a good example), the depth samplers can be used incorrectly.

For example, if the Coliwasa tube is inserted into a drum of liquid waste too quickly, the fill level of the tube will be lower than the surface of the liquid. When insertion is completed, the remaining space in the tube will be filled by taking liquid from near the bottom of the tube. Obviously, this procedure rejects material from the upper portion of the liquid and produces a high bias to material from the lower portion.

8.4 Preparation Error

Preparation error *PE*, while still classified as one of the seven sampling errors, is the only error that is nonselective. The errors described thus far are selective, and it is important to distinguish between the two kinds of errors. Preparation stages render the material into an appropriate form and an appropriate location for analytical operations.

While preparation errors are very important and can significantly alter the sample (typically introducing a bias), the discussion in this section will be brief. Much attention has been given to the topic of preparation in the literature and by USEPA. Therefore, only a summary of the concepts will be presented.

Preparation stages can be placed in a variety of categories. Some of these include:

- Comminution stages, such as grinding, crushing, and pulverizing. These operations assist in reducing the particle size *d*, contributing to a lower fundamental error *FE*, and increasing the number of particles.
- Screening (wet or dry), which often accompanies comminution.
- Drying of solids to remove moisture.
- Filtration to separate solids from liquids.

Several categories of preparation errors will be discussed. These include contamination errors, loss errors, change in chemical composition, unintentional mistakes, fraud, and sabotage.

8.4.1 Contamination Errors

It is nearly impossible to eliminate all forms of contamination. Contamination can, however, be minimized by proper procedures. In addition, some contaminants are not as detrimental as others, depending on the purpose of the sampling. Knowing which contaminants are detrimental requires skilled operators and specialists.

Contamination by dust is a problem for those working with fine and dry particles. Free falls of material should be minimized, and hoods with slight negative pressure should be employed. Decontamination of equipment used in previous operations is also important. Operations involving abrasion, including grinding, pulverizing, crushing, and, even to some extent, screening, can introduce unwanted materials.

8.4.2 Loss Errors

These errors often occur during crushing and pulverizing operations and are exaggerated by dedusting systems using strong vacuums. Loss errors are often significant for volatile organic compounds when improper collection, storage, and handling procedures are not employed.

8.4.3 Alteration of Chemical Composition

For environmental sampling, this is often related to loss error. Improper field handling and implementation of preservation techniques, especially for volatiles, can promote degradation of the sample material.

8.4.4 Unintentional Mistakes

Even though sampling protocols are often detailed in writing, the possibility still exists for inappropriate human interpretation of the intentions. This can lead to unintentional errors on the part of the operator. This is, of course, exacerbated by vague and ambiguous sampling protocols that are all too common. Other mistakes arise simply from carelessness or awkwardness.

Some common types of unintentional mistakes include dropping samples and recovering them incompletely, mixing fractions belonging to different samples, mixing labels, poor maintenance of sampling equipment, contamination, and losses. This category of unintentional mistakes again points out the advantage of having skilled operators and supervisors.

8.4.5 Sabotage and Fraud

Because sampling is a mystery to many people, opportunities for fraud exist. Incorrect but plausible sampling can be used to voluntarily or involuntarily alter sample results. This tactic can be used by industry to demonstrate that contamination is not a problem or by regulatory agencies to condemn a site that actually is below regulatory limits.

Closely related are the errors, often in the unintentional category, that occur in the balance room where a sample instantly becomes a specimen via materialization errors. The knowledge exists to produce equitable samples; one can only appeal to samplers to act in a responsible manner.

Conclusion

There are multiple opportunities to introduce sampling errors during the sampling and subsampling process. These errors are most frequently introduced unintentionally because of a lack of knowledge on the part of the sampler or laboratory technician. This does not diminish their effect, however, and can lead to samples that are nowhere near representative of the lot. Fortunately, each type of error has identifiable causes and solutions that can be implemented to minimize and manage the overall sampling error OE. Management of these errors at the sampling stage is important as these sampling errors are carried forward to the statistical analysis and geostatistical appraisal (GA) phases of GEM. Ultimately, these samples may represent areas or volumes on a map that need to be scheduled for, or pardoned from, remedial activities. It has been shown that sampling correctness does not necessarily imply immense cost; in fact, the opposite is sometimes true. The cost of remedial activities is generally far beyond that of sampling. Therefore, one should carefully weigh the cost of correct sampling against the cost of incorrect decisions leading to excessive remediation.

PART 4 Geostatistical Appraisal

Chapter 9

Bivariate Distributions

SYNOPSIS: Chapters 4 through 8 dealt with issues concerning univariate distributions. Univariate distributions contributed error management tools for items such as population statistics, probabilities, distributional models, and sampling errors. Chapter 9 takes us into the area of bivariate distributions. Bivariate tools expand the available error management measures, allowing for more advanced forms of estimation and prediction. The concept of correlation will be examined, along with ways to model it and apply it.

KEY WORDS: Univariate Distribution, Bivariate Distribution, Quantile, *q-q* Plot, Independence, Correlation, Scatter Plot, Regression, Covariance, Conditional Expectation

9.1 INTRODUCTION

Chapters 4 through 8 presented a number of tools and techniques that applied to univariate distributions. In the univariate case, only individual variables were of interest. Examples of univariate variables are aquifer thickness, average concentration of chromium in soil, and the variability of volatile organic compounds (VOCs) in groundwater.

While important and interesting in itself, the univariate view is fairly limited in its potential for analysis. Commonly, the relationships and interdependencies between variables provide more stimulating insight into the nature of the phenomena. The variables, in fact, may take many forms. Of interest may be the relationship between two contaminants such as lead (Pb) and arsenic in soil.

Similarly, we may investigate the relationship between pH and Hg concentrations in groundwater. Other examples include the relationship between exposure to adverse environmental conditions (radiation, chemicals, etc.) and the development of cancer, the relationship between porosity and permeability in an aquifer, and the concentration of arsenic in soil around a smelter with increasing depth from the surface. Bivariate possibilities are almost limitless and can be used in a wide variety of applications.

Bivariate distributions also are useful in the processes of inference and prediction. For example, there is often a relationship between the levels of heavy metals in a stream and the number and types of fish or other aquatic life that inhabit the stream. While heavy metals exist in almost all surface waters, the question of concern becomes one of defining the levels at which the fish population begins to exhibit adverse effects. Once known, the impact of additional metals introduced by a planned mining operation may be evaluated and predicted.

Bivariate techniques are also very useful in comparing two distributions to determine if they are different. Univariate statistics can provide insights regarding blatant differences through means, standard deviations, and histograms. Bivariate tools offer more sophisticated ways to evaluate and compare data sets.

Two distributions are often compared in the environmental field. A notable example is the comparison of background levels of a contaminant to the levels in an area allegedly impacted by a contaminant of concern (COC). Typically, the two domains are established representing (1) the background area and (2) the allegedly impacted area. Data from each are collected. If significant differences are found between the distributions, an environmental impact may be inferred or impacted boundaries determined.

One of the most fundamental relationships for the GEM process is that between variance and distance. This relationship is quantified by means of the variogram, which incorporates both variables into the analysis. The insight gained from bivariate correlation analysis in this chapter will be expanded in Chapter 10 to cover the nature of spatial correlation and the variogram.

9.2 Comparing Distributions

In Chapter 5, cumulative frequency plots were shown as a means of comparing a data set to the standard normal distribution. The idea was to compare the quantiles of a particular data set to those of the standard normal distribution. If a linear relationship was found in the data set using probability paper, the data were declared to be normally distributed. A similar exercise was performed for data alleged to be distributed lognormally.

It is not always necessary or advisable to match a data set with a formal distributional model. What may be of more importance is comparing the equivalence of two data distributions. One comparison technique is accomplished by means of a *q-q* plot using techniques similar to those for cumulative frequency plots.

A *q-q* plot graphs the quantiles from two distributions, one on the *X* axis and one on the *Y* axis. The first step is to organize the data sets into quantiles. Quantiles are simply regular percentage divisions of the data sets. For instance, quartiles represent data broken into one quarter of the distribution, deciles into tenths, percentiles into hundredths.

Two nearby lead (Pb) smelters (DMC and RSR) and a nonimpacted reference area (REF) are shown in Figure 9-1. Quantile data information (5% quantiles) for all areas have been summarized and appear in Table 9-1. Because each smelter data set had an equal number of samples, the 204 data values could be sorted in ascending order and the pairs plotted. To make a less cluttered graph, however, only the 5% quantiles were plotted. The resultant graph comparing smelter data is shown in Figure 9-2a.

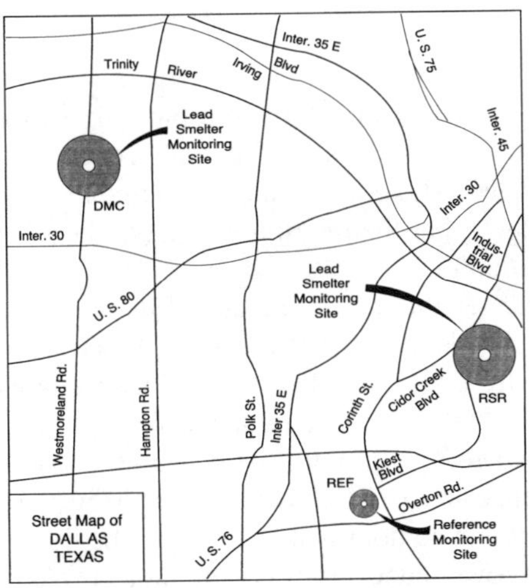

Figure 9-1. Location map of Dallas Lead smelters and Reference area. (From Geostatistical sampling and contamination analysis of lead values surrounding industrial smelters, by J.C. Myers, Master's thesis, Colorado School of Mines, 1985.)

Table 9-1. Quantile data for the DMC and RSR smelters.

	Quantile		
Cumulative Frequency	**DMC**	**RSR**	**REF**
0.05	41.4	32.2	24.4
0.10	52.2	45.2	40.3
0.15	61.9	52.9	50.2
0.20	71.4	60.4	53.0
0.25	79.8	71.6	65.4
0.30	91.8	83.1	75.4
0.35	102.0	97.1	83.0
0.40	121.0	113.0	100.0
0.45	131.0	126.0	120.0
0.50	150.0	141.0	137.0
0.55	161.0	154.0	145.0
0.60	185.0	168.0	155.0
0.65	216.0	206.0	175.0
0.70	238.0	239.0	207.0
0.75	275.0	298.0	220.0
0.80	349.0	344.0	252.0
0.85	435.0	447.0	300.0
0.90	674.0	605.0	340.0
0.95	2130.0	920.0	365.0

If two distributions are identical, the *q-q* plot will be the straight line $x = y$. As with cumulative frequency plots, perfection is never attained and some departures should be expected. The *q-q* plot may be a line but may not follow the plot of $x = y$. This indicates that the distributions have the same shape yet differ in their spread and location.

Figure 9-2a indicates that the DMC and RSR smelter distributions for lead (Pb) in soil are nearly identical until the last (95%) quantile. At this point, a significant departure from the $x = y$ line exists. This result was not unexpected as the summary statistics (Table 9-1) are quite similar.

Figure 9-2b shows the *q-q* plot for the distributions of the DMC smelter and a nonimpacted reference area (REF) located nearby (Fig. 9-1). In the case of DMC and REF, the distributions do not track as closely. The line in Figure 9-2b departs from the $x = y$ line at lower concentrations and indicates that the data in these two locations follow different distributions. The nonequivalence is expected as the summary statistics show distinct differences (Table 9-1).

Figures 9-3 and 9-4 show other examples of *q-q* plots. In Figure 9-3a, pH

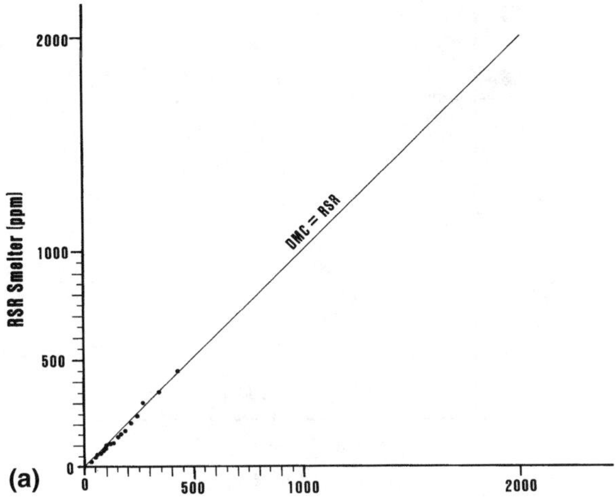

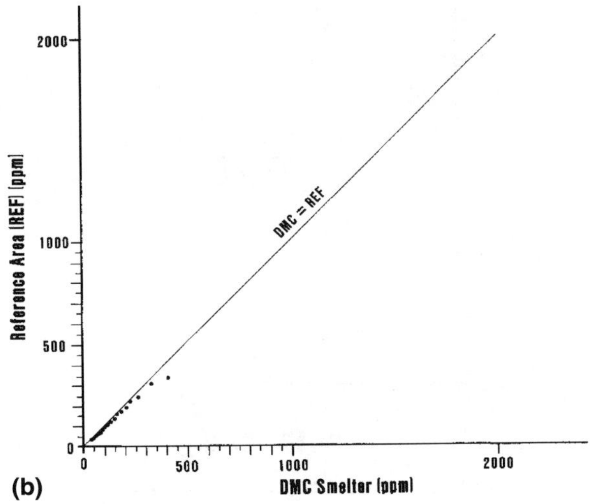

Figure 9-2. *q-q* plots of the Dallas Lead (Pb) data: (a) DMC vs. RSR data; (b) DMC vs. REF data.

values in regraded spoil overburden compare well in overlying vertical layers at the same site. When the same layers (0–1 ft depth) are compared at different sites (Fig. 9-3b), distributions differ markedly. Similarly, acid-base account (ABA) values for regraded spoil overburden compare fairly well for overlying layers (Fig. 9-4a), while the same layers (0–1 ft) depth) differ greatly between sites (Fig. 9-4b).

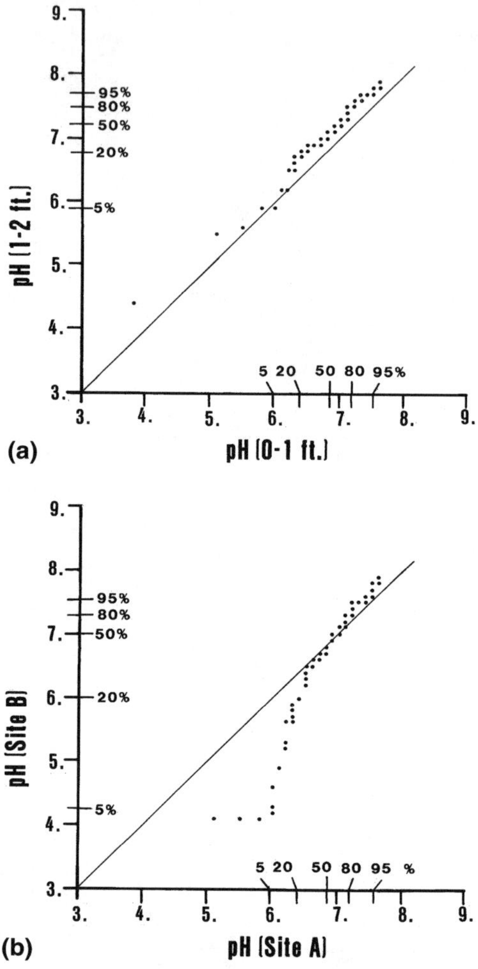

Figure 9-3. *q-q* plots of pH and ABA in Colorado coal mine overburden: (a) Data in the 0–1 ft interval plotted vs. the 1–2 ft interval; (b) data in the surface interval (0–1 ft) at Site A plotted vs. the surface interval at Site B.

9.3 Independence

Chapter 4 stated that two variables are independent if knowing the value of one variable does not provide insight regarding the other. While intuitively a good working definition, a more formal description can be developed using the idea of conditional probability.

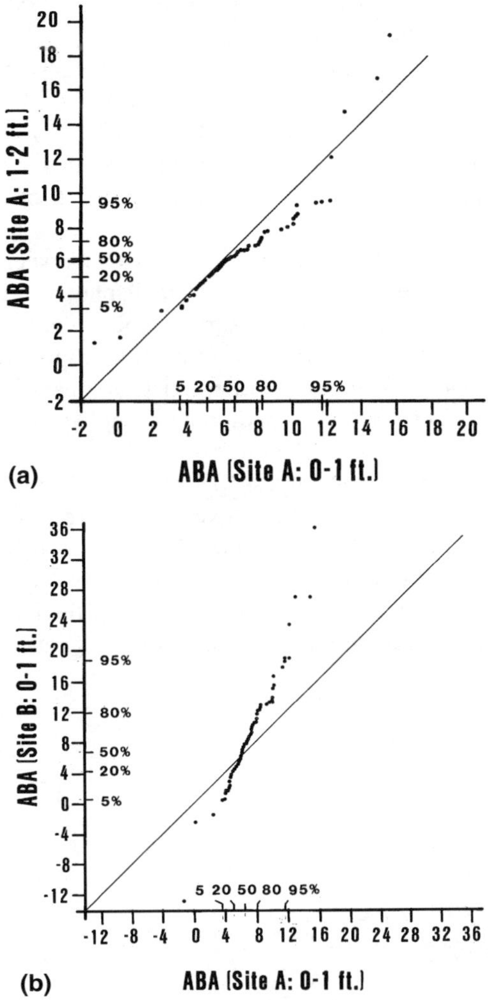

Figure 9-4. *q-q* plots comparing ABA distributions at Site A for (a) surface soil (0–1 ft) vs. depth (1–2 ft), and (b) in surface soils at Site A vs. Site B.

Conditional probability assumes that one event has already occurred and attempts to place a probability on the occurrence of a second event, given the knowledge of the first outcome. A typical statement with regard to conditional probability would be "the probability of X given that Y has happened." As a practical example, this is equivalent to asking, "What is the probability of an outcome of 6 with a second roll of a die, given that the first roll was also a 6?"

Similarly, one could ask, "Given a TOC content of 90,000 ppm, what is the expected PCB concentration of the sample?"

Mathematically, the general form of the statement is written $P\{X|Y\}$ and reads "the probability of X given that Y has occurred." If two variables are independent, knowledge of Y is irrelevant; thus, $P\{X|Y\} = P\{X\}$ which is simply the probability of X. For the above example, knowing that a 6 has just occurred does not assist in predicting the outcome of the next roll of the die. Therefore, $P\{X|Y\} = 1/6$.

Visual tools provide a way to gain a qualitative understanding as to whether two variables are independent. A common way to display bivariate data is through the use of a scatter plot or correlation diagram. One variable is plotted on the X axis and the other on the Y axis. If the data pairs plot in a scatter, we may say that, for practical purposes, no correlation exists. If the variables show some degree of linearity or other type of nonrandom grouping, some degree of correlation exists.

Figure 9-5 shows an example of a scatter plot. The X axis shows the concentration of lead (Pb) in soil, and the Y axis shows the corresponding arsenic concentration. The random "shotgun" pattern indicates that little if any correlation is present between the lead (Pb) and arsenic values. Thus for this example, the data are considered independent; that is, knowing the concentration of lead (Pb) does not assist in predicting the corresponding concentration of arsenic, and vice versa.

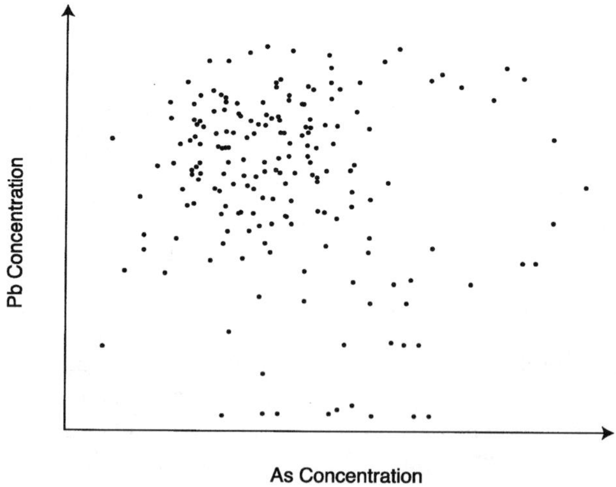

Figure 9-5. Scatter plot of lead (Pb) vs. arsenic concentrations, showing no correlation.

9.4 Correlation

9.4.1 Graphical Approach

While not a part of Figure 9-5, correlation is often present between two random variables and can easily be seen when it is present. Figure 9-6 shows a scatter plot of field electric conductivity measurements (EC), a rough measure of soil salinity, plotted against laboratory EC measurements for the same samples. The graph in Figure 9-6 shows a distinct linear relationship between field and laboratory EC data. In other words, the larger the field EC measurement, the larger the corresponding laboratory EC result, in general.

Such a graph may be used as a primitive prediction model; that is, knowing the analytical result using a field-portable device provides an idea of the "actual" associated EC value (as determined by the laboratory). With a correlation coefficient of .888, use of field EC measurements as a predictive tool may be attempted, with two caveats: (1) a higher degree of correlation is preferred for predictive purposes, and (2) the field EC measurements are biased low.

The correlation seen in Figure 9-6 is referred to as positive correlation. Positive correlation takes its name from the fact that the slope of the data is positive. Correlation may also be negative, as in Figure 9-7, which shows a scatter plot of coal quality data from an east Texas lignite mine where ash

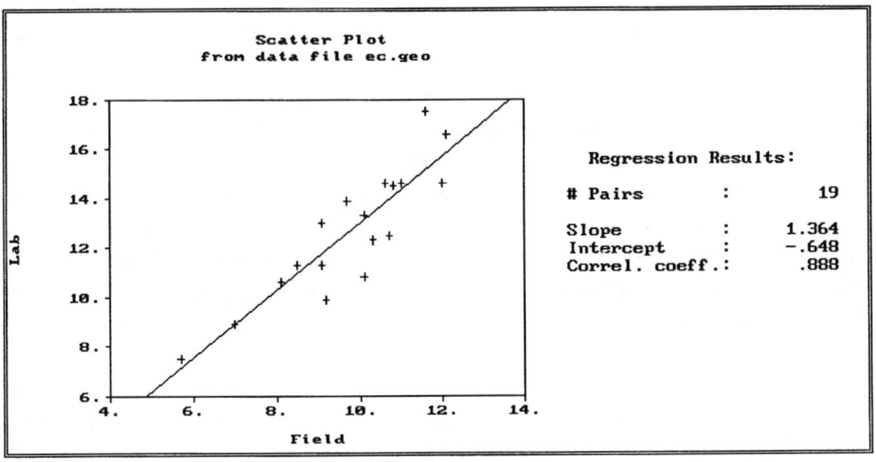

Figure 9-6. Scatter plot of field-derived electrical conductivity (EC) values vs. laboratory EC measurements for the same samples.

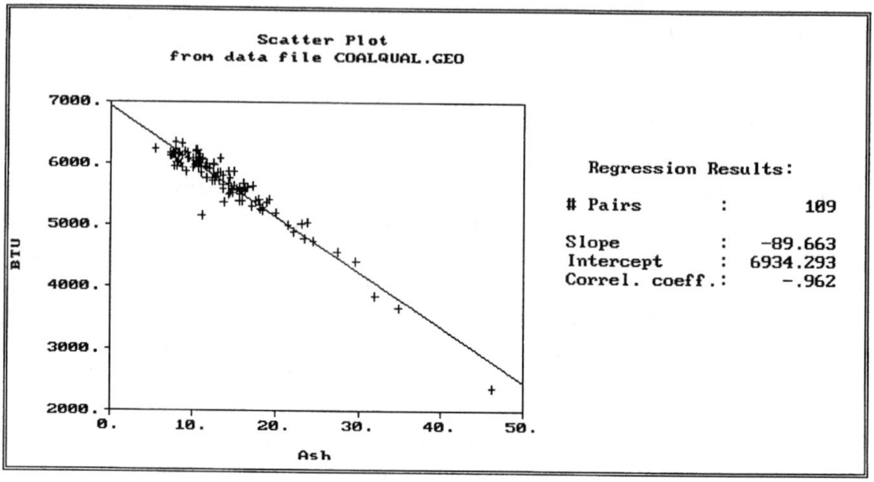

Figure 9-7. Scatter plot of ash content vs. heating value (BTUs) in lignite coal.

content vs. heating value (BTUs/lb) are plotted. In this case, as the ash content increases, the heating value (BTUs/lb) decreases. A plot of variables such as the total vegetative mass per acre vs. annual rainfall or arsenic concentration vs. the distance from the smelter would also be expected to show a negative correlation.

In general, data are positively correlated, negatively correlated, or uncorrelated. These three basic conditions have been summarized visually in Figures 9-5 through 9-7.

Occasionally, the scatter plots show more than one type of correlation structure on the graph. This often indicates two data populations and is analogous to a bimodal histogram. Figure 9-8 shows a scatter plot of aqueous bromide concentrations vs. chloride concentrations, where both a positive linear correlation and an intersecting negative correlation exist. The positive trend indicates the mixing of ocean bay water with native groundwater that is flowing through the aquifer into the bay. The negative trend indicates ground water mixing with an industrial process brine solution that contains concentrations of mercury.

Figure 9-9 shows another example of two populations on a scatter plot of PCB and TOC concentrations. While these two variables often show good correlation, on this site, they do not. Two distinct clouds of points exist, however, one lower and one higher, separated by a sparse zone. This effect is equally as pronounced when the scatter plot graphs the PCB concentration vs. the natural logarithm of TOC (Fig. 9-10).

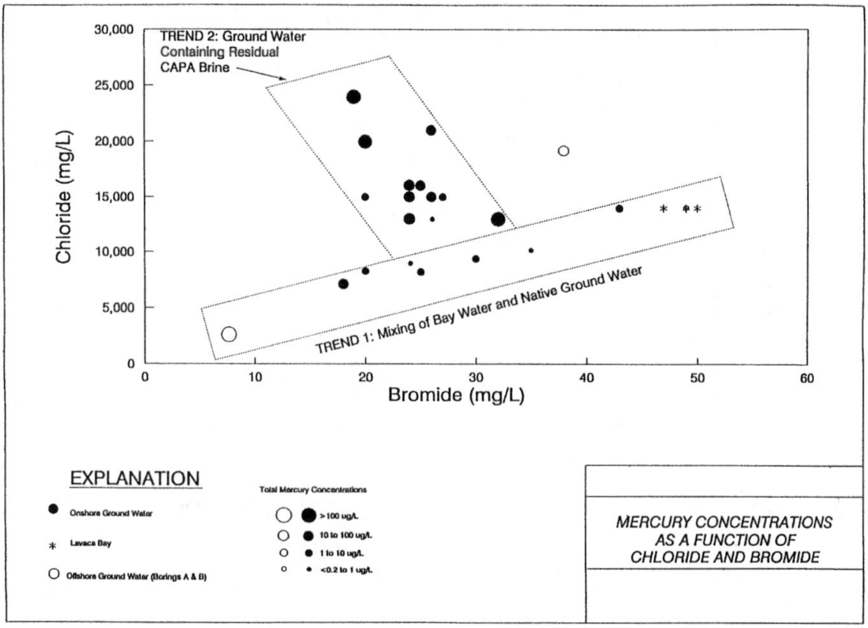

Figure 9-8. Scatter plot showing two populations: bromide vs. chloride.

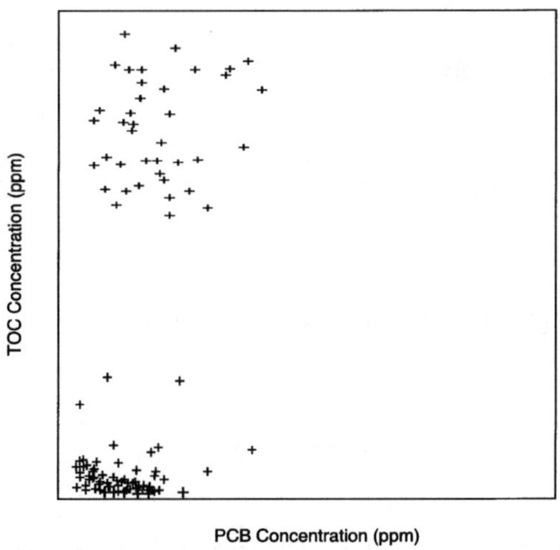

Figure 9-9. Scatter plot showing two populations: PCB vs. TOC.

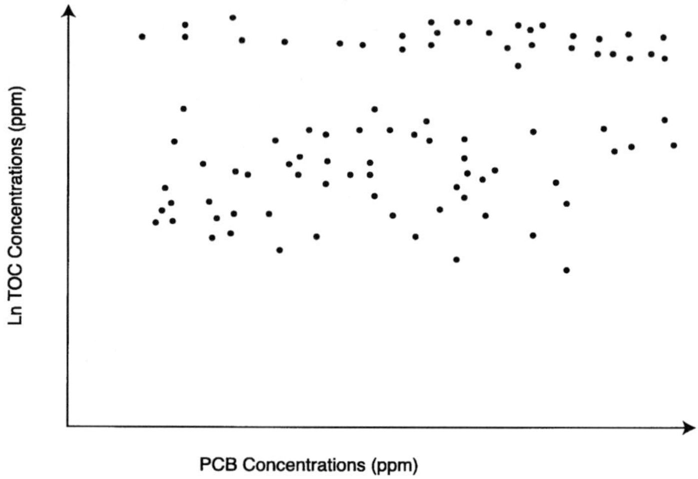

Figure 9-10. Scatter plot showing two populations: PCB vs. Ln (TOC).

9.4.2 Relationship of Correlation and Covariance

Mathematically, the covariance can be used to assess the mutual relationship (correlation) between two random variables. The covariance is written

$$\mathrm{COV}\,(X, Y) = E[(X - \overline{X})(Y - \overline{Y})] \tag{9.1}$$

or, for practical computation,

$$\mathrm{COV}\,(\mathrm{X}, \mathrm{Y}) = \frac{1}{n} \sum_{i=1}^{n} (x_i - \overline{x})(y_i - \overline{y}) \tag{9.2}$$

Note that the covariance is commonly written using each of the following notations:

$$\mathrm{COV}\,(X, Y) = \alpha\,(X, Y) = \alpha_{XY} \tag{9.3}$$

These definitions show that the covariance is the expected product of the two variables X and Y around their respective mean values. Similar to the variance in that it is a product, the covariance differs in that the product may be negative (a nonexistent condition for variances). When X and Y grow together in either the positive or negative direction, the product is always positive. When one grows while the other decreases, the product is negative. If no correlation exists, some products are positive while others are negative, canceling each other to produce a zero average.

Using the formula for the covariance, the correlation coefficient r may be calculated. The correlation coefficient is commonly used to summarize the relationship between two variables. It has this convenient property: r can range only between 1 and -1. This property is achieved by dividing the covariance by the standard deviations of X and Y. The correlation coefficient r is calculated

$$r = \frac{\frac{1}{n}\sum_{i=1}^{n}(x_i - \bar{x})(y_i - \bar{y})}{\sigma_X \sigma_Y} \tag{9.4}$$

9.5 Linear Regression

Beyond just determining whether two variables are independent or what their degree of correlation is, it is possible to use these tools to make predictions about one variable by knowing the value of the other. It was noted earlier that a linear relationship often exists between two variables. This provides a convenient framework for modeling and prediction.

A straight line can be described mathematically by an equation of the form:

$$y = mx + b \tag{9.5}$$

where m is equal to the slope of the line and b is the y intercept. The equation $y = 5 - 2x$ describes a straight line with a y intercept of 5 and a slope of -2. Similarly, $y = 4 + 1x$ has an intercept of -4 and a slope of 1. Both equations are plotted in Figure 9-11.

Linear regression uses the same basic format to provide an optimized line. In the case of regression, the slope m is defined as

$$m = r\frac{\sigma_y}{\sigma_x} \tag{9.6}$$

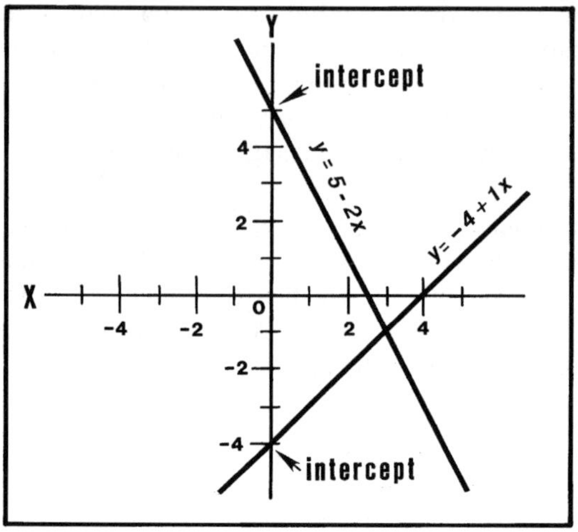

Figure 9-11. Plots of straight lines.

and the y intercept b as:

$$b = \overline{Y} - m\overline{X} \tag{9.7}$$

Notice that the slope m is nothing else than the correlation coefficient multiplied by the ratio of the standard deviations of X and Y. Once the slope has been calculated, the mean values of X and Y can be used to determine the constant b.

Figure 9-12 shows the linear regression model for the variables chromium (Y) and cadmium (X) in soils around a settling pond for a metal-plating facility. The statistics produced $\overline{X} = 45.1$, $\overline{Y} = 142.5$, $r = .735$, $\sigma_x = 22.78$, and $\sigma_y = 24.74$. Using equations 9.6 and 9.7 along with the results from the statistics, it is now possible to predict a value for chromium (Y) given only a value for cadmium (X). Inserting the values $m = .80$ and $b = 173.2$ along with a cadmium concentration of 51.8 ppm produces

$$y = 173.2 + (.8)(51.8) \cong 215 \text{ ppm} \tag{9.8}$$

Linear regression is sometimes called *least-squares regression* or a *minimum variance regression.* This stems from the fact that the regression minimizes the sum of squared deviations of the Y values around the regression line. These deviations are vertical line segments created by drawing a line from the observed

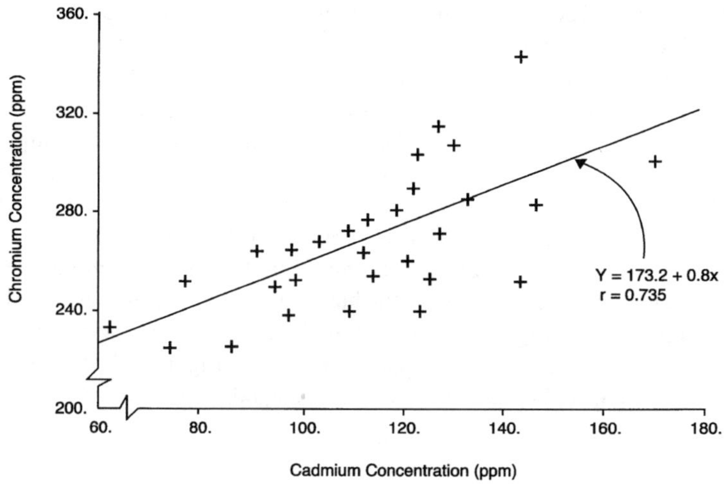

Figure 9-12. Linear regression for chromium (Y) on cadmium (X).

value of Y to the regression line. Examples can be seen in Figure 9-13. Smaller deviations will obviously produce a smaller sum of squares along with a tighter and better-looking fit.

The correlation coefficient attempts to quantify how closely the observed values follow the straight line. If the observed points were a perfect fit for the line, the scatter plot would also be a straight line with $r = 1$ and a positive slope. Conversely, with $r = -1$ the scatter plot would be a straight line with a negative

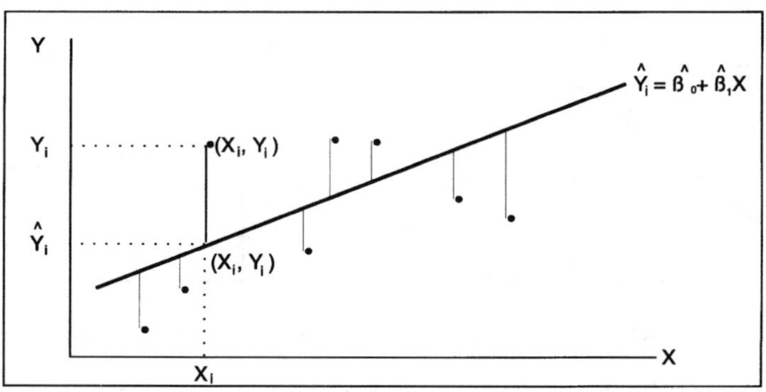

Figure 9-13. Deviations of observed points from the regression line. (From *Applied Regression Analysis*, by D.G. Kleinbaum and L.L. Kupper. Copyright © 1988, 1978 by Wadsworth Publishing Company. Reprinted by permission of Brooks/Cole Publishing Company, Pacific Grove, CA, a division of International Thomson Publishing Inc.)

slope. As r approaches zero, any linear attributes of the scatter plot tend to disappear, looking much like the "shotgun" scatterplot in Figure 9-5. A summary of these three described conditions appears in Figure 9-14.

The quantity r (or equivalently r^2) is commonly used erroneously to measure two entities: (1) the magnitude of the slope of the regression line and (2) the appropriateness of the straight-line model. Figure 9-15 shows two examples of perfect linear regression; in each case, the regression line exactly fits the data points. The slopes of the lines, however, are quite different.

Figure 9-16 shows examples of the appropriateness of a straight-line fit. In Figures 9-16a and 9-16b, both situations have a value of r^2 equal to zero. Because no correlation exists in Figure 9-16a and the plot is a random scatter, neither a line nor other type of curve seems an appropriate fit. In Figure 9-16b, a linear fit is inappropriate because the data are horseshoe-shaped.

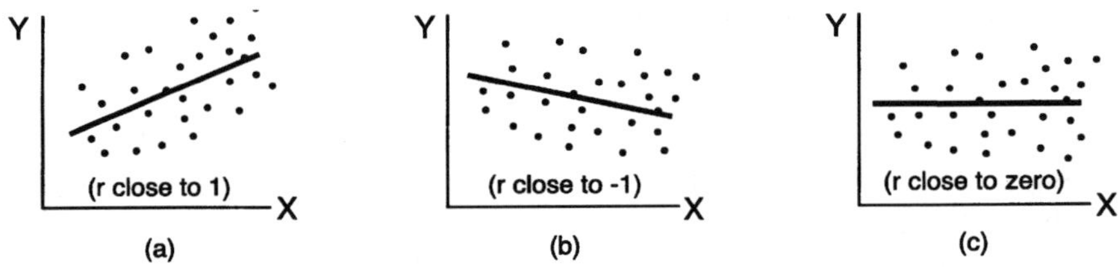

Figure 9-14. Summary of positive, negative, and zero correlation graphs showing the correlation coefficient as a measure of association. (From *Applied Regression Analysis*, by D.G. Kleinbaum and L.L. Kupper. Copyright © 1988, 1978 by Wadsworth Publishing Company. Reprinted by permission of Brooks/Cole Publishing Company, Pacific Grove, CA, a division of International Thomson Publishing Inc.)

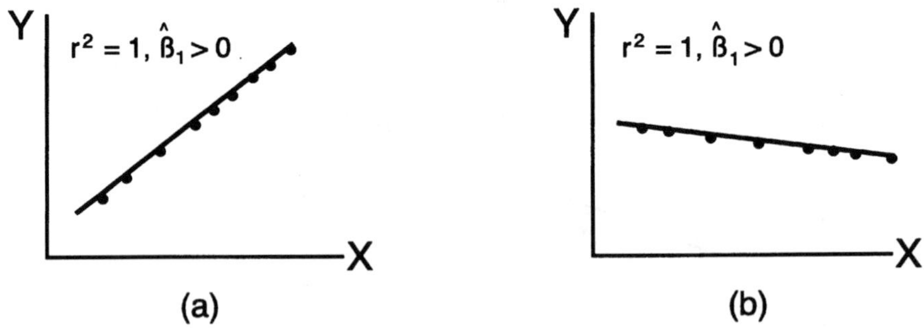

Figure 9-15. Perfect linear regression for two lines with different slopes. (From *Applied Regression Analysis*, by D.G. Kleinbaum and L.L. Kupper. Copyright © 1988, 1978 by Wadsworth Publishing Company. Reprinted by permission of Brooks/Cole Publishing Company, Pacific Grove, CA, a division of International Thomson Publishing Inc.)

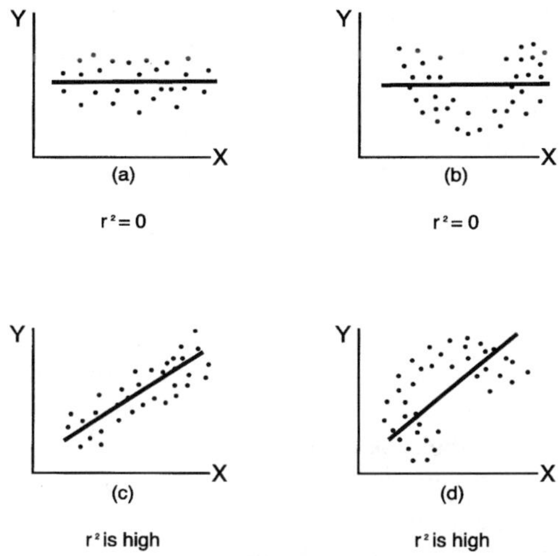

Figure 9-16. Appropriateness of straight-line fit for linear regression based on the calculated r^2 value. (From *Applied Regression Analysis*, by D.G. Kleinbaum and L.L. Kupper. Copyright © 1988, 1978 by Wadsworth Publishing Company. Reprinted by permission of Brooks/Cole Publishing Company, Pacific Grove, CA, a division of International Thomson Publishing Inc.)

Figure 9-16c shows a situation where r^2 is expected to be high and is, in fact, high. Figure 9-16d shows an arch-shaped scatter plot that has an unexpectedly high correlation for a linear fit. Figure 9-16d would be better modeled using a nonlinear approach.

In actual practice, there is no guarantee that the relationship between the two random variables will be linear. The scatter plot may show a parabolic shape, an arch, and *S* shape, or other type of nonlinear structure. In these cases, the correlation coefficient may be very low despite a noticeable correlation. Also, the linear fit will be poor and inappropriate, as seen in Figure 9-16.

While nonlinear regression techniques are available, they are more difficult and beyond the scope of this book. Two fairly simple alternatives exist to deal with the problem of nonlinear correlation. They are rank correlation and conditional expectation.

9.5.1 Rank Correlation

The linear regression approach uses sample data to fit a straight line to the *XY* scatter plot. *Rank correlation* makes a simple departure from this approach for nonlinear situations. Instead of using the original data values, the correlation coefficient is calculated using ranked data. Equation 9.4 becomes

$$r_{\text{RANK}} = \frac{\frac{1}{n}\sum_{i=1}^{n}(Rx_i - R\overline{x})(Ry_i - R\overline{y})}{\sigma_{Rx}\sigma_{Ry}} \tag{9.9}$$

Rx_i is the rank of x_i in relation to all other x values. Ranking is generally preferred in ascending order. The smallest value of x is given a rank of 1, while the largest value receives the rank of n. Ry_i is performed in the same manner. The mean values for Rx and Ry are simply $n/2$. This mean is used to calculate the standard deviations σ_{Rx} and σ_{Ry}.

Rank correlation often dampens the effect of extreme pairs on the scatter plot. This, in turn, may significantly improve the correlation coefficient r_{rank}. The location of the extreme pairs on the scatter plot may also influence results. Under proper conditions, the r value may be higher than that for r_{rank}.

9.5.2 Conditional Expectation

The linear regression approach and the *q-q* plot both relate closely to straight line functions: regression fits a line to the scatter plot and the *q-q* plot compares a plot of the data to the $X = Y$ line. Linear functions are mathematically simple to understand and use and provide continuous functions for estimation.

Conditional expectation offers a discrete stepwise way to fit nonlinear data on a scatter plot. This can often be satisfying where scatter plots are not linear or where correlation decreases, forcing a straight line through what more and more resembles a cloud of points as concentrations increase.

The conditional expectation approach calculates the mean value of Y for discrete ranges of X. Each XY pair is assigned to an X category. Within the X class, the mean of all Y values is then calculated. The expected value of Y then becomes constant over the entire range of the particular X class in question. An example of this type of classification has been performed for arsenic and cadmium data. Classifications appear in Table 9-2.

The mean of the Y values plots on the scatter plot as a horizontal line within the confines of the X category. This can be seen in Figure 9-17, where a conditional expectation curve has been plotted for chromium and cadmium data. The predicted value of Y is constant within the X category. The expected values are conditional as they are only valid within the specified range. Note that the expected value of Y does not have to be a strictly increasing function. The mean

Table 9-2. Mean values of chromium within classes defined for cadmium.

Class	Number of Pairs	Mean of Chromium
60–80	3	235.6
80–100	6	254.8
100–120	8	267.4
120–140	8	279.7
140–160	3	292.6
160–180	1	307.5

value of Y in one category of X may be less than the mean value of Y in the previous category of X.

Various class sizes may be selected for categorization. As with histograms, the size of the category will slightly alter the trace of the graph. Optimally, we would wish for a large data set so that a large number of classes could be defined, with each class containing a sufficient number of samples to be used in calculating the class mean. This would produce an appealing graph in that the steps would be smoother than those for graphs with only a few categories of X. With small data sets, we encounter the practical problem of minimal pairs for extreme classes (lowest and lowest). This can cause the graph to become erratic.

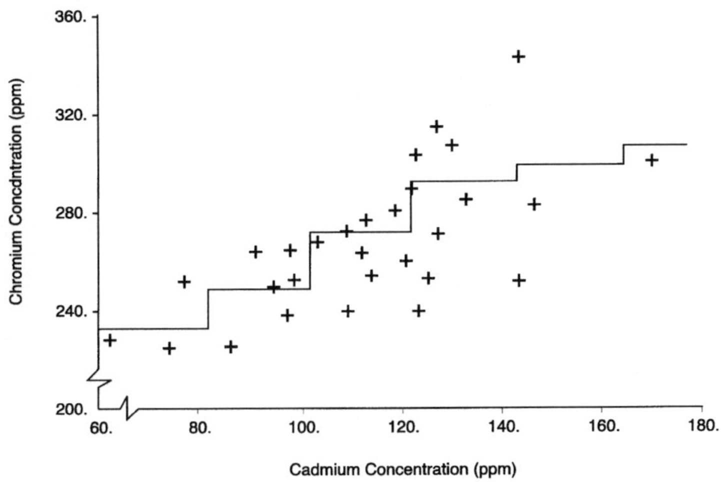

Figure 9-17. Conditional expectation curve for cadmium and chromium data.

CONCLUSION

The tools offered by the bivariate distribution can be of great use in the GEM process. Comparison of distributions can isolate contaminated areas from unaffected or background areas. The degree of correlation between two variables can be determined and regressed to serve as a predictive model. The linear regression approach introduced here serves as the underlying process for the kriging algorithm, a key error management tool of geostatistical appraisal (GA), to be discussed in later chapters.

Chapter 10

Variograms: Quantification of Spatial Continuity

SYNOPSIS: Perhaps no other entity of geostatistics embodies the creative uniqueness of the variogram function. This twist on the calculation of a classical variance allows the quantification of uncertainty vs. distance. It can unravel the unique spatial continuity structure for each contaminant of concern (COC) at each site, something intuitively known to exist but heretofore unquantifiable. Like a skilled detective, the variogram can unlock the important secrets surrounding the spatial continuity that reflect site geology and history. This is particularly impressive because the variogram calculation is little more than a subtract-and-square technique. Kriging, while an important and powerful geostatistical counterpart to the variogram, is a modified regression "machine" that cranks through its process with predictable precision. Kriging is dependent on the COC and site- specific distance-weighting function provided by the variogram. The discovery of the variogram for a COC is always an individual experience, offering opportunities for the geostatistician to employ artfulness, creativity, and geologic intuition. This chapter will explore the calculation, interpretation, modeling, and geologic implications of the geostatistical function known as the variogram.

KEY WORDS: Variogram, Spatial Correlation, Continuity, Nugget Effect, Zone of Influence, Range, Sill, Variogram Model, Isotropy,

Anisotropy, Angular Window, Angular Direction, Lag Distance, Outliers, Relative Variogram, Logarithmic Variogram, Indicator Variogram, Cross-Variogram, Weak Stationarity, Intrinsic Hypothesis

10.1 Introduction to Variograms

Variograms are fundamental tools in geostatistical appraisal and the GEM process. As described in Chapter 1, the intuitive concept of the variogram has been known to earth scientists for a long time. It was frequently observed that nearby sample data had similar values. Allowing for a broad interpretation as to the definition of *nearby*, almost all spatial data display this similarity characteristic.

The definition of independence and the insight provided by the concept of correlation demonstrate that there is a statistical basis for the earth scientist's intuition. By modifying existing statistical tools, the variogram provides a quantitative measure of the intuition as well as the statistical correlation.

Ironically, the modifications made in order to calculate a variogram (and later kriging) are so insignificant mathematically that, initially, many statisticians were not interested. In fact, at times one could literally see the yawn of boredom! This lack of interest may also have been rooted in the fact that most statisticians deal exclusively with independent variables. Consequently, a minor change to the math in an area of minor interest (correlated variables) did not make an impact on statisticians. To those endeavoring to quantify spatial relationships, however, the development of geostatistical tools was a revolutionary and long-awaited development.

To this point, the discussion has referred to the similarity between nearby samples. The variogram and statistics, in general, focus on the dissimilarity. For example, the variance σ^2 looks at the difference between values for individual samples and the mean value $(x_i - m)^2$ for univariate data. For bivariate data, the covariance σ_{XY} measures the differences between each of the variables X and Y as compared to their respective means.

The variance provided by the variogram uses a slightly different twist on the variance or covariance approach. First, the variogram compares pairs of sample data $(Z_x - Z_{x+h})^2$ at different locations and at distance h, not a datum to the mean $(x_i - m)^2$. Pairing of samples is done so that a second feature may be included: the distance between the two samples. Mathematically, the difference of two values $Z(x)$ and $Z(x+h)$ is determined and then squared to produce a variance.

Having the squared difference of one pair of points is not sufficient; rather, it is necessary to obtain the average value of all such pairs of points at that distance. Thus, the variogram function may be defined:

$$\gamma(h)=\frac{1}{2N(\mathrm{h})}\sum_{i=1}^{N}[Z(x-h)]^2 \tag{10.1}$$

It is important to notice that the variogram is a vector function, that is, a function of both distance and direction. The variogram quantifies how values of the variable *Z(x)* differ (on the average) as distance *h* increases along a given direction.

Figure 10-1 illustrates a typical pairing scheme for the calculation of an experimental variogram. Samples are distributed on a regular grid, and pairs have been formed in the east-west direction at a distance interval *h*. The average of these squared pairs can be computed using equation 10.1 to obtain $\gamma(h)$, the variance at that distance.

Figure 10-2 shows such a data set where arsenic concentrations have been sampled on a regular grid. Using the pairing scheme shown in Figure 10-1, 99 sample pairs of distance $1h$ ($h = 10$ ft) can be developed. Figure 10-3 presents a complete listing of these computation pairs, along with the calculated $\gamma(10)$ value of 43.9 ppm^2.

Figure 10-4 illustrates the next step, which is to calculate $\gamma(2h)$ or $\gamma(20)$ for the arsenic data. Here the distance between sample pairs, called the *lag*, is twice

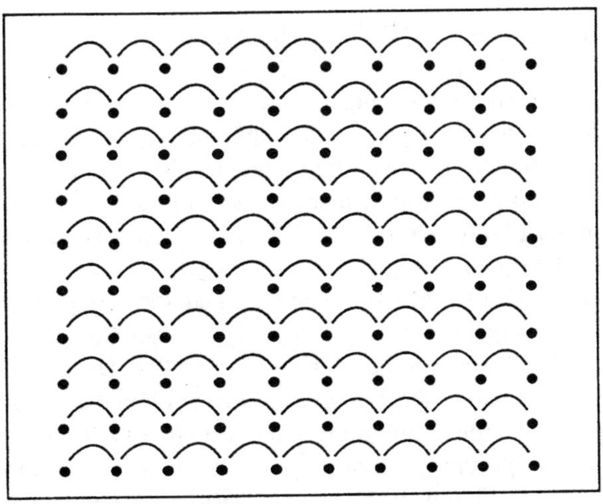

Figure 10-1. East-west pairing for variogram calculation.

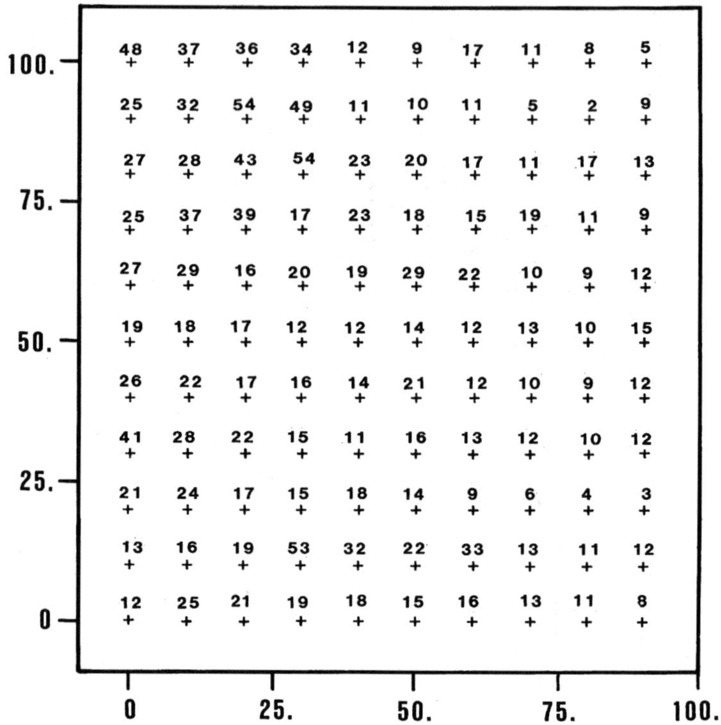

Figure 10-2. Data set of arsenic concentrations on a regular grid.

the initial distance. Figure 10-5 lists the 88 pairs for the lag of $2h$, or 20 ft, along with the calculated $\gamma(20)$ value of 86.5 $(\text{ppm})^2$.

The process of increasing the lag distance by h (h, $2h$, $3h$, etc.) continues until such time as nh is close to or greater than the largest distance between any two samples. Figure 10-6 plots the variogram graph for the first four lag distances (10, 20, 30, and 40 ft). With the grid of arsenic data, a maximum lag of $9h$ (90 ft) is possible looking east-west or north-south, although only 11 sample pairs would result for these directional variograms.

Once all the values of $\gamma(h)$ have been obtained, they are plotted as a function of distance to produce the variogram graph. An example of an experimental variogram for PCBs in soil can be seen in Figure 10-7. Distance h is plotted as an increasing function along the X axis, and $\gamma(h)$ is plotted along the Y axis. The reader is reminded that this graph represents only the east-west direction. Figure 10-8a shows another grouping of variogram pairs at distance h, this time in the north-south direction. Figure 10-8b shows variogram pairs in the northwest-southeast direction.

$$
\begin{aligned}
\gamma(10) = {} & (48-37)^2 + (37-36)^2 + (36-34)^2 + (34-12)^2 + (12-9)^2 + \\
& (9-17)^2 + (17-11)^2 + (11-8)^2 + (8-5)^2 + (25-32)^2 + \\
& (32-54)^2 + (54-49)^2 + (49-11)^2 + (11-10)^2 + \\
& (10-11)^2 + (11-5)^2 + (5-2)^2 + (2-9)^2 + (27-28)^2 + \\
& (28-43)^2 + (43-54)^2 + (54-23)^2 + (23-20)^2 + \\
& (20-17)^2 + (17-11)^2 + (11-17)^2 + (17-13)^2 + \\
& (25-37)^2 + (37-39)^2 + (39-17)^2 + (17-32)^2 + \\
& (23-18)^2 + (18-15)^2 + (15-19)^2 + (19-11)^2 + (11-9)^2 + \\
& (27-29)^2 + (29-16)^2 + (16-20)^2 + (20-19)^2 + \\
& (19-29)^2 + (29-22)^2 + (22-10)^2 + (10-9)^2 + (9-12)^2 + \\
& (19-18)^2 + (18-17)^2 + (17-12)^2 + (12-12)^2 + \\
& (12-14)^2 + (14-12)^2 + (12-13)^2 + (13-10)^2 + \\
& (10-15)^2 + (26-22)^2 + (22-17)^2 + (17-16)^2 + \\
& (16-14)^2 + (14-21)^2 + (21-12)^2 + (12-10)^2 + (10-9)^2 + \\
& (9-12)^2 + (41-28)^2 + (28-22)^2 + (22-15)^2 + (15-11)^2 + \\
& (11-16)^2 + (16-13)^2 + (13-12)^2 + (12-10)^2 + \\
& (10-12)^2 + (21-24)^2 + (24-17)^2 + (17-15)^2 + (15-18)^2 + \\
& (18-14)^2 + (14-9)^2 + (9-6)^2 + (6-4)^2 + (4-3)^2 + \\
& (13-16)^2 + (16-19)^2 + (19-53)^2 + (53-32)^2 + \\
& (32-22)^2 + (22-33)^2 + (33-13)^2 + (13-11)^2 + \\
& (11-12)^2 + (12-25)^2 + (25-21)^2 + (21-19)^2 + \\
& (19-18)^2 + (18-15)^2 + (15-16)^2 + (16-13)^2 + \\
& (13-11)^2 + (11-8)^2 \div (2)(99) \\
= {} & 43.9\ (\text{ppm})^2
\end{aligned}
$$

Figure 10-3. Variogram pairs for arsenic data at a lag of 1*h* (10 ft).

Figure 10-9 presents an idealized variogram graph and the variogram components of interest. This figure plots a mathematical model that will be used to represent the spatial continuity in the kriging program. The ideal variogram rises from the origin and then reduces its rate of increase until it eventually levels off. The distance at which the graph flattens is called the *range*. The height at which this plateau is reached is called the *sill*. In practice, the height of the sill is

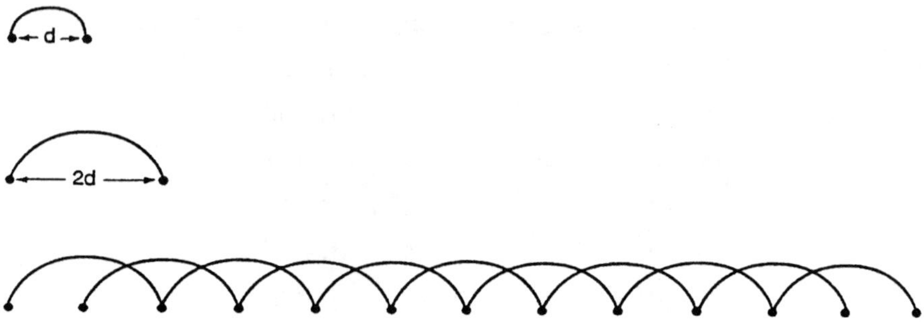

Figure 10-4. Pairing with lag of $2h$.

$$
\begin{aligned}
\gamma(20) = {} & (48-36)^2 + (37-34)^2 + (36-12)^2 + (34-9)^2 + (12-17)^2 + \\
& (9-11)^2 + (17-8)^2 + (11-5)^2 + (25-54)^2 + (32-49)^2 + \\
& (54-11)^2 + (49-10)^2 + (11-11)^2 + (10-5)^2 + (11-2)^2 + \\
& (5-9)^2 + (27-43)^2 + (28-54)^2 + (43-23)^2 + (54-20)^2 + \\
& (23-17)^2 + (20-11)^2 + (17-17)^2 + (11-13)^2 + \\
& (25-39)^2 + (37-17)^2 + (39-23)^2 + (17-18)^2 + \\
& (23-15)^2 + (18-19)^2 + (15-11)^2 + (19-9)^2 + (27-16)^2 + \\
& (29-20)^2 + (16-19)^2 + (20-29)^2 + (19-22)^2 + \\
& (29-10)^2 + (22-9)^2 + (10-12)^2 + (19-17)^2 + (18-12)^2 + \\
& (17-12)^2 + (12-14)^2 + (12-12)^2 + (14-13)^2 + \\
& (12-10)^2 + (13-15)^2 + (26-17)^2 + (22-16)^2 + \\
& (17-14)^2 + (16-21)^2 + (14-12)^2 + (21-10)^2 + (12-9)^2 + \\
& (10-12)^2 + (41-22)^2 + (28-15)^2 + (22-11)^2 + \\
& (15-16)^2 + (11-13)^2 + (16-12)^2 + (13-10)^2 + \\
& (12-12)^2 + (21-17)^2 + (24-15)^2 + (17-18)^2 + (15-14)^2 + \\
& (18-9)^2 + (14-6)^2 + (9-4)^2 + (6-3)^2 + (13-19)^2 + \\
& (16-53)^2 + (19-32)^2 + (53-22)^2 + (32-33)^2 + \\
& (22-13)^2 + (33-11)^2 + (13-12)^2 + (12-21)^2 + \\
& (25-19)^2 + (21-18)^2 + (19-15)^2 + (18-16)^2 + \\
& (15-13)^2 + (16-11)^2 + (13-8)^2 + (2)(88) + \\
= {} & 86.5\ (\text{ppm})^2
\end{aligned}
$$

Figure 10-5. Variogram pairs for arsenic data at a lag of $2h$ (20 ft).

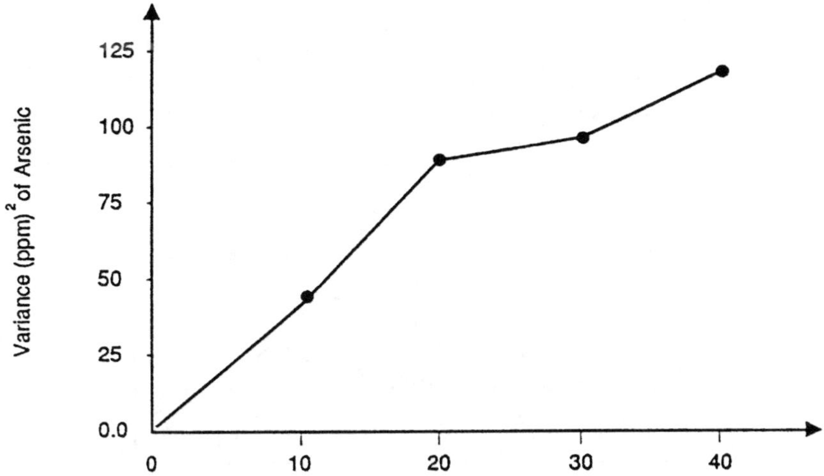

Figure 10-6. Variogram graph of arsenic data for lag distances of $1h$, $2h$, $3h$, $4h$.

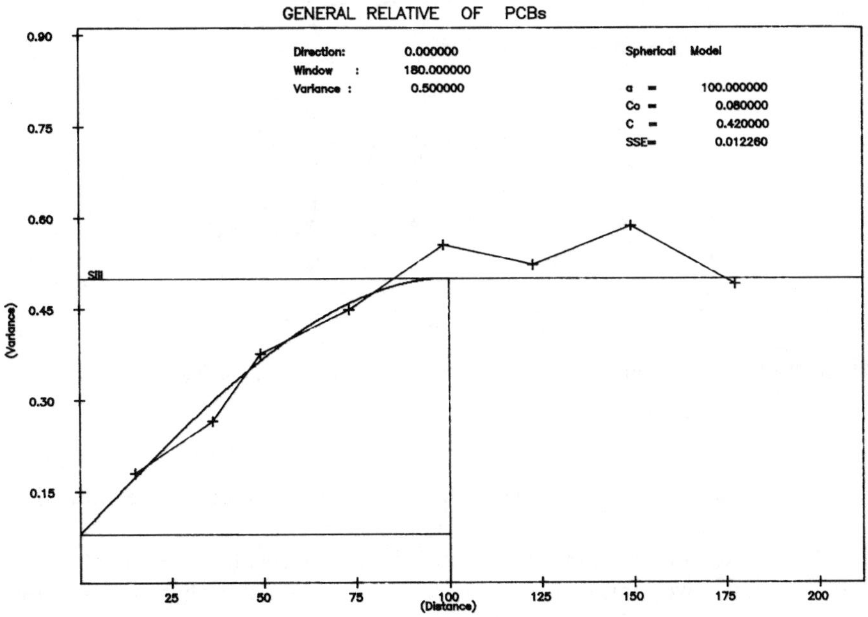

Figure 10-7. Example experimental variogram: PCBs in soil.

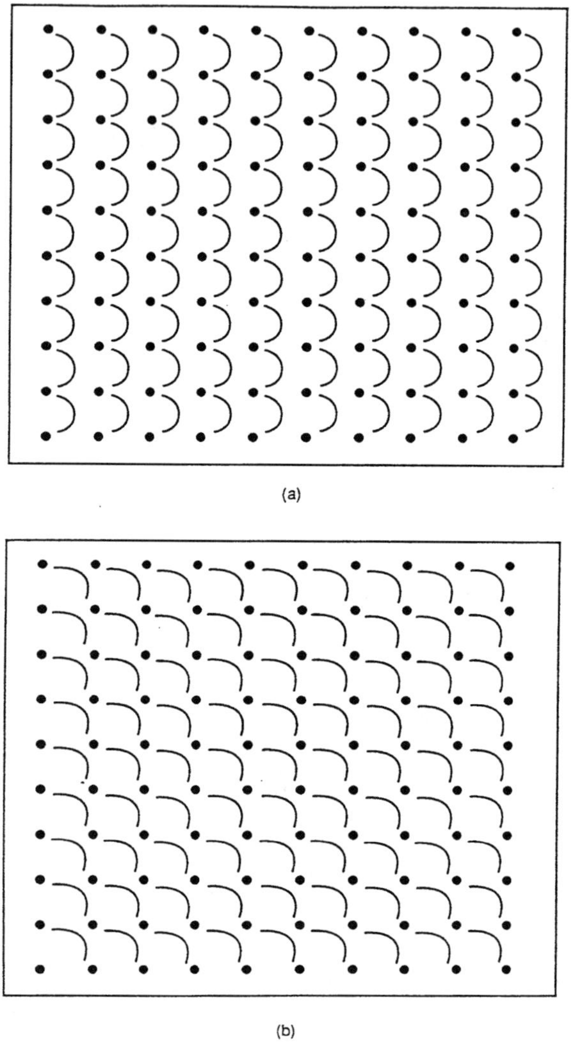

Figure 10-8. North-south and northwest-southeast variogram pairing.

generally equal or close to the population variance s^2. In theory, the sill and the population variance are equivalent (Barnes 1991).

The variogram graph embodies three basic concepts that were qualitatively known to those working with earth science or spatial data. Thse are continuity, zone of influence, and anisotropies. Each is related to the components of the idealized variogram, and each will be discussed individually below.

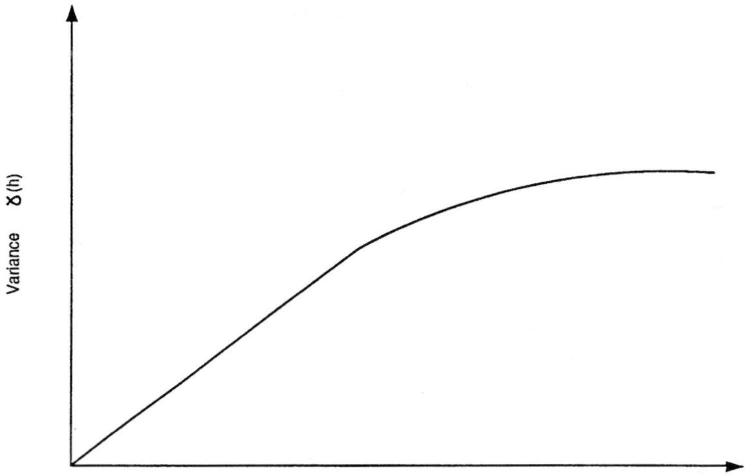

Figure 10-9. Idealized variogram graph.

10.1.1 Continuity

Continuity refers to the smoothness of the transition between very closely spaced samples. Quantitatively, this is measured by the rate at which $\gamma(h)$ grows near the origin of the variogram graph. By definition, $\gamma(0)$ is zero. However, the experimental variogram graph may show a *y* intercept at some value greater than zero. An example of this behavior in soil lead (Pb) concentrations is shown in Figure 10-10.

The implication is that a variance exists at zero distance. While contrary to the definition, this is a common practical occurrence. This phenomenon is called the *nugget effect* C_0 and stems from the mining industry, where mineralization often occurrs as nuggets. The nugget effect is often extremely pronounced for gold because, as a result of gold's high malleability, nuggets of gold are not reduced in size during comminution. Other metals may have large nugget effects as well.

In environmental situations, erratic variables such as PCB concentrations in soil often show nugget effects. Because the oily substances to which PCBs adhere tend to cluster and not disseminate evenly, rapid variations in concentration frequently occur over short distances.

The nugget effect is a picturesque name for a very troublesome phenomenon that immediately introduces error into an estimation process. The higher the nugget, the higher the error. The randomness introduced makes prediction of unsampled locations more difficult.

The nugget effect can result from causes other than nuggets. In many cases, the nugget effect is nothing else than a reflection of the fundamental error *FE* or the short-range heterogeneity fluctuation error CE_1 ($FE + GE$). By minimizing

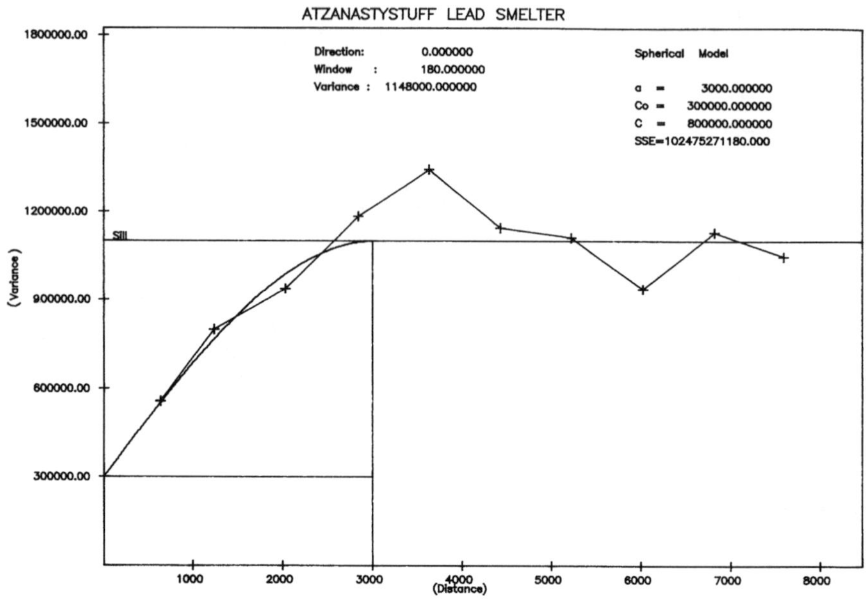

Figure 10-10. Nugget effect for lead (Pb) in soil.

the *FE* and *GE* during sampling, the nugget effect C_0 can be minimized during the variogram analysis.

Figure 10-11 shows an experimental variogram graph for copper concentrations in soils, with no nugget effect or discontinuity at the origin. This is one limit case for the nugget effect that sometimes occurs in practice. It is also the desirable limit case. Figure 10-12 shows another limit case in which all that exists is a nugget effect for lead in soil. This variogram is totally devoid of any spatial continuity and indicates that the data are independent.

This situation is termed the pure nugget effect. Variability is essentially constant for all distances *h*. This implies that the data are independent and that geostatistical methods cannot decrease the error when predicting the concentration of unsampled locations in comparison to classical statistical approaches. In other words, the two approaches are equivalent when a pure nugget effect exists, notwithstanding the declustering contribution made by kriging.

10.1.2 Zone of Influence

A zone of influence around a sample is an intuitive concept understood by most earth scientists. It is generally assumed that a sample will predict con-

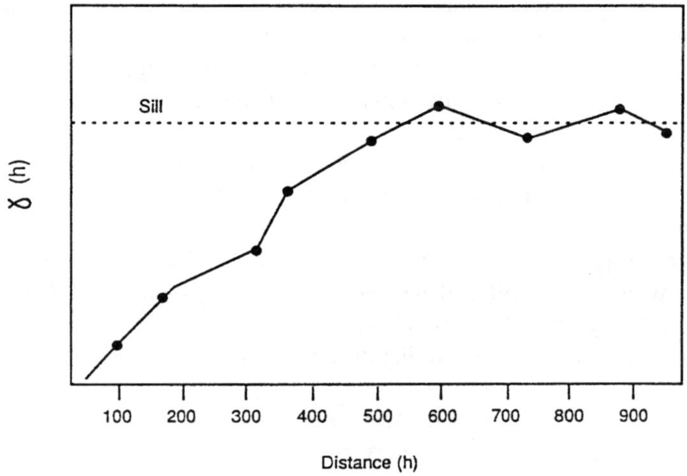

Figure 10-11. Experimental variogram for copper concentrations in soil, exhibiting no nugget effect.

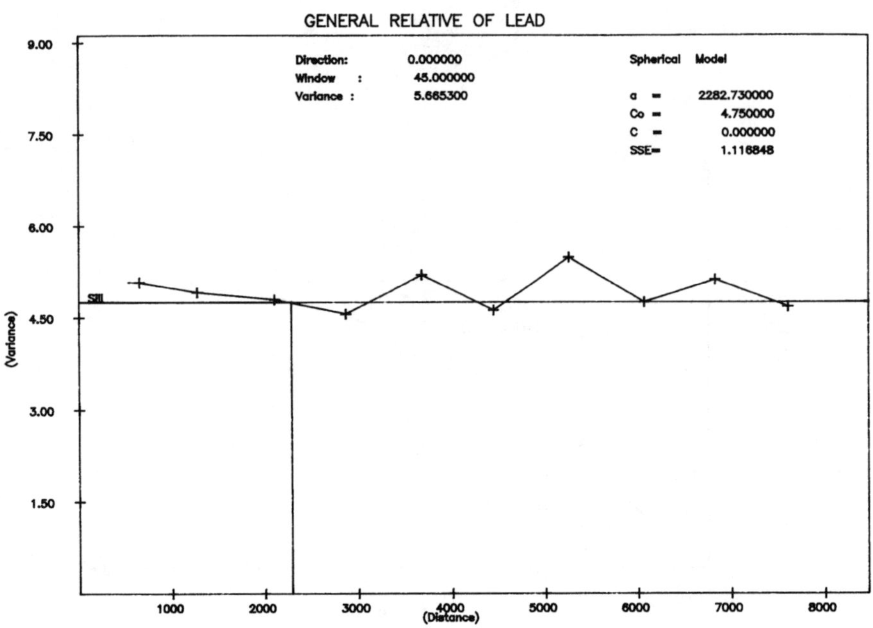

Figure 10-12. Pure nugget effect variogram for lead (Pb) in soil.

centrations for a certain distance from the sample location with some degree of accuracy. The range of the variogram quantifies the zone of influence by providing a distance within which similarity between samples is noted. Beyond that zone, data are independent. This concept was described in Chapter 3 as the neighborhood.

The variogram can serve as a rough measure of the level of accuracy for predictions when only one sample datum is used. Figure 10-13 shows an ideal variogram. As distance increases, the variance also increases. The counterpart to the variance (Fig. 10-13) is the covariance or correlation. The covariance is the difference between the sill and the model variance as shown in Figure 10-13. As the variance grows, the correlation component shrinks. Since correlation is information, we can see that prediction becomes more difficult as distance increases.

Figure 10-14 shows how the variogram may be used to quantify the error in prediction. The contaminant concentration—say, 50 ppm—is known at a sample location *A*. The known concentration can be extended to the unknown location *B*, giving *B* a value of 50 ppm as well. When this extension is performed, an error is likely to result.

The distance between *A* and *B* is known to be 15 ft. The value of $\gamma(15)$ from the variogram in Figure 10-14 is approximately 44 $(\text{ppm})^2$ or a standard deviation of around 6.7 ppm. Assuming a normal distribution of the errors, we can say that the value at point *B* is 50 ppm ± 6.7 ppm at a 68% level of confidence.

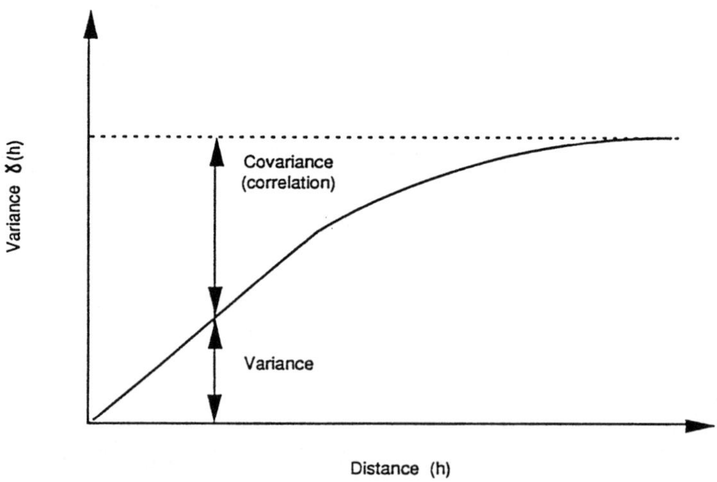

Figure 10-13. Ideal variogram showing correlation and variance components.

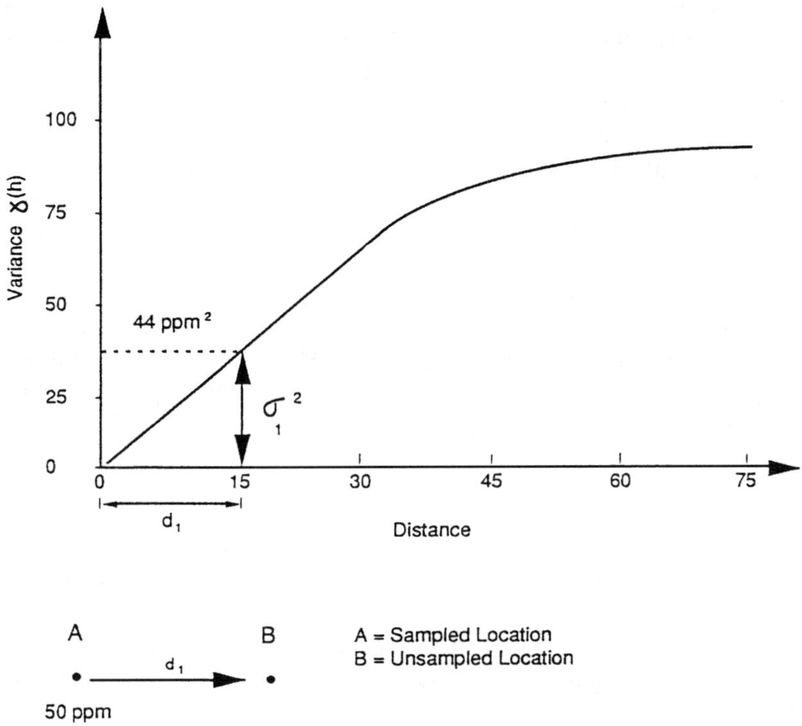

Figure 10-14. Calculation of the expected prediction error at an unsampled location using the variogram.

10.1.3 Variogram Models

A well-behaved experimental variogram provides a clear picture of the spatial continuity present at a site for samples of a given support. It is limited in practical usefulness, however, in that $\gamma(h)$ values are available for only those specific distances corresponding to the average separation distances between samples derived in the calculation of the experimental variogram. The number of average separation distances is almost always less than 20. The kriging process will require the knowledge of the variogram value between all sample locations and the locations at which estimates need to be made. It is, therefore, extremely unlikely that the 20 or less h distances available will satisfy the multitude of distances required during the kriging process.

The solution comes in the form of a mathematical model for the experimental variogram. A number of functions are commonly used to model variograms and will be described in this section. Whereas a suite of models is available, not all mathematical functions are appropriate. Only positive definite functions may be used in order to avoid problems in the kriging matrices. The models discussed in this section all qualify as positive definite functions.

Although the few models presented here may appear to be limiting, these models may be combined to provide new opportunities. Experience in practical situations also has proved that these models or combinations of models are sufficient for the vast majority of situations encountered. The interested reader is referred to the thesis by Matheron (1965) where a detailed discussion of admissible variogram functions is presented.

Models are loosely grouped into two categories: (1) those with sills and (2) those without. Those with sills or plateaus, sometimes called *transitive models*, include the spherical, exponential, Gaussian, and hole effect models. Those without sills include the nugget effect and the linear model. Both types of models will be discussed below.

10.1.3.1 Spherical Model

The spherical model is, in all probability, the most frequently used function for modeling variograms. Its shape is presented in Figure 10-15. The beginning portion of the graph, indicating small separation distances, displays a linear component. This linear behavior is lost at larger distances, and the curve flattens, eventually reaching a plateau. A line tangent at the origin will intersect the sill at two-thirds the range.

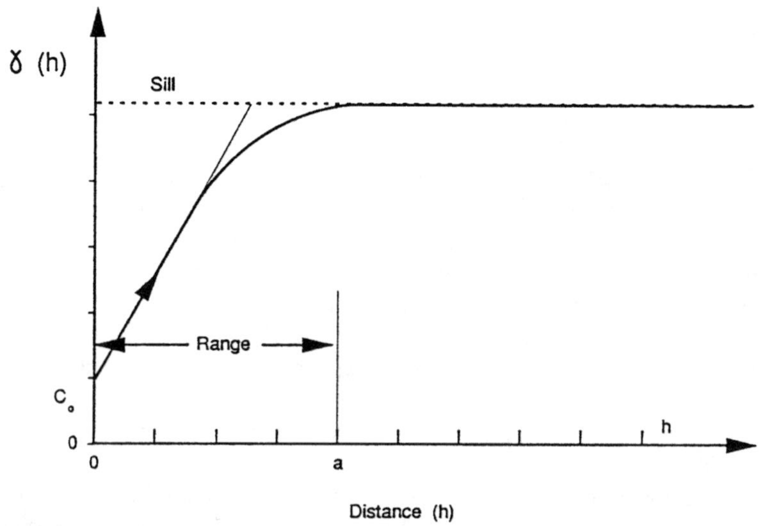

Figure 10-15. Spherical variogram model.

The equation for the spherical model is

$$\gamma(h)=C\left[\frac{3}{2}\frac{h}{a}-\frac{1}{2}\frac{h^3}{a^3}\right]+C_o \qquad h\leq a \tag{10.2}$$

$$\gamma(h)=C-C_o \qquad h>a \tag{10.3}$$

$$\gamma(0)=0 \tag{10.4}$$

where a is the range, C_o is the nugget effect, and $C + C_o$ the sill.

10.1.3.2 Exponential Model

The exponential model is visually similar to the spherical model (Fig. 10-16) but represents a more continuous process. It rises more steeply from the origin than the spherical model but then flattens at a more leisurely rate. In fact, it approaches a sill only asymptotically, with a practical range usually defined to be the variogram value representing 95% of the sill. As with the spherical model, it is linear at the origin. With the exponential model, the linear component exists

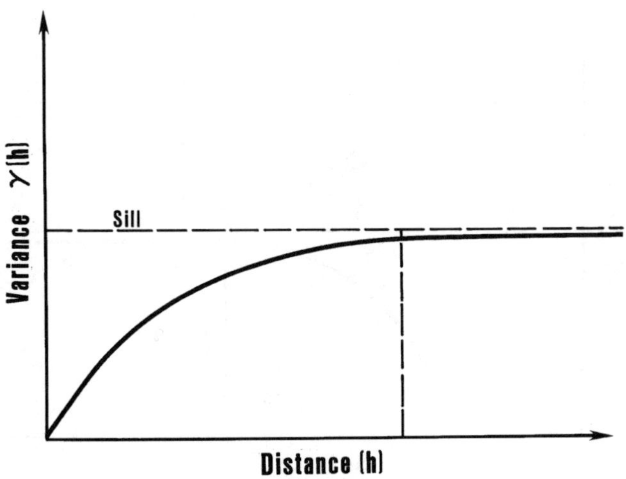

Figure 10-16. Exponential variogram model.

only for very small distances. A line tangent at the origin will intersect the sill at one-fifth the range. The equation for the exponential model is

$$\gamma(h) = C_o - C\left[1 - \exp\left(-\frac{|h|}{a}\right)\right] \tag{10.5}$$

10.1.3.3 Gaussian Model

The Gaussian model is typically used to model processes that show extremely low variation at short distances, to the extent that the model is parabolic at the origin (Fig. 10-17). As with the exponential model, the Gaussian model asymptotically approaches its sill, the practical range again defined as the distance at which the variogram value achieves 95% of the sill. The equation is

$$\gamma(h) = C_o + C\left[1 - \exp\left(-\frac{3h^2}{a^2}\right)\right] \tag{10.6}$$

The exponential model should be used with caution as it is prone to causing large negative weights during kriging.

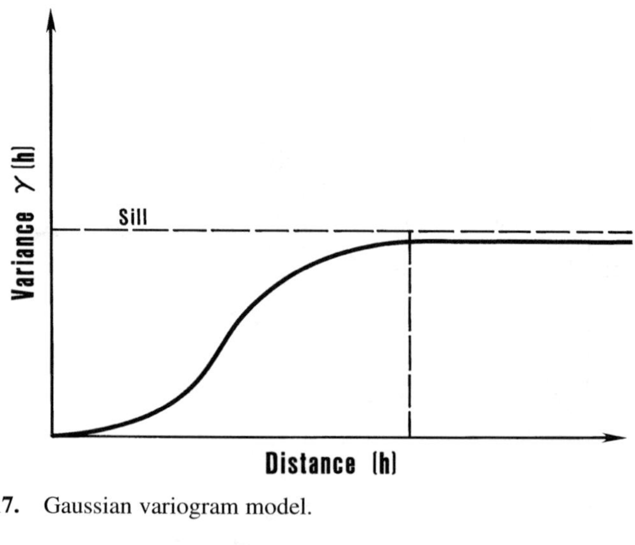

Figure 10-17. Gaussian variogram model.

10.1.3.4 Hole Effect Model

The hole effect model is used to model periodic behavior. This type of phenomenon is encountered where a series of high- and low-concentration areas exist in succession—in an area of multiple spills, in lagoons, or in process areas, for example. The graph is slightly parabolic at the origin, although not so much as for the Gaussian model. After reaching a maximum, the graph oscillates up and down in a sine wave of ever-decreasing amplitude (Fig. 10-18). The equation for the hole effect model is a combination of the exponential model with a sine model:

$$\gamma(h) = C\left(1 - \frac{\sin ah}{ah}\right) \tag{10.7}$$

10.1.3.5 Nugget Effect Model

As seen in Figure 10-10, many variogram graphs do not begin at the origin. Rather, a vertical discontinuity exists at distance zero. By definition, $\gamma(0)$ is always zero, but it is known that variation may exist (fundamental error, analytical error, etc.). Also, variation may exist at very small distances. The nugget effect is defined as a constant C_o and is simply added to the other variogram model or models being used. For purposes of estimation, $C_o = 0$ when $h = 0$.

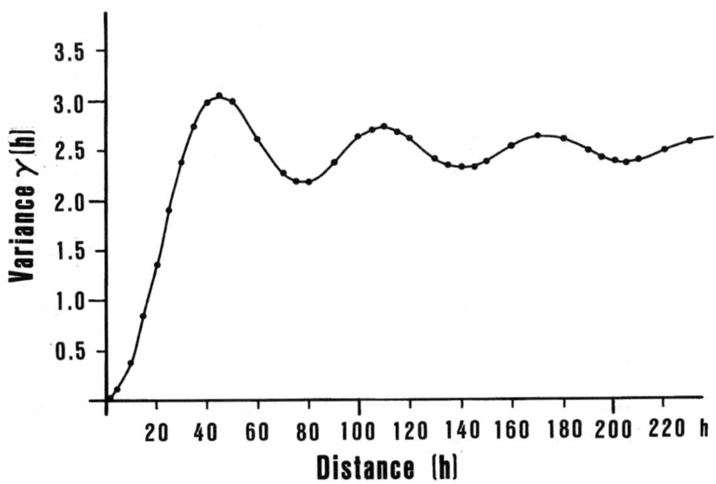

Figure 10-18. Hole effect variogram model.

10.1.3.6 Linear Model

Aside from the nugget effect model, the linear model is the simplest of the variogram models. The linear model does not reach a sill; instead, it grows in a linear fashion as a function of distance *h* (Fig. 10-19). Its equation is written

$$\gamma(h) = ah + B \tag{10.8}$$

10.1.4 Anisotropy

Sample data, along with the intuition of the geologist, have often suggested that it is easier to predict the value of an unsampled location in one direction than in another. This may correspond to values along strike, down dip, downwind of a smelter, along the axis of a groundwater plume, and so on.

By calculating variograms in different directions and evaluating the associated ranges, the anisotropy can be determined. Figure 10-20 shows four directional variograms plotted on the same graph. Although there is some variation in the placement of the points for the curves, each reaches the sill at a distance of 100 feet. This is a simple case and shows no anisotropy because all ranges are equal. Equal ranges are also termed *isotropic ranges.*

Perhaps a more common case is where ranges differ according to direction. For example, as shown in Figure 10-21, the north-south range is clearly longer than the east-west range, in this case by a factor of 2. This gives an anisotropy

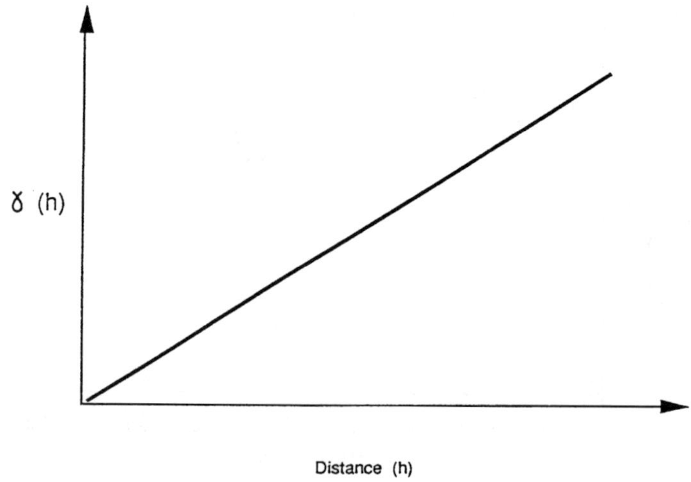

Figure 10-19. Linear variogram model.

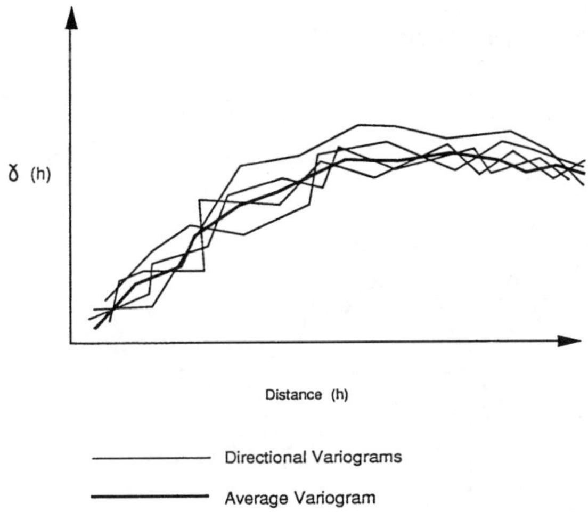

Figure 10-20. Four directional experimental variograms: isotropic situation.

ratio of 2:1. This also means that if we try to predict the value of a location 20 ft to the north or south of a known location, the expected error will be the same as that for an unsampled location only 10 ft east or west of the known location. Thus, there is more information available along the axis of greater continuity. This information can be incorporated into a sampling program that uses the budget efficiently, as fewer samples are needed in the north-south direction.

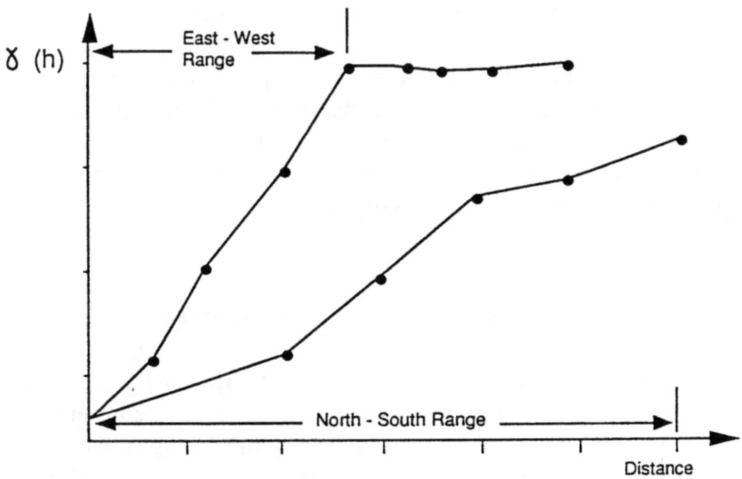

Figure 10-21. Anisotropic variogram ranges.

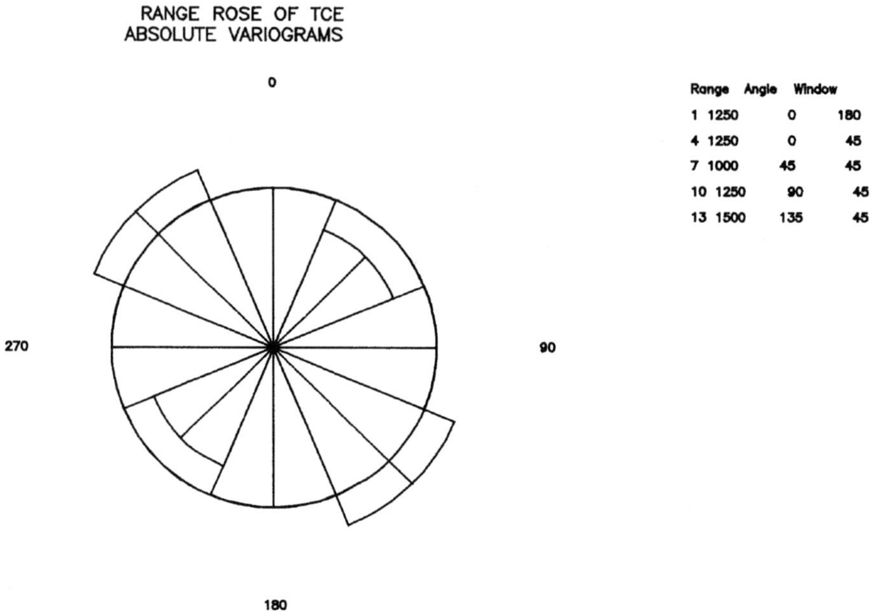

Figure 10-22. Anisotropic range rose diagram.

There are two commonly used methods for displaying and interpreting anisotropy in practice. The most common is through a range rose diagram; examples are shown in Figures 10-22 and 10-23. Anisotropic variograms stand out as an ellipse (Fig. 10-22) with isotropic situations forming a circle (Fig. 10-23).

A second method is to use a contour map of the experimental variogram surface. The contour map is centered at a pair separation distance of zero. Pairs are grouped according to the separation distances to form a grid and then contoured. The result can be seen in Figure 10-24. The contour map rises from the center of the grid (zero distance) in a pattern similar to that of a cone of depression for groundwater. An anisotropy can easily been seen.

10.1.5 Types of Anisotropy: Geometric vs. Zonal

In practice, two types of anisotropy exist: (1) geometric anisotropy and (2) zonal anisotropy. The anisotropy described thus far refers to geometric anisotropy. As shown in Figure 10-21, the range of the variogram changes according to the direction while the value of the sill remains constant. For example, if the horizontal ranges are 50 ft and the vertical range is 10 ft, the anisotropy ratio k is 50/10

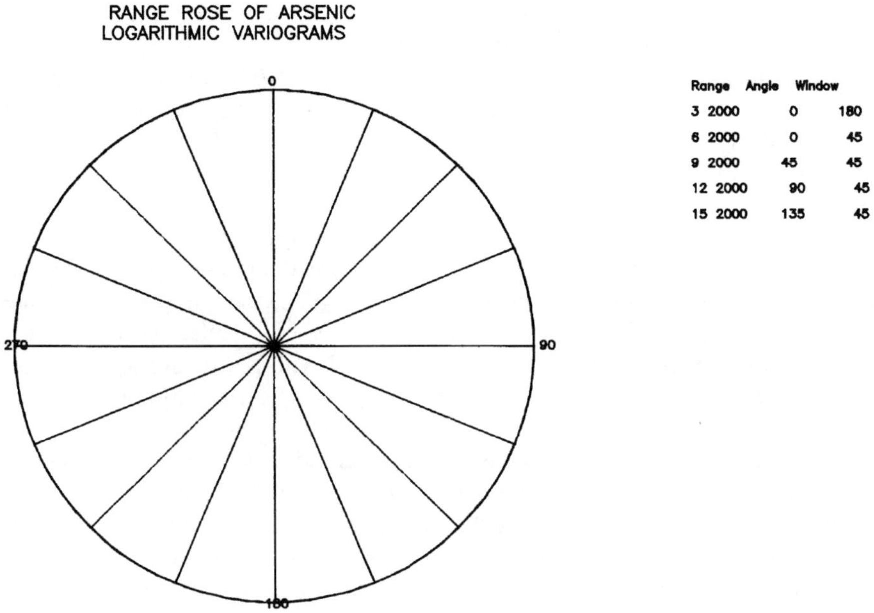

Figure 10-23. Isotropic range rose diagram.

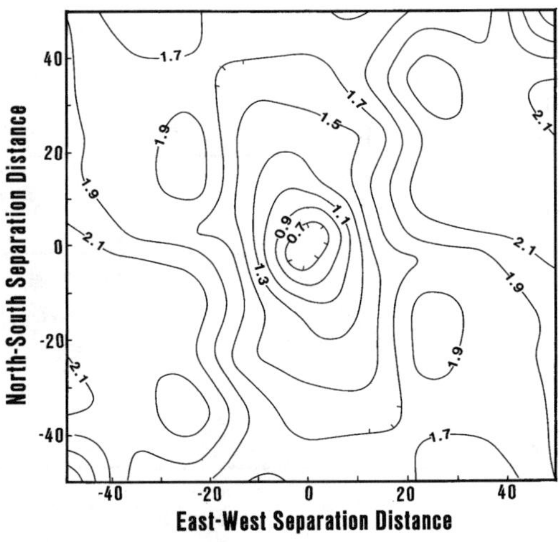

Figure 10-24. Contour map of variogram values.

= 5. Thus, two samples 20 ft apart horizontally show the same amount of similarity as two samples 4 ft apart vertically.

The anisotropy ratio k is characteristic of a given COC at a site for a sample of a given size. The conceptualization is quite simple: The anisotropy can be thought of as an ellipsoidal zone of influence. The ellipsoid, once a sphere, has been altered by the anisotropy ratio k, which modifies the distance units on one or more axes.

Zonal anisotropy differs from geometric anisotropy in that the ranges of the variograms remain constant while the sill changes with direction. This phenomenon is commonly equated with layering of some type, although the layering may be more conceptual than real as no layering may be visible. In any event, a direction of maximum variation exists contrasted with a direction of minimum variation. Figure 10-25 shows a visual display of zonal anisotropy for the case of the spherical variogram.

Once the axes of anisotropy have been identified, it is necessary to construct a model that reflects the changes in the variogram due to distance and direction. Isaaks and Srivastava (1989) provide a useful method for combining all directional models into a single directional model that has a range of 1. This method transforms the separation distance to a standard model such that the variogram value $\gamma(h)$ is identical to any directional model at that particular value of h.

Figure 10-26 shows two spherical variograms with different ranges and the same sill. The shorter-range variogram can be thought of as a unit variogram with a range equal to 1. The other variogram has a range equal to an arbitrary distance h that is greater than 1. The variogram value at distance h/a for the unit variogram

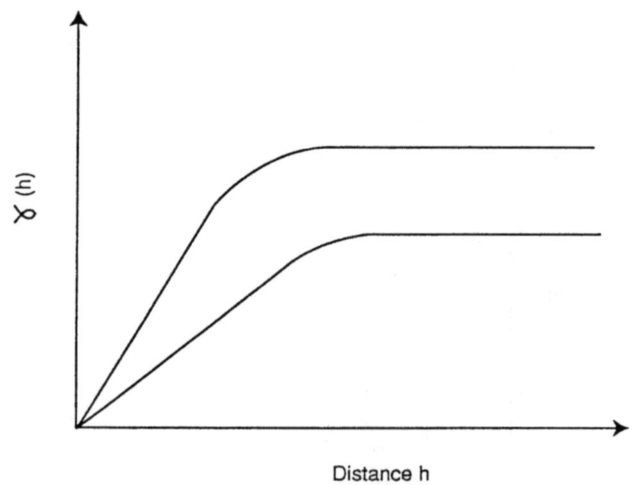

Figure 10-25. Zonal anisotropy for the spherical variogram model.

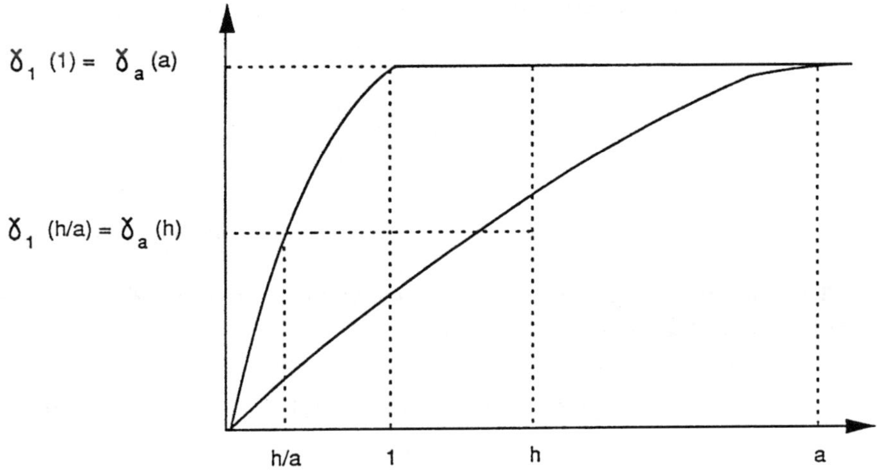

Figure 10-26. Variogram models for the reduced distance technique. (From *An Introduction to Applied Geostatistics*, by E.H. Isaaks and R.M. Srivastava. Copyright © 1989 by Oxford University Press, Inc. Used by permission of Oxford University Press, Inc.)

(range = 1) is exactly equal to the value for the variogram of range a at distance h. This reduces the range a model to an equivalent model based on a range of 1, which, in turn, is based on the reduction of the separation distance h to h/a. This can be expressed as

$$\gamma_1\left(\frac{h}{a}\right)=\gamma_a(h) \qquad \text{or} \qquad \gamma_1(h)=\gamma_a(ah) \tag{10.9}$$

or, if $h_1 = h/a$, as

$$\gamma_1(h_1)=\gamma_a(h) \tag{10.10}$$

This shows that, simply by replacing the separation distance h by h/a, any directional model with range a can be reduced to a standard model with a range of 1. Extending this concept to two dimensions, a_x and a_y become the ranges in the x and y directions, respectively. This can be expressed as

$$\gamma(h)=\gamma(h_x, h_y)=\gamma_1(h_1) \tag{10.11}$$

with the reduced distance given by

$$h_1 = \sqrt{\left(\frac{h_x}{a_x}\right)^2 + \left(\frac{h_y}{a_y}\right)^2} \tag{10.12}$$

where h_x is the component of h along the x axis and h_y the component along the y axis.

The final step is to expand the anisotropic variogram model to three dimensions, with ranges a_x, a_y, and a_z expressed as

$$\gamma(h) = \gamma(h_x, h_y, h_z) = \gamma_1(h_1) \tag{10.13}$$

and the reduced distance h_1 given by

$$h_1 = \sqrt{\left(\frac{h_x}{a_x}\right)^2 + \left(\frac{h_y}{a_y}\right)^2 + \left(\frac{h_z}{a_z}\right)^2} \tag{10.14}$$

The directional variogram models along all three axes of a three-dimensional geometric anisotropy can be seen in Figure 10-27. Note that linear and other models that do not reach a sill can make use of the equivalent model approach. For example, equations 10.2 through 10.8 can be applied to linear directional models with slopes

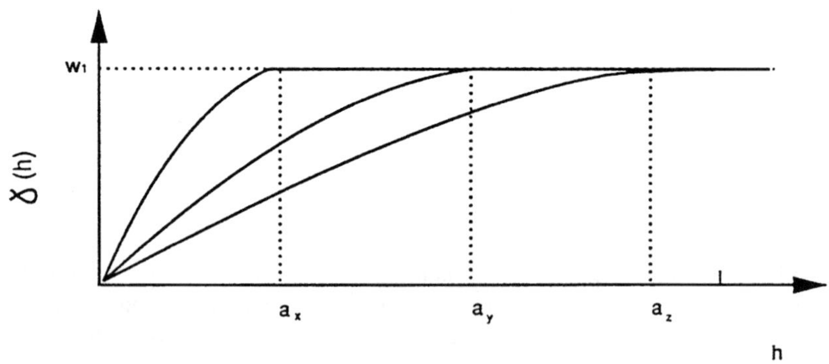

Figure 10-27. Directional variogram models along three axes of three-dimensional geometric anisotropy. (From *An Introduction to Applied Geostatistics*, by E.H. Isaaks and R.M. Srivastava. Copyright © 1989 by Oxford University Press, Inc. Used by permission of Oxford University Press, Inc.)

a_x, a_y, and a_z with the reduced distances being $a_x h$, $a_y h$, and $a_z h$. The equivalent isotropic model with slope 1 is given by $\gamma_1(a_x h) = \gamma_{ax}(h)$ and so forth.

10.1.6 Other Features

It has been shown that the variogram can quantify the geologic concepts of continuity, zone of influence, and anisotropy. In addition to these standard situations, the variogram can provide other interesting insights into the nature of the contamination.

Figure 10-28 shows a variogram of lead (Pb) concentrations in soils surrounding the DMC smelter at the Dallas Lead site. Initially, the variogram graph rises, as is typical. At a distance of approximately 500 ft, the rate of increase in the variance decreases although the variance still continues to rise. Eventually, the graph levels off at around 5000 ft.

Figure 10-28 is an example of a *nested* variogram structure. Such an occurrence suggests that two different phenomena are influencing the contaminant distribution and, subsequently, the variogram. For the lead (Pb) smelter, it may indicate two different domains, perhaps related to wind, particle size, or human activity.

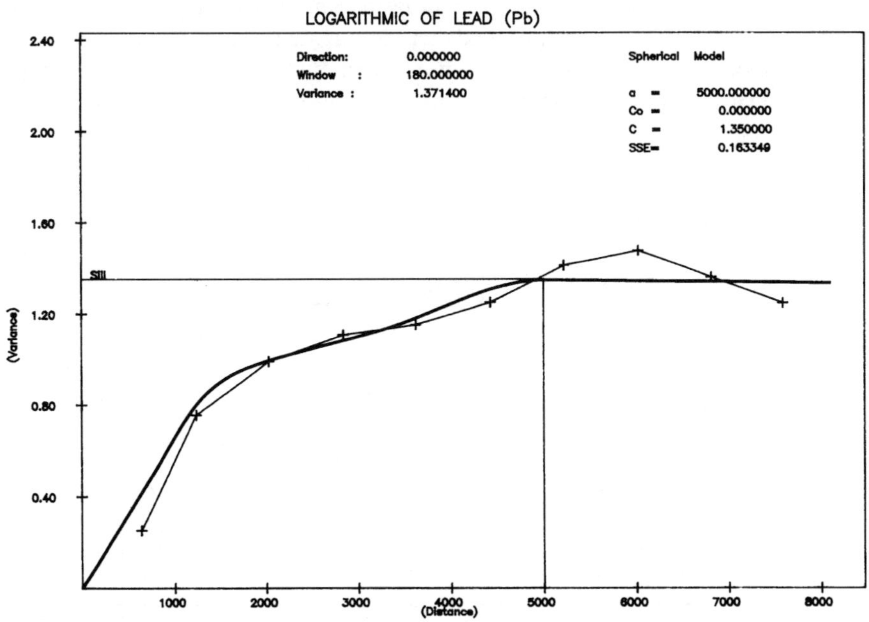

Figure 10-28. Nested variogram structure for lead (Pb) values. (From Geostatistical sampling and contamination analysis of lead values surrounding industrial smelters, by J.C. Myers, Master's thesis, Colorado School of Mines, 1985.)

Figure 10-29 shows another interesting situation that can occur in variogram analysis. The variogram rises from the origin, reaches a maximum (not really a sill since a sill implies some form of stability or constancy), and then plummets toward zero variance.

Conventional wisdom states that the sill, once achieved, is maintained. The graph in Figure 10-29 tells us two things. One is that the entire contaminant plume has been bridged by the lag distances. The second is that the contamination is generally high in the center of the plume and decreases systematically away from the center, eventually reaching background concentrations.

Consequently, the maximum variance is reached when samples from the background area on one side of the plume (say the west side) are compared with the high-concentration samples in the center of the plume (source area). At larger lag distances, the west-side samples are compared to samples further east that decrease in concentration away from the source, and the variance decreases. Eventually, west-side background samples are compared to east-side background samples. Since both sets of background samples have very low concentrations, the variance is correspondingly low.

A variation on this theme is shown in Figure 10-30. In this case the variogram reaches a maximum, then declines, then rises again, and repeats this pattern on a smaller and smaller scale. This indicates that a number of small plumes or extensions of plumes may be present. As the variogram crosses the transition zones between plumes, the variance falls, and then rises again as concentrations rise. This was described in Section 10.1.3.4 as the hole effect. Figure 10-31 shows an example of cyclical variogram behavior in temporal air monitoring (Buxton 1994).

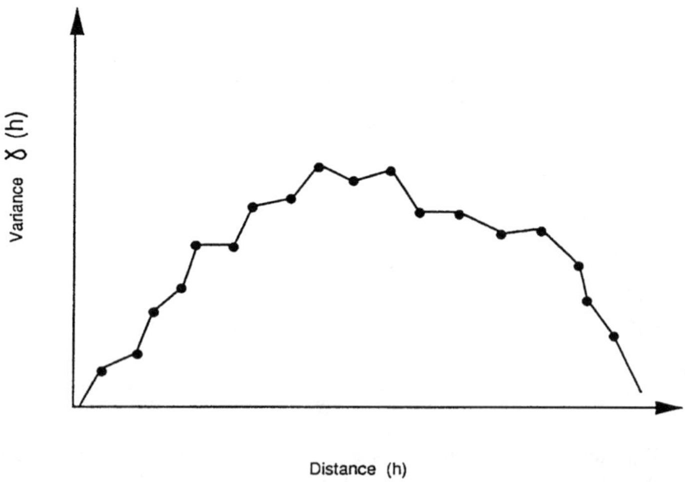

Figure 10-29. Variogram arch for soil pH data.

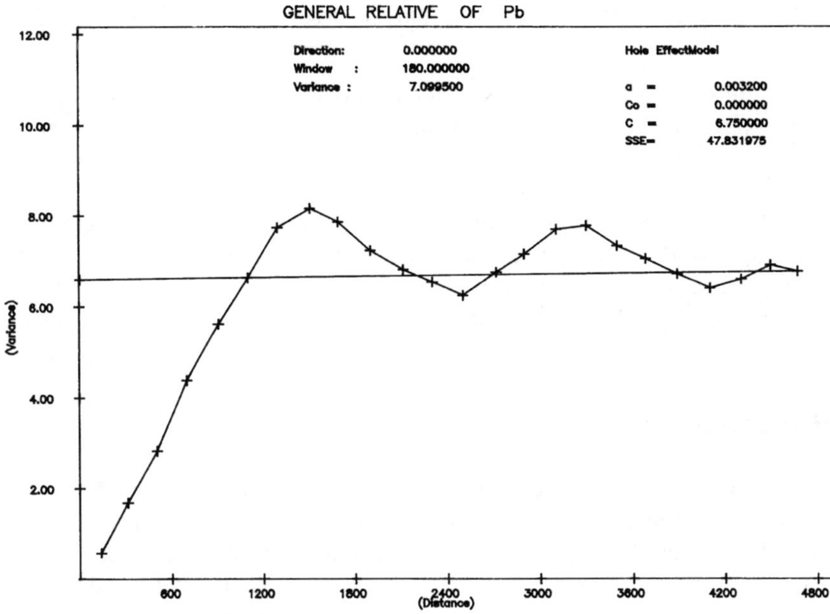

Figure 10-30. Hole effect for experimental variogram calculation.

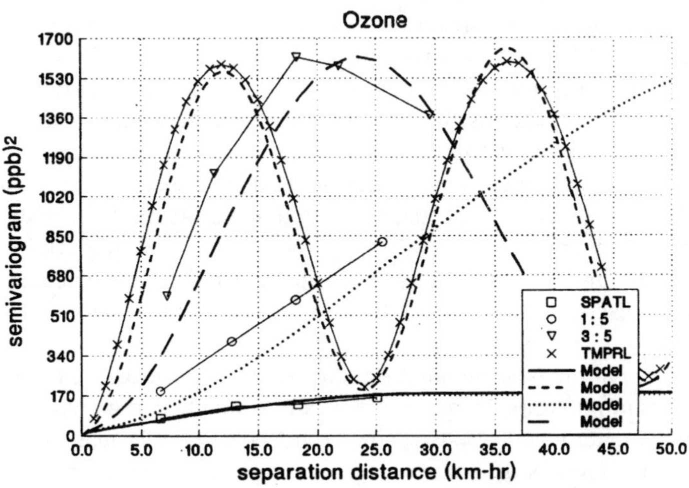

Figure 10-31. Variograms showing temporal and joint spatial-temporal oscillations in ozone concentrations. (Reprinted by permission of Kluwer Academic Publishers from Joint-temporal spatial modeling of concentrations of hazardous pollutants in urban air, by B.E. Buxton and A. Pate, in *Geostatistics for the Next Century*, Vol. VI, ed. R. Dmitrakopoulos, Fig. 2, 82. Copyright © 1994 Kluwer Academic Publishers.)

10.2 Calculation of Experimental Variograms

Section 10.1 assumes that the involved detail work of calculating the experimental variogram has been performed so that the best structure has been found and the anisotropy axes have been determined. There are a number of practical considerations in the process of calculating experimental variograms that significantly affect the experimental graphs. Among these are the angular window and direction, the lag distance, the vertical window, the influence of outlier data, and the data domains. Each can have a significant impact on the results of the experimental variograms and can help unlock and/or incorporate the site geology and history.

10.2.1 Angular Window and Direction

The convenient square grid of samples shown in Figures 10-1 and 10-8 for variogram pairing and calculation is a limit case that seldom occurs in practice. The usual case yields sample data that are taken at unequal intervals scattered across the site. Even data planned on a regular grid are inevitably altered to adjust to practical realities: lakes, buildings, roads, steep terrain, lack of permission from landowners, even survey precision.

With scattered data, samples are not aligned in perfect north-south or east-west patterns. As such, it becomes necessary to allow an angular tolerance for sample pairs in variogram calculation. It would be highly unusual for two samples to be aligned exactly north-south or east-west from each other. Therefore, without an angular tolerance, it would be impossible to calculate a variogram.

Figure 10-32 illustrates the process for capturing sample pairs in angular windows. The first step is to specify the angular direction to search for pairs, in this case, east-west. In addition, an angular window is specified. In Figure 10-32, the angular window is 45°. The angular direction bisects the 45° window, leaving a window of 22.5° on either side of the east-west directional line. All samples that fall within the angular window of 45° will be accepted for calculation of the experimental east-west variogram. All samples that fall outside this window will be rejected.

It is necessary to have an understanding of the angular reference system used by a variogram program. Many programs are available commercially and in the public domain, but not all use the same reference system. In many programs, the east-west direction has an angular window of 0°, corresponding to a Cartesian coordinate system commonly used by mathematicians. Thus, the north-south direction would have an angular direction of 90°.

Other packages, however, use an "azimuth" system for directional orientation, corresponding to the system used by most field geologists and engineers. In this case, 0° is north and 90° is east. Failure to understand the program default

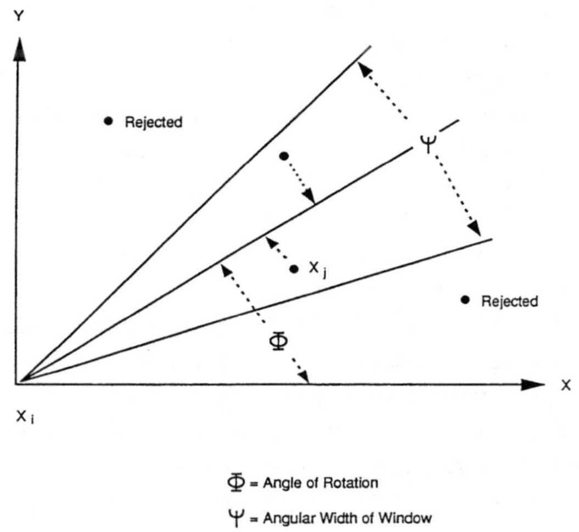

Figure 10-32. Angular window to capture variogram pairs.

orientation can result in a misinterpretation of anisotropy relationships. Examples of the Cartesian and azimuth orientation approaches can be seen in Figures 10-33a and 10-33b respectively.

The angular window may be as large or as small as desired, although some practical limitations do arise. For very small angular windows, it may not be

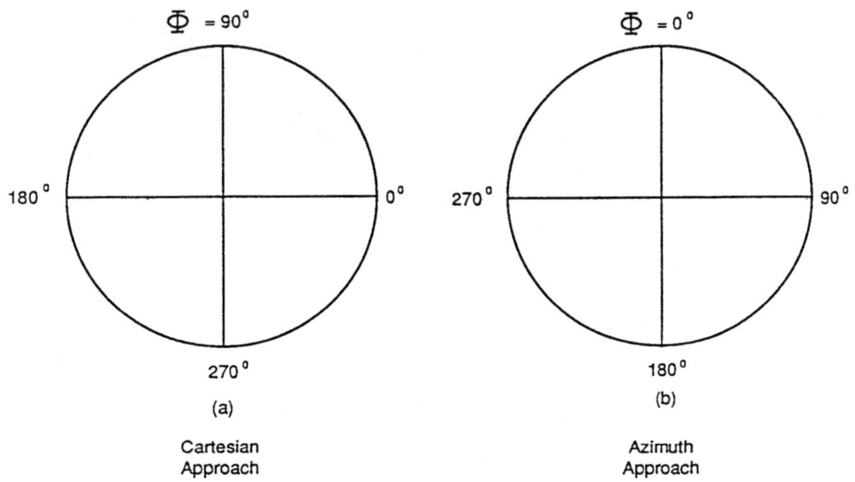

Figure 10-33. Approaches to experimental variogram orientation: (a) Cartesian and (b) azimuth.

possible to collect a sufficient number of sample pairs to calculate the variogram. In fact, in small data sets, no sample pairs may be a common result.

For large windows, the concept of direction becomes blurred, although finding sample pairs is not generally a problem. An angular window of 45° is usually a reasonable starting point for experimental variogram analysis. An illustration is shown in Figure 10-34. Four angular directions are used: north-south, northeast-southwest, east-west, and northwest-southeast, or 0°, 45°, 90°, and 135°, respectively, using the azimuth orientation system. These are often referred to as the cardinal directions for experimental variogram calculation.

The system shown in Figure 10-34 allows all sample data to be captured in one and only one window. All data are utilized, yet none overlap in two windows, and, thus, cause a potential for interference.

A similar convention issue is of concern for the definition of angular windows. Some programs define a 45° window by having the user input a value of 22.5°; one-half the total window. Other programs require an input value of 45° for a 45° window.

Commonly, an omnidirectional variogram is calculated along with the directional variograms. For an omnidirectional variogram, the angular window is set to a large enough value (e.g., 180°) to allow all data to be evaluated without regard to angular direction. This can be viewed as a type of "average" variogram and will generally plot in the middle of the four (or more) directional variograms. This can be seen in Figure 10-35. It should be noted, however, that this is not a strict average since certain portions of the site may be overrepresented by denser sampling.

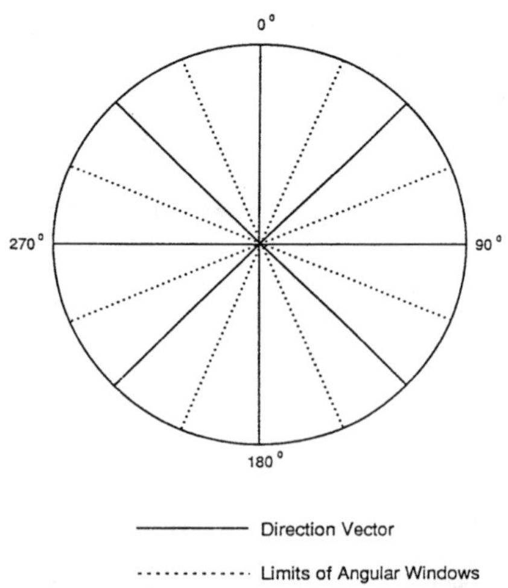

Figure 10-34. Cardinal directions and window for variograms.

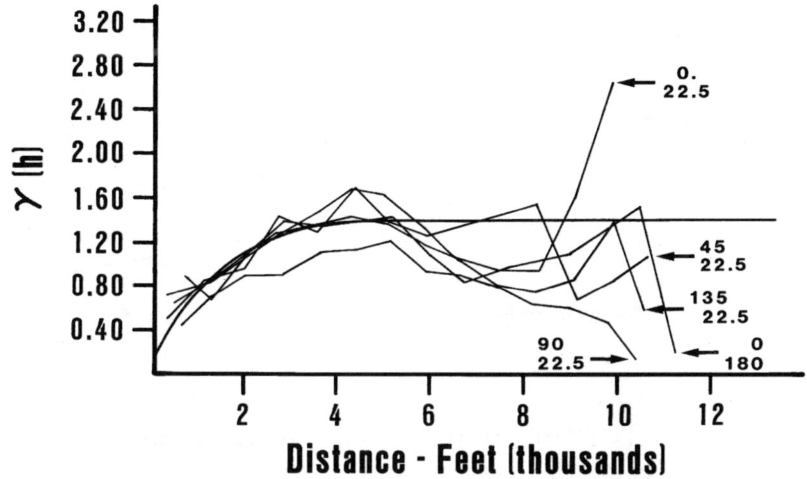

Figure 10-35. Average variogram with four directional variograms plotted for Dallas Lead (Pb) soil data, RSR smelter. (From Geostatistical sampling and contamination analysis of lead values surrounding industrial smelters, by J.C. Myers, Master's thesis, Colorado School of Mines, 1985.)

The omnidirectional variogram is often a good starting point for experimental variogram analysis. Because it uses all the data, the omnidirectional variogram will have the most sample pairs. As such, it is more likely to produce a definable structure. If no structure is present, it is not likely that the directional variograms will produce good structures.

10.2.2 Lag Distance

Scattered data also create the need for a type of distance tolerance in experimental variogram analysis. Because all data are at different distances, all data within a specified distance interval or "lag distance" are treated together.

For example, a lag distance of 100 ft (h_1) may be specified. Based on this, all sample data between 0 and 100 ft are analyzed together. Similarly, all data between 100 and 200 ft are grouped for analysis, as are all data between 200 and 300 ft, and so forth. An example is shown in Figure 10-36.

Data may fall anywhere within the lag distance and still contribute to the variogram for that distance. It is common in practice, however, to plot the $\gamma(h)$ value at the average distance for all pairs in the specified distance interval. For example, the average distance between sample pairs in the 100- to 200-ft category may be 165 ft. The $\gamma(h)$ value would then be plotted at 165 ft instead of one-half

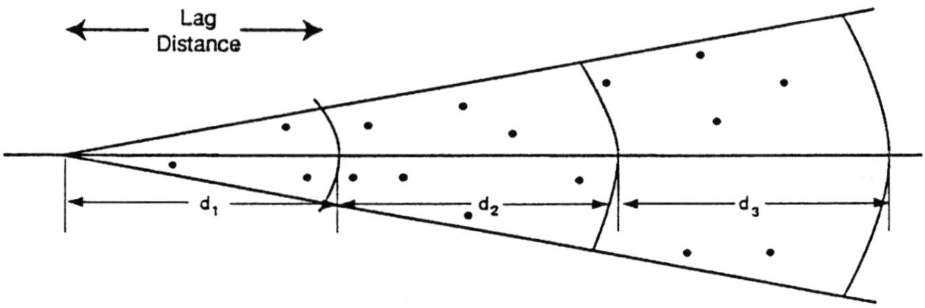

Figure 10-36. Data grouped by lag distance.

the distance interval (150 ft). This is essentially plotting the variance $\gamma(h)$ at the center of gravity for the lag distance interval.

Without the benefit of a standardized regular sample grid, it becomes necessary to select an appropriate lag distance for variogram analysis. Lag distances that are too large mask the spatial structure, whereas inappropriately small lags do not capture sufficient samples. Inappropriate selection of lag distances can even produce pure nugget variograms, suggesting no structure or continuity.

A basic rule of thumb for selecting an appropriate lag distance is to use the "average" or median distance between sample data points. We can visually examine the data and determine a distance between samples that is common or representative; we need not actually measure and calculate an average.

During the variogram analysis, the average or median distance provides a good starting point. Other lag distances, both larger and smaller, should also be used to determine the sensitivity of the lag distance to variogram results. Typically, lag distances slightly less than the average or median distance provide good results. They also can be helpful in providing a good indication of the *Y* intercept on the variogram graph (nugget effect). It is possible to analyze separately a single lag with a very short distance to investigate closely spaced or duplicate samples.

10.2.3 Vertical Window

For three-dimensional data, it is also necessary to define a vertical tolerance for sample pairs. Ideally, a three-dimensional sample search would be conical in shape (Fig. 10-37). This type of search is rather difficult and rapidly consumes computer time. As a good approximation, a "pie-slice" approach is often used to speed up the search process.

Figure 10-38 illustrates the pie-slice approach. A tolerance is given to the program for the vertical direction. For each sample location, this vertical tolerance

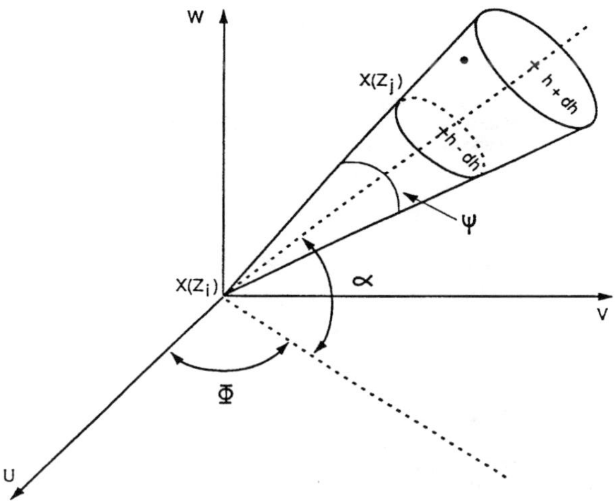

Figure 10-37. Conical search for variogram computation in three dimensions.

is allowed for samples falling within the specified lag distance and angular direction.

The vertical window can greatly affect the experimental variogram results. While good continuity and structure often exist horizontally, concentrations tend to change rapidly in the vertical direction. Thus, small vertical windows are generally recommended.

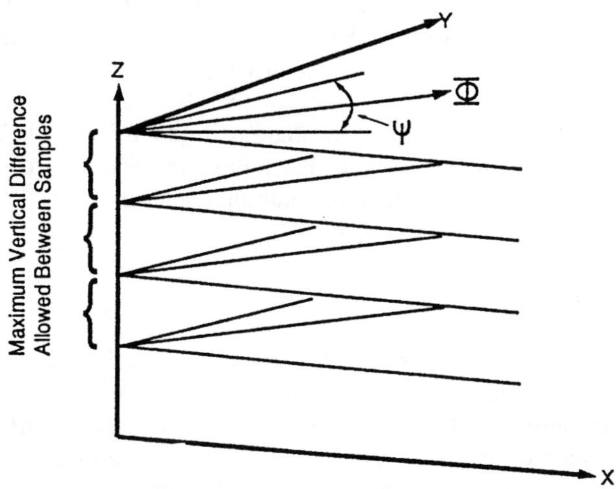

Figure 10-38. Vertical window "pieslice." (Courtesy of Geostat Systems International.)

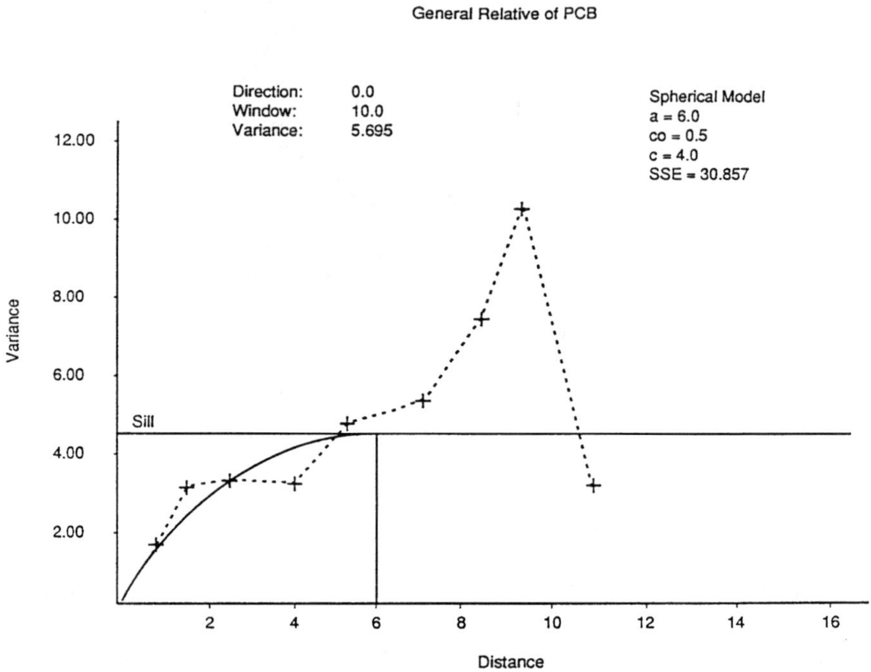

Figure 10-39. Vertical variogram for PCBs in soil at the ReSolve site.

Although the vertical window introduces a vertical component to the analysis, it does not constitute a vertical variogram. Instead, a separate variogram should be performed in the vertical direction ("down the hole"). Note that the lag distance in the vertical direction will typically be much smaller than that for the horizontal direction.

Figure 10-39 shows a vertical variogram for PCBs in soil. The 1-ft lag distance for analysis matches the vertical sampling interval of 1-ft. samples. The corresponding lag distance for the horizontal direction was 40 ft.

10.2.4 Outlier Data

Because the variogram uses a squared function, large differences in concentration between the sample data composing a variogram pair become exponentially large. This can cause the shape of the experimental variogram curve to become erratic and difficult to interpret. With sample concentrations often spanning several orders of magnitude in environmental situations, erratic behavior often arises.

One way to attempt to improve the appearance of the variogram is to remove the samples containing the highest concentrations. The idea is that these have the greatest potential to produce large squared differences. Two problems exist with this approach. First, this assumption is not necessarily true. The samples with high concentrations may be grouped with others of similar magnitude, producing relatively small differences. Thus, removing them will not necessarily solve the problem.

The second problem is how to determine how many samples with high concentrations to remove. There are no exact rules, numbers, or percentages. And as just stated, these may not be significant in improving the variogram structure.

It is possible to eliminate the particular pairs of data that contribute most to the erratic nature of the experimental variogram. Although this seems to be a good solution, it often has little effect on the variogram results. In any situation, removing samples is generally not satisfying and, with the absence of objective rules, any procedure may be interpreted as arbitrary. We must also consider that the data are part of the true continuity structure that should not be ignored.

Despite these problems, removal of problem data or data pairs can dramatically improve the appearance of the experimental variogram for many data sets. It is not possible to tell beforehand whether data removal will be effective.

Difficulty obtaining good spatial continuity may indicate that the statistical domains have been improperly defined. This is a good point at which to check the assumptions used to develop the domains. Also, poor sampling (i.e., high *FE*, *GE*, etc.) may be influencing the variogram results. Unfortunately, the opportunity to correct sampling problems has usually been lost, yet the penalty will continue to be paid.

10.3 Types of Variograms

The erratic nature of the variogram for certain data sets may be addressed in more sophisticated ways. The use of various mathematical transformations can be effectively employed to achieve better results.

10.3.1 Relative Variograms

It has often been found in earth science applications that the variance of a population grows as the mean value of the parameter of interest grows. Correspondingly, the sill of the variogram grows as the mean increases.

This phenomenon indicates a relationship between the mean and the variability of the data. In fact, a linear relationship often exists between the mean and the

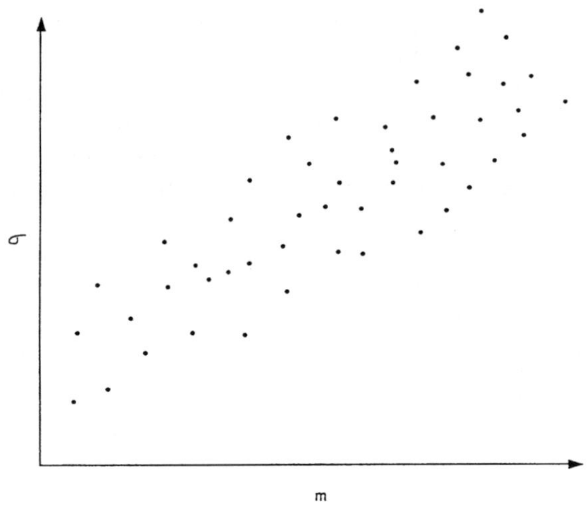

Figure 10-40. Proportional effect: graph of σ vs. *m*.

standard deviation, implying that the coefficient of variation σ/m is constant. This is known as the *proportional effect* (David 1977). An example for TCE data is shown in Figure 10-40.

For variogram computation, the idea is to divide the variance of $\gamma(h)$ by the square of the mean value. This makes the variance proportional to the mean. It also tempers the effects of the outlier values since the larger squared differences are divided by the larger squared mean. Figure 10-41a shows a variogram of PCB concentrations using raw data. The structure is limited, and the behavior is quite erratic. Figure 10-41b shows the same variogram using relative techniques. This time, the structure is better defined and has less erratic behavior.

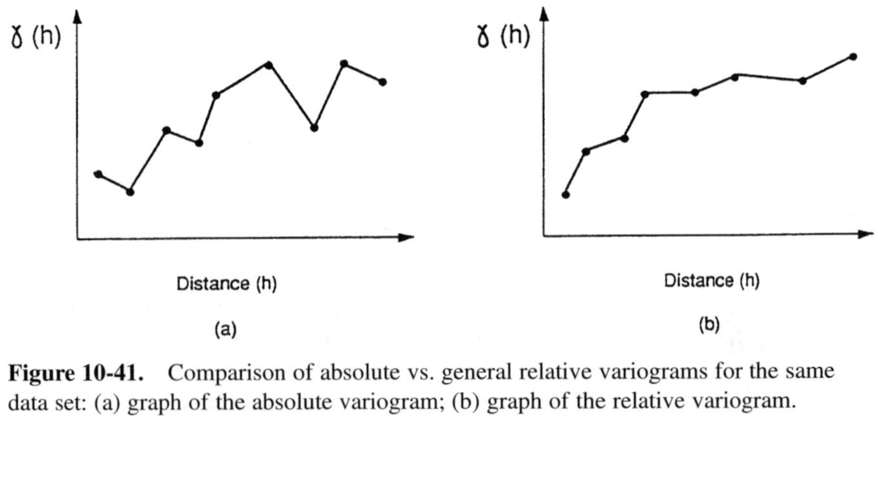

Figure 10-41. Comparison of absolute vs. general relative variograms for the same data set: (a) graph of the absolute variogram; (b) graph of the relative variogram.

10.3.1.1 Local Relative Variograms

There are three ways to use relative variogram techniques: local relative variograms, general relative variograms, and pairwise relative variograms. Local relative variograms account for the effect of the local mean by defining separate regions. Data within each region are analyzed separately using the local mean and number of sample pairs. While clearly a logical approach, the local relative variogram often becomes erratic because of a lack of data in some regions. The equation for the local relative variogram is

$$\gamma_{\mathrm{LR}}(h) = \frac{\sum_{i=1}^{n} N_i(h) \frac{\gamma_i(h)}{m_i^2}}{\sum_{i=1}^{n} N_i(h)} \tag{10.15}$$

10.3.1.2 General Relative Variograms

A commonly used alternative is the general relative variogram. Instead of a local mean, the population mean is used, in other words, the mean of all the data used to calculate the variogram. This is expressed as

$$\gamma_{\mathrm{GR}}(h) = \frac{\gamma(h)}{m(h)^2} \tag{10.16}$$

or

$$\gamma_{\mathrm{GR}}(h) = \frac{\dfrac{\sum_{i=1}^{n} [x_i - x_{i+h}]^2}{2N(h)}}{\dfrac{\sum_{i=1}^{n} [x_i + x_{i+h}]^2}{2N(h)}} \tag{10.17}$$

The general relative variogram is usually the most stable relative variogram and is recommended over the other two types in most situations.

Use of relative variograms in kriging provides an additional benefit. The kriging variance is a relative variance and is expressed in percent error. Such a format is often easier for communicating results than standard deviations.

10.3.1.3 Pairwise Relative Variograms

Pairwise relative variograms use the average of the two sample data comprising the variogram pair as the local mean value. The equation is

$$\gamma_{\mathrm{PR}}(h)=\frac{1}{2n}\sum_{(i,j)|h_{ij}\approx h}^{n}\frac{(x_i-x_{(i+h)})^2}{\left(\frac{(x_i-x_{(i+h)})^2}{2}\right)} \tag{10.18}$$

This formulation reduces the influence of very large values on a local basis. The pairwise relative variogram can become unstable, however. If the two data values are both equal to zero, the denominator is also zero and $\gamma_{\mathrm{PR}}(h)$ is infinite. As a practical matter, one can set a lower bound for the denominator or simply discard pairs having a denominator of zero.

10.3.2 Logarithmic Variograms

Another way to reduce the erratic behavior of an experimental variogram is by transforming the raw data by means of logarithms. This can be performed by using the following equation:

$$\gamma_n(h)=\frac{\sum_{i=1}^{n}[\ln(x_i)-\ln(x_{i+h})]^2}{2n} \tag{10.19}$$

Use of logarithms is a very powerful transformation and may be likened to a "sledgehammer" approach. Data concentrations several orders of magnitude apart suddenly appear similar when one considers that:

$$\begin{aligned}\ln(1)&=0\\ \ln(10)&=1\\ \ln(1000)&=3\\ \ln(10000)&=4\\ \ln(100000)&=5\end{aligned}$$

Whereas the transformed data appear alike, they still retain their original magnitudes. However, what is actually happening inside the variogram equation with the log-transformed data is unclear. The subtraction process is actually division, division is actually taking the square root, and squaring the quantity is somewhat obscure.

Logarithmic variograms can be used only in the lognormal kriging process. Lognormal kriging can be quite sensitive and problematic and will be discussed later. While this type of kriging performs very well for some cases, it is not recommended for most situations.

The logarithmic variogram may be used as a qualitative tool for experimental variogram analysis. When the erratic behavior is calmed, an idea of the range of the variogram and anisotropy ratios may be obtained.

10.3.3 Cross-Variograms

Occasionally, it is desired or necessary to predict the value of one variable at a location using the value of a second variable at the same location. For instance, radiometric assays are cheaper and faster than chemical assays. If a good correlation exists between the radiometric and chemical assays, the radiometric assays may provide a cost-effective alternative to expensive chemical sampling.

It is not always this easy, however, as both variables may not be available at the time of measurement. For example, before a surface mining operation is begun, it may be important to know what the expected pH values will be in the soil after the reclamation work is performed. Undisturbed overburden generally has a low pH but may have a high acid-base potential, indicating that large amounts of acid may be produced subsequent to mining.

In this case, it may be possible to determine the yearly needs for lime in the neutralization process in advance, using a secondary variable. In agricultural applications, knowing the concentration of specific trace elements in the soil may be a predictor of crop yields.

The process by which this estimation is performed is called cokriging. Cokriging will be discussed in Chapter 14. In order to perform cokriging, however, it is necessary to determine the simultaneous variability between a primary variable at distance h and the variability of a secondary variable at the same distance.

Measurement of covariability is performed by means of the cross-variogram. The spatial structure of the cross-variogram is estimated by using the following formula:

$$\lambda_{\mathrm{PS}}(h)=\frac{\sum_{i=1}^{n}([Z_p(x_i)-A_p(x_i+h)][Z_s(x_i+h)])}{2n(h)} \tag{10.20}$$

where $Z_P(x)$ represents the primary variable at location x_i, and Z_S is the secondary variable at location x_i. $N(h)$ represents the number of sample pairs of $Z_P(x_i)$ and $Z_S(x_i)$ in a given lag interval. The linear model of coregionalization requires that the variance of any possible linear combination of the two variables be positive. This can be assured if the cross-variogram function satisfies the Cauchy-Schwartz inequality given by (Myers 1984):

$$|\gamma_{Z_{PS}}(h)| \leq |[\gamma_{Z_P}(h)][\gamma_{Z_S}(h)]|^{\frac{1}{2}} \tag{10.21}$$

As a more practical demonstration, a portion of a pH and acid-base potential (ABA) data file from regraded overburden at a Colorado surface mining operation is presented in Table 10-1. Using these data, pairings for the east-west cross-variogram between pH and ABA would be the following:

$$\begin{aligned}\gamma_{\text{pH}-\text{ABA}}(1) = &[(5.80-6.00)][(2.54-0.21)]\\ &+[(6.00-6.50)][(0.21-4.13)]\\ &+[(6.50-6.30)][(4.13-3.95)]\\ &+\ldots+[(6.90-6.90)][(5.90-5.44)]\ /\ (2)\quad(30)\end{aligned} \tag{10.22}$$

10.3.4 Indicator Variograms

As discussed in Chapter 5, sample data are commonly transformed into indicator values based on some type of action level or cutoff. This transformation dramatically reduces the influence of outlier data and skewed data distributions.

The indicator transformation also provides the opportunity to produce better-behaved experimental variograms. Indicator variograms use the 0/1 data of the threshold-based transformation to calculate the spatial continuity of the 0/1 data. Even the simple on-off status of data will generally produce a recognizable spatial structure.

The simplistic nature of the transformed data provides a simplified array of indicator variogram sample pairs. The only indicator pairing options are:

1. $(0 - 0)^2$
2. $(0 - 1)^2$
3. $(1 - 0)^2$
4. $(1 - 1)^2$

Because of this limited array of outcomes and the relatively small difference between any possible sample pair, indicator variograms tend to be very well

Table 10-1. Coordinate, elevation, depth, pH, and acid-base potential data from regraded spoil overburden to be used in the calculation of cross variograms.

X	*Y*	*Z*		pH	ABA	Depth
0.	0.	1.	1.	5.80	2.54	0–1′
1.	0.	1.	1.	6.00	0.21	0–1′
2.	0.	1.	1.	6.50	4.13	0–1′
3.	0.	1.	1.	6.30	3.95	0–1′
4.	0.	1.	1.	6.40	6.04	0–1′
5.	0.	1.	1.	7.20	11.40	0–1′
6.	0.	1.	1.	6.30	4.96	0–1′
7.	0.	1.	1.	6.30	6.14	0–1′
8.	0.	1.	1.	7.10	10.30	0–1′
9.	0.	1.	1.	6.10	5.78	0–1′
0.	−1.	1.	1.	6.00	7.88	0–1′
1.	−1.	1.	1.	7.40	6.42	0–1′
2.	−1.	1.	1.	7.10	10.30	0–1′
3.	−1.	1.	1.	7.00	6.97	0–1′
4.	−1.	1.	1.	7.10	6.47	0–1′
5.	−1.	1.	1.	7.20	8.21	0–1′
6.	−1.	1.	1.	7.10	8.53	0–1′
7.	−1.	1.	1.	6.70	10.10	0–1′
8.	−1.	1.	1.	7.20	7.42	0–1′
9.	−1.	1.	1.	7.20	11.70	0–1′
0.	−2.	1.	1.	6.20	6.26	0–1′
1.	−2.	1.	1.	6.60	3.70	0–1′
2.	−2.	1.	1.	6.20	4.90	0–1′
3.	−2.	1.	1.	6.50	5.36	0–1′
4.	−2.	1.	1.	6.40	10.10	0–1′
5.	−2.	1.	1.	6.80	5.45	0–1′
6.	−2.	1.	1.	6.60	5.91	0–1′
7.	−2.	1.	1.	6.60	6.31	0–1′
8.	−2.	1.	1.	6.90	5.90	0–1′
9.	−2.	1.	1.	6.90	5.44	0–1′

behaved and to show good structure. Spatial continuity structure may vary, however, at different action level cutoffs. Generally there is only a single action level for environmental situations but, occasionally, multiple action levels occur. In this case, experimental indicator variograms should be calculated at each action level.

If the action level for the COC places an overwhelming proportion of the distribution in either the zero or one category, problems may occur in obtaining good variogram structure. Experience has shown that the median of the distribution, with half the data transformed to zero and half to one, generally provides satisfying variograms that approximate the structure at the desired action level. Note, however, that median-based indicators are only an approximation of the true spatial continuity at the actual threshold, though, in some cases, a good approximation.

10.4 Geostatistical Assumptions

Calculation of the experimental variogram depends on sample data collected at the site. According to the definition of a regionalized variable, a sample $Z(x)$ at location x is but one realization of a random function (RF). This essentially turns a uniquely defined sample value into a single realization of a random process. This implies other potential results at the same location. The chapters on STP demonstrate how this is possible.

Although only one realization of the random function is available, one nevertheless wants to determine the characteristics of the RF $Z(x)$ so that unknown points may be estimated. To overcome this limitation, assumptions and hypotheses have been introduced into the geostatistical literature regarding the type of random function under consideration.

The first assumption, the *weak-stationarity assumption*, often referred to as *second-order stationarity*, states two conditions:

1. The expected value of the regionalized variable $Z(x)$ is the same over the entire field of interest.
2. The spatial covariance of the regionalized variable $Z(x)$ is the same over the entire field of interest.

The weak-stationarity assumption is well named in that this condition seldom exists in earth science applications. It is nevertheless necessary to state the assumption, and geostatistics remains an effective tool despite common violations of the assumption.

The second assumption is the intrinsic hypothesis. Early geostatistical work on ore deposits showed that a finite variance did not exist. Second-order stationarity requires that the variance be finite. This can be circumvented by considering the variations of the grades or concentrations rather than the grades or concentrations themselves. Thus, if one considers increments of the function, $Z(x) - Z(x+h)$, as opposed to the function itself, the following assumptions can be made:

$$E[Z(x+h)-Z(x)]=0 \tag{10.23}$$

$$\mathrm{VAR}[Z(x+h)-Z(x)]=2\gamma(h) \tag{10.24}$$

Equation 10.22 is nothing more than the definition of the variogram. The variogram implies that the variability between any two samples depends only on their separation vector and not on their specific locations.

This chapter will end with a few words about a long-standing source of confusion in terminology. Although equation 10.24 defines the quantity $2\gamma(h)$, the quantity $\gamma(h)$ is more frequently used. The net result is that $\gamma(h)$ is frequently called the semivariogram, which is acceptable terminology. Nevertheless, for the sake of simplicity, the GEM process will refer to the quantity $\gamma(h)$ as the variogram.

CONCLUSION

Any interpolation or estimation method such as kriging, inverse distance, spline, multiquadrics, or minimum tension, depends on the use of a distance weighting function, such as the variogram. Often, nonkriging interpolators are used to contour data of varying types in varying locations. What many users fail to recognize is that, when they do this, the same "variogram" or distance-weighting function is being used for every COC at every site. With the insights gained into how much geologic, temporal, historical, and other types of useful site-specific information is to be gleaned by variogram analysis, using an identical distance-weighting function for any situation now appears ludicrous and naive, perhaps even reckless. The variogram provides the opportunity to customize the distance-weighting function for each COC at every site. In addition, it provides the kriging system with information needed to determine the statistical accuracy of the estimation, something no other interpolator can accomplish. On the darker side, the variogram is not a panacea. Numerous practical difficulties arise in its calculation and modeling. As with almost any good tool, the variogram can be misused or misinterpreted. The practitioner must recognize that there always remains a degree of uncertainty surrounding the variogram model as a result of poor-quality sample data, improper variogram analysis and modeling, and assumptions of stationarity and domains.

Chapter 11

The Volume-Variance Relationship

SYNOPSIS: In Chapters 7 and 8 (STP), the critical relationship between sample volume or weight and the fundamental error FE was demonstrated. This provided a measure of a volume-variance relationship at a small-scale or local level that is important to the GEM process. In this chapter, a larger-scale volume-variance support relationship is shown to have similar ramifications on a geostatistical or regional level (GA). While originally developed for mining applications, the volume-variance relationship has a direct influence on environmental decision making. All volume-variance relationships must be understood prior to the DQO process in order that the effects of volume may be accounted for in an appropriate manner.

KEY WORDS: Volume-Variance, Fundamental Error, Support, Block Variance, Change of Support, Grade-Tonnage Curve, Concentration-Volume Curve, Support Correction Factors, Affine Correction, Indirect Lognormal Correction, Dispersion Variance, Krige's Relationship, Discretization Grid, Gamma Bar, F-Function

11.1 Introduction

Previous chapters have emphasized the impacts associated with samples having different support volumes. These include different shapes of sample distributions and associated population statistics as seen in Chapters 4, 5, and 6 as well as significant variation in the fundamental error *FE* shown in Chapter 8. In addition to the support volume of the sample data used, kriged block estimates have an associated area or volume. As with both the variance of the fundamental error *FE* and the variance of samples s^2, the magnitude of the block variance is related to the support volume of the block.

As has been shown, the variance of the fundamental error FE and the sample variance can be determined. The method for calculating the block variance is the focus of this chapter. Block variance has been tracked in mining situations for many years as a means of determining expected ore recovery and associated profit. Block variance is directly related to a mining problem called "vanishing tons" (David 1977) where some of the mine areas anticipated to have high concentrations turn out to have much lower concentrations at the time of extraction. To experience the concept of vanishing tons can mean severe economic hardship to a mine.

In the case of environmental situations, "profit" is defined in the context of health and safety. This new perspective changes some of the issues involved, including the concept of recovery. For the mine, all material above an economic cutoff must be extracted for profitability. In the environment, all material exceeding a level deemed appropriate for health and safety must be remediated. This volume may change at varying levels of support. This chapter will examine a number of the volume-variance support relationships, starting with a mining perspective and then relating the concepts to environmental situations.

11.2 Change of Support

11.2.1 Grade-Tonnage Curves

Sample data collected during the initial phases of a mining operation are similar to environmental data from remedial investigations (RIs) and other phases in that a wide range of concentrations is usually encountered. In each situation, an action level will have importance for the engineering operation, be it mineral extraction or contaminant remediation. For the mine, all material above an economic cutoff will be extracted and processed in hopes of earning a profit. In the environment, all soil above a regulatory or risk-based threshold limit will be subject to remedial action in anticipation of health and safety benefits.

Based on the sample data, it is usually desired in either situation to determine the amount of material that will need to be excavated. This provides an estimate of the cost and level of effort and the anticipated duration of the operation.

At the limit, a cutoff of zero means that all material will be treated. As the cutoff is moved positively from zero and is continually increased, less and less of the material will be treated. Figure 11-1, known as a grade-tonnage curve, shows the nature of the declining tonnage or volume to be treated as the action level is increased. On the *X* axis, the cutoff grade is plotted. The cutoff grade is analogous to the action level. On the *Y* axis, the percent of the metal recovered is plotted, analogous to the percent of the COC removed.

With a cutoff of zero, all material is mined and all metal is recovered. As the cutoff grade increases, the amount of material mined decreases sharply, with the quantity of metal declining more slowly. The environmental analogue is that, with higher and higher threshold limits, smaller volumes of material are processed during remediation, and a greater percentage of the COC remains at the site, although at levels below the action threshold.

The grade-tonnage or concentration-volume (C-V) curve shows an intuitive result for a hazardous waste site: As the action level for lead (Pb) in soil around a smelter goes from 500 to 1000 ppm, a smaller volume will be treated and more lead will remain after the cleanup ceases.

What is not so obvious is the effect that block size has on the final remedial volume. Figure 11-1 shows the C-V curve for sample points. The assumption inherent in this graph is that one will be able to differentiate and remediate on a very small scale (i.e., remediate on sample-sized support). This is not a practical reality. Instead, kriging estimation and subsequent remedial activities

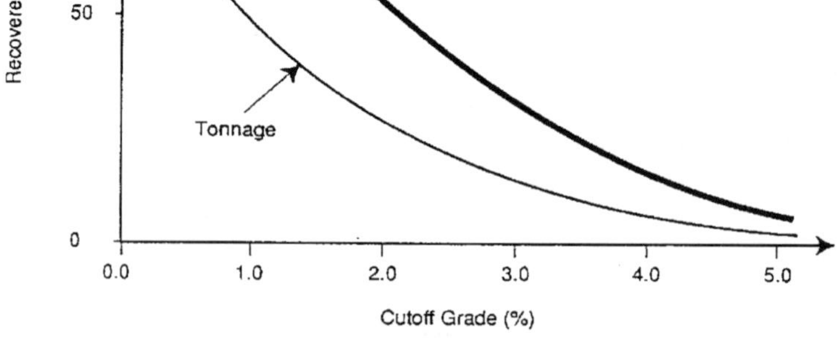

Figure 11-1. Recovery curves (tonnage, quantity of metal) as functions of cutoff grade for point samples and blocks (grade-tonnage curves).

will be based on block units that, as shown in Chapter 4, follow a different distribution.

11.2.2 Relationship of the Mean to the Threshold Concentration

As part of the volume-variance analysis, it is also important to be aware of the relationship between the mean of the sample data and the action level. For action levels below the mean, increasing the size of the block generally results in an increase in the amount of remedial volume. For action levels above the mean, increasing the block size generally decreases the remedial volume.

This relationship between the mean and the increase or decrease of material with a support change can easily be shown using a normal distribution. Figures 11-2a and 11-2b show the effects in both cases. In both parts of Figure 11-2, the sample and block distributions are shown, along with the action level concentration. In Figure 11-2a, the threshold cutoff x_{c1} is below the mean. The solid area combined with the shaded area to the left of the cutoff indicates the expected volume of material that will not require remediation based on the sample distribution. The solid area shows the actual amount of clean material based on block estimates.

It is important to remember that remediation is performed on blocks and not on samples. This fact can have serious consequences for remedial contractors. During the bidding process, it is tempting to estimate the amount of volume to be moved based on the sample distribution since sample data are readily available. As shown in Figure 11-2a, such an approach can lead to serious underestimation and significant cost overruns.

Figure 11-2b shows that the opposite effect occurs when the cutoff x_{c2} is above the mean. Again, the actual volume (number of blocks) shrinks because of the smaller variability of blocks vs. samples, with reversed but nevertheless serious impacts. If a certain volume is required for profitability to a remedial contractor, a reduced volume may cause the project to fall short of profitable revenue levels. For the miner, this is always a problem because vanishing tons can ruin the economic viability of a mine.

An exercise using the normal distribution can also be used to illustrate the effect of sample vs. block variability. Consider a site where the samples of lead (Pb) in soil follow a normal distribution, with a mean of 738 ppm, a standard deviation of 225 ppm for samples, and a standard deviation of 195 ppm for blocks. Total potential remedial volume at the site is estimated to be 50,000 yd^3 of soil. From this information, proportions of the total volume above any cutoff can be determined.

Table 11-1 shows the results using the sample data distribution and three cutoffs above the mean (900, 1000, 1100 ppm) and three cutoffs below the mean

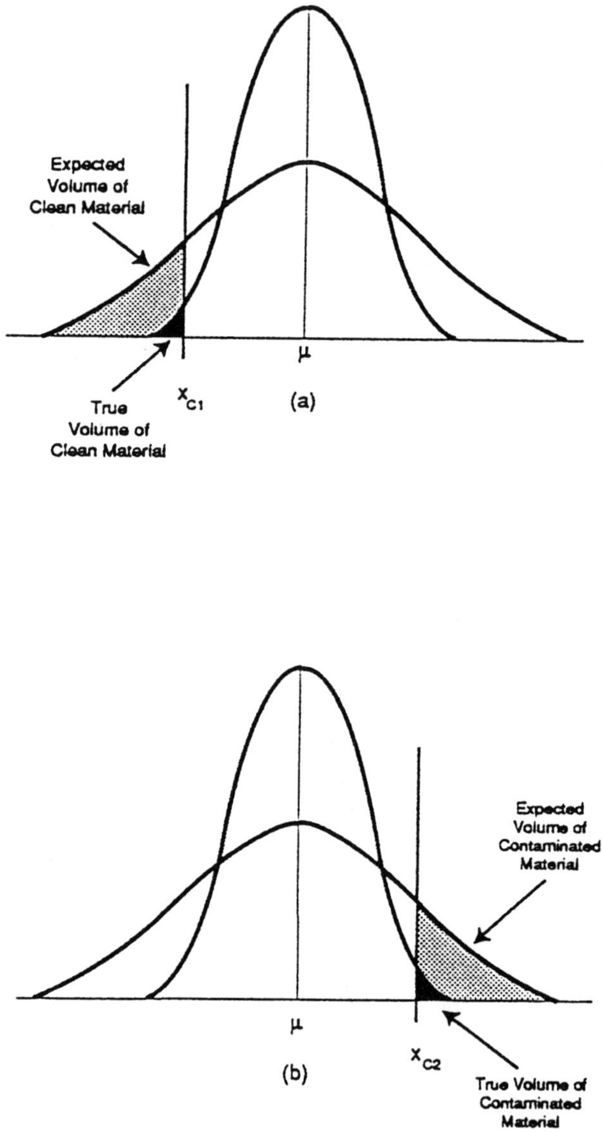

Figure 11-2. Relationship between increase/decrease in volume with a change of support for thresholds above and below the mean.

Table 11-1. Predicted vs. actual volume with a change of support using a normal distribution.

Cutoff (ppm)	Reduced Normal Variable	Proportion Above Cutoff (%)	Volume
Samples			
400	400–738/225	93.3	46,650
500	500–738/225	85.5	42,750
600	600–738/225	72.9	36,450
Mean = 738			
900	900–738/225	23.6	11,800
1000	1000–738/225	12.3	6,150
1100	1100–738/225	5.4	2,685
Blocks			
400	400–738/195	95.8	47,900
500	500–738/195	88.9	44,450
600	600–738/195	76.1	38,050
Mean = 738			
900	900–738/195	20.3	10,150
1000	1000–738/195	9.0	4,500
1100	1100–738/195	3.1	1,550

(400, 500, 600 ppm). Table 11-1 shows that a higher proportion of blocks will be above the cutoff when samples and blocks are compared for cutoffs below the mean. The reverse is also true: A higher proportion of samples are above the cutoff for cutoffs above the mean. Note that basing volume estimates on samples produces a shortfall of 1700 yd^3 for a cutoff of 500 ppm and an overage of 1650 yd^3 for a cutoff of 1000 ppm.

It is clear from Table 11-1 that regulatory agencies may incur unpleasant and unexpected costs if projections are made using sample data. Underestimating the total disposal volume by 1700 yd^3 may translate into additional costs running into the hundreds of thousands of dollars.

Although the normal distribution does a good job of illustrating the change of support concept, most environmental data sets are highly skewed to the right. For environmental data, the C-V curve can be used effectively for most data sets. Figure 11-3 shows the C-V curve for a PCB site. The *X* axis has two simultaneous scales: the PCB concentration and the natural logarithm of the PCB concentration. The volume of material as a percentage of the total site volume is plotted on the *Y* axis. Similarly, the quantity of recovered PCB is plotted as a percentage of the total mass of PCBs at the site.

The graphs show a significant difference between the estimated values for declustered sample and block distributions. Based on sample values, the mass of PCBs at the site is considerably overestimated. In accordance with the results

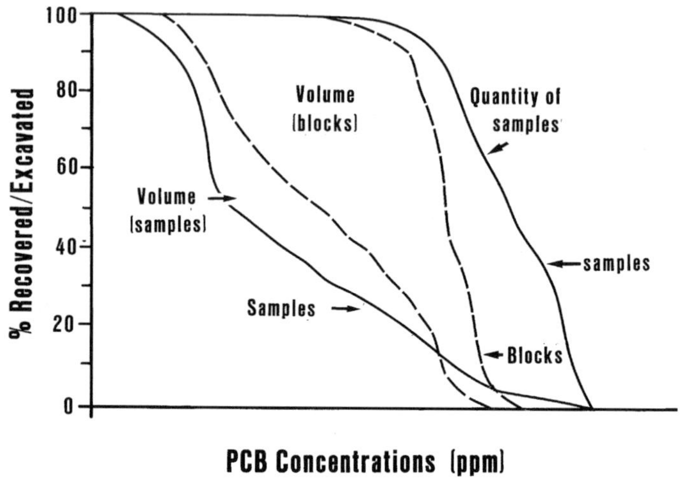

Figure 11-3. Concentration-volume curves for a PCB site.

shown in the exercise using the normal distribution, the volume of contaminated material is underestimated for low cutoffs and overestimated for high cutoffs when the sample distribution is used.

For example, at a cutoff of 5 ppm, the expected volume based on samples is approximately 45% of the total volume. This compares with an actual volume equating to almost 65% of the total based on blocks. The reverse effect can be seen at high cutoffs but have little practical significance for PCBs since cutoffs are usually 25 ppm or less.

Note that the block size plotted on the graph is 10 ft × 10 ft. Differences would be even greater for blocks with larger support. Even at this support, an action level anywhere between zero and approximately 500 ppm would significantly underestimate the volume to be addressed during remediation. In terms of the quantity of PCB recovered, however, there is little difference in the amount recovered below a cutoff of 50 ppm. This is satisfying from a risk perspective because most of the COC is being removed, no matter which distribution is used. However, if mass balance calculations are being performed, sample distributions could overestimate the quantity of COC.

Figures 11-1 and 11-3 show the volume of contaminated material as a function of cutoff. These graphs are in fact cumulative distribution curves that show the cumulative distribution above a range of cutoff values. A cumulative tonnage graph similar to the cumulative frequency graphs shown in Figures 5-6 and 5-14 are shown in Figure 11-4. This graph confirms the results seen in Figures 11-1 and 11-2 and Table 11-1: For low cutoffs, samples underestimate remedial volume and, for high cutoffs, samples overestimate remedial volume.

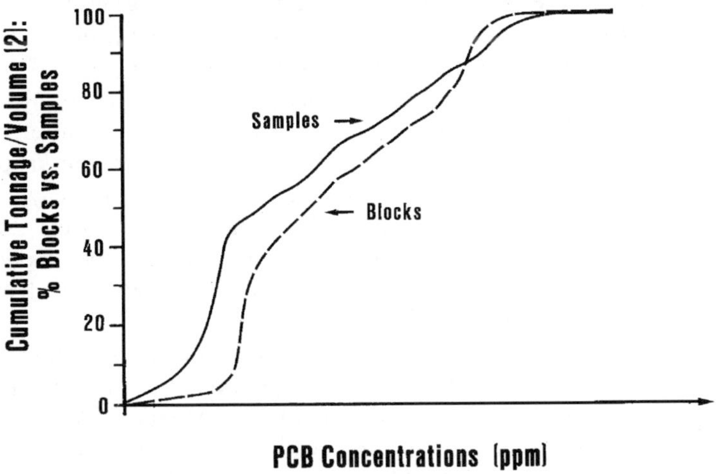

Figure 11-4. Cumulative volume curve for a PCB site.

11.3 Support Correction Factors

11.3.1 Introduction

The graphs in Figures 11-2 through 11-4 share one other thing in common. Conveniently, all these graphs had a known block distribution or block standard deviation. This information is generally not available from standard statistical packages. Therefore, it is often necessary to calculate the effect of a support change using other methods.

One method, used in Figures 11-3 and 11-4, is to krige the site using a desired block size. This, of course, still poses some risk because the actual size of the remedial block may change before actual remediation begins. Kriging, however, is not a reliable change of support tool. This is true especially when block sizes smaller than the sampling grid are used. If samples are on a 100 × 100 ft grid, for example, there will be no significant difference between a kriged histogram for 50 × 50 ft blocks and 1 × 1 ft blocks.

Other support correction factors for sample distributions are available that use alternative approaches. Two of these techniques will be discussed: the *affine correction* and the *indirect lognormal correction*. These techniques have two features in common. The first is that they do not alter the mean value of the distribution. This is clearly a desirable feature in that the constancy of the mean is one of the few safe assumptions in support correction. The variance, for example, is generally less well known than the mean.

The second feature is that both methods reduce the variance by a factor f called the *variance adjustment factor*. This reduces the spread of the histogram distribution and allows sample and block distributions to be compared visually.

In addition to variance reduction, another phenomenon may occur during a support correction. The effect is known as *symmetrization*. which refers to the degree to which the block distribution becomes less skewed and, thus, approaches a symmetric normal distribution. The two correction factors impact symmetry in different ways. The affine correction assumes that there is no corresponding increase in symmetry with the reduction in variance because of the support effect. Conversely, the indirect lognormal correction lowers skewness and increases symmetry.

Increase in symmetry is also related to spatial continuity. The more homogeneous a data set, the more rapidly the increase in symmetry will occur. Homogeneity implies randomness and is thus related to the variogram. Where extreme values are clustered in spatial groups (common in environmental data sets), the reduction in variance and in asymmetry occurs more slowly than in data sets where the extreme values are scattered about.

The variogram is thus a useful tool in evaluating the effect of change of support. The variogram does have limitations in this regard, however. It is not able to decipher the nature of the connectivity of extreme values (Isaaks and Srivastava 1989). The degree of symmetrization to be expected thus falls into the realm of professional judgment. Qualitative information about the site, along with past experience and variogram results, must all be used to choose an appropriate procedure. If extreme values are judged to be generally well connected, a technique that does not induce symmetry may be preferred. If entropy or randomness is pervasive, a method that increases symmetry may be better suited to the task.

11.3.2 The Affine Correction

One of the more pleasing aspects of the affine correction is its simplicity. This technique reduces the variance of a distribution by compressing the entire distribution closer to the mean. As stated above, there is no alteration to the mean value nor any increase in symmetry. The amount of compression is directly related to the variance correction factor f selected by the user.

The affine correction is achieved by using the following equation:

$$x' = \sqrt{f} \cdot (x - m) + m \tag{11.1}$$

where x is a value (or quantile) of the original distribution, x' a value (or quantile) of the transformed distribution, m the mean, and f the variance correction factor.

The transformation affects the data or quantile values themselves. Working like a vise, it squashes both sides of the distribution toward the mean. With a simple algorithm, an entire set of data values may be quickly transformed by a short computer program or spreadsheet. Figures 11-5a and 11-5b show the before and after histograms of lead (Pb) concentrations for the Dallas Lead DMC smelter

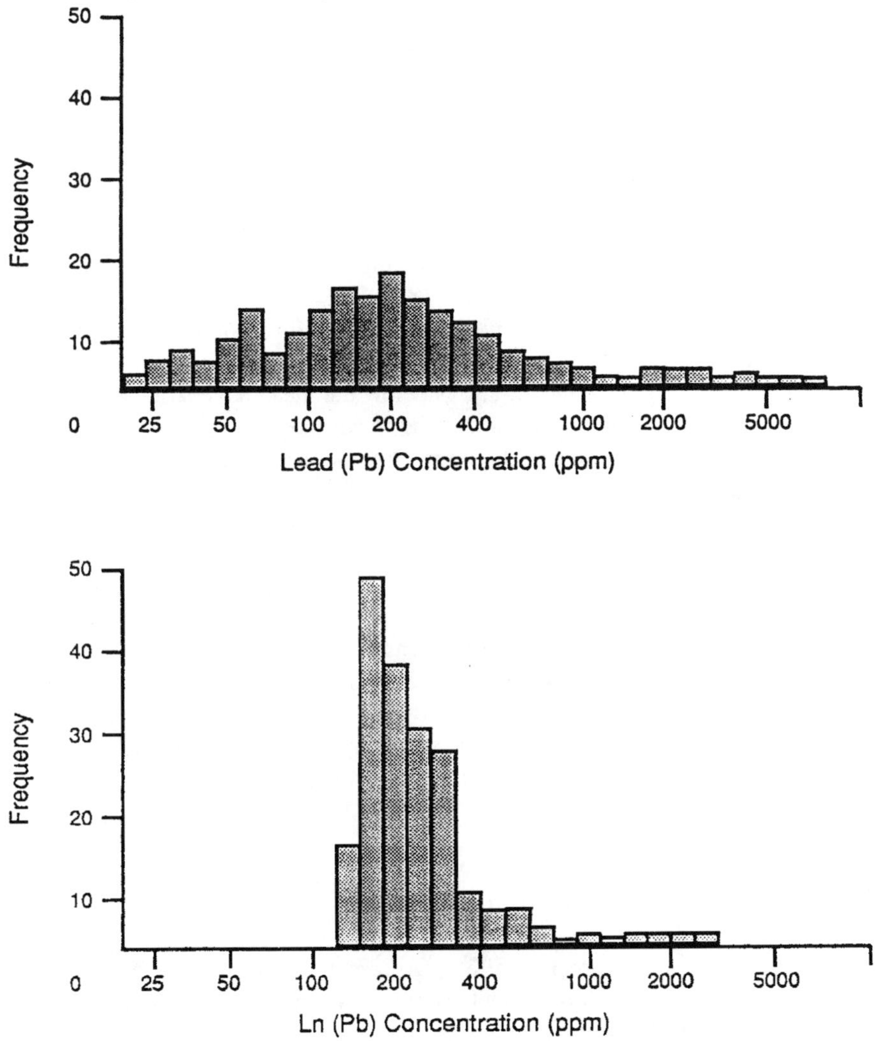

Figure 11-5. Histograms of Ln transformed Dallas Lead (Pb) DMC data: (a) Data prior to affine correction; (b) data after the correction ($f = 0.7$).

Table 11-2. Application of the method of affine correction to quantile data.

Cumulative Frequency	Quantile Concentration	Correction Equation	Corrected Value
.10	223.7	(223.7–311.8) (.7) + 311.8	250.1
.20	245.1	(245.1–311.8) (.7) + 311.8	265.1
.30	267.6	(267.6–311.8) (.7) + 311.8	280.9
.40	289.7	(289.7–311.8) (.7) + 311.8	296.3
.50	319.9	(319.9–311.8) (.7) + 311.8	317.3
.60	344.3	(344.3–311.8) (.7) + 311.8	334.6
.70	360.8	(360.8–311.8) (.7) + 311.8	346.1
.80	376.9	(376.9–311.8) (.7) + 311.8	357.4
.90	408.1	(408.1–311.8) (.7) + 311.8	379.2

area for a value of $f = 0.7$. The effect is dramatic as the adjusted histogram is much narrower.

The data used in the affine correction can also correspond to quantile values as shown in Table 11-2.

Note that all original concentrations below the mean have a higher corrected value and that all concentrations above the mean have a lower corrected value. Also, note that the magnitude of the correction increases as concentrations move away from the mean.

The simple compression offered by the affine correction may be inappropriate for many environmental situations. Specifically, the correction may provide a minimum value that is too high. At environmental sites, the regulatory threshold cutoff is often very low with respect to the overall distribution of sample values. Thus, it is possible to have a corrected minimum above the threshold limit, indicating that all blocks at the site need to be remediated. Whereas this may be the reality, there are generally at least a few "clean" blocks at the site, most notably at the margins. As a rule of thumb, if the variance adjustment factor is .7 or greater and the threshold is relatively close to the mean, the affine correction is often appropriate.

11.3.3 The Indirect Lognormal Correction

The indirect lognormal correction offers an alternative to the affine correction that is often better suited to environmental data. It offers a more prudent approach when extreme cutoffs are to be applied. This is commonly the case when, for example, the PCB limit may be between 1 and 25 ppm and the sample data range into the thousands or tens of thousands.

The indirect lognormal correction has the effect of increasing symmetry, which creates a more normal-like appearance in the block distribution. The booming values in the upper tail are compressed toward the mean, lowering skewness while simultaneously allowing the minimum to remain at zero.

This correction is a bit more complex in that it requires a two-step process for the transformation. The initial equation to be applied is

$$x' = ax^b \tag{11.2}$$

where the coefficient a and the exponent b are given by the following equations:

$$a = \frac{m}{\sqrt{f \cdot CV^2 + 1}} \left(\frac{\sqrt{CV^2 + 1}}{m} \right)^2 \tag{11.3}$$

$$b = \sqrt{\frac{\ln(f \cdot CV^2 + 1)}{\ln(CV^2 + 1)}} \tag{11.4}$$

where f is the variance adjustment factor and CV is the coefficient of variation.

A practical problem frequently arises with the direct application shown in equation 11.2. If the data values are not exactly lognormally distributed, the mean of the distribution is not necessarily maintained during the transformation. As stated earlier, preserving the mean is important and, thus, a mean correction is applied by the following equation:

$$x'' = \frac{m}{m'} \cdot x' \tag{11.5}$$

where m' is the mean of the distribution after transformation by equation 11.2. For data distributions close to lognormal, the scaling factor m/m' will be close to 1.

11.4 Dispersion Variance and Krige's Relationship

11.4.1 Introduction

Both the support correction factors presented require the use of a variance adjustment factor. The examples shown in Figure 11-5 and Table 11-2 for the affine correction conveniently use a prechosen variance adjustment factor of .7. Thus far, however, no means of selecting such a factor has been presented. It is not appropriate to select such a factor arbitrarily. The variogram, however, may be used effectively to determine an appropriate variance adjustment factor.

Before discussing the specifics of the role played by the variogram in determining the variance adjustment factor, two related and more general concepts will be presented. They are the dispersion variance and Krige's relationship.

11.4.2 Dispersion Variance

The dispersion variance and Krige's relationship are themselves closely related. The dispersion variance is the variance of a smaller support unit within a larger unit and specifically states the support units of the mean. Examples could be the variance of samples at the entire site, the variance of the distribution of 20 × 20 ft blocks at a site or in an exposure unit, or the variance of samples in an EU. This can be stated mathematically as

$$\sigma^2(a, b) = \frac{1}{n} \sum_{i}^{n} (x_i - m_{a_i})^2 \tag{11.6}$$

and

$$x_b = m_{a_i} \tag{11.7}$$

where a is a specific support size, and b is a second specific support size containing n units of size a.

11.4.3 Krige's Relationship

Krige's relationship provides a general statement of the relationship between any support size for a site. It is given by the equation

$$\sigma^2(a, c) = \sigma^2(a, b) + \sigma^2(b, c) \tag{11.8}$$

where

$\sigma^2(a,c) =$ total variance at the site
$\sigma^2(a,b) =$ variance within blocks
$\sigma^2(b,c) =$ variance between blocks

A simple adjustment to the notation may bring a bit more meaning to equation 11.8. Substituting:

(.) for a (support of a sample data point)
v for b (support of small blocks)
V for c (support of large blocks, i.e., the site in most cases)

the equation may be rewritten as

$$\sigma^2_{(./V)} = \sigma^2_{(./v)} + \sigma^2_{(v/V)} \tag{11.9}$$

where:

$\sigma^2_{(./V)} =$ variance of samples at the site
$\sigma^2_{(./v)} =$ variance of samples within blocks
$\sigma^2_{(v/V)} =$ variance of blocks at the site

11.4.4 Calculating the Dispersion Variance

An example of dispersion variance calculations and Krige's relationship can be seen in Figure 11-6. Figure 11-6a shows 16 sample data points on a regular grid spacing. Calculation of the mean results in a value of 96 ppm. The variance of the sample values can be calculated using the standard formula:

$$\begin{aligned}\sigma^2_{\text{point values}} = {} & [(72-96)+(75-96)+(83-96)+(90-96)\\ & +(81-96)+(89-96)+(91-96)+(87-96)\\ & +(101-96)+(98-96)+(100-96)+(109-96)\\ & +(111-96)+(104-96)+(113-96)+(132-96)\\ & \div 16 = 230 \text{ ppm}^2\\ & = \sigma^2_{(./V)}\end{aligned}$$

Figure 11-6b shows a change of support with the samples grouped into four blocks, each containing four sample points. Associated with each sample is the

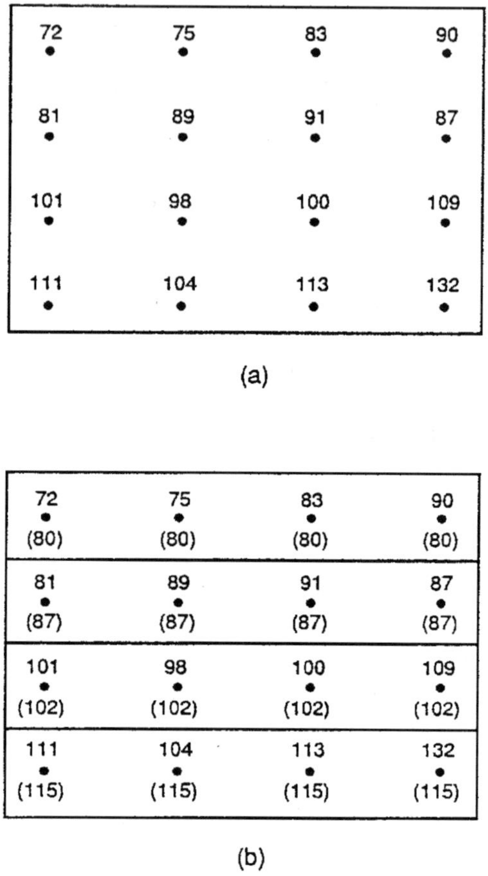

Figure 11-6. Grid of 16 data used to calculate the dispersion variance. (From *An Introduction to Applied Geostatistics,* by E.H. Isaaks and R.M. Srivastava. Copyright © 1989 by Oxford University Press, Inc. Used by permission of Oxford University Press, Inc.)

mean of the particular block (shown in parentheses). By applying such a block grouping, a variance of a new support that is equal to the size of the block may be calculated using the four block averages:

$$\begin{aligned}\sigma^2_{\text{block values}} &= [(80-96)+(87-96)+(102-96)+(115-96)\div 4 \\ &= 184 \text{ ppm}^2 \\ &= \sigma^2_{(v/V)}\end{aligned}$$

The final support variance to be calculated is the variability of points within the block support unit. This variance is determined by the following:

$$\begin{aligned}\sigma^2_{\text{point in block}} = {} & [(72-80)+(75-80)+(83-80)+(90-80) \\ & +(81-87)+(89-87)+(91-87)+(87-87) \\ & +(101-102)+(98-102)+(100-102)+(109-102) \\ & +(111-115)+(104-115)+(113-115)+(132-115) \\ & \div 16 = 46 \text{ ppm}^2 \\ & = \sigma^2_{(./v)}\end{aligned}$$

Substituting these calculated values for the three supports shown in Figure 11-6 into Krige's relationship gives

$$\sigma^2_{(./V)} = \sigma^2_{(./v)} + \sigma^2_{(v/V)} \tag{11.10}$$

$$230 \text{ ppm} = 46 \text{ ppm} + 184 \text{ ppm} \tag{11.11}$$

A bit of simple algebra can be used to rearrange equation 11.10 to produce

$$\sigma^2_{(v/V)} = \sigma^2_{(./V)} + \sigma^2_{(./V)} \tag{11.12}$$

$$184 \text{ ppm} = 230 \text{ ppm} - 46 \text{ ppm} \tag{11.13}$$

Equation 11.12 shows that, once any two terms of the equation are known, the third can be easily found by subtraction. The objective stated in change of support calculations is to determine the variability of blocks at the site (block variance). The variance $\sigma^2_{(./V)}$ of the sample distribution is known. Thus, the within-block variance $\sigma^2_{(./v)}$ needs to be determined before the block variance $\sigma^2_{(v/V)}$ can be obtained. It is by means of the variogram that the within-block variance may be determined.

11.4.5 Calculating the Within-Block Variance

11.4.5.1 Discretization Grids

The dispersion variance $\sigma^2_{(./v)}$ averages the variance of the data values contained in a support block, as shown in Figure 11-6. If we divided the site into block-sized units, calculations similar to those shown previously could be performed. This would be a very tedious process. In addition, many blocks would contain only one sample, thus making calculation of a variance impossible.

The variogram is used in practice to overcome this dilemma. The dispersion variance $\sigma^2_{(./v)}$ averages the squared differences of all pairs contained in the block. The variogram averages squared differences of pairs at a distance *h*. The dispersion variance is thus a type of variogram calculation in that pairs are considered, provided that the separation vector h_{ij} is contained inside the block.

What we need, then, is the average value of the variogram within the confines of the block. This average variogram value is referred to as *gamma bar*. While the value of gamma bar can most accurately be estimated by an integral, practitioners use a discrete approximation to the integral as an acceptable and expedient alternative.

Figure 11-7 shows an example of how the discretization process works. A grid size is selected and inserted into the block. Each grid node has a distance to each other grid node that can be easily calculated by the computer. Once the distance is determined, the value of the variogram can be calculated for that distance by

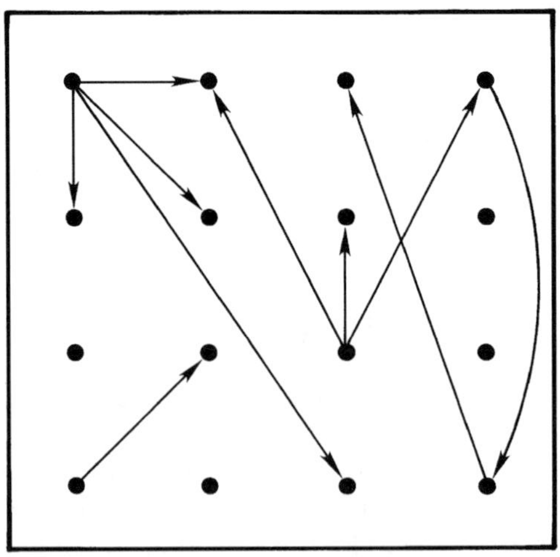

Figure 11-7. Internal discretization grid: calculation of averages differences (gamma bar).

using whatever variogram model has been selected. The n^2 values are then averaged to come up with a value for gamma bar.

Gamma bar is an average and, as is true for most averages, the more samples used in the calculation, the better. The size of the discretization grid therefore has an effect on the accuracy of the estimate for gamma bar. There is also a type of spatial component that affects the accuracy. Figure 11-8 shows a fixed block size with 2×2, 3×3, 4×4, and 5×5 grid node discretizations.

The spatial coverage in the block by the grid nodes becomes better and better as the grid size increases and grid nodes encroach on the edge of the block. The better the coverage, the better the gamma bar estimate. Remember that the value of gamma bar is an estimate as there are always at least the uncertainties about the variogram parameters themselves.

Table 11-3 shows how the number of grid nodes internal to the block affects the gamma bar value for a typical two-dimensional block using a spherical variogram model. As the number of grid nodes increases, the value of gamma bar also increases as the grid is pushed closer and closer to the edge of the block, allowing for greater distances and, thus, higher variogram values. The rate of increase declines rapidly, however.

With nine grid nodes (3×3 grid), the value of gamma bar has reached about 94% of its potential, a good approximation. With 25 nodes, 98% of the potential has been reached and should be considered an excellent approximation. In general, the number of internal grid nodes should be kept to a reasonable minimum. Chapter 12 will show that the calculation of gamma bar is performed only once for determining the estimation variance in kriging. However, Chapter 12 will also

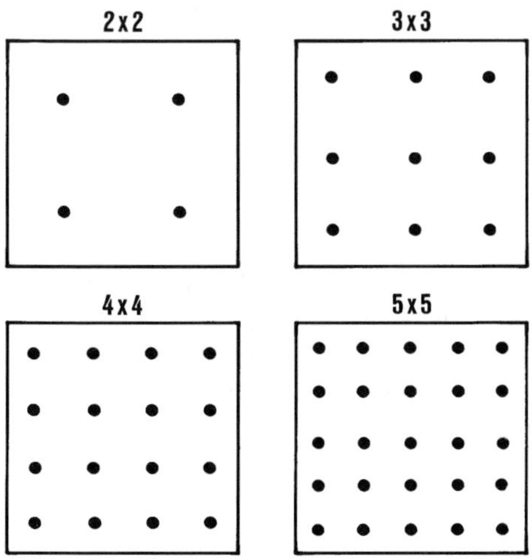

Figure 11-8. Fixed block size with various-sized discretization grids.

Table 11-3. Number of discretized internal grid Nodes with corresponding gamma bar value.

Internal Grid Size	Gamma Bar	% Total (Gamma Bar)
1 × 1	0.00	0.0
2 × 2	3.11	82.9
3 × 3	3.51	93.6
4 × 4	3.63	96.8
5 × 5	3.68	98.1
6 × 6	3.71	98.9
7 × 7	3.72	99.2
8 × 8	3.73	99.5
9 × 9	3.74	99.7
10 × 10	3.74	99.7

show how the time it takes to krige a grid of blocks increases rapidly as the internal grid size increases.

Note also that, for blocks that are not square or for blocks where an anisotropic variogram is used, the internal grid may need to be modified. For example, for a 10 × 20 ft block or a 10 × 10 ft block with a 2:1 anisotropy, the internal grid may need to be 3 × 6, 4 × 6, 4 × 7, and so forth.

11.4.5.2 Use of Charts

Calculation of gamma bar may also be performed using charts and the F-function (David 1977). The block variance may be obtained by

$$\sigma^2_{(v/V)} = \sigma^2_{(./V)} - \sigma^2_{(./V)} = F(V) - F(v) \tag{11.14}$$

For two- or three-dimensional cases, the so-called $F(S)$ function can be used. This assumes that the block of interest is a rectangle or parallelepiped whose sides are parallel to the main axes of variogram anisotropy, and two of the block dimensions are assumed proportional to the variogram model (Figs. 11-9 and 11-10). The function charted in Figure 11-10 is actually

$$\frac{1}{C} F\left(\frac{h}{a}, \frac{h}{a}, \frac{1}{a}\right) \tag{11.15}$$

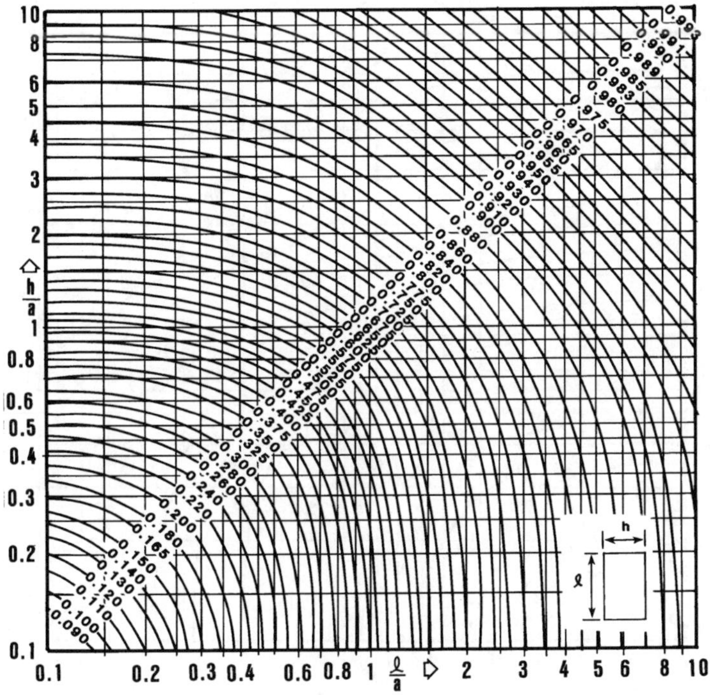

Figure 11-9. Variance $\sigma^2(./v)$ (O/S) of a point within a rectangle (h,l) for the spherical model, function $(1/C)\ F(h/a,l/a)$. (From *Geostatistical Ore Reserve Estimation*, by M. David, Elsevier, 1977. Courtesy of Michel David.)

Consider that the size of the remedial blocks proposed are 20 × 10 × 2 ft, with variogram ranges of 120, 60, and 5 ft, and with a sill C of 500 ppm^2. Reading from Figure 11-10 and incorporating the anisotropies, we get, for the small block,

$$\frac{h}{a}=\frac{20}{120}=\frac{10}{60}=0.167 \tag{11.16}$$

$$\frac{1}{a}=\frac{2}{5}=0.4 \tag{11.17}$$

$$\frac{1}{C}\,F(0.167,0.4)\cong 0.255 \tag{11.18}$$

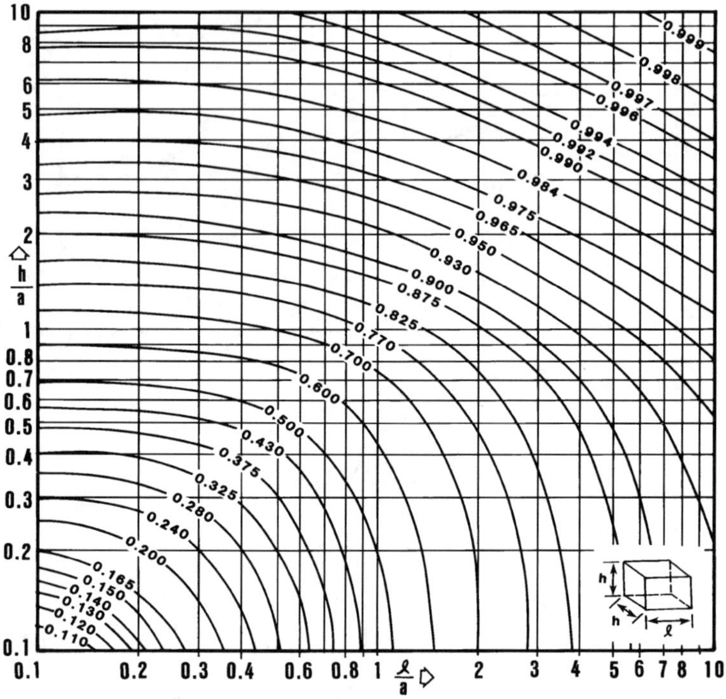

Figure 11-10. Variance $\sigma^2(./v)$ (O/V) of a point within a parallelpipedic block (h, h, l), function $(1/C)\ F(h/a, h/a, l/a)$. (From *Geostatistical Ore Reserve Estimation*, by M. David, Elsevier, 1977. Courtesy of Michel David.)

$$F(V) = 500\,(0.255) = 127.5 \tag{11.19}$$

and, for the site,

$$\frac{h}{a} = \frac{360}{120} = \frac{180}{60} = 3.0 \tag{11.20}$$

$$\frac{l}{a} = \frac{10}{5} = 2.0 \tag{11.21}$$

$$\frac{1}{C}\,F(3.0, 1.0)\ \cong 0.965 \tag{11.22}$$

$$F(V) M 5500(0.965) = 482.5 \tag{11.23}$$

Thus, the block variance can be calculated:

$$\sigma^2(v\ /\ V) = 500(0.965 - 0.255) = 355\ \text{ppm}^2 \tag{11.24}$$

Having obtained gamma bar and, subsequently, the value of the block variance, it is now possible to calculate the variance adjustment factor described previously. The variance adjustment factor is simply the ratio of the block variance $\sigma^2_{(v/V)}$ to the point variance $\sigma^2_{(./V)}$ and can be shown as

$$f = \frac{\sigma^2(v\ /\ V)}{\sigma^2(./V)} \tag{11.25}$$

$$= \frac{\sigma^2(./V) - \sigma^2(./v)}{\sigma^2(./V)} \tag{11.26}$$

$$= 1 - \frac{\sigma^2(./v)}{\sigma^2(./V)} \tag{11.27}$$

The value of $\sigma^2_{(./V)}$ is commonly taken from the variogram model. The numerator in equations 11-14 through 11-16 is estimated by using the variogram model and its associated sill. Thus, if the denominator also uses the variogram model, we compare variance models of equal magnitude. This approach is acceptable when the variogram has a clear sill and relatively short range. The sample variance, declustered if necessary, is better.

Conclusion

The volume-variance relationship quantifies the variability associated with blocks of a specified volume. Consistent with the volume properties seen in the fundamental error *FE*, the block variance decreases as the volume increases. This shared result emphasizes that the volume for estimation, be it sample or block, is always an important factor and should in no case be ignored. With the development of a block variance, the block distribution may be obtained. Careful analysis of the block distribution is recommended because key insights may be obtained regarding remedial volumes that can be integrated into the DQO process. The block variance is also a starting point for additional error and uncertainty quantification that will be presented in Chapters 12 and 13, which deal with estimation variance and kriging.

Chapter 12

Estimation Variance

SYNOPSIS: GEM recognizes that numerous mathematical functions allow one to estimate the concentration of a COC at an unsampled location using a single or multiple samples. Kriging is unique, however, in that it provides a measure of the uncertainty associated with the estimate. To determine the level of confidence associated with an estimate, the estimation variance simultaneously incorporates several items: the spatial correlation information embodied in the variogram; the amount of sample data available; the closeness of the samples to the point, area, or volume being estimated; the closeness of the samples one to another; and the size of the block. This chapter will present both the intuitive and mathematical sides of the concept known as estimation variance.

KEY WORDS: Estimation Variance, Kriging, Covariance

12.1 Introduction

In Chapter 3, the estimation error was defined using a simple formula where the estimated value was subtracted from the true known value. Mathematically, this can be stated

$$e_o = Z(x_o) - Z(x_o)^* \tag{12.1}$$

where $Z(x_0)$ equals the true but unknown concentration and $Z(x_0)$ equals the estimate Z^* at location x_0.

Immediately, we realize that the true value is not known, for if it were, there would be no need to incur an estimation error. Instead, a random function model will be used along with the random variables (samples, unknown true value, estimated value, estimation error). These quantities will provide a quantification of the variance of the estimation error.

The discussion on estimation variance will begin with an intuitive and visual approach to the problem. This is intended to offer a more tangible foundation to those readers who need mathematical reinforcements. This will be followed by a detailed derivation of the formula for the estimation variance, which will relate the visual concepts to the mathematics.

12.2 Relationships of Blocks and Samples

In a sense, the estimation variance is a bridge between dispersion variance and kriging. The estimation variance draws on the dispersion, or block, variance seen in Chapter 11 and adds the contribution of sample data to create a formula that describes the variability of the estimates. As will be shown in Chapter 13, kriging minimizes the estimation error to produce an optimized and statistically unbiased estimate.

The formula for the estimation variance has three components. The first is the block, or dispersion, variance. This is a needed component because it establishes the baseline difficulty in predicting the value of an unknown area or volume based on sample data. The concentration of a COC in a 1 × 1 ft block will be more difficult to estimate than the concentration in a 25 × 25 ft block. Small blocks are more difficult to estimate than large blocks because the larger support imparts smaller amounts of variability (Fig. 12-1). The analogy: It would be easier to predict the average temperature for the month of March than to guess the average temperature for any particular day because months have a larger temporal support than days.

The second term in the estimation variance equation describes the contribution made by the sample data in relation to the block to be estimated. Logically, the closer the samples are to the block being estimated the better the estimate. The variogram plays a pivotal role here as it reveals the expected variability for any distance from a sample to the block. Figure 12-2 is a visualization of this term.

Figure 12-2 brings up an interesting and important question: Just how far is a sample from a block? Since a block is a two- or three-dimensional object, a given sample point will be simultaneously closer and further from different parts of the block. The solution to the problem comes in using the identical internal

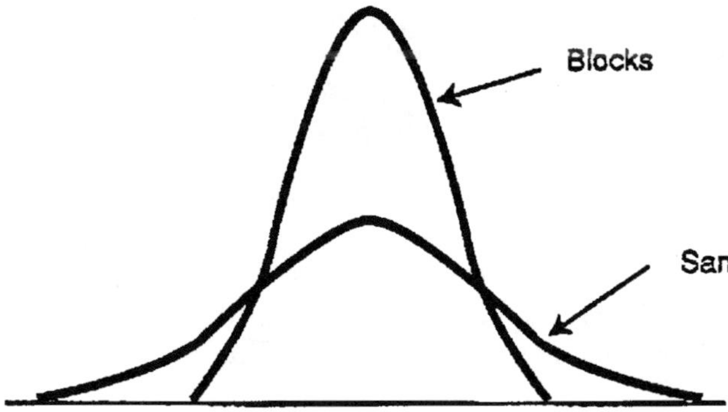

Figure 12-1. Block distribution and variance contrasted with sample distribution and variance.

discretization grid that was used to calculate gamma bar for the within-block variance in Chapter 11.

First, the distance from a given sample to the n discretization points is determined. Then, for each individual distance, a value of the expected variability can be calculated using the variogram model. The n variances are then averaged to obtain the average amount of variability between the sample and the block. This average is commonly called the point-to-block variance. Note that, for three-dimensional blocks, the discretization grid is repeated for as many layers as necessary within the block.

The final term is very important because it provides a key correction factor to the equation that is not found in other popular estimators. It is also a term that surprises newcomers to geostatistics. It quantifies the relationship between samples with regard to how close they are, one to another.

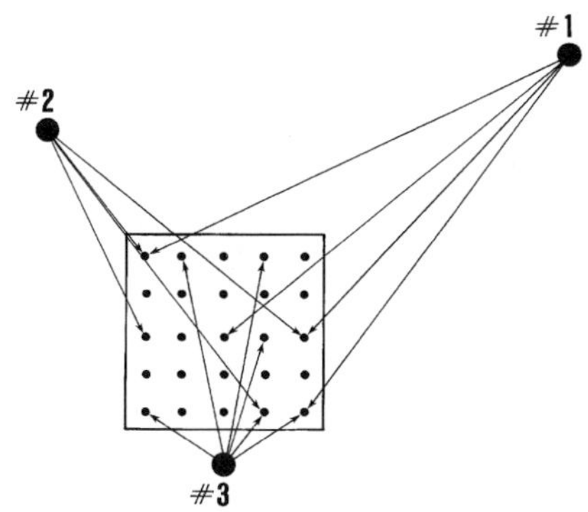

Figure 12-2. Illustration of the method used to calculate the distance of sample data points from a block to be estimated.

Reflecting on the concept of spatial correlation quickly reveals how this makes perfect sense. Spatial correlation states that nearby samples are alike. This is good but, as two samples draw closer and closer, we realize that, at some point, they become redundant for estimation purposes. In other words, a single sample (either one, in fact) could adequately represent the concentration for the area they share. This confirms the intuitive notion that sampling too close together is a waste of money. Chapter 13 will show how elegantly kriging handles this potentially dangerous and common estimation problem involving clustered samples. A visual expression of this concept is shown in Figure 12-3.

On a simplistic level, the equation for the estimation variance σ_e^2 can be given as follows:

$$\sigma_e^2 = \sigma_{v/V}^2 - \mathrm{COV}_{\text{samples2block}} + \mathrm{COV}_{\text{samples2samples}} \tag{12.2}$$

where:

$\sigma^2_{(v/V)}$ = block or dispersion variance (Fig. 12-1)

$\mathrm{COV}_{\text{samples2block}}$ = covariance (correlation) between the samples and the block being estimated (Fig. 12-2)

$\mathrm{COV}_{\text{samples2samples}}$ = covariance of sample data points one to another, that is, clustering (Fig. 12-3)

Estimation variance sets the stage for the next step, which is optimization through kriging. Kriging will take these concepts and minimize the equation to

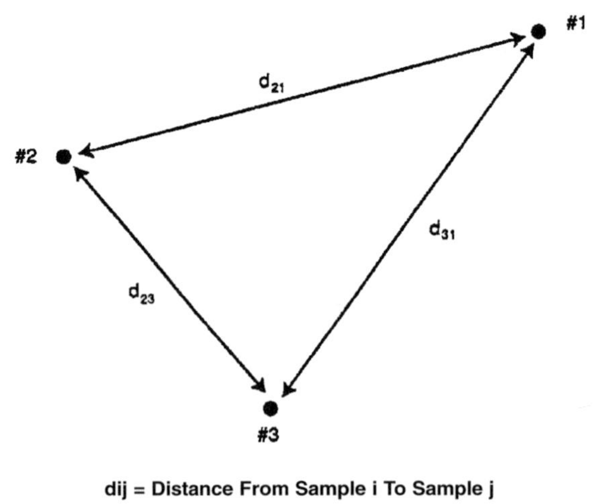

Figure 12-3. Distance relationship of sample data points among themselves.

come up with a minimum variance and an unbiased estimate. Kriging cannot, however, use the equation in this form. The next section will derive the equation in detail (after Barnes 1984) and come up with a more usable form for kriging.

12.3 Derivation of the Estimation Variance Equation

The definition of the estimation variance equation was given earlier as

$$e_o = Z(x_o) - Z(x_o)^* \tag{12.3}$$

Because the true value is generally not known, all that can be calculated is the distribution of the possible errors. Thus, given N samples at locations

$$X_1, X_2, X_3, \ldots, X_n$$

with associated known concentrations

$$Z(X_1), Z(X_2), Z(X_3), \ldots, Z(X_n)$$

we wish to estimate the unknown concentration

$$Z(X_o)$$

at location X_o. This estimated concentration will be denoted

$$Z(X_o)^*$$

Logically, this estimate should be a function (F) of the n known sample concentrations available. This can be written

$$Z(X_o)^* = F\,[Z(X_1), Z(X_2), Z(X_3), \ldots, Z(X_n)]$$

If the true concentration were known, the estimation error could be calculated as

$$e_o = Z(X_o) - Z(X_o)^*$$

Typically, a large number of estimates are made over the area of a site. Thus, we can view this estimation error as an element of a population of errors from similar estimates, and the distribution of errors can be investigated.

Before the investigation begins, it is reasonable to establish the desirable characteristics of such a distribution of errors. Intuitively, what makes a "good" estimate? Previous discussion has indicated the following:

1. There should not be any regular bias (always too high or always too low). (*Unbiasedness!*)
2. The spread of the error distribution should be as small as possible. (*Minimum Variance!*)

Another very common decision is to limit the estimation to a "linear" estimator. That is

$$F(*) = \sum_{i=1}^{n} a_i Z(X_i) \tag{12.4}$$

where

a_i = the weights given to the i samples.

The most important reason for the choice of a linear model is that the mathematics for linear estimators are tractable; however, there is a theoretical reason. It can be shown that, if the underlying distribution is normal, the best (minimum variance) estimator is necessarily linear (Barnes 1985).

Thus, the estimator of interest for the estimation variance is

$$Z(X)* = \sum_{i=1}^{n} a_i Z(X_i) \tag{12.5}$$

The first requirement for a "good" estimator as mentioned above was unbiasedness. What is required for unbiasedness, and how is this condition introduced? First, unbiasedness means

$$E(e_o) = 0 \qquad (\textit{No systematic error}) \tag{12.6}$$

$$= [E(ZX_o) - Z(X_o)^*] \qquad \text{(by definition of } e_o\text{)}$$

$$= E[Z(X_o)] - \sum_{i=1}^{n} a_i E[Z(X_i)]$$

$$= 0 - \sum_{i=1}^{n} a_i M_i \tag{12.7}$$

where the M_i terms are the mean values. Under the assumption of stationarity, we can say that

$$M_0 = M_1 = M_2 = \ldots = M_n$$

Thus,

$$M - \sum_{i=1}^{n} a_i M = 0 \tag{12.8}$$

Factoring out M yields

$$M(1 - \sum_{i=1}^{n} a_i) = 0 \tag{12.9}$$

Dividing both sides of the equation by M gives

$$1 - \sum_{i=1}^{n} a_i = 0 \tag{12.10}$$

Rearranging the equation results in the unbiasedness condition

$$\sum_{i=1}^{n} a_i = 1 \tag{12.11}$$

where the sum of the weights used in the estimate equals one. If the weights sum to a value greater than one, a positive bias will be introduced. Likewise, weight sums less than one will produce a negative bias.

So far, the definition of the error e_o has been established and the unbiasedness condition derived. What is left now is to determine the variance of e_o. By definition,

$$\begin{aligned} \sigma_o^{\,2} &= E[(e_o)^2] \\ &= E(e_o^{\,2}) \end{aligned}$$

since $E(e_o^{\,2}) = 0$. Substituting in the definition of e_o

$$= E[(Z(X_o) - Z(X_o)^*)^2]$$

Subtracting the mean from both terms yields

$$= E[((Z(X_o) - M) - (Z(X_o)^* - M))^2]$$

Expanding the square gives

$$= E[(Z(X_o)\text{-}\ M)^2 - 2(Z(X_o) - M)(Z(X_o)^* - M) + (Z(X_o)^* - M)^2]$$

And, finally,

$$\sigma^2_{Z(X_o)} - 2\ \mathrm{COV}\ [Z(X_o),\ Z(X_o)^*] + \sigma^2_{Z(X_o)}\ ^* \tag{12.12}$$

Note the similarity between equation 12.12 and equation 12.2.

The terms in equation 12.12 can be examined one at a time to determine how the variogram and sample data influence them. First,

$$\sigma^2_{Z(X_o)} \tag{12.13}$$

This is the variance of the block being estimated. In other words, this is the block variance $\sigma^2_{(v/V)}$ (Fig. 12-1). If the variability of the block increases, it follows that the estimation becomes more difficult and the variability of the estimation error should rise.

Second, we must examine the term

$$- 2 \text{ COV } [Z(X_o), Z(X_o)^*] \tag{12.14}$$

This term should intuitively make sense: As the covariance (similarity) between the unknown concentration and its estimate goes up, the quality of the estimate goes up and the variance of the error goes down.

It is also useful to express this second portion of the estimation variance equation in terms of the sample grades:

$$= E\left[(Z(X_o) - M)\left(\sum_{i=1}^{n} a_i Z(X_i) - M\right)\right] \tag{12.15}$$

$$= E\left[(Z(X_o) - M)\left(\sum_{i=1}^{n} a_i Z(X_i) - M\right)\right] \tag{12.16}$$

$$E\left[\sum_{i=1}^{n} a_i (Z(X_i) - M)\ (Z(X_o) - M)\right] \tag{12.17}$$

$$\sum_{i=1}^{n} a_i E[(Z(X_i) - M)\ (Z(X_o) - M)] \tag{12.18}$$

$$\sum_{i=1}^{n} a_i \text{ COV } [Z(X_i),\ Z(X_o)] \tag{12.19}$$

Again, this should intuitively make sense: As the covariance between samples X_i and the unknown concentration X_o goes up (i.e., the samples get closer to the block being estimated), the quality of the estimate should go up with a corresponding reduction in the estimation error. Indeed, since this term is negative in sign, it subtracts a portion of the error from the block variance.

Third, consider the quantity

$$\sigma^2_{Z(X_o)}{}^* \tag{12.20}$$

This term indicates that, as the variability of the estimator increases, the variance of the estimation error increases. Again, it is informative to re-express this using the sample data values:

$$E\,[(Z(X_o)\,^* - E\,(Z(X_o)^*)^2\,] \tag{12.21}$$

$$E\,[(Z(X_o)\,^* - M^2\,] \tag{12.22}$$

$$= E\left[\left(\sum_{i=1}^{n} a_i(Z(X_i) - M)^2\right)\right] \tag{12.23}$$

$$= E\left[\left(\sum_{i=1}^{n} a_i(Z(X_i) - M)^2\right)\right] \tag{12.24}$$

$$= E\,[(a_1\,(Z(X_1) - M) + a_2\,(Z(X_2) - M) + \cdots + a_n\,(Z(X_n) - M)] \tag{12.25}$$

$$= E\left[\sum_{i=1}^{n}\sum_{j=1}^{n} a_i a_j (Z(X_i) - M)(Z(X_j) - M)\right] \tag{12.26}$$

$$= \sum_{i=1}^{n}\sum_{j=1}^{n} a_i a_j\ E\,[(Z(X_i) - M)(Z(X_j) - M)] \tag{12.27}$$

$$= \sum_{i=1}^{n}\sum_{j=1}^{n} a_i a_j\ \mathrm{COV}\ (Z(X_i),\ Z(X_j)) \tag{12.28}$$

Equation 12.28 shows that, as the covariance between the samples goes up, the samples are each giving less new information. This redundancy causes an increase in the estimation variance. This means that an interesting and important tensional relationship exists between the samples and the block: The samples should be as close to the block as possible but, simultaneously, the samples should be as far apart from one another as possible. This situation brings maximum sample coverage and proximity along with minimum clustering.

Summarizing, the estimation variance can be expressed as

$$\sigma_e^2 = \sigma_{Z(X_o)}^2 - 2\sum_{i=1}^{n} a_i \text{ COV } [Z(X_i), Z(X_o)] + \sum_{i=1}^{n}\sum_{j=1}^{n} a_i a_j \text{ COV } [Z(X_i), Z(X_j)] \tag{12.29}$$

12.4 Extension Variance

12.4.1 Definition

Equation 12.29 is a generalized formulation for determining the estimation variance and is typically solved using a kriging program and multiple sample data points. It is sometimes useful, however, to determine the estimation variance for a single sample within a block. This is called the *extension variance*.

As described in Chapter 3 and shown in Figure 3-4, an error is made when the concentration of a central sample is extended to be representative of the entire block area. The magnitude of the expected error could be determined using equation 12.29 with the appropriate variogram model or by means of a kriging program.

12.4.2 Use of Charts

For situations that involve the extension of a single sample, it is often easier to make use of charts that have already been developed (Matheron 1971). Figure 12-4 is such a chart, which can be used to calculate the estimation variance of the concentration of a square-based block that has been estimated from a central sample intersection. Note the similarity of Figure 12-4 to Figures 11-9 and 11-10, which allowed the computation of the block variance.

As with the charts for block variance, Figure 12-4 gives the estimation variance using standardized variograms. Standardized variograms have a zero nugget effect, a range of one, and a sill equal to one. In order to obtain the proper value of σ_e^2 it is necessary to multiply chart readings by C, the sill of the actual variogram model for the COC of interest.

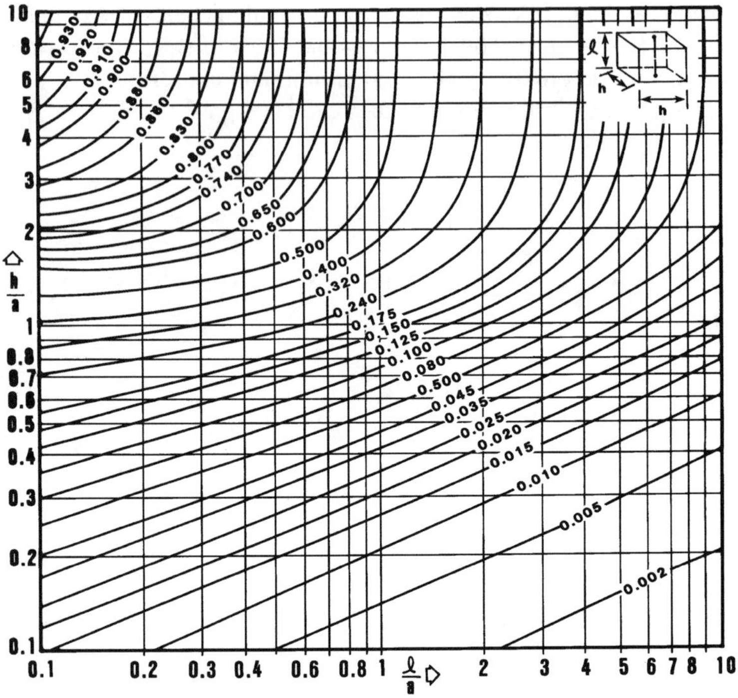

Figure 12-4. Chart to determine the estimation variance of a concentration in a square-based block using a central sample and a spherical variogram model. (From *Geostatistical Ore Reserve Estimation*, by M. David, Elsevier, 1977. Courtesy of Michel David.)

As an example, consider an arsenic site with a spherical variogram model with sill C equal to 500 (ppm)^2 and an isotropic range of 100 ft. The estimation variance of a central sample is desired for a block measuring 20 × 20 × 5 ft. Using Figure 12-4, the values for h/a and ℓ/a are found to be .2 and .05, respectively. Extrapolating for the value of ℓ/a this yields

$$\frac{1}{C}\sigma_e^2 = .002 \tag{12.30}$$

Thus, with $C = 500$, the answer is

$$\sigma_e{}^2 = (12{,}500)(.002) = 25 \text{ ppm}^2 \tag{12.31}$$

and it equates to a standard deviation of 5.0 ppm arsenic. This standard deviation is sometimes referred to as the standard error of the mean. This is, in fact, the nonindependent form of the well-known formula

$$\sigma_m = \sqrt{\frac{\sigma^2}{n}} \tag{12.32}$$

which is the equation for the standard error of the mean for independent samples. For the sake of brevity, the term *standard error of the mean* is often used for spatial data. The extension variance would be more correctly defined as "the variance of the average error made in estimating the average concentration of a block by using the concentration of a central sample" (David 1977). The reason for the short form should be obvious.

12.4.3 Composition of Extension Variances

The errors involved in repeated extensions of sample data to blocks can be considered as independent for practical purposes. Although this is not strictly true, discrepancies are always very small (David 1977). This implies that the variances associated with successive extensions may be added.

Adding successive extension errors can provide a basis for an estimate of global precision. In other words, we may want to estimate the error associated with the mean of the entire site, perhaps in anticipation of a risk assessment. It must be stressed that, for this technique to be valid, each extension must be to the same-sized block. The formula for the standard error of the mean at a site with spatially correlated data is, thus,

$$\sigma^2_{m_e} = \frac{1}{n}\sigma^2_{\text{ext}} \tag{12.33}$$

For global estimation of precision where extensions are made to volumes of unequal sizes, weighted averages must be used. The following formula can be used for extensions of unequal volumes or areas:

$$\sigma^2_{m_{\text{ext}}} = \frac{\dfrac{(Vol)_1(\sigma^2_{\text{ext}})_1}{TV} + \dfrac{(Vol)_2(\sigma^2_{\text{ext}})_2}{TV} + \cdots + \dfrac{(Vol)_n(\sigma^2_{\text{ext}})_n}{TV}}{n} \tag{12.34}$$

CONCLUSION

Obtaining a formula for the precision of the estimated values is a major step in the GEM process. The next step is to find a way to minimize the variance of the error. Once minimized, confidence intervals can be placed on blocks, contour lines, even the mean concentration for the entire site for input into risk assessment models. To quantify the error of an estimate means that decision risk can be better evaluated as part of the GEM or DQO process. The magnitude of the error relates to the number of samples needed to characterize a site, where the samples should be placed, and what an appropriate block estimation size might be for accurate forecasting of remedial volumes.

Chapter *13*

Optimizing Estimation: Kriging

SYNOPSIS: Quantification of the spatial continuity through the variogram allows the development of the block variance and, later, the estimation variance, both important components in the GEM quest to evaluate and quantify uncertainty and error. A valid goal of any statistical analysis is to minimize the error; this is what kriging brings to the GEM process. This chapter will begin by deriving the kriging equations, followed by a look at how the matrix formulation operates. Then we will examine important properties of kriging and how the variogram parameters influence the kriging results.

KEY WORDS: Block Kriging, Estimation Variance, Covariance, Weights, Partial Derivative, Lagrange Multiplier, Unbiasedness, Constraint, Matrix, Point Kriging, Variogram Model, Declustering, Inverse Power Distance, Additivity, Screen Effect, Negative Weights, Exact Interpolation, Smoothing Effect, Conditional Unbiasedness, Sill, Nugget Effect, Range, Anisotropy, Shape

13.1 INTRODUCTION

In Chapter 12, considerable effort was expended to formulate the problem of determining the error involved in estimating a block using sample data of known concentration. It was shown that a linear combination using the variogram function and sample weights could define the precision with which an estimated location is known.

Given a quantification of the expected error, it is possible to find a set of sample weights that will minimize the magnitude of the error variance. This, of course, is *kriging*. Using the time-honored method of taking the derivative with respect to the appropriate variables, the estimation variance equation produces a series of equations that, when solved, will produce the minimum-variance optimal estimate.

Despite this optimality, the task kriging attempts to accomplish is challenging. We will never know the true value of the mean average error m_e or its variance σ_e^2. Thus, we cannot be assured that m_e will exactly equal zero, nor that σ_e^2 will be minimized. Given that the variogram has been developed on a limited set of sample data that we hope is representative, even the error variance model is subject to uncertainty. Kriging will, however, provide a model that assures that these quantities will be optimized for the chosen error model. Kriging has jokingly been described as "rude, crude, and socially unacceptable," but it nevertheless offers the most powerful approach to error quantification and minimization for spatially correlated data.

13.2 The Kriging Equations

Starting with the definition of the estimation variance,

$$\sigma_e^2 = \sigma_{Z(X_o)}^2 - 2\sum_{i=1}^{n} a_i \mathrm{COV}[Z(X_i), Z(X_o)] + \sum_{i=1}^{n}\sum_{j=1}^{n} a_i a_j \mathrm{COV}[Z(X_i, Z(X_j))] \tag{13.1}$$

we wish to determine the set of weights a_i $(i = 1,...,n)$ that will allow the weighted average

$$Z^* = \sum_{i=1}^{n} a_i Z(X_i) \tag{13.2}$$

to be the minimum variance estimator of *Z(V)*, where *Z(V)* is the true unknown concentration of block *V*, and $Z(X_i)$ represents a series of samples $(i = 1,...,n)$ with known concentrations. In addition, all of this is subject to the constraint that the estimate be unbiased, with the weights summing to one:

$$\sum_{i=1}^{n} a_i = 1 \tag{13.3}$$

For notational simplicity, equation 13.1 can be rewritten as

$$\sigma_e^2 = \sigma_v^2 - \sum_{i=1}^{n} a_i \sigma_{VX_i} + \sum_{i=1}^{n} \sum_{i=1}^{n} a_i a_j \sigma_{X_i X_j} \tag{13.4}$$

where

$$\sigma_V^2 \tag{13.5}$$

is the variance of the blocks being estimated (block variance);

$$\sigma_{VX_i}^2 \tag{13.6}$$

is the covariance COV[$Z(V)$,$Z(X_i)$] of the concentration of block V and the concentration of sample X_i (sample-to-block covariance);

$$\sigma_{X_i X_j} \tag{13.7}$$

is the covariance COV[$Z(X_i)$,$Z(X_j)$] of the concentrations of sample X_i and sample X_j (sample-to-sample covariance).

The n sample weights a_i (i = 1,...,n) will produce n equations with n unknowns when the n partial derivatives are taken. One problem still remains, however. The unbiasedness constraint (sum of the weights = 1) adds an additional equation yet does not contribute an additional unknown. Thus, the system has $n + 1$ equations and n unknowns, a very difficult solution to obtain.

The Lagrange multiplier may be used to circumvent this problem. A new unknown μ is introduced to stand for the estimated variance of the errors. The Lagrange principle states that, when a constraint C (note: $C = 0$) is present, the function

$$F = Q + 2\mu C \tag{13.8}$$

must be minimized. Combining equations 13.4 and 13.8, the short form of the equation becomes

$$F=\sigma_e^2+2\mu\left(\sum_{i=1}^{n} a_i-1\right) \qquad (13.9)$$

or, more fully,

$$F=\sigma_V^2-2\sum_{i=1}^{n} a_i\sigma_{VX_i}+\sum_{i=1}^{n}\sum_{j=1}^{n} a_ia_j\sigma_{X_iX_j}+2\mu\left(\sum_{i=1}^{n}-1\right) \qquad (13.10)$$

Note that the equality of the equation is not affected by adding the Lagrange parameter. Because

$$\sum_{i=1}^{n} a_i-1=0 \qquad (13.11)$$

it also follows that

$$2\mu\left(\sum_{i=1}^{n} a_i-1\right)=0 \qquad (13.12)$$

Thus, zero is added to Q in the general equation 13.8 and to the estimation variance in equation 13.9, causing no net effect to the equality.

To obtain the error variance for the model, it is necessary to take the $n + 1$ partial first derivatives corresponding to the n sample weights and μ. This yields $n + 1$ equations with $n + 1$ unknowns. In addition, the unbiasedness condition is fulfilled by setting the partial first derivative with respect to μ equal to zero. The derivatives can be written as follows:

$$\frac{\delta F}{\delta a_i}=-2\sigma_{VX_i}+\sum_{i} a_j\sigma_{X_iX_j}+2\mu=0 \qquad (13.13)$$

for each of the $i = 1,...,n$ sample weights corresponding to the n samples to be used in the estimate, along with the unbiasedness constraint

$$\frac{\delta F}{\delta \mu} = \sum_i a_i - 1 = 0 \tag{13.14}$$

The more usual form of these equations can be given by

$$\sum_j a_j \sigma_{X_i X_j} + \mu = \sigma_{VX_i} \tag{13.15}$$

and

$$\sum_i a_i = 1 \tag{13.16}$$

Expanding equation 13.15 produces the n equations needed for each of the n sample weights:

$$\sum_j a_j \sigma_{X_1 X_j} + \mu = \sigma_{VX_1} \tag{13.17}$$

$$\vdots$$

$$\sum_j a_j \sigma_{X_2 X_j} + \mu = \sigma_{VX_2} \tag{13.18}$$

$$\vdots$$

$$\sum_j a_j \sigma_{X_n X_j} + \mu = \sigma_{VX_n} \tag{13.19}$$

This system of equations can be written in matrix form:

$$[\mathbf{\Sigma}][\mathbf{A}] = [\boldsymbol{D}] \tag{13.20}$$

and is generally known as the ordinary kriging system. Using σ_{ij} as shorthand for σ_{XiXj}, the $\mathbf{\Sigma}$, $\mathbf{A}$, and $\mathbf{D}$ matrices can be expanded respectively, as

$$\Sigma = \begin{vmatrix} \sigma_{11} & \sigma_{12} & \cdots & \sigma_{1n} & 1 \\ \sigma_{21} & \sigma_{22} & \cdots & \sigma_{2n} & 1 \\ \cdot & & & & \\ \cdot & & \ddots & & \\ \cdot & & & & \\ \sigma_{n1} & \sigma_{n2} & \cdots & \sigma_{nn} & 1 \\ 1 & 1 & \cdots & 1 & 0 \end{vmatrix} \tag{13.21}$$

$$\boldsymbol{A} = \begin{vmatrix} a_1 \\ a_2 \\ \vdots \\ a_n \\ \mu \end{vmatrix} \tag{13.22}$$

$$\boldsymbol{D} = \begin{vmatrix} \sigma_{VX_1} \\ \sigma_{VX_2} \\ \vdots \\ \sigma_{VX_n} \\ 1 \end{vmatrix} \tag{13.23}$$

The $\mathbf{\Sigma}$ matrix is symmetric and contains information dependent solely on the known sample concentrations and the covariances between them (sample-to-sample covariance) This type of relationship was shown previously in Figure 12-3. The $\mathbf{D}$ matrix depends on both the block and the sample data (point-to-block covariance), shown previously in Figure 12-2. The $\mathbf{A}$ matrix contains the n as yet unknown sample weights and the Lagrange multiplier.

To solve for the weights, the equation becomes

$$[\mathbf{A}] = [\mathbf{\Sigma}]^{-1}\,[\boldsymbol{D}] \tag{13.24}$$

where $[\mathbf{\Sigma}]^{-1}$ is the inverse of the $[\mathbf{\Sigma}]$ matrix. Note that the value for every σ_{ij} and σ_{VXi} can be determined from the selected variogram model. An example of this process will be given in Section 13.4.

With the weights determined, the kriging estimate becomes simply a matter of calculating the weighted average of the sample data used. This gives a minimum variance and unbiased estimate for the block in question. All that remains to be done is to calculate the value of the uncertainty associated with the kriging estimate.

The weights obtained by multiplying the $[\boldsymbol{\Sigma}]^{-1}$ can be substituted into equation 13.1 since the kriging weights were obtained by minimizing the estimation variance. This somewhat tedious process can be simplified by using the following formula:

$$\sigma_K^2 = \sigma_V^2 - \sum_i a_i \sigma_{VX_i} - \mu \tag{13.25}$$

where $\sigma_K{}^2$ is the minimized error variance, which will now be called the *kriging variance*. The kriging variance may also be expressed in matrix form by

$$\sigma_K^2 = \sigma_V^2 - \boldsymbol{A} \cdot \boldsymbol{D} \tag{13.26}$$

Since all values on the right side of equation 13.26 are available in the kriging system, the kriging variance is easily obtained.

13.3 Point Kriging

The reader should recall that the above equations relate to the concept of block kriging, which is a two- or three-dimensional process. In order to build the **D** matrix of sample to block covariances, a discretization procedure was used to determine the average distance of a sample to that block (Fig. 12-2).

In practice, it is often desired simply to estimate a point value instead of a block value. This is commonly done for purposes of creating a contour map. This greatly simplifies the distance calculation as

$$\sigma_{VX_i} \rightarrow \sigma_{X.X_i} = \sigma_{\cdot i} \tag{13.27}$$

if $X_.$ is the point to be estimated. The Σ matrix remains unchanged since it still depends only on the samples X_i and their relationship one to another. However, the **D** matrix becomes much easier to calculate. Figure 13-1 shows a visual example. Figure 13-1a shows the multiple distances (25) *per sample* and associated covariance determination required for block kriging. This is contrasted with

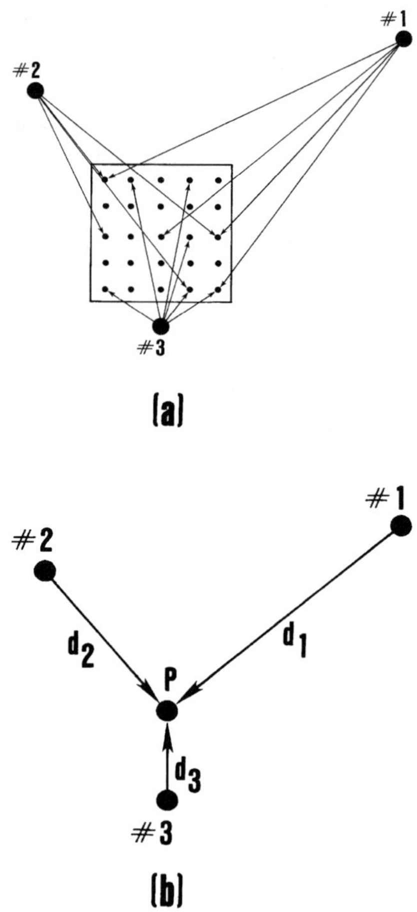

Figure 13-1. Comparison of block and point kriging with respect to the **D** matrix determination.

Figure 13-1b, which shows that three calculations, one per sample, are needed for point kriging.

13.4 Application of Kriging Equations

Even though the derivation of the equations for estimation variance and kriging is important from a conceptual perspective, it can be a dry and tedious exercise for the practitioner. What is of great benefit is seeing the samples and the variogram used actually to krige a block. This section will present an example of two-dimensional kriging using sample data and a block to be estimated. The kriging process will be exposed in great detail in a step-by-step manner. For simplicity in demonstration, all examples will use the point kriging method.

Figure 13-2 shows a sample configuration from a PCB site. Six sample data labeled 1 through 6 are present. The point to be estimated is the location labeled *P* located at coordinate $X = 27.5$ and $Y = 37.5$. Table 13-1 lists the *XY* coordinates of the samples, the sample concentrations, and the distance from the samples to point *P*.

With kriging the estimator of choice, it is necessary to define a variogram function by which the covariances may be determined. For this example, the commonly used spherical model will be employed, with a nugget effect of zero ($C_o = 0$), a sill of 500 ($C = 500$), and an isotropic range of 100 ft in all directions.

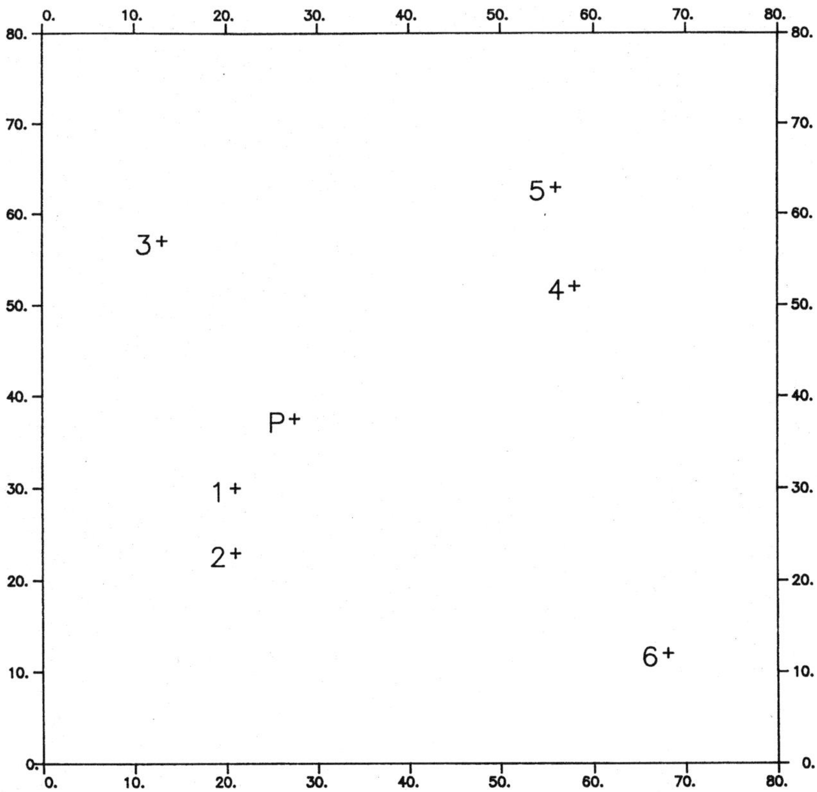

Figure 13-2. Sample data configuration and point to be estimated in the kriging example.

Table 13-1. Sample data, sample concentrations, and sample distances from the point *P* to be estimated.

Sample Number	*X* Coord.	*Y* Coord.	Sample Concentration	Distance from Point *P*
1	21.	30.	551.	9.9
2	21.	23.	675.	17.7
3	13.	57.	197.	24.3
4	56.	52.	395.	33.8
5	58.	63.	322.	38.2
6	68.	12.	137.	47.9

With sample data and a covariance function, the [**Σ**] and [**D**] matrices may be determined. Starting with the [**Σ**] matrix, it is necessary first to calculate the distances between all sample pairs. These values are given in Table 13-2.

Using the formula for the spherical variogram model

$$\gamma(h) = C_o) + C\left[(0.5)\frac{h}{a} - (.5)\frac{h^3}{a^3}\right] \tag{13.28}$$

the distances listed in Table 13-2 may be plugged in, with the covariances calculated by means of the following formula:

$$\sigma_{ij} = C_o + C - \gamma(h_{ij}) \tag{13.29}$$

Table 13-2. Distances between all sample points for use in calculating the [**Σ**] matrix for kriging.

	Distance						
Location	***P***	**1**	**2**	**3**	**4**	**5**	**6**
P	0.0	9.9	15.9	24.3	33.8	38.2	47.9
1	9.9	0.0	7.0	28.2	43.1	48.1	50.3
2	15.9	7.0	0.0	34.9	47.0	53.2	48.3
3	24.3	28.2	34.9	0.0	45.3	43.4	71.1
4	33.8	43.1	47.0	45.3	0.0	11.2	41.2
5	38.2	48.1	53.2	43.4	11.2	0.0	52.4
6	47.9	50.3	48.3	71.1	41.2	52.4	0.0

which is simply the sill of the variogram minus the $\gamma(h)$ value for the distance between samples i and j. Since $C_o + C = 500$ for the chosen variogram model, combining equations 13.28 and 13.29 gives

$$\sigma_{12} = \sigma_{21} = 500 - [\gamma(7.0)] \tag{13.30}$$

$$\sigma_{12} = \sigma_{21} = 500 - \left[500\left[(1.5)\frac{7.0}{100} - (.5)\frac{(7.0)^3}{(100)^3}\right]\right] \tag{13.31}$$

$$\sigma_{12} = \sigma_{21} = 447.5 \tag{13.32}$$

for the covariance between samples 1 and 2. Other ij sample pairs are calculated in a similar manner. Note that the covariance between a point and itself ($i = j$) is equal to the sill value, implying complete correlation. The complete matrices of σ_{ij} combinations and corresponding covariance values for the distances listed in Table 13-2 appear below:

$$\Sigma = \begin{vmatrix} \sigma_{11} & \sigma_{12} & \sigma_{13} & \sigma_{14} & \sigma_{15} & \sigma_{16} & 1 \\ \sigma_{21} & \sigma_{22} & \sigma_{23} & \sigma_{24} & \sigma_{25} & \sigma_{16} & 1 \\ \sigma_{31} & \sigma_{32} & \sigma_{33} & \sigma_{34} & \sigma_{35} & \sigma_{36} & 1 \\ \sigma_{41} & \sigma_{42} & \sigma_{43} & \sigma_{44} & \sigma_{45} & \sigma_{46} & 1 \\ \sigma_{51} & \sigma_{52} & \sigma_{53} & \sigma_{54} & \sigma_{55} & \sigma_{56} & 1 \\ \sigma_{61} & \sigma_{62} & \sigma_{63} & \sigma_{64} & \sigma_{65} & \sigma_{66} & 1 \\ 1 & 1 & 1 & 1 & 1 & 1 & 0 \end{vmatrix} \tag{13.33}$$

$$\Sigma = \begin{vmatrix} 500 & 448 & 294 & 197 & 167 & 154 & 1 \\ 448 & 500 & 249 & 173 & 139 & 166 & 1 \\ 294 & 249 & 500 & 184 & 195 & 57 & 1 \\ 197 & 173 & 184 & 500 & 416 & 208 & 1 \\ 167 & 139 & 195 & 416 & 500 & 143 & 1 \\ 154 & 166 & 57 & 208 & 143 & 500 & 1 \\ 1 & 1 & 1 & 1 & 1 & 1 & 0 \end{vmatrix} \tag{13.34}$$

The **D** matrix, whose distances are summarized in Table 13-1, appears below in covariance form:

$$D = \begin{vmatrix} \sigma_{\cdot 1} \\ \sigma_{\cdot 2} \\ \sigma_{\cdot 3} \\ \sigma_{\cdot 4} \\ \sigma_{\cdot 5} \\ \sigma_{\cdot 6} \\ 1 \end{vmatrix} = \begin{vmatrix} 426 \\ 382 \\ 321 \\ 256 \\ 227 \\ 168 \\ 1 \end{vmatrix} \tag{13.35}$$

Taking the inverse of the **Σ** matrix yields:

$$\Sigma^{-1} = \begin{vmatrix} .01191 & -.00935 & -.00201 & -.00069 & .00085 & .00005 & -.06194 \\ -.00935 & .10203 & .00003 & .00005 & .00006 & -.00101 & .22638 \\ -.00201 & .00003 & .00303 & .00001 & -.00108 & .00002 & .27491 \\ -.00069 & .00005 & .00001 & .00750 & -.00561 & -.00127 & .02894 \\ .00009 & .00006 & -.00108 & -.00561 & .00651 & .00003 & .25978 \\ .00005 & -.00101 & .00002 & -.00127 & .00003 & .00218 & .32981 \\ -.06194 & .22628 & .27491 & -.02894 & .25978 & .32981 & -239.74 \end{vmatrix} \tag{13.36}$$

Multiplying the $\mathbf{\Sigma}^{-1}$ matrix by the **D** matrix provides the set of weights that yields an unbiased and minimum variance estimate for the unsampled point *P*:

$$\boldsymbol{A} = \begin{vmatrix} a_1 \\ a_2 \\ a_3 \\ a_4 \\ a_5 \\ a_6 \\ \mu \end{vmatrix} = \Sigma^{-1} \cdot \boldsymbol{D} \begin{vmatrix} 0.642 \\ 0.015 \\ 0.165 \\ 0.137 \\ 0.019 \\ 0.021 \\ 15.72 \end{vmatrix} \tag{13.37}$$

With the weights determined, the estimated value for the point may be calculated using the sample data:

$$Z(X_.)^* = \sum_{i=1}^{n} a_i X_i \tag{13.38}$$

$$= (.642)(551) + (.015)(675) + (.165)(197) + (.137)(395) + (.019)(322) \quad (13.39)$$

$$= 460 \text{ ppm} \quad (13.40)$$

Again, using the weights, the corresponding minimum error variance can be calculated:

$$\sigma_K^2 = \sigma_V^2 - \sum_i a_i \sigma_{VX_i} - \mu \quad (13.41)$$

$$500 - (.642)(426) + (.015)(382) + (.165)(321) + (.137)(256) + (.019) \quad (13.42)$$

$$= 109 \text{ ppm}^2 \quad (13.43)$$

13.5 Properties of Kriging

13.5.1 Declustering

Deriving the kriging equations provides an informative look at the source and development of the kriging system. Similarly, plugging in sample data and working through the process in a step-by-step manner reveals the mechanics of kriging in intimate detail. Yet a single example has limitations. We may ask: How does kriging compare to other estimators? or What happens if the variogram model changes? These are important questions regarding the properties of the kriging estimator. These properties will be explored in this section.

Kriging uses two matrices, the **C** and the **D**, to develop the optimal and minimum variance weights for the samples. Understanding the relationship of these matrices helps to develop an appreciation and understanding of the uniqueness of kriging.

To start, the **D** matrix provides a set of distances from the available sample data to the point or block being estimated. This, of course, is useful information but information that other estimating techniques use as well. For instance, inverse power distance (IPD) methods use a **D**-type matrix to calculate a weighted point or block estimate. Inverse distance weights are limited, however, to the form $|h|^{-p}$, where h is the distance and p the power used in the weighting.

Kriging may use this type of inverse distance function but is not restricted to it. While the **D** matrix uses a distance-weighting function, an important distinction between the IPD method and kriging is that the **D** matrix uses a "statistical" distance rather than a geometric distance. Of far greater importance than statistical vs. geometric distance is the contribution provided by the **Σ** matrix. It is the **Σ** matrix that accounts for the declustering property of kriging.

Multiplying $\mathbf{\Sigma}^{-1}$ by **D** rescales the covariances in **D** so that they sum to one. Yet more than this is happening. The **Σ** matrix embodies the statistical distance between each sample and all other samples to be used in the estimate. This yields information regarding the degree of clustering among the sample data. Closely spaced samples report a large covariance value to the **Σ** matrix; distant samples report smaller covariances. Multiplying $\mathbf{\Sigma}^{-1}$ by **D** accounts for redundant sample information and restructures the weights in **D** so that the influence of clustered data is minimized.

Returning to the kriging example (Fig. 13-2 and the σ_{ij} matrix in equation 13.33), we see that sample 1 is very close to sample 2 and that both are relatively close to the point P being estimated. They are, in fact, the two closest points. The **Σ** matrix confirms that the covariance is high between the two samples, returning a value of 448 out of a possible 500.

Equation 13.37 reveals, however, that sample 2 gets a weight of only .015 vs. a weight of .642 for sample 1. In fact, samples 3, 4, and 5 all return larger weights than sample 2 despite their greater distances from point P. This demonstrates that the final sample weights are influenced differently by the **Σ** and **D** matrices. The **D** matrix accounts for distance from the point or block; the **Σ** matrix accounts for clustering.

Note that another cluster exists, samples 4 and 5, and that the weights for these samples are subject to the declustering effect. The covariance between samples 4 and 5 is high; a value of 416 is seen in the **Σ** matrix. The weight for sample 4 is much higher, however, than for sample 5 (.137 vs. .019). The cluster total for the weights is less than for the samples 1 and 2 cluster because of the increased distance, but a similar declustering process has taken place.

As a comparison, Table 13-3 shows a listing of the samples and weights for kriging, inverse distance ($p = 1$), and inverse distance ($p = 2$). The kriging weights come from equation 13.37. The IPD weights have been calculated using the formula:

$$Z(X)^*_{\mathrm{IPD}} = \frac{\sum_{i=1}^{n} \frac{1}{d_i^p} v_i}{\sum_{i=1}^{n} \frac{1}{d_i^p}} \tag{13.44}$$

Table 13-3. Kriging and inverse power distance sample weights for example data configuration.

Sample	Distance	Kriging Weight	$1/d_i$	IPD (p = 1) Weight	$1/d_i^2$	IPD (p = 2) Weight
1	9.9	.642	.1010	.367	.01020	.597
2	17.7	.015	.0565	.205	.00319	.187
3	24.3	.165	.0412	.150	.00169	.099
4	33.8	.137	.0296	.108	.00088	.047
5	38.2	.019	.0262	.095	.00069	.040
6	47.9	.021	.0209	.076	.00044	.026

Comparing the kriging weights to the IPD (p = 1) weights shows significant differences. The range for the weights (difference between largest and smallest) is more than double, .623 versus .291 respectively. The IPD weight for sample 1 is far less than that for kriging, .367 vs. .642. For sample 2, however, the inverse distance weight is far greater than for kriging, .205 vs. 015. This illustrates the fact that the IPD method cannot perform a declustering function. Note that kriging detected another cluster composed of samples 4 and 5. Sample 4 received the bulk of the weight for the cluster while sample 5 received a very small weight.

Other differences exist as well. The sample 1 and 2 cluster received 65.7% of the weight in kriging compared to only 57.2% of the weight for IPD with p = 1. Kriging attributes a greater value to the cluster than IPD, which values the more distant samples (as a group) more than the close-in cluster. Note that this result is a function of the distance-weighting function used for both estimators. Using a different distance-weighting function will alter the result.

For IPD (p = 2), results differ from kriging as well as from IPD (p = 1) because the parameters of the weighting function have changed. The weight for sample 1 is almost as large as that for kriging (.597 vs. .642). Unlike kriging but like IPD (p = 1), sample 2 returns a relatively high weight. The weights of both IPD estimators for sample 2 are much higher than the kriging weight, which determines a need to apply declustering to sample 2. This also results in an even greater sum of weights for samples 1 and 2 for IPD than that for kriging (.784 vs. .657). Almost 80% of the weight has been assigned to the sample 1 and 2 cluster.

IPD also fails to decluster samples 4 and 5. Sample 4 still receives a higher weight than sample 5 since sample 4 is a bit closer to P yet, without the ability to decluster, both receive almost equal weight (.047 vs. .040). Sample 6 now receives the smallest weight by virtue of the fact that it is at the greatest distance from point P.

When the weights from the three estimators are multiplied by the sample concentrations, the results are:

Kriging	460 ppm
IPD ($p=1$)	456 ppm
IPD ($p=2$)	515 ppm

The IPD (p = 1) estimate is almost identical to that of kriging, and the IPD (p = 2) estimate results in a higher estimated concentration than that of kriging. As the power p of the IPD estimate continues to rise, the weight given to sample 1 will also increase. As p approaches ∞, all the weight is given to the closest sample point and IPD becomes a polygonal estimator. Conversely, as p approaches zero, the weights become more nearly equal, and IPD becomes a simple averaging technique.

Figures 13-3a through 13-3d show a generalized example for the differences in declustering abilities of kriging compared to inverse power distance. In each case, a sample point will be estimated using three equidistant data locations or two equidistant locations and an equidistant cluster. All estimations assume an isotropic variogram.

In Figure 13-3a, all three samples receive an equal portion of the weight for both kriging and IPD. In Figure 13-3b, location 1 contains a cluster of two samples. With kriging, both samples receive one-sixth of the weight or a total of

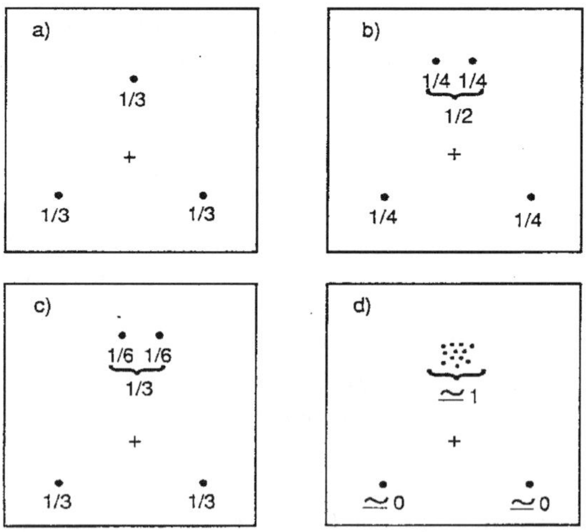

Figure 13-3. Generalized example of the declustering of samples by kriging vs. the nondeclustered weights resulting from IPD method. (From *Geostatistical Reserve Estimation*, by M. David, Elsevier, 1977. Courtesy of Michel David.)

one-third for the cluster. Given the redundancy of their information contribution, this weighting is appropriate.

In actual practice, we would expect kriging to give just a bit more than one-third of the weight to the cluster with the other two locations receiving slightly less than one-third. After all, the additional sample in the cluster does provide confirmation of the original sample concentration and thus should contribute a small amount.

The IPD weights, however, give each sample in the cluster a weight of 1/4, or a total of 1/2 to the entire cluster (Fig. 13-3c). The redundancy is not accounted for and the cluster receives too much weight. Taken to the extreme, Figure 13-3d shows a cluster of a dozen samples. Kriging will again apportion the weights in an appropriate manner; IPD, however, will assign virtually all the weight to the cluster of samples.

The ability of kriging to provide a measure of the expected error at each estimated location is no small advantage. In practice, this uncertainty value is not always used in subsequent applications. Given this, declustering joins the estimation variance as a very important attribute of kriging. Given a choice between kriging and another interpolation method in a situation for which no spatial continuity exists, kriging stands out as the logical choice simply because it can appropriately decluster nearby samples. This rule does not apply to pure nugget effect variogram models, in which no declustering occurs.

13.5.2 Additivity

An alternative to block kriging is often cited in the literature (see, e.g., David 1977, Isaaks and Srivastava 1989). This alternative method involves discretizing the block to be kriged into many point locations, perhaps 100. Point kriging can then be employed to estimate each of the 100 point locations if it is assumed that the same samples will be used in all 100 estimates. Then, the arithmetic average of the 100 point estimates will be assigned as the concentration for the block.

This simple approach has one major drawback: Computationally, it is extremely expensive. Block kriging is much more efficient. However, the average of the 100 point estimates and the block estimate will be the same. This is a property that is unique to kriging. It can also have some practical applications.

Consider a hazardous waste site that was kriged on 10 $\times$ 10 ft blocks, based on excavation constraints. If the remedial equipment changes and a new block size of 20 $\times$ 20 ft is needed, then blocks may simply be added and averaged to calculate the new block estimate.

This simple approach may be used to avoid rekriging locally or globally. Regrettably, this technique cannot be used for precisions or variances because the variances of the blocks are not independent. The relationship between the blocks to be averaged involves all the weights and covariances and thus makes simple averaging impossible.

13.5.3 Screen Effect

A sample is said to be screened if another sample comes between it and the point or block to be estimated. Figure 13-4 shows a modified arrangement of Figure 13-2, where sample 2 has been moved so that sample 1 is almost directly between sample 2 and point *P*.

Because both these samples have the same vector orientation with respect to point *P*, it is logical to give more weight to the sample that is not being screened, that is, the closer sample. This is, in fact, accomplished during the multiplication

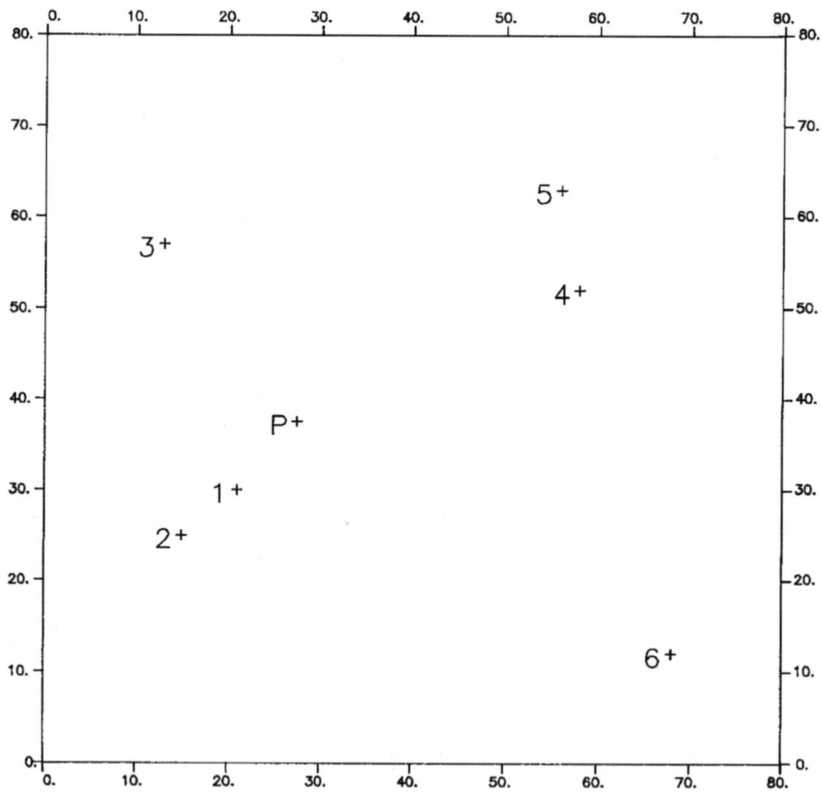

Figure 13-4. Modified example data set to demonstrate the screen effect of kriging and negative weight assignments. (From *An Introduction to Applied Geostatistics*, by E.H. Isaaks and R.M. Srivastava. Copyright © 1989 by Oxford University Press, Inc. Used by permission of Oxford University Press, Inc.)

by the Σ^{-1} matrix. The amount of screening depends on the model of spatial continuity chosen.

Figure 13-4 also shows the weights assigned to the samples in the new configuration. Note that sample 2 is so strongly screened that it has received a negative weight. Negative weights are unique to the kriging estimator. Estimators that are limited to weights falling between zero and one can produce only estimated values that are bounded by the high and low sample concentrations.

There are pros and cons surrounding the issue of negative weights. The advantage is that weights larger than one or less than zero can be allocated. Note, however, that the sum of the weights still must add up to one. It would be an unusual sampling program that captured both the highest and lowest concentrations present at a site. True values may exist beyond the recorded sample data values, and negative weights provide an opportunity to estimate above or below the concentration range of sample data encountered.

Extrapolating beyond the range of observed data is an even riskier endeavor than simple estimation. It may be more prudent to take additional samples if you suspect that the full distribution of sample values has not been encountered during sampling.

Another drawback to negative weights is that negative estimated concentrations may also occur. This is a distinct possibility at many environmental sites where closely spaced sample data quickly span several orders of magnitude. Even a very small negative weight, perhaps .005, could cause an estimated block to be negative.

For example, PCBs often have action levels around 25 ppm. Consider, then, a sample of concentration 5000 ppm, not an unusual value for PCBs. If this sample is used to krige a block and receives a weight of $-.005$, the block estimate will be negative if the weighted average of all the other sample concentrations is 25 ppm or less. If the sample screening the 5000 ppm sample has a concentration close to zero, one may reasonably question the validity of using the 5000 ppm sample at all.

Note that, in certain special instances, negative concentrations may be totally valid. For instance, acid-base account for soil reclamation issues takes on a negative value when the neutralizing potential of the soil is high. Thus, the data may support the existence of negative estimated values. This, however, should not be confused with the issue of negative weights.

Some kriging programs offer options whereby negative weights can be eliminated if detected in the kriging system. It is generally recommended to use such a removal option if available.

13.5.4 Exact Interpolation

An estimator can be termed *exact* when it returns the sample datum value for the estimated value. For instance, if kriging attempts to estimate the concentration of a known point, the concentration will be equal to the sample value at that point even if other samples are available in the kriging matrix. This implies that the weight for the sample point will be one whereas all others are zero.

Exact interpolation is a pleasing feature for the production of contour maps. One common problem occurs when a contour line passes on the "wrong" side of a data point. Although this situation may, in fact, be reasonable or explainable, it is aesthetically displeasing and often is not tolerated. With exact interpolation, occurrences of "wrong" contour lines are reduced. They are not entirely eliminated, however, as sample points seldom fall exactly at the point location to be estimated.

13.5.5 Smoothing Effect

This is a topic related to the concepts presented in Chapter 11. The dispersion of estimated concentrations is less than that for sample concentrations. Correlation diagrams confirm that high concentrations tend to be underestimated, whereas low concentrations tend to be overestimated. The result is a smoothing effect where estimates are less variable than original data.

13.5.6 Conditional Unbiasedness

Conditional unbiasedness, while daunting as a phrase, arguably describes the fundamental property of kriging (David 1977). It states that, on the average, all blocks Z that are predicted to have a concentration of Z_o will indeed have that concentration. Mathematically, this is written

$$E(Z|\, Z^* = Z_o) = Z_o \tag{13.45}$$

This condition is very difficult to achieve in practice. It is precisely true, however, only when the Zs are normally distributed and when the average concentration at the site is known. These conditions will not, of course, occur in practice. Experience, however, has shown that even strong departure from the assumptions will not seriously affect the equality in equation 13.45.

Most estimators are biased over some range of estimated values even if they are globally unbiased. The smoothing effect is an example.

Conditional unbiasedness is in most cases of more benefit in mining scenarios than in environmental contexts. This property indicates that if one mines based on the predicted grades, the expected grade will be achieved in the end. Thus, as long as other economic factors are correct, the mine will be profitable.

The objectives are different for the environment, however. One does not try to achieve the recovery of an average concentration. This objective implies that material with concentrations above the action level will go unremediated. Rather, it is more important to recover all material above a specified concentration to minimize health and safety risks. Unbiasedness for below-threshold concentrations is thus more important than for other portions of the distribution.

13.6 Influence of the Variogram on Kriging

The variogram has been touted as the means by which the unique spatial continuity characteristics can be quantified and input to the site model. Fitting an appropriate model to the experimental variograms for the site COCs is a very important step. This section will outline to the practitioner how the selection of model parameters influences the kriging system and, hence, the kriging weights, kriging estimate, and estimation variance.

13.6.1 Influence of the Sill

Two spherical variogram models are shown in Figure 13-5. Both have the same range but differ in the height of the sill. The model $\gamma_a(h)$ is the same one used in the kriging example in Section 13.4, with a nugget effect of 0, a sill of 500, and an isotropic range of 100:

$$\gamma(h) = 0 + 500\left[(1.5)\frac{h}{100} - (0.5)\frac{h^3}{(100)^3}\right] \tag{13.46}$$

Model $\gamma_b(h)$ is identical to $\gamma_a(h)$ except that the sill is twice as large:

$$\gamma(h) = 0 + 1000\left[(1.5)\frac{h}{100} - (0.5)\frac{h^3}{(100)^3}\right] \tag{13.47}$$

Figures 13-6a and 13-6b show the outcomes when the models $\gamma_a(h)$ and $\gamma_b(h)$ are input to the kriging system. Each sample location shows the sample concen-

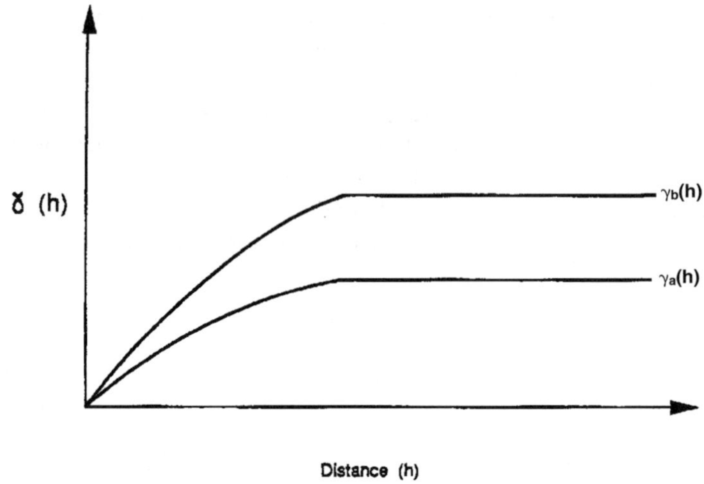

Figure 13-5. Two variograms that differ only in the height of the sill. (From *An Introduction to Applied Geostatistics*, by E.H. Isaaks and R.M. Srivastava. Copyright © 1989 by Oxford University Press, Inc. Used by permission of Oxford University Press, Inc.)

tration along with the kriging weight. Point *P* displays the kriged estimate along with the kriging variance.

Both models yield exactly the same kriging weights and estimated value for point *P*. When the height of the sill is changed, no corresponding effect has been realized on the estimate. However, the estimation variance has doubled as a result of the higher sill in model $\gamma_b(h)$. This result is a rule in rescaling: The estimate remains unaffected while the estimation variance increases by the factor used to rescale the variogram sill.

13.6.2 Influence of the Nugget Effect

Figure 13-7 presents two variogram models: the model used in the kriging example (eq. 13.46) and the same model with a nugget effect equal to 50% of the sill. Figure 13-8 shows the differences in the kriging weights for these two models.

The weights associated with the nugget effect model differ significantly from the original model. The weights from the nugget effect model are more alike than those for the original model. The largest weight has been reduced from .642 to .271 (original vs. nugget) and the smallest weight has been increased from .015 to .094. The declustering effect has been highly dampened, with the weights in the sample 1 and 2 cluster and the sample 3 and 4 cluster becoming quite similar.

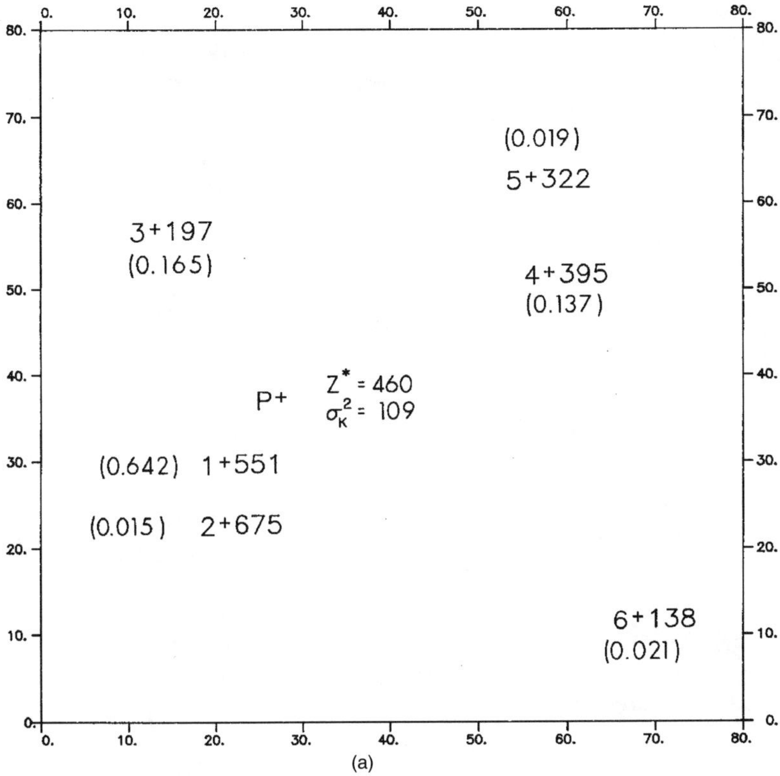

Figure 13-6. Kriging weights, kriging estimate, and estimation variance for two variogram models differing only in the height of the sill. (From *An Introduction to Applied Geostatistics*, by E.H. Isaaks and R.M. Srivastava. Copyright © 1989 by Oxford University Press, Inc. Used by permission of Oxford University Press, Inc.)

The nugget effect influences the kriging system so that the estimate produced is closer to a simple average than it would be without a nugget value. Taken to the limit, a pure nugget effect will, in fact, simply average the available data. The resulting weights are all equal—nice for the kriging system but not good for the kriging variance. Figure 13-8 shows that $\sigma_K{}^2$ goes from 109 ppm^2 in the original model to 362 ppm^2 in the nugget model. Using additional samples in the pure nugget case will slowly bring down the kriging variance, as shown in the following equation:

$$\sigma_K^2 = \sigma^2 + \frac{\sigma^2}{n} \tag{13.48}$$

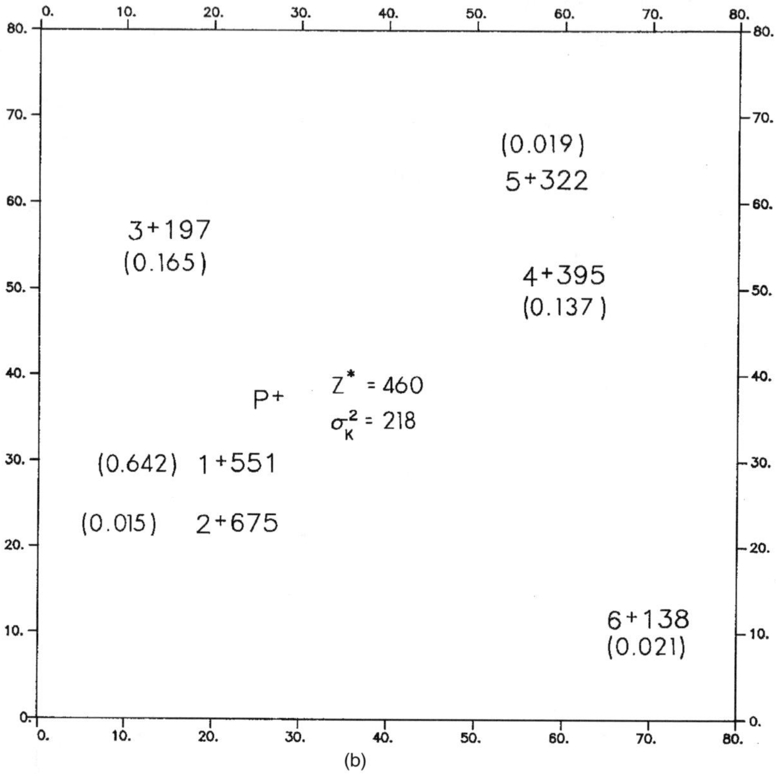

Figure 13-6. (*continued*)

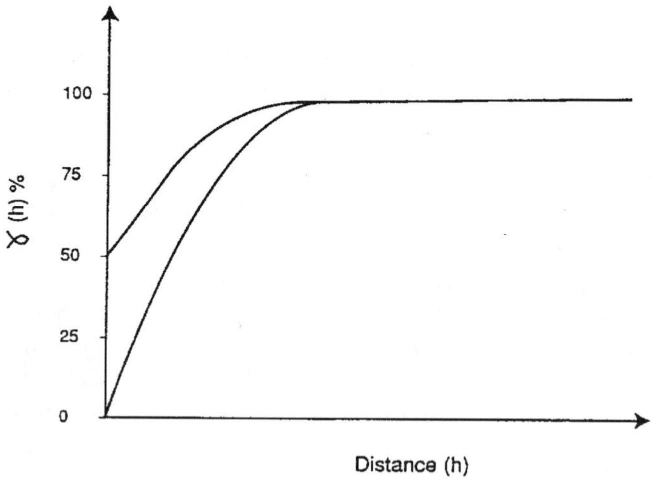

Figure 13-7. Two variograms that differ only in the nugget effect. (From *An Introduction to Applied Geostatistics*, by E.H. Isaaks and R.M. Srivastava. Copyright © 1989 by Oxford University Press, Inc. Used by permission of Oxford University Press, Inc.)

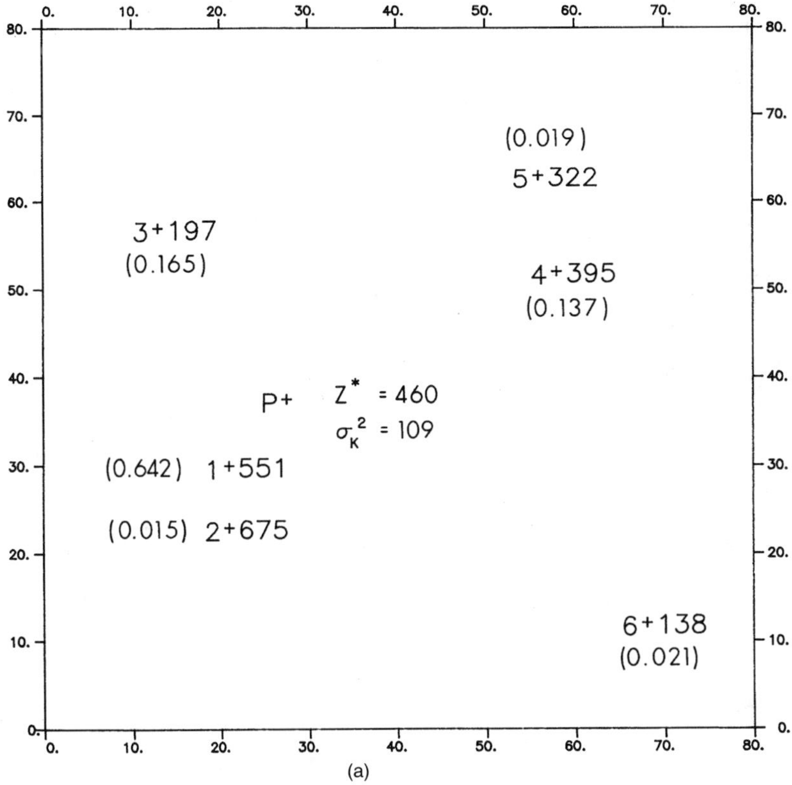

Figure 13-8. Kriging weights, kriging estimates, and estimation variances for two variogram models differing only in the nugget effect. (From *An Introduction to Applied Geostatistics*, by E.H. Isaaks and R.M. Srivastava. Copyright © 1989 by Oxford University Press, Inc. Used by permission of Oxford University Press, Inc.)

where the first term is the variance of the sample data and the second term is the standard error of the mean.

13.6.3 Influence of the Range

Figure 13-9 shows the original variogram model and a model in which the range has been doubled (200 ft). Figure 13-10 shows the results for both models after kriging. Despite the increased spatial continuity, little change is noticed in either the weights or the value of the point estimate. The most noticeable change is to the kriging variance, which has been cut approximately in half. This is to be

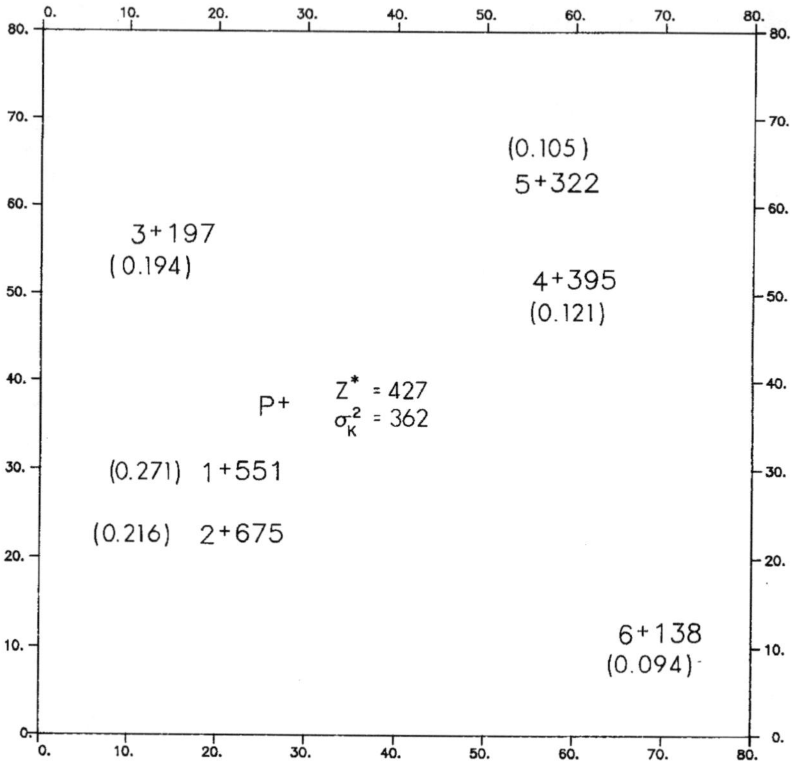

(b)

Figure 13-8. (*continued*)

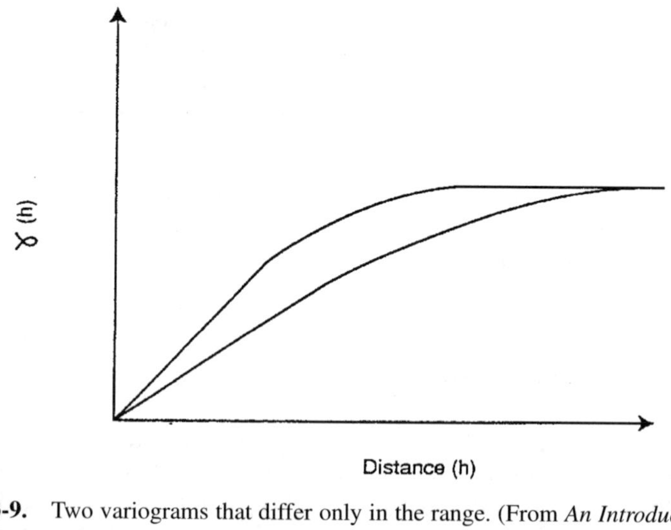

Figure 13-9. Two variograms that differ only in the range. (From *An Introduction to Applied Geostatistics*, by E.H. Isaaks and R.M. Srivastava. Copyright © 1989 by Oxford University Press, Inc. Used by permission of Oxford University Press, Inc.)

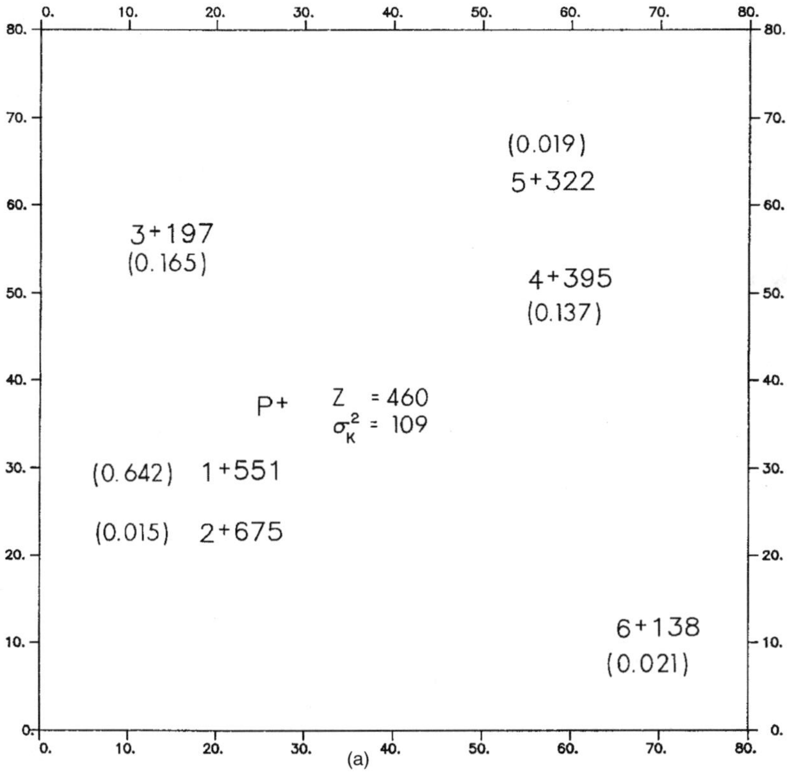

Figure 13-10. Kriging weights, kriging estimates, and estimation variances for two variogram models differing only in the range. (From *An Introduction to Applied Geostatistics*, by E.H. Isaaks and R.M. Srivastava. Copyright © 1989 by Oxford University Press, Inc. Used by permission of Oxford University Press, Inc.)

expected as, in terms of statistical distance, the samples now appear twice as close as before. The opposite becomes true for shorter ranges, and samples all appear equally far away. This situation simulates a pure nugget effect, and the weights approach $1/n$ as the range approaches zero or the variogram approaches a pure nugget effect model.

13.6.4 Influence of Anisotropy

Many spatial variables exhibit continuity features that are not isotropic in nature, as discussed in Chapter 10. Often, contamination is more continuous in one

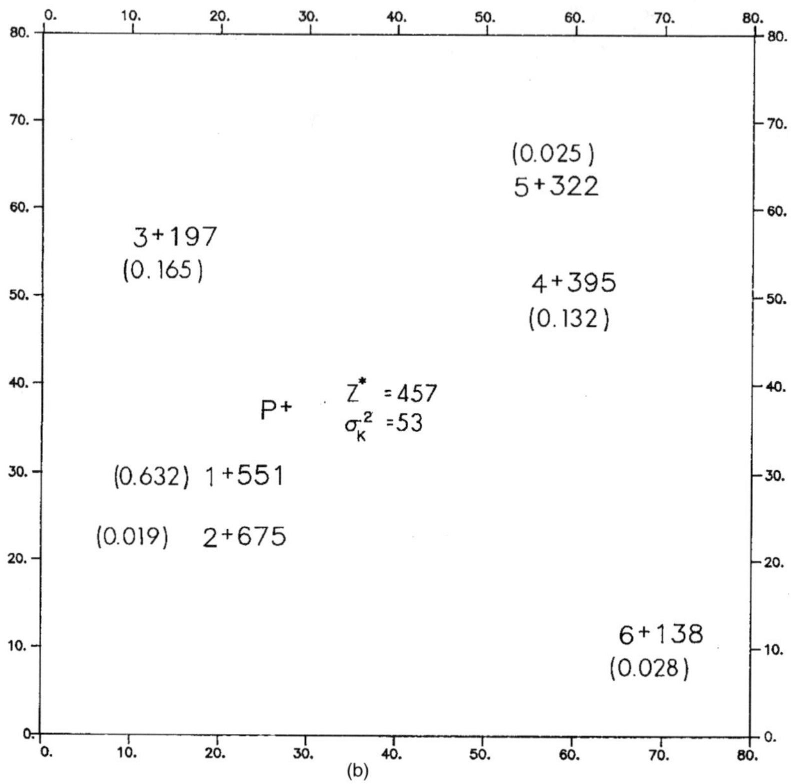

Figure 13-10. (*continued*)

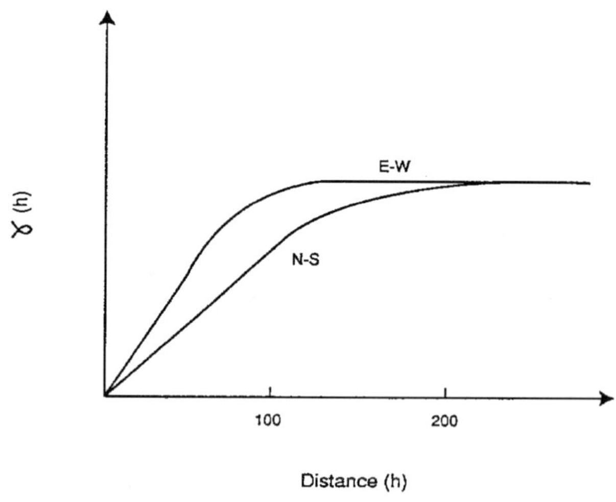

Figure 13-11. Two variograms that differ only in the anisotropy. (From *An Introduction to Applied Geostatistics*, by E.H. Isaaks and R.M. Srivastava. Copyright © 1989 by Oxford University Press, Inc. Used by permission of Oxford University Press, Inc.)

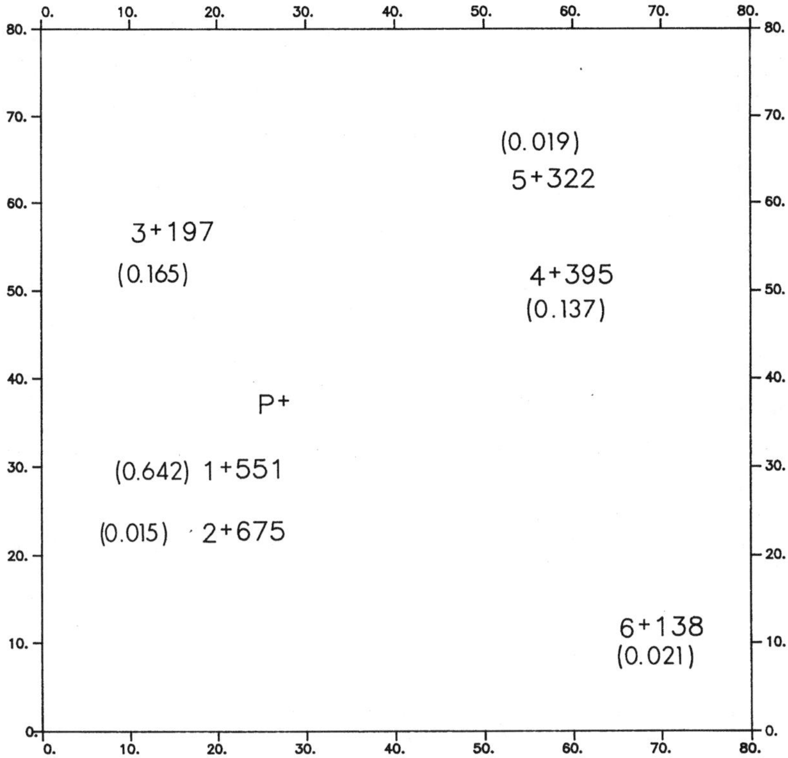

(a)

Figure 13-12. Kriging weights, kriging estimates, and estimation variances for two variogram models differing only in the anisotropy. (From *An Introduction to Applied Geostatistics*, by E.H. Isaaks and R.M. Srivastava. Copyright © 1989 by Oxford University Press, Inc. Used by permission of Oxford University Press, Inc.)

direction than another. Figure 13-11 shows a variogram for which the range in the north-south direction is twice that of the east-west direction, 200 ft vs. 100 ft, resulting in an anisotropy ratio of 2:1.

Using an anisotropic variogram in kriging will significantly influence the results as the data in the direction of the long axis suddenly become statistically closer than in the isotropic case. Figure 13-12 shows the results of kriging using both an isotropic and anisotropic model (Figs. 13-12a and 13-12b, respectively). The anisotropic model gives a much higher weight to sample 2, a higher weight to sample 5, and lower weights to samples 1, 3, 4, and 6.

One result is that the sample 1 and 2 cluster weight increases from .657 to .700. Since both these samples also have high concentrations, the estimated value

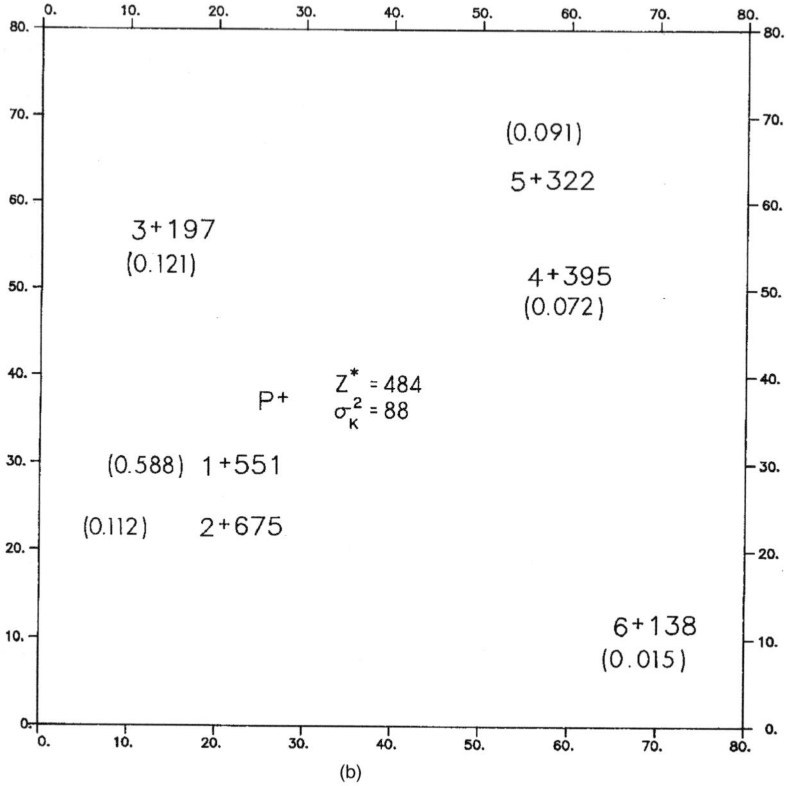

Figure 13-12. (*continued*)

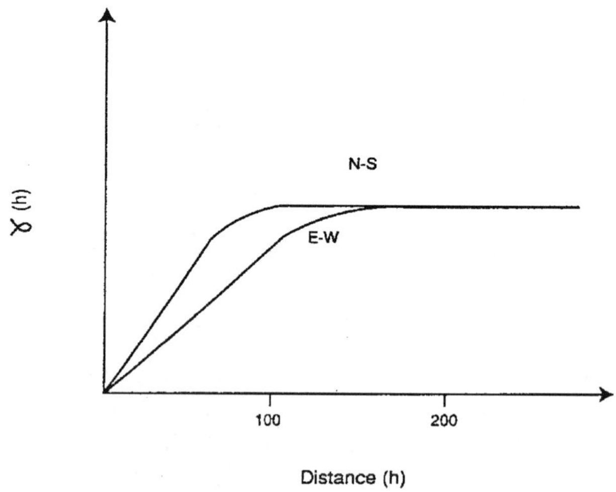

Figure 13-13. Two variograms that differ only in the anisotropy. (From *An Introduction to Applied Geostatistics*, by E.H. Isaaks and R.M. Srivastava. Copyright © 1989 by Oxford University Press, Inc. Used by permission of Oxford University Press, Inc.)

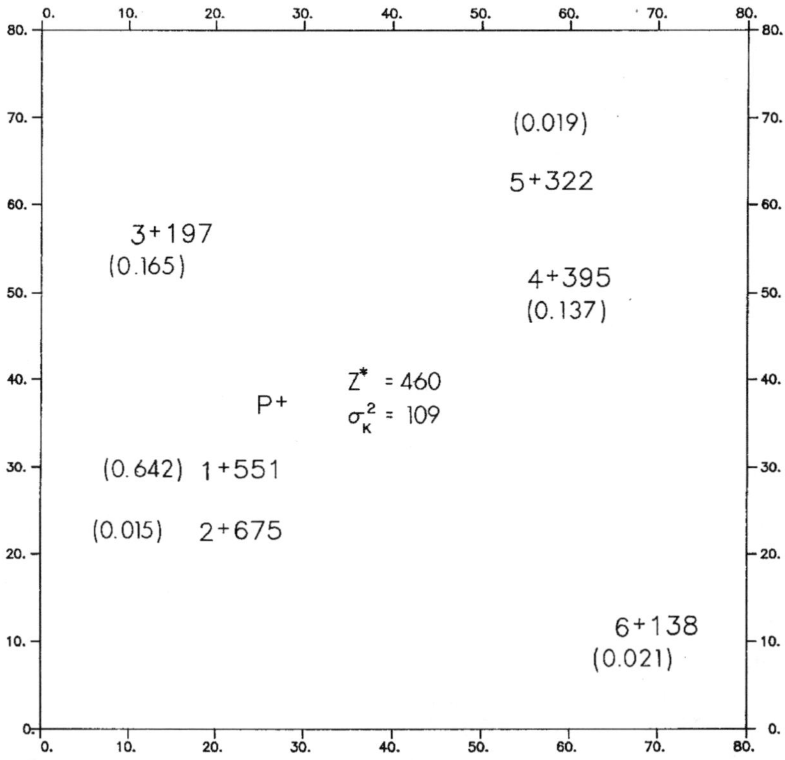

Figure 13-14. Kriging weights, kriging estimates, and estimation variances for two variogram models differing only in the anisotropy. (From *An Introduction to Applied Geostatistics*, by E.H. Isaaks and R.M. Srivastava. Copyright © 1989 by Oxford University Press, Inc. Used by permission of Oxford University Press, Inc.)

of point *P* goes up. The kriging variance has also been influenced. Because of the longer range in the north-south direction, the variance drops from 109 to 88 ppm^2.

Figure 13-13 shows another anisotropic variogram with a 2:1 ratio but, this time, the long axis of continuity is in the east-west direction. Figure 13-14 shows the kriging results for the isotropic model (Fig 13-14a) and anisotropic model (Fig. 13-14b). This time, the weights for samples 1, 3, and 4 increase, weights for 2 and 5 become negative, and 6 remains the same. The net result is that the estimate of point *P* drops only slightly. Also, the kriging variance drops from 109 to 84 ppm^2, about the same reduction as for the north-south anisotropy.

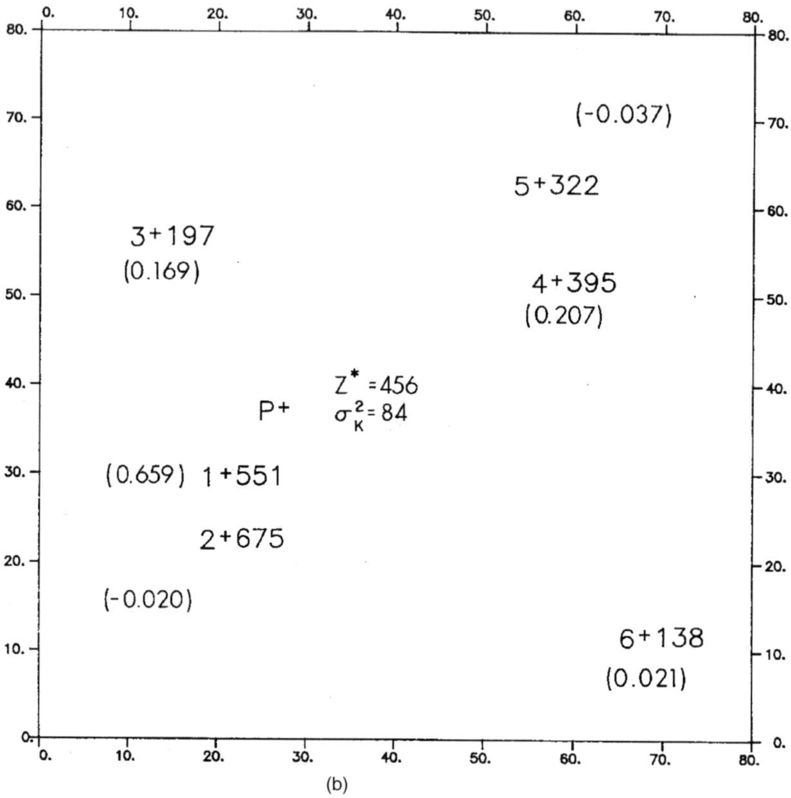

Figure 13-14. *(continued)*

The anisotropy affects not only the estimate at point *P* but all other points in the region that may need to be estimated, perhaps for contouring. Figure 13-15 shows three such contour maps. Figure 13-15a shows the contours developed using the variogram model with the north-south anisotropy. Figure 13-15b shows the contours for the east-west anisotropy.

Visually, it is easy to see the influence. For both maps, the contours trend in the direction of anisotropy. For anisotropy ratios higher than 2:1, the elongations noticed in the direction of maximum continuity will be even more pronounced. As a final contrast, Figure 13-15c shows the resultant contours with an isotropic variogram essentially an average between the two anisotropic maps.

A visualization of the variogram model surface can help us understand what happens when kriging is performed with an anisotropy in the variogram model. Figure 13-16 shows a three-dimensional fishnet representation of an isotropic model surface. This features a circular cone of depression in contrast to Figure

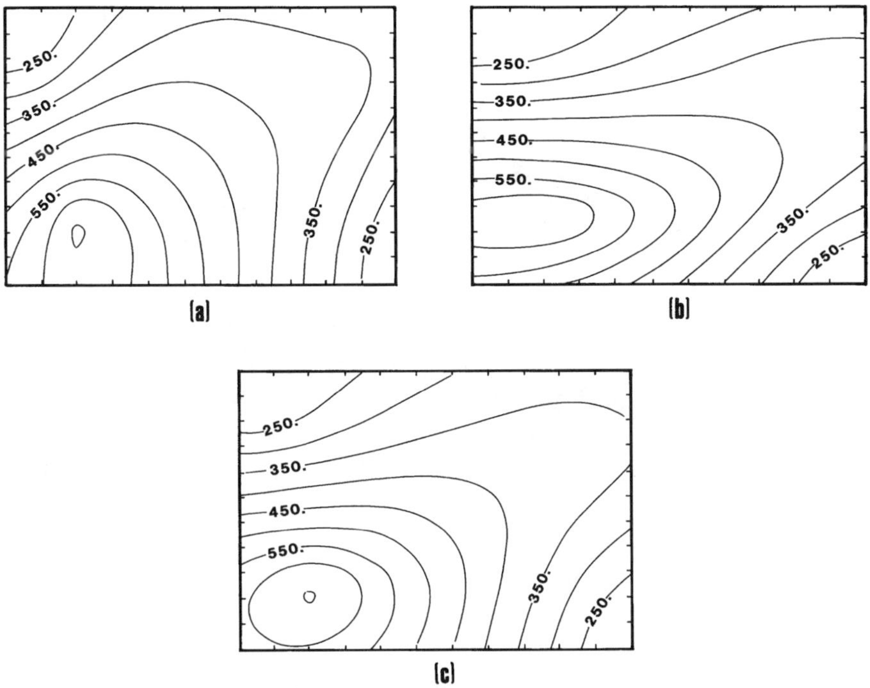

Figure 13-15. Contour maps of the same data set using different anisotropies in the variogram models during kriging.

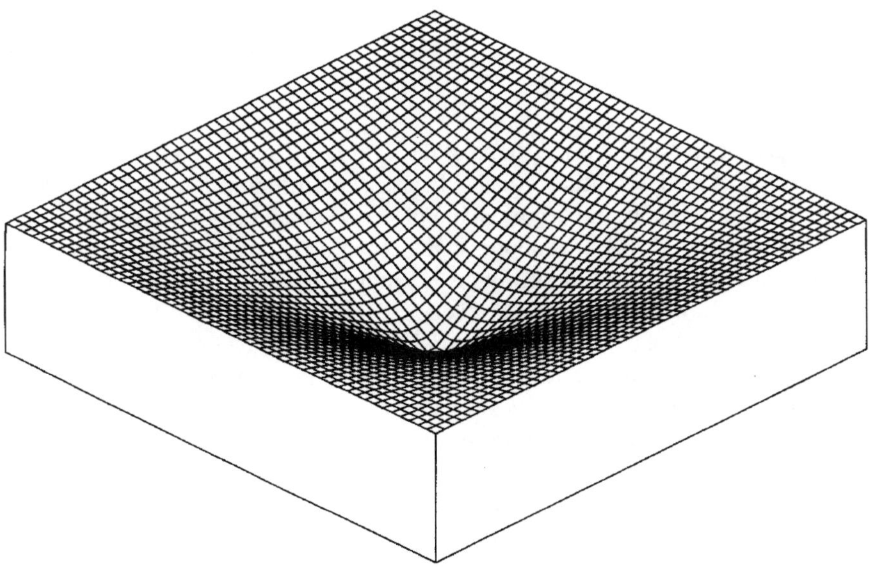

Figure 13-16. Three-dimensional fishnet diagram of an isotropic variogram surface.

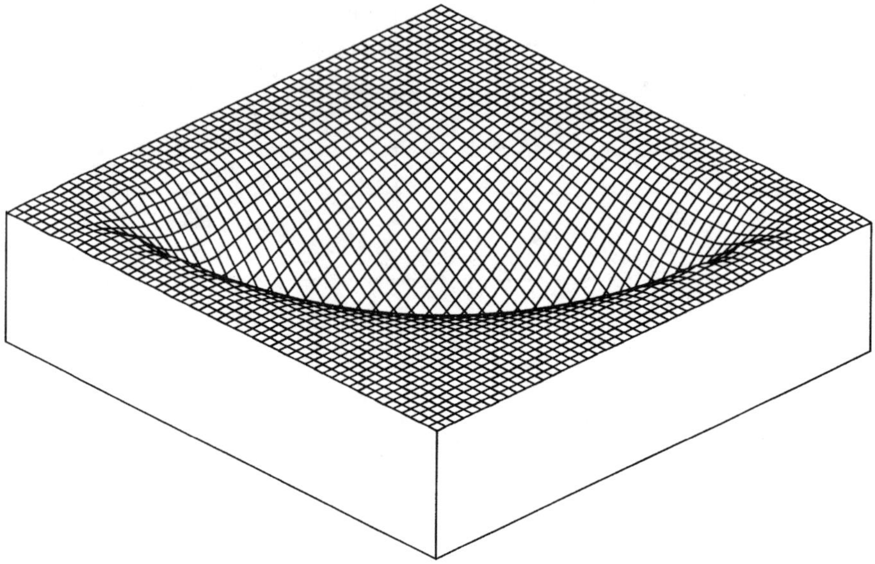

Figure 13-17. Three-dimensional fishnet diagram of an anisotropic variogram surface.

13-17, where an anisotropy is present, and the cone has been elongated in the direction of the long axis of continuity.

13.6.5 Influence of Shape

This final section addresses the shape of the variogram model. Figure 13-18 shows two variogram models with identical sill values but different shapes. The spherical model used throughout the kriging example is compared to the Gaussian model. The Gaussian model has a parabolic shape at the origin of the graph. This means that the variance rises very slowly at the beginning and more steeply later on. Such behavior is seen in variables that are very continuous.

Figure 13-19 shows the results of kriging using a Gaussian model. The weights assigned are dramatic. Sample 1 receives an enormous weight of 1.870. To balance this, sample 2 receives a weight of −.978 with a net weight to the cluster of .892. With the exception of sample 4, most other weights are close to zero and are both positive and negative.

This type of behavior is the reason that practitioners generally refrain from using Gaussian and other models that are parabolic near the origin. Although a phenomenon may indeed be continuous, the negative weights produced during estimation may cause erratic results. For example, we may speculate that these

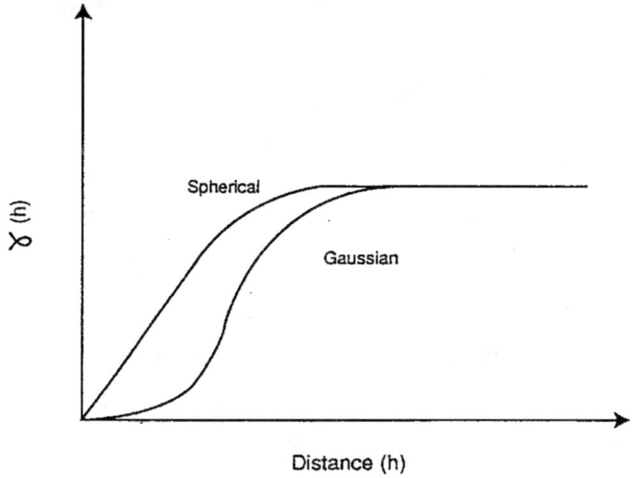

Figure 13-18. Two variograms that differ only in shape.

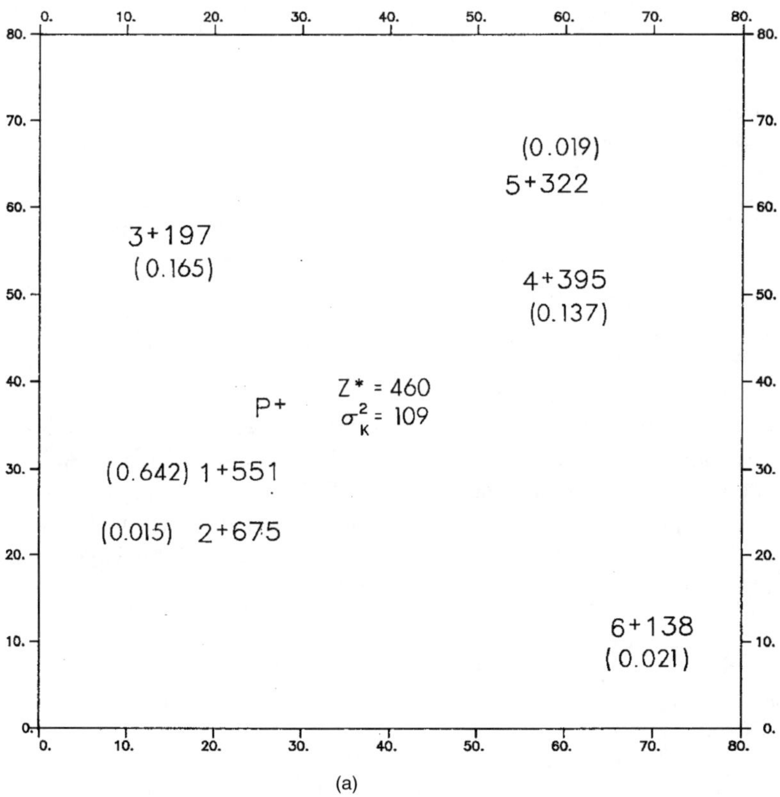

Figure 13-19. Kriging weights, kriging estimates, and estimation variances for two variogram models differing only in shape.

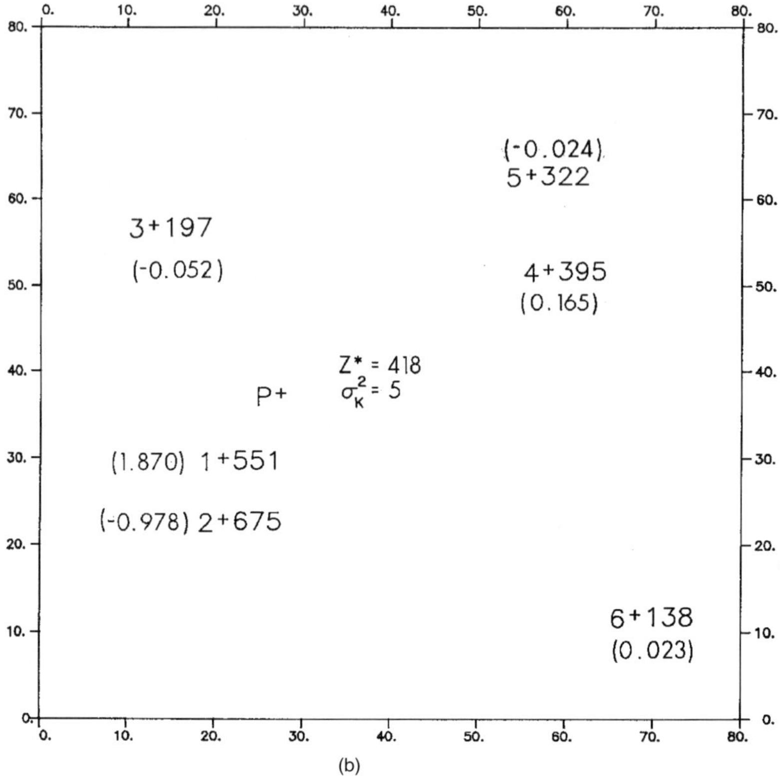

Figure 13-19. (*continued*)

data reflect a point source in the southwest portion of the map. As we draw closer to the source, concentrations are likely to increase. Consider a single additional sample—sample 7 in Figure 13-20— that is closer to the source than sample 2 and with a concentration at least that of sample 2. If a negative weight is assigned to this new sample, the estimate of point *P* could be less than any observed concentration or even negative in value. Contour maps may exhibit large areas with concentrations below any observed data value.

CONCLUSION

Kriging achieves the objectives of being a minimum-variance unbiased estimator by minimizing the estimation variance equation. Mechanically, it uses the

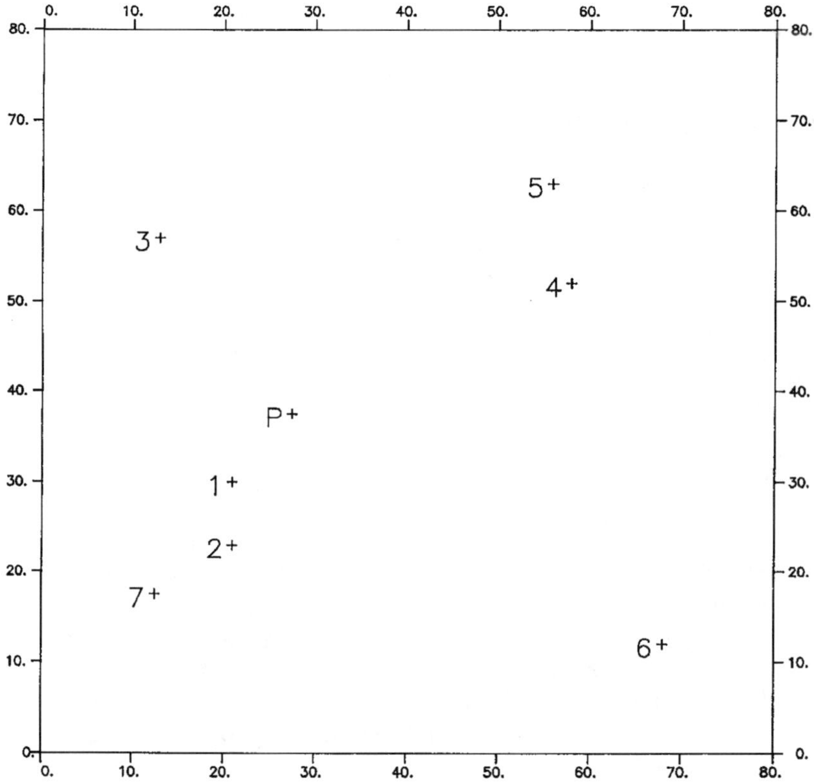

Figure 13-20. Addition of new sample 7 could cause an inappropriately low or negative value to be estimated at point *P*.

variogram model almost countless times to derive the covariances needed to fill the kriging matrices and to solve for the estimate and its variance. As an estimator, it distinguishes itself with key properties such as declustering, additivity, and exact interpolation. Whereas the mechanics of kriging are straightforward, kriging is sensitive to the parameters of the variogram model are critical, and their impact on the final results should be expressly understood.

Chapter **14**

Practical Aspects of Kriging

SYNOPSIS: Understanding the mathematical mechanics of kriging provides a good basis for practical application. When evaluating the kriging system, we tend to think in terms of kriging a single block. For practical situations, the focus shifts to a "mass production" or "assembly line" operation, where hundreds or thousands of blocks (or more) are estimated in a single computer program run to produce a block model. These massive kriging runs form the basis for two- and three-dimensional map visualizations. This chapter will emphasize the guidelines necessary to implement successful production of a kriged model and detail some of the pitfalls that await the practitioner.

KEY WORDS: Block Model, Bounding Techniques, Contour Constraints, Horizontal Boundaries, Search Neighborhood, Sample Sufficiency, Excess Samples, Volumetric Calculations, Extraction Volumes, Floating Cone, Cross Validation, Error Distribution, Typecasting of Error, Number of Neighbors, Search Radius

14.1 Concept of a Block Model

The discussion on kriging has thus far been limited to two-dimensional situations, mainly with regard to contouring problems. This is an important function, but a three-dimensional perspective is needed as well.

Figure 14-1a shows the locations of a sample data set along with a point grid

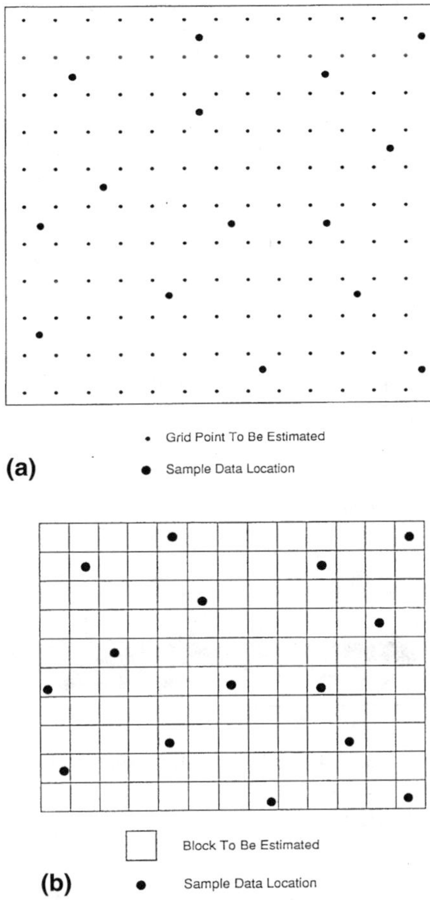

Figure 14-1. (a) Two-dimensional contour grid and (b) two-dimensional block model.

to be estimated by kriging in anticipation of contouring. Figure 14-1b shows the same samples with a block grid to be estimated for contouring or block mapping. Both represent two-dimensional estimation problems. Figure 14-1b can be called a two-dimensional block model.

The two-dimensional block model can be extended into three dimensions as shown in Figure 14-2. In this case, a series of two-dimensional block models are essentially stacked one atop the other to form a three-dimensional block model. We can conceive of this as a block of sugar cubes forming a three-dimensional shape. The block model should be of sufficient size to encompass all areas and depths that need to be estimated at the site.

As with the contour grid, the dimensions of the blocks may vary in the X, Y,

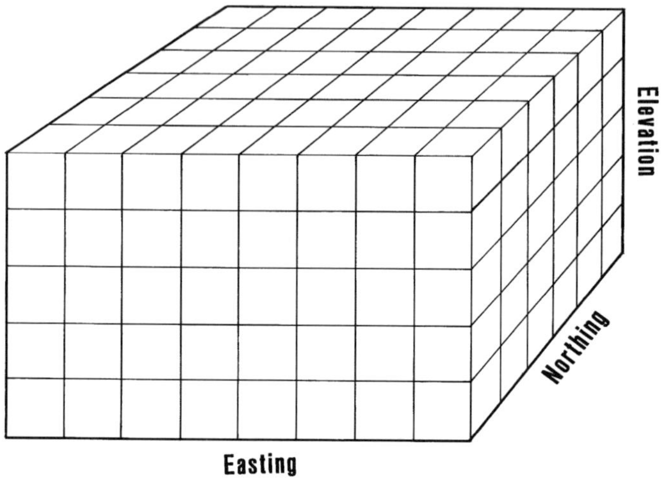

Figure 14-2. Three-dimensional block model.

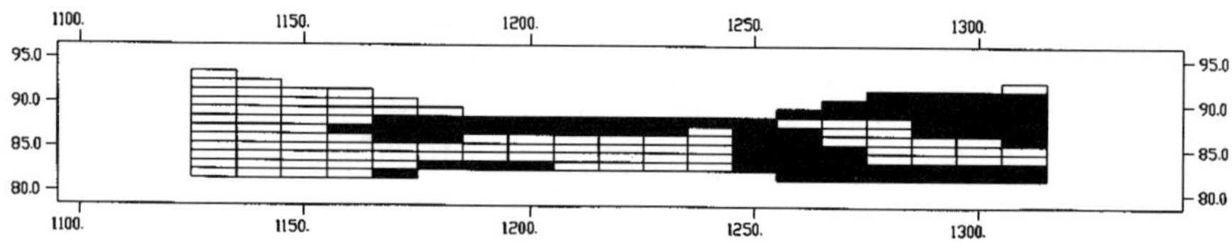

Figure 14-3. Vertical cross section of a three-dimensional block model; vertical exaggeration is 20:1. (From A geostatistical analysis of PCB concentrations in soils at the ReSolve site, by J.C. Myers, 1992.)

and Z directions. In plan view, the blocks may be square or rectangular. For environmental situations, the Z direction of a block (depth) is typically much smaller than the XY dimensions. Figure 14-3 shows a block cross section through a PCB site, where the block dimension is smaller in the Z direction.

14.2 Bounding Techniques

It would be a rare event for a parallelepipedic block model such as the one shown in Figure 14-2 to fit the boundaries of a hazardous waste site exactly. In most situations, several factors influence the shape of the final block model. A few of the most important aspects of bounding the block model will be discussed.

14.2.1 Contour Surface Constraints

Contour surfaces are often needed to restrict the limits of kriging estimation to practical, engineering, or political areas of interest. Among these, include the ground surface, water table surface, soil/bedrock contact, maximum depth of excavation, faults, and others. At some point before, during, or after the kriging process, each block in the model must be checked against the contour surface constraints to see whether the block falls inside or outside the defined area. Only blocks inside the area of interest should become candidates for estimation; inappropriate blocks must rejected.

14.2.1.1 Ground Surface Topography

An obvious boundary for the block model is the surface of the earth. It is inappropriate to estimate blocks above the ground surface because it may cause confusion as well as provide potential for excessive volume to be calculated. Figure 14-4 shows a plan view of a block map from the ReSolve site where large numbers of blocks have been excluded from the model based on the imposition of a surface topography model.

14.2.1.2 Lower Bounding Surfaces

Lower bounding surfaces may take many forms as described above. When appropriate, they should be used simultaneously with the ground surface topographic constraint. Figure 14-5 shows one layer in a three-dimensional block model (ReSolve), where the blocks in the western portion of the map have been screened out by imposing the seasonal low groundwater table contours.

Figure 14-6 shows a three-dimensional representation of the ReSolve surface topography along with the water table surface below. These two surfaces form a type of "topographic sandwich" for estimation and will restrict kriging to appropriate areas for the block model. Note that the upper and lower topographic surfaces may be the structure contour of the top and bottom, respectively, of an aquifer.

14.2.2 Horizontal Boundary Constraints

Horizontal boundary constraints are highly useful for limiting estimation to desired areas, such as the site boundary, a subportion of the site, a zone of special engineering interest, and the boundary of a waste lagoon.

Boundaries may be inclusionary or exclusionary. Figures 14-4 and 14-5 show inclusionary boundaries. Figure 14-7 shows both inclusionary and exclusionary

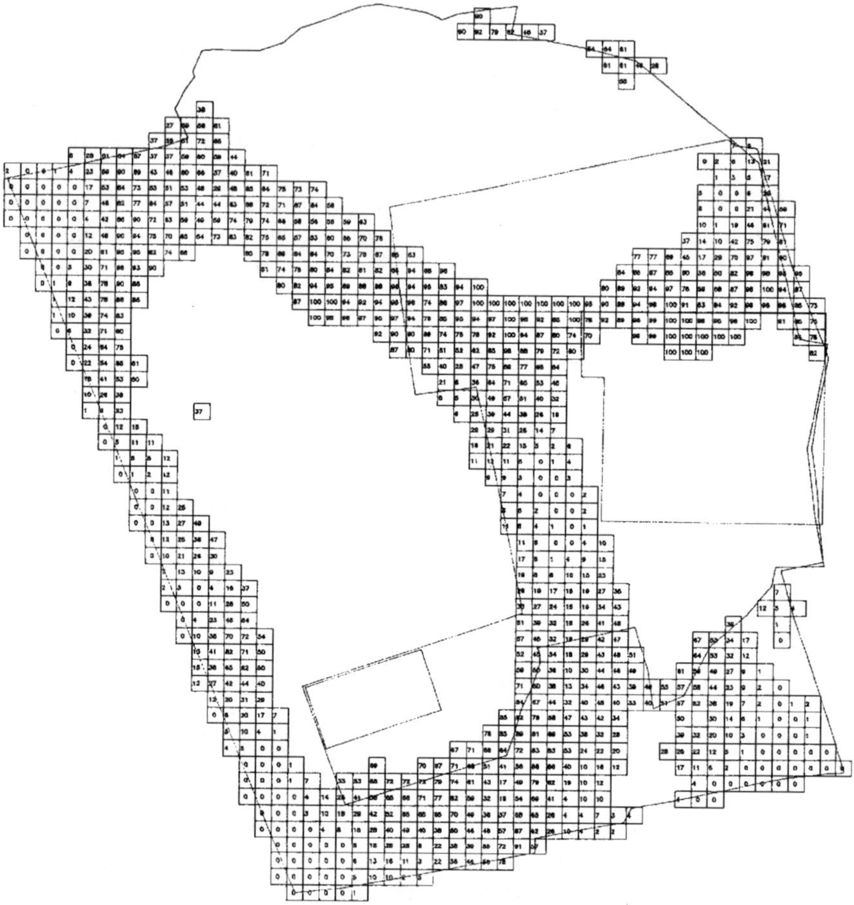

Figure 14-4. Plan block map (level elevation 89 ft above sea level) showing block screening resulting from the use of a surface topography model. Blank areas indicate that actual surface elevations are below this level, signifying a depression in the ground surface.

boundaries. In Figure 14-7, most of the area of the inclusionary boundary is filled with blocks. Two areas, however, show missing blocks based on the imprint of buildings planned for contaminant processing.

The horizontal boundary is usually identical for each estimated level in environmental situations, although this is not necessary. When combined with upper and lower topographic surfaces, a three-dimensional kriging envelope is formed. Figure 14-8 shows such a kriging envelope.

Figure 14-5. Plan block map showing block screening resulting from the use of a groundwater table topography model. Blocks in the western portion of the map fall below the groundwater table surface.

14.3 The Search for Samples

The kriging example in Chapter 13 conveniently used all six available sample data points to estimate an unknown concentration. Unfortunately, in actual practice, conditions are not so straightforward. Many options regarding the search method must be considered before deciding on an appropriate search strategy. The type of

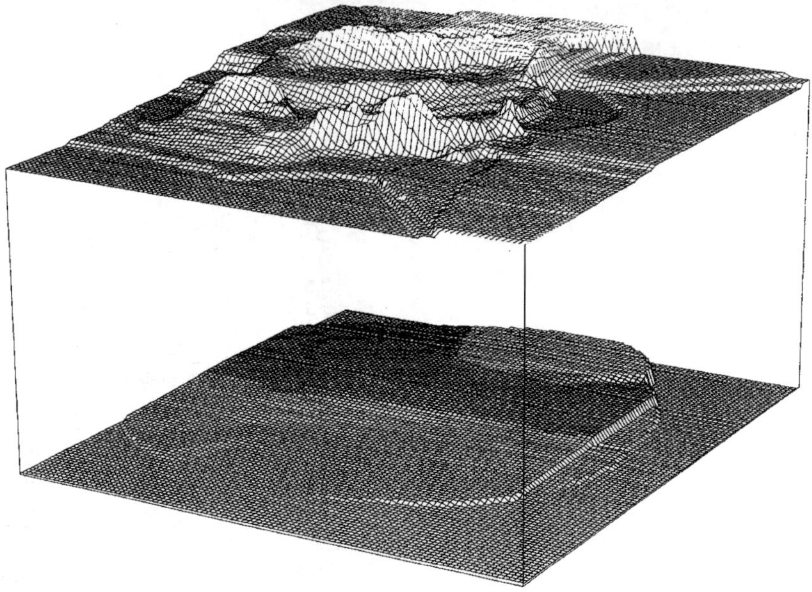

Figure 14-6. "Topographic sandwich" formed by combining surface topography and groundwater elevation contours.

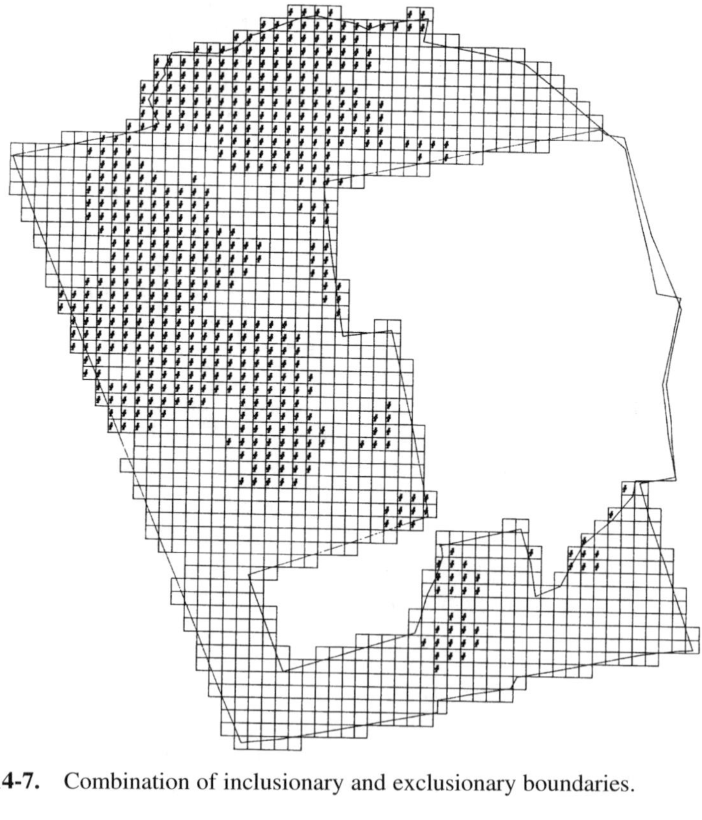

Figure 14-7. Combination of inclusionary and exclusionary boundaries.

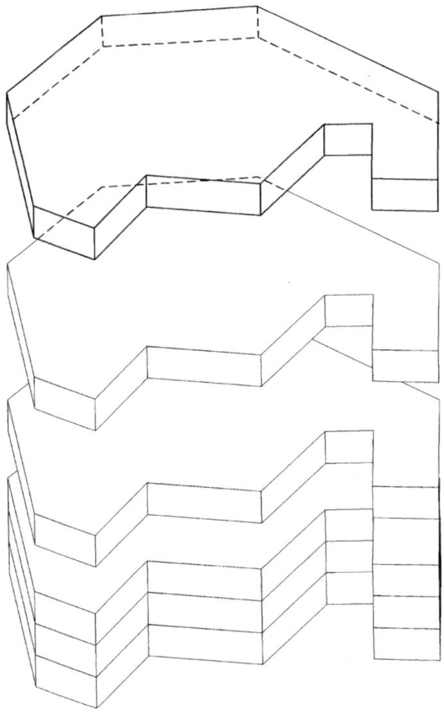

Figure 14-8. Example of a kriging envelope.

search parameters chosen will have a significant impact on the final estimates, no matter what type of interpolation method is chosen.

14.3.1 Defining the Search Neighborhood

14.3.1.1 Simple Search

A simple search approach uses only sample data that are "close" to the point or block being estimated. Close is defined as those samples that fall within an ellipse or ellipsoid centered on the point or block to be estimated. The ellipse can be either isotropic or anisotropic in shape (Fig. 14-9). In general, the shape of the ellipse is consistent with the anisotropy of the variogram parameters. If the spatial continuity is anisotropic, the search ellipse is also. If spatial continuity is isotropic, so is the search ellipse.

For anisotropic search ellipses, the long axis of the ellipse is usually oriented in the same direction as the long axis of spatial continuity. There is no strict need for this, and most kriging programs offer the user the option to select a different

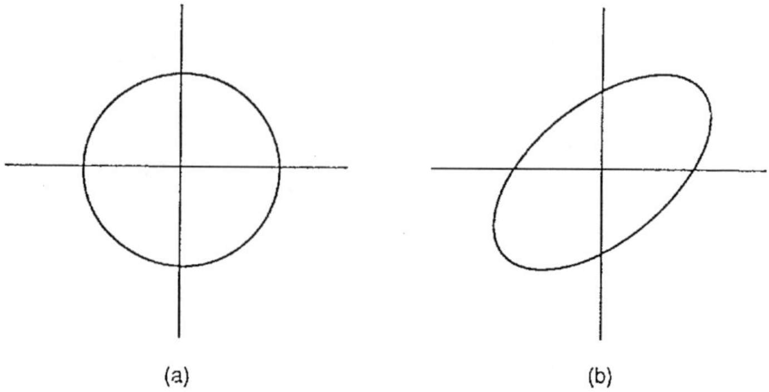

Figure 14-9. (a) Isotropic search ellipse vs. (b) anisotropic search ellipse.

orientation for the search. However, it is logical to want to use samples that are much more continuous in the preferred direction.

14.3.1.2 Partitioned Search

If a data set has clusters of sample points, a simple search procedure will inevitably capture sample clusters for some of the areas to be estimated. An example is given in Figure 14-10, where samples are clustered to the southeast of the point being estimated. Although kriging is able to decluster effectively such a sample configuration, other estimators, such as IPD, will assign an inappropriately large weight to this cluster.

The solution consists of using a partitioned search tactic. Most commonly, the search area is divided into quadrants or octants (Fig. 14-11). For each quadrant or

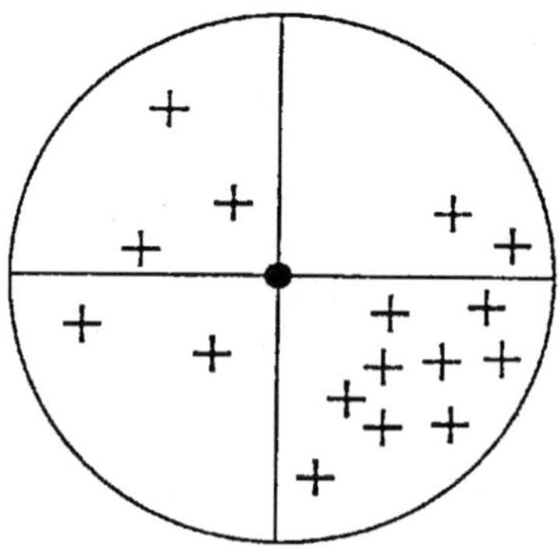

Figure 14-10. Sample data clusters captured during a typical sample search operation for kriging.

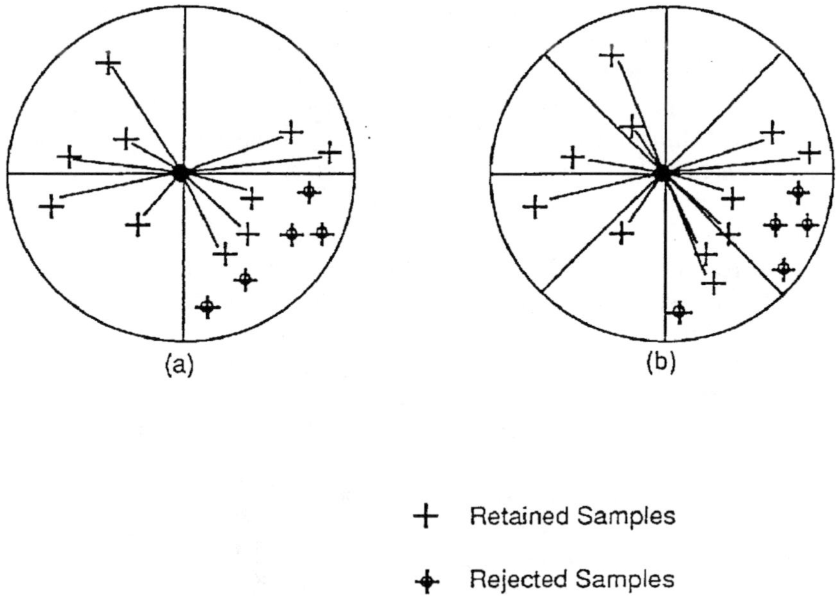

Figure 14-11. Sample search strategies to decluster data: (a) a quadrant-oriented search strategy is used; (b) an octant approach is employed.

octant, the maximum number of samples allowed is specified for the program. In Figure 14-11, a maximum of three samples per quadrant and two samples per octant was allowed. If more than three samples are found in a quadrant, only the closest three are used for estimation. If three samples or fewer are found, then all samples in the quadrant are retained. Figure 14-11b also shows that some sectors may contain no samples.

Figure 14-11a shows that for the quadrant search, 10 samples were retained for estimation. This compares with Figure 14-11b where eleven samples were kept. The octant search kept the same 10 that were retained for the quadrant search but added one additional sample. By using a slightly different set of samples, the estimation produced by the octant search will be slightly different from that for the quadrant search. However, since the new sample is at a fairly large distance, its weight will be small.

14.3.2 Sample Sufficiency

The problem of sample declustering by partitioning assumes that an excess number of samples have been captured during the search process. For relatively sparse data sets (such as often occur in groundwater studies), a lack of sample data may be the problem. In this case, it is necessary to select a search radius of sufficient size to capture at least one sample, without which estimation is not possible.

Generally, sparse data sets are also irregularly gridded. Once the area or boundary for estimation has been determined, it is a simple matter to measure the maximum distance between a grid node and the closest sample point. This can be considered a minimum search radius to complete the grid. It may be appropriate to use a longer radius if this maximum distance is similar for a number of samples. In this case, a single sample could be used for a large number of grid nodes, resulting in a mixture of polygonal and interpolation estimation.

Another consideration is the length of the search radius compared to the range of the variogram. It is commonly—and wrongly—assumed that the search radius should not exceed the variogram range. The size of the search radius may simply be a function of the sparse sample geometry and of a need to collect multiple sample data. Note that the need for search radii longer than the range applies not only to contouring grids but to kriging envelopes as well.

On the other hand, this does not give the practitioner license to reach out to any distance for samples. For extremely large search windows, the assumption of stationarity may be violated, and we should question whether such samples can appropriately represent the area to be estimated. This is an example of where the properties of kriging may lure one down a questionable path: Additional samples will always reduce the kriging variance of the estimate. Whether extremely distant samples will improve local estimation is a different matter altogether.

14.3.3 Excess Samples

For densely sampled sites or for sites where samples are clustered in some areas but sparse in others, it is possible to capture too many samples during the search operation. Two considerations need to be addressed in this regard. First, the number of samples used in kriging dramatically affects the computational effort required. The time it takes to solve the kriging matrix is proportional to the cube of the number of samples. For example, it takes eight times longer to solve a matrix with 10 samples than one with 5.

This situation is easy to fix; an example of the solution is shown in Figure 14-12. This figure shows the same sample data as those used for the clustering and quadrant/octant examples. Although the search radius has found 16 samples, the program can be instructed to retain only the closest 6. This effectively reduces the size of the search radius in densely sampled areas, as shown in Figure 14-12.

The second consideration relates to distant samples. As discussed previously, we should always question the validity of using a distant sample for a local estimation.

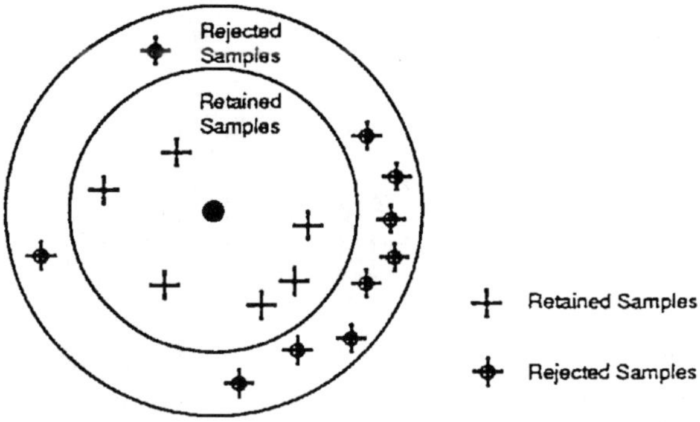

Figure 14-12. Technique for eliminating excess samples. Only those samples closest to the point being estimated are retained; all other candidate samples are rejected.

14.4 VOLUMETRIC CALCULATIONS

14.4.1 Summaries of Block Model Contents

Most kriging programs also have a corresponding reporting program that will summarize the estimated concentrations for the block model. The reporting program will generally be used in one of two ways: (1) to obtain a simple two-category summary of the volume of material estimated to be above and below the threshold concentration ("dirty" and "clean"), and (2) as a detailed summation by concentration or quantile category to enable a block distribution to be calculated.

For either type of report, it is important to impose the topographic surfaces on the block model so that partial volumes are calculated at the surface interfaces. This will prevent volumes from being exaggerated.

14.4.2 Extraction Volumes

The simple tabulation of clean and dirty volumes may be misleading depending on the nature of the contamination at the site. If the contamination is in a single, continuous mass, a simple tabulation may be appropriate. If contamination is layered or in lenses, a simple tabulation may be misleading as to the level of effort required during the remedial process.

14.4.2.1 Contaminated Volumes

Figure 14-13 shows a block cross section from a PCB site. In the diagram, any block that has a symbol inside is above the action limit and requires remedia-

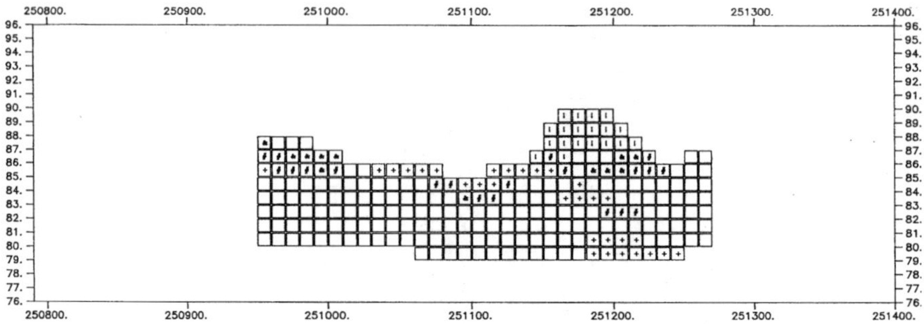

Figure 14-13. Block cross section showing clean volumes requiring excavation.

tion. Any block without a symbol is considered clean. At this site, any block that has been classified as dirty will be excavated and treated. This means that there is only one category of dirty block.

14.4.2.2 Clean Volumes

Unlike dirty blocks, clean blocks come in two categories. The first category is clean blocks that require no excavation. Examples of such blocks can be seen in the left group of blocks in Figure 14-13.

The second category is clean blocks that require excavation. On the right side of Figure 14-13, a number of clean blocks are positioned vertically between dirty blocks. This means that the upper dirty blocks must first be excavated, the clean blocks removed, and the lower dirty blocks removed.

This three-step process has serious economic implications. After the first set of dirty blocks is removed, appropriate decontamination procedures must be performed to workers and equipment. This has an associated cost. Then, once the clean blocks have been removed, the material must be stockpiled in a clean zone. This implies the necessity of a staging area on a site that may already have space restrictions. Also, it means that this clean material must be rehandled at an additional cost. Alternatively, if it is less expensive, "clean but excavated" blocks may be processed as "dirty." Understanding the nature of the contaminated volume can be valuable for bidding on, costing, and planning a remedial effort.

14.4.2.3 Floating Cone Techniques

The amount of clean material that must be excavated can be calculated by using block plan and block cross-section maps. This is a time-consuming endeavor, however. An easier approach is to use a computer-automated method known as the *floating cone*.

The floating cone method was developed in the mining industry to determine

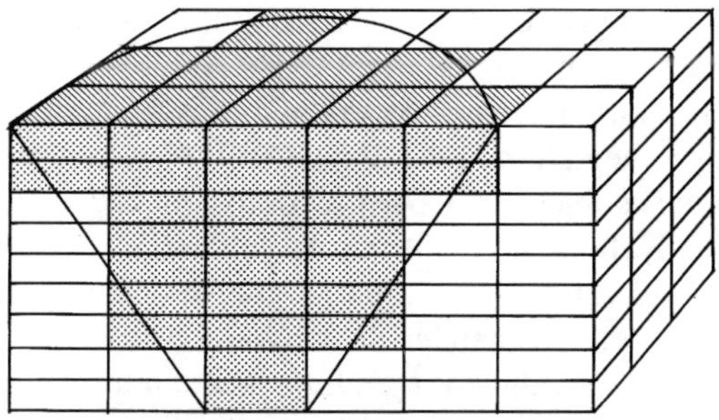

Figure 14-14. Floating-cone algorithm for volumetric calculation.

the limits of an excavation pit based on economic parameters. Each block in the three-dimensional model is assigned a value based on the metal content, current or projected selling price of metal, extraction cost, hauling cost, processing cost, and so forth. If a block has a positive value, it exerts an upward buoyant force; if negative, the block force is downward.

If the value of a block is sufficiently positive to lift overlying barren material at a profit, the block is mined. If not, the block is left in place. The floating cone algorithm looks at each positively valued block in the model to see if it should be removed. When the process is complete, a summary is compiled of (1) the amount of profitable material to be mined, (2) the amount of waste material to be mined, and (3) the elevation of the surface for the final pit. An example is given in Figure 14-14.

This same principle may be applied, with some slight modifications, to a hazardous waste site. It is possible for a miner to have a marginally positive block overlaid by a large amount of waste material, making this block uneconomic to mine. The floating cone will leave this block in place. In many environmental situations, any block having estimated concentrations above the action limit must be removed, regardless of the amount of overlying clean material. The cost function applied must be sufficient to lift all blocks that are contaminated. This can easily be accomplished by assigning a zero cost of excavation to clean blocks.

14.5 Cross Validation

14.5.1 Introduction

In Chapter 3, kriging and inverse squared distance were used to estimate the concentration of a location where the concentration was known. By comparing the estimates with the known concentrations, useful information was gleaned regarding the performance of each estimator.

In practice, we do not have access to the true concentration of an estimated location, even after remediation. Yet it would still be of interest to know beforehand how the selected estimator will perform or to compare several estimators and use the one that yields the best results. Cross validation provides a type of dress rehearsal for the actual estimation. Although this does not guarantee success, it can at least warn of severe problem areas to avoid.

The technique of cross validation provides useful information based solely on the information contained in the sample data. In addition to comparing different estimators, we may evaluate the sensitivity of other parameters, such as search radius, variogram model, and number of neighbors. Whereas the mechanics of the cross validation process are simple, the interpretation of results can be more difficult.

14.5.2 Mechanics of Cross Validation

The process of cross validation is quite straightforward. Starting with a sample data set, we remove a single sample location. Then, using an estimator and distance-weighting function, the concentration at the vacant sample location is estimated. Two values now exist for the selected location: the true value and an estimate. The error of estimation can be calculated by subtracting the estimate from the true value. Undoubtedly, we will make an error in estimation. By comparing the estimate to the true value, the magnitude, sign, and classification may be determined (Fig. 14-15).

On completion of the estimation/error-determination process, the sample is returned to the data set. This procedure of removal/estimation is repeated using each of the remaining samples in the data set. Statistical tools may then be used to assess the performance of the selected estimator and its associated parameters.

14.5.3 Assessing Performance: Traditional Methods

No single estimator is likely to perform best for all samples in the cross validation process. It is therefore necessary to develop types of statistical criteria against which we can measure the performance.

14.5.3.1 Distribution of Estimates

Starting with a distribution of sample data, the cross validation process develops a corresponding distribution of estimated values on a sample-by-sample basis. An estimator that performs well yields a distribution of estimates that is as

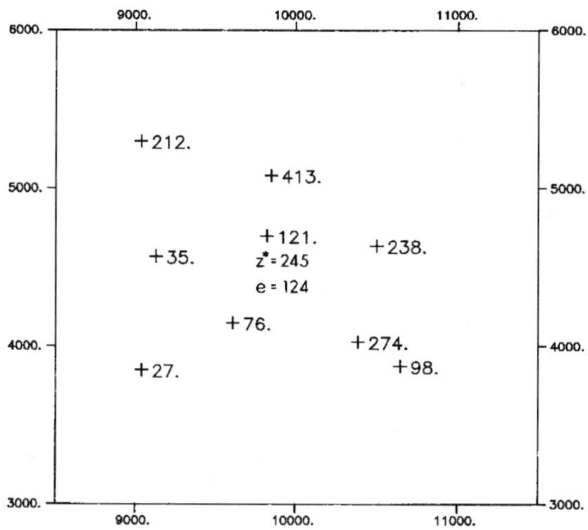

Figure 14-15. Example showing the result of a cross validation experiment for a single sample location.

close to the original sample distribution as possible. This implies that the mean values should be close, providing a summary measure of performance.

For a more complete comparison of the distributions, a *q-q* plot may be used. Identical distributions will plot as a straight line; deviations from this can be informative. Figure 14-16 shows a hypothetical *q-q* plot from a cross validation exercise with a typical result. Values in the lower quantiles of the distribution tend to be overestimated, whereas values in the upper quantiles of the distribution tend to be underestimated, a result of the smoothing effect inherent in most interpolation procedures.

14.5.3.2 Distribution of Errors

The cross validation process results in an error value for each sample location in the data set. These errors are often referred to as *residuals*. In general, a mean error of zero is desired as a nonzero mean indicates a bias. Figure 14-17 shows three common situations. Figure 14-17a shows a mean less than zero indicating a negative bias. Figure 14-17b shows a positive mean, indicating a positive bias. Figure 14-17c depicts a distribution with a mean of zero, suggesting no bias.

All elements of interest are not captured in the mean value, however. We would also like the median and the mode to be close to the mean. As an average value, the mean is composed of numerous elements. If the mean is composed of numerous insignificant underestimates and a handful of very large overestimates, a zero value may still result. Such a case is shown in Figure 14-18a. In general a more symmetric distribution is preferred (Fig. 14-18b).

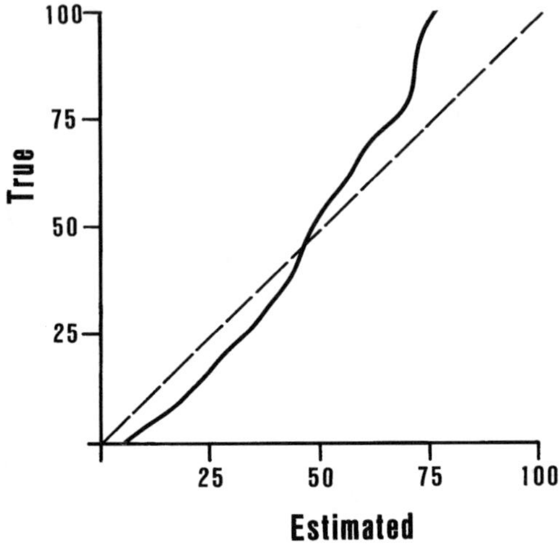

Figure 14-16. A *q-q* plot of a typical result comparing the sample distribution and cross validation estimates. (From *An Introduction to Applied Geostatistics,* by E.H. Isaaks and R.M. Srivastava. Copyright © 1989 by Oxford University Press, Inc. Used by permission of Oxford University Press, Inc.)

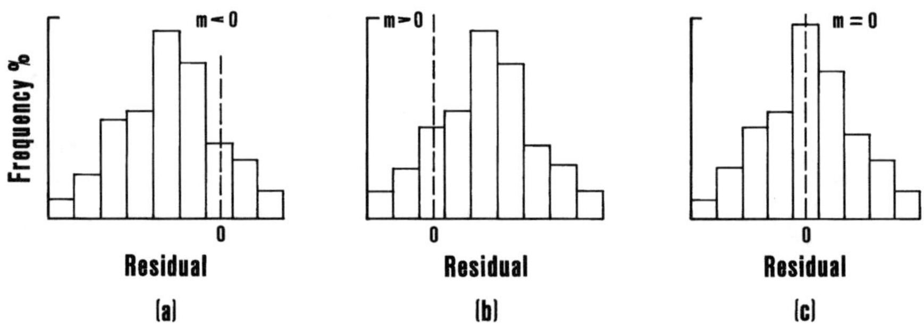

Figure 14-17. Cross validation error distributions showing negative, positive, and zero bias. (From *An Introduction to Applied Geostatistics,* by E.H. Isaaks and R.M. Srivastava. Copyright © 1989 by Oxford University Press, Inc. Used by permission of Oxford University Press, Inc.)

Even symmetric distributions with zero means may not be equivalent. Figure 14-19 shows two such distributions. In Figure 14-19a, the variance of the distribution is much greater than in Figure 14-19b. The narrower distribution would be preferred.

So far, positive attributes of the error distribution have been minimum variance and a mean of zero. Some error distributions result in a mixture of the positive and negative aspects. Figure 14-20 shows two distributions where the

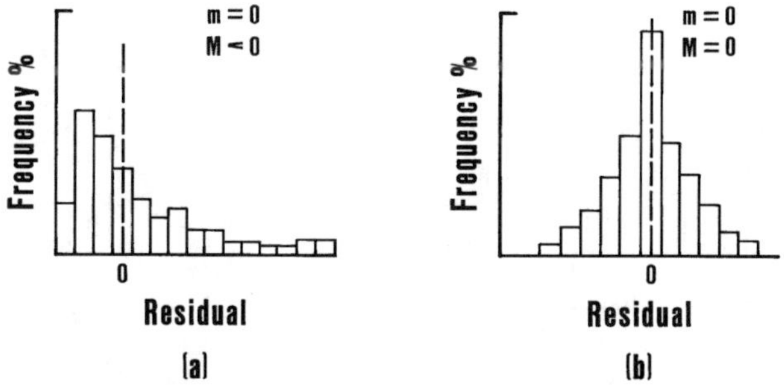

Figure 14-18. Cross validation error distributions with zero means: (a) A skewed distribution compared to (b) a symmetric distribution. (From *An Introduction to Applied Geostatistics,* by E.H. Isaaks and R.M. Srivastava. Copyright © 1989 by Oxford University Press, Inc. Used by permission of Oxford University Press, Inc.)

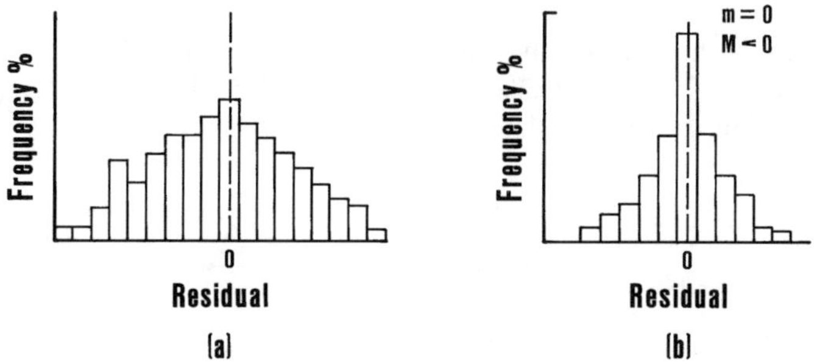

Figure 14-19. Cross validation error distributions with zero means: (a) A distribution with a large variance contrasted to (b) a distribution with a small variance. (From *An Introduction to Applied Geostatistics,* by E.H. Isaaks and R.M. Srivastava. Copyright © 1989 by Oxford University Press, Inc. Used by permission of Oxford University Press, Inc.)

preferred choice is not clear. In Figure 14-20a, the errors are normally distributed, with a mean of zero, as desired. The spread of the distribution, however, is quite large, indicating a significant amount of large estimation errors.

Figure 14-20b shows another normal error distribution but, this time, the variance is quite small, which is a desirable feature. The problem is that the mean carries a positive value, indicating a persistent bias in the estimation. Given the choice between the two estimators represented by these error distributions, we must make an awkward decision: Is it more acceptable to allow a small bias while receiving less variable estimates, or is unbiasedness more critical than error variability? The answer is subjective and may not be the same for all situations or all stakeholders.

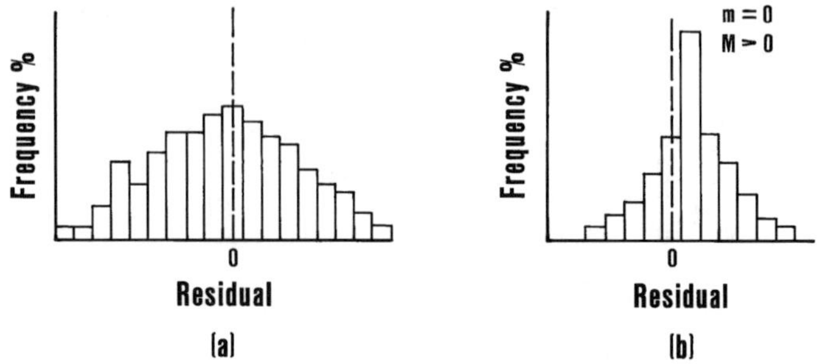

Figure 14-20. Cross validation error distributions where positive and negative assessment criteria are mixed. (From *An Introduction to Applied Geostatistics,* by E.H. Isaaks and R.M. Srivastava. Copyright © 1989 by Oxford University Press, Inc. Used by permission of Oxford University Press, Inc.)

Several statistical measures may be applied to help us interpret the mean of the cross validated residuals. The mean of the error distribution is calculated simply by summing the individual errors e_i and dividing by n:

$$\text{Mean error} = \frac{1}{n}\sum_{i=1}^{n} e_i \tag{14.1}$$

The bias and the spread of the errors are incorporated into the summary statistics of the mean absolute error and the mean squared error as given by the following equations:

$$\text{Mean absolute error} = \frac{1}{n}\sum_{i=1}^{n} |e_i| \tag{14.2}$$

$$\text{Mean squared error} = \frac{1}{n}\sum_{i=1}^{n} e_i^2 \tag{14.3}$$

Note that the mean squared error (MSE) includes both variance and bias components in the following way:

$$\text{MSE} = \text{variance} + \text{bias}^2 \tag{14.4}$$

In the analysis of the entire distribution of estimates, a *q-q* plot is useful. Similar techniques may be applied to the distribution of errors. One particular item of interest is conditional unbiasedness. If the summary statistics indicate that the bias is low, we also hope that the bias will be low for the entire error distribution. For environmental situations, we hope, at least, that the errors will be conditionally unbiased near the action level threshold.

Figure 14-21 shows a visual method for investigating the possibility of conditional bias. The approach plots the error values as a function of the estimated values. For conditionally unbiased estimates, the errors should plot both above and below the horizontal zero line at roughly equal distances for particular estimated values or ranges of estimated values. An example of this type of outcome is shown in Figure 14-21b.

Figure 14-21a shows a plot of conditionally biased estimates. The lowest estimated concentrations tend to be too high, whereas the highest estimated concentrations tend to be too low. This again demonstrates how combining several samples to develop an estimate smooths the distribution.

Another method for investigating conditional bias is using scatter plots of true vs. estimated concentrations as shown in Figure 14-22. As described in Chapter 3, perfect estimation would result in a 45° line, indicating that true and estimated concentrations match exactly. This limit case does not occur in practice, where a cloud of points appears instead. "Thinner" clouds are preferred, and the results shown in Figure 14-22b are more desirable than those illustrated in Figure 14-22a.

Fitting a conditional expectation curve to the scatter plot is an additional way to check for conditional bias. If no conditional bias is present, the conditional

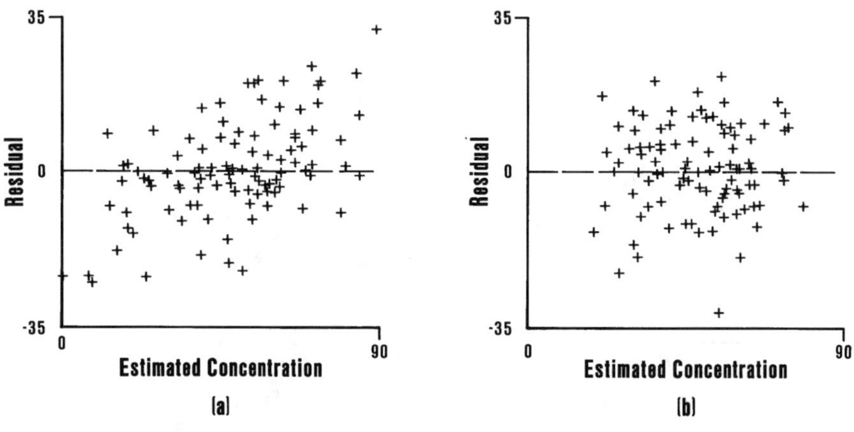

Figure 14-21. Plot of estimated concentrations vs. residuals to investigate residual bias. (From *An Introduction to Applied Geostatistics,* by E.H. Isaaks and R.M. Srivastava. Copyright © 1989 by Oxford University Press, Inc. Used by permission of Oxford University Press, Inc.)

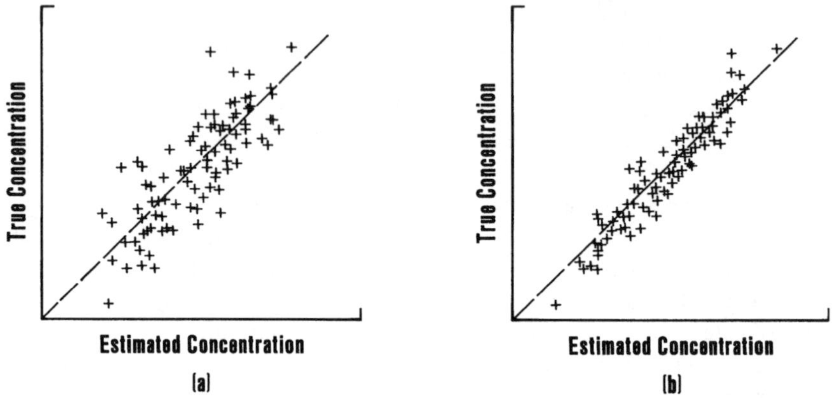

Figure 14-22. Scatter plots of estimated vs. true concentrations showing (a) larger and (b) smaller variability. (From *An Introduction to Applied Geostatistics,* by E.H. Isaaks and R.M. Srivastava. Copyright © 1989 by Oxford University Press, Inc. Used by permission of Oxford University Press, Inc.)

expectation curve will plot on the 45° line for all ranges of concentrations. Figure 14-23 shows two sets of scatter plots and conditional expectation curves. In Figure 14-23b, conditional bias is seen in the lower and upper concentration ranges, as evidenced by the departures of the conditional expectation curve from the 45° line. This is in spite of statistics for these data that indicate a global absence of conditional bias.

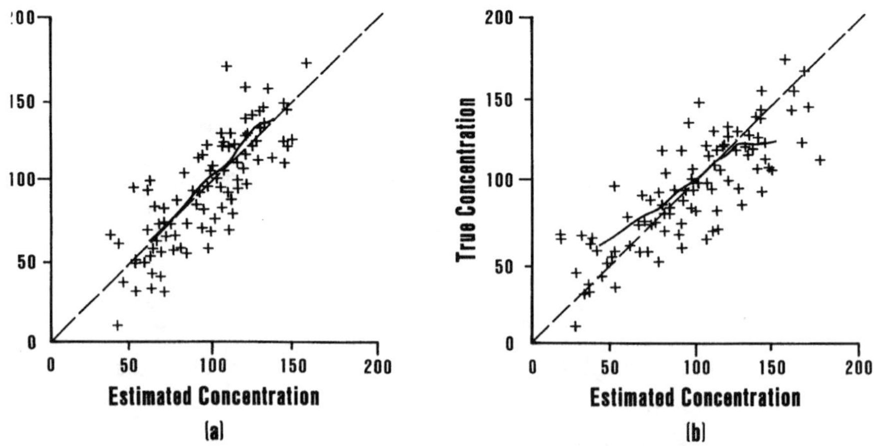

Figure 14-23. Scatter plots with conditional expectation curves showing (a) no conditional bias and (b) conditional bias. (From *An Introduction to Applied Geostatistics,* by E.H. Isaaks and R.M. Srivastava. Copyright © 1989 by Oxford University Press, Inc. Used by permission of Oxford University Press, Inc.)

In Figure 14-23a, there is little deviation of the conditional expectation curve from the 45° line over any range of concentrations. No conditional bias is present in this case, and the estimates are also globally unbiased.

14.5.4 Assessing Performance: Typecasting Error

Summary statistics of estimates and errors, along with distribution analysis of estimates and errors derived from the cross validation process, provide a general basis for decision making. However, as Figure 14-20 demonstrates, the results of different statistical criteria used to measure performance may be in conflict, once again relegating decisions regarding preferred estimators and appropriate variogram parameters to the subjective realm. Modified approaches have been proposed (Clark 1986, David 1988), but the conclusions again indicated that subjectivity could not be eliminated from the cross validation process.

The simplified decision rules applicable to environmental remediation offer an opportunity to remove the subjectivity from cross validation studies. Once an action level has been selected for remediation, all soil estimated to contain concentrations equal to or greater than this value will be subject to remediation. For remedial purposes, it does not matter whether the concentration of lead (Pb) in soil is 501 or 50,000 ppm; both will be treated when the action level is 500 ppm. If a block concentration is incorrectly estimated, the magnitude of the error is far less important than the classification of clean or dirty. One possible exception is PCBs, for which incineration may be required at levels above 500 ppm in soil.

In Chapter 3, the misclassification ellipse was introduced, along with the process of typecasting error. Figure 14-24 shows a diagram of the misclassifica-

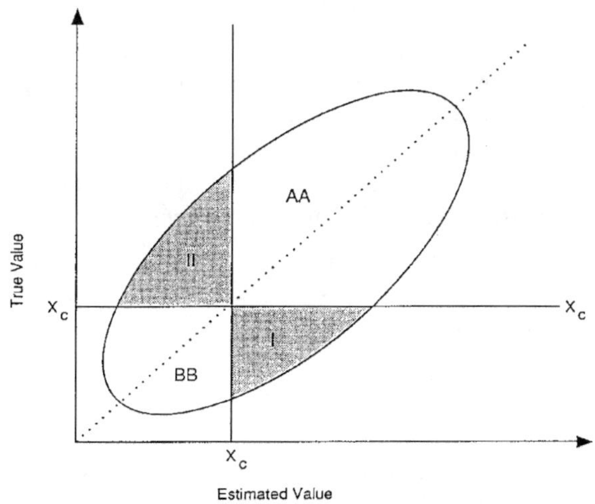

Figure 14-24. Misclassification ellipse.

Table 14-1. Typecasting of error cross-validation results for a PCB project.

Threshold concentration, ppm:	25	
Total number of estimates:	665	
	Number	*Percent*
Occurrences of Type I error:	86	12.93
Occurrences of Type II error:	63	9.47
Occurrences below threshold:	318	47.82
Occurrences above threshold:	198	29.77
Total misclassifications:	149	22.41
Total correct decisions:	516	77.59

tion ellipse. Two general categories are illustrated by the ellipse: (1) correct classification in the lower left and upper right quadrants (BB and AA), and (2) incorrect classification (Types I and II) in the lower right and upper left quadrants, respectively.

By plotting the action threshold limit on a scatter plot of true concentrations vs. concentrations estimated by cross validation, the misclassification ellipse is formed. Then, either by counting points on a scatter plot or by computer-generated summary, the number of correct classifications and the number of errors for each category may be tabulated. This comparison of correct vs. erroneous decisions is called *typecasting of error* (Myers 1991).

Using this process, a cross validation estimate falls into only one of four possible classes. The results of the classification are then totaled and summarized. An example result of a typecasting cross validation for a PCB project is presented in Table 14-1.

The objective of the typecasting approach is to minimize the misclassification errors associated with a given estimator or given set of estimation parameters. This will help minimize decision errors during remediation. Goals can be set during the DQO process according to the needs of the stakeholders. PRPs will benefit by reducing the occurrences of Type I errors, and thus minimizing false positive results and costly overremediation. Regulatory agencies will benefit by minimizing Type II errors, thereby helping to assure health and safety.

14.5.5 Typecasting Error as a Decision Tool

14.5.5.1 Selecting An Estimator

Many computerized interpolation methods are available for gridding data or creating block models. Among the commonly available routines are inverse power distance, triangulation, radial basis functions, Shepard's method, minimum curva-

1 LINE LONG

ture, polynomial regression, splines, trend surface analysis, and, of course kriging. Kriging also comes in many "flavors": ordinary, simple, disjunctive, multi-gaussian, universal, indicator, probability, and so on.

It is unlikely that a single estimator will be optimal for all situations. Therefore, we may wish to test various interpolation methods in order to determine which method performs best for a particular data set or a particular COC. The typecasting error approach allows us to compare estimators based on environmentally relevant objectives.

14.5.5.2 Selecting Sample Locations

The results from a typecasting cross validation lend themselves to informative visual display. Figure 14-25 shows one level of the ReSolve site, a multilevel

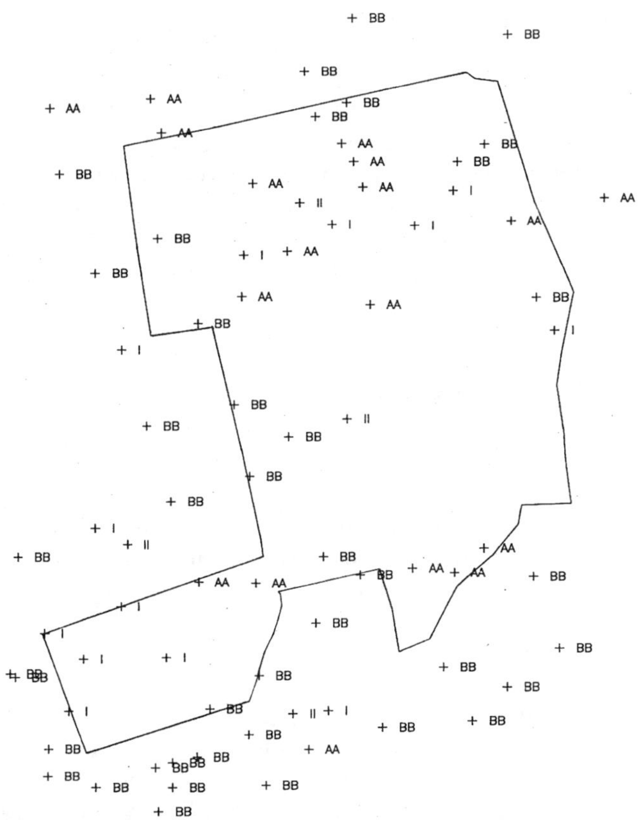

Figure 14-25. Typecasting of error location map for one level of the ReSolve PCB site. (From A geostatistical analysis of PCB concentrations in soils at the ReSolve site, by J.C. Myers, 1992.)

site contaminated with PCBs. As a group, the classification results show roughly the same classification percentages as those listed in Table 14-1.

The spatial configuration of the results is also of interest. BB (below/below) results (from locations where both the true and estimated concentrations are below the threshold, or Below/Below) tend to be at the margins of the site, as might be expected. Similarly, AA (above/above) results tend toward the interior portions of the site. Type I errors occur in both interior and exterior locations. They also cluster in the southwest portion of the map, indicating a consistent difficulty in appropriate estimation. Type II errors are few and widely dispersed, showing no consistent pattern.

Once the original sample data are returned to the data set, we would expect local estimation to improve, eliminating some of the misclassifications. The next step, however, is to estimate block concentrations instead of sample locations. It is reasonable to assume that the areas that were correctly classified without the benefit of the removed sample will continue to be correctly classified. Whether the estimates given to the transition areas between the Type I and II errors and AA and BB classifications will result in correct classifications is still in doubt.

Transition areas provide good targets for additional samples. This may be possible if the site is in the early stages of investigation or if additional sampling programs are already planned. PRPs wishing to reduce the cost of remediating clean soil classified as dirty may wish to invest some money in attempts to reduce the remedial cost associated with Type I error. If no additional characterization sampling is planned, transition Type II areas provide good targets for postremedial excavation sampling. These areas are known to be difficult to classify and have the highest probability of false negative classifications.

14.5.6 Optimizing Interpolation Parameters

Cross validation and typecasting error techniques may be used for more than simply distinguishing between different estimation techniques. They may also be used to optimize the parameters used for a particular estimator, some of which are extremely sensitive. For simplicity, only kriging parameters will be discussed in this section; the same or similar methods may be used for other interpolators.

14.5.6.1 Number of Neighbors

Experience has shown that the parameter most sensitive to typecasting error for kriging is the number of neighboring samples used in the cross validation process (Myers 1991). One reason for this is that environmental data often span several orders of magnitude. An extreme concentration can dramatically influence the estimated concentration, even at great distances.

Figure 14-26 shows a plot of the misclassification error as a function of the

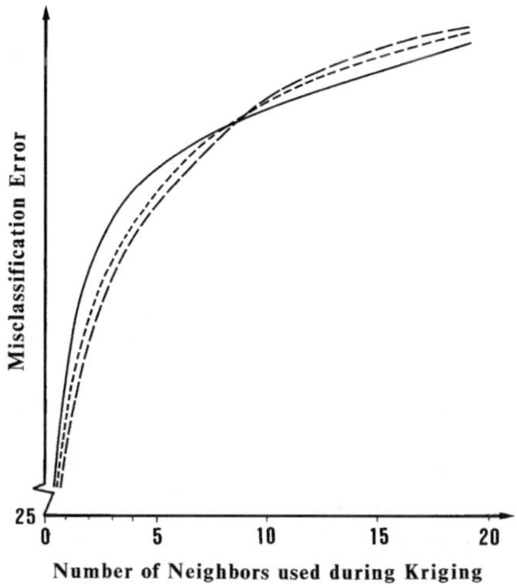

Figure 14-26. Plot of the total misclassification error vs. the number of neighbors used in typecasting of error cross validation (sample concentrations) for the ReSolve PCB site. (From A geostatistical analysis of PCB concentrations in soils at the ReSolve site, by J.C. Myers, 1992.)

number of neighbors for a soil PCB site using three different sets of spherical variogram parameters. The graph depicts a steady and rapid increase in total misclassifications from one sample to approximately 10, with a slower rate of decline beyond. Based on this curve, the optimum number of neighbors for cross validation is one. This implies that, for the actual kriging, two samples will be used because the original sample will be replaced in the data set.

Note that all three variogram models performed in a surprisingly similar fashion. This does not imply that they are not important. It does, however, suggest a reasonable approach to optimizing the kriging parameters. A preliminary variogram analysis should be run and a reasonable model fit to the experimental data. This preliminary model may then be used to determine the optimal number of neighbors. Once that number has been determined, it can be held constant while the sensitivity of other kriging parameters is investigated.

The results presented in Figure 14-26 suggest an almost polygonal approach to estimation. Such results are not necessarily typical. Figure 14-27 shows another graph of the same type as Figure 14-26, this time for the indicator values of the same data. Instead of a continuous rise with increasing neighbors, the graph shows a consistent drop in error until a minimum is reached for 15 neighbors. Beyond this minimum, errors begin a slow but fluctuating rise.

Figure 14-28 shows another example of misclassification error behavior. Error values fall sharply at first, reaching a minimum at four. Error then begins a

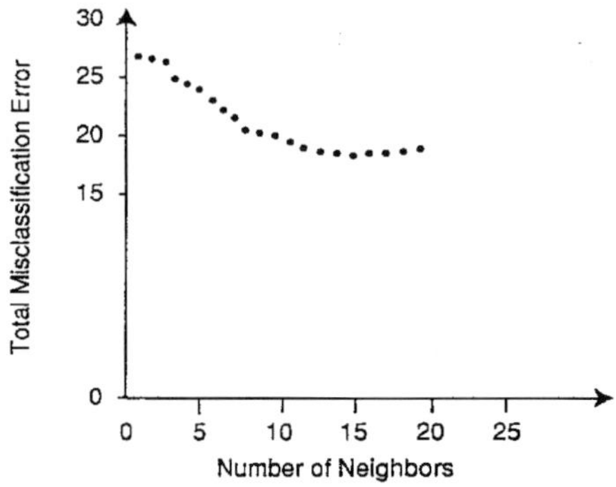

Figure 14-27. Plot of the total misclassification error vs. the number of neighbors used in typecasting of error cross validation (indicator values) for the ReSolve PCB site. (From A geostatistical analysis of PCB concentrations in soils at the ReSolve site, by J.C. Myers, 1992.)

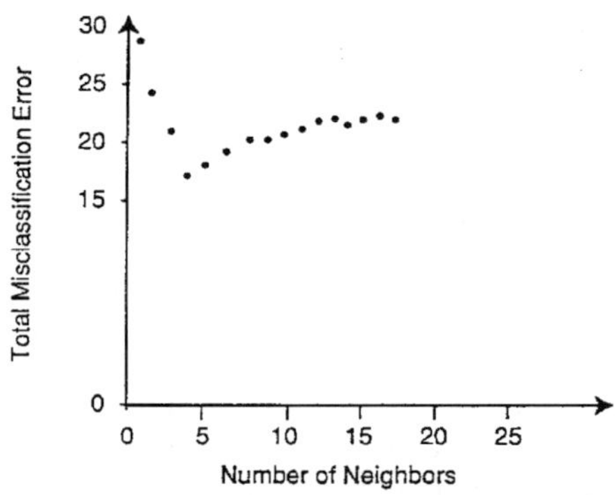

Figure 14-28. Plot of the total misclassification error vs. the number of neighbors used in typecasting of error cross validation of sample concentrations for a VOC in groundwater. (From A geostatistical analysis of PCB concentrations in soils at the ReSolve site, by J.C. Myers, 1992.)

slow but mostly consistent rise beyond the minimum. As a reminder, five samples should be used during kriging as it is appropriate to include the original sample and all four optimal neighbors.

14.5.6.2 Search Radius

As discussed earlier in this chapter, the final estimate is influenced by the search parameters used. Cross validation results are also influenced by the search parameters and are somewhat related to the number of neighbors.

In order to obtain a large number of neighbors, a reasonably large search radius will be needed in many instances. One must remember that, whereas 15 neighbors may be specified as a maximum, an insufficient radius may never capture this many sample points. This must be balanced by the fact that samples at extreme distances may not be appropriate for local estimation.

14.5.6.3 Variogram Range

The kriging example in Chapter 13 demonstrated that the range of the variogram model has an effect on the kriging weights and, consequently, the final estimate. While typecasting error results are not as sensitive to the range as to the number of neighbors, different variogram ranges and anisotropies (both ratios and orientations) should be investigated. The model that provides the best fit to the experimental variogram may not, in fact, be the most reliable estimator.

14.5.6.4 Nugget Effect

Typecasting error results are quite sensitive to the influence of a nugget effect. As the nugget effect increases, sample weights become more alike and estimates draw closer to a simple average. In general, the addition of a nugget effect has a detrimental effect on the typecasting error results. The effect of various nugget-to-sill ratios should be investigated during a cross validation study.

14.5.6.5 Variogram Sill

As discussed in Chapter 13, the height of the sill does not affect the kriging weights nor the block or point estimate. Therefore, it is not necessary to investigate different sills during cross validation.

14.5.7 Caveats for Cross Validation

14.5.7.1 Problems in Two Dimensions

Figure 14-29a shows a regular unit grid of samples at distance d ($d = 1$), with an unsampled point P to be estimated. In this case, for any unsampled point,

$$\text{Maximum distance} = \frac{d}{\sqrt{2}} \tag{14.5}$$

or approximately 0.7.

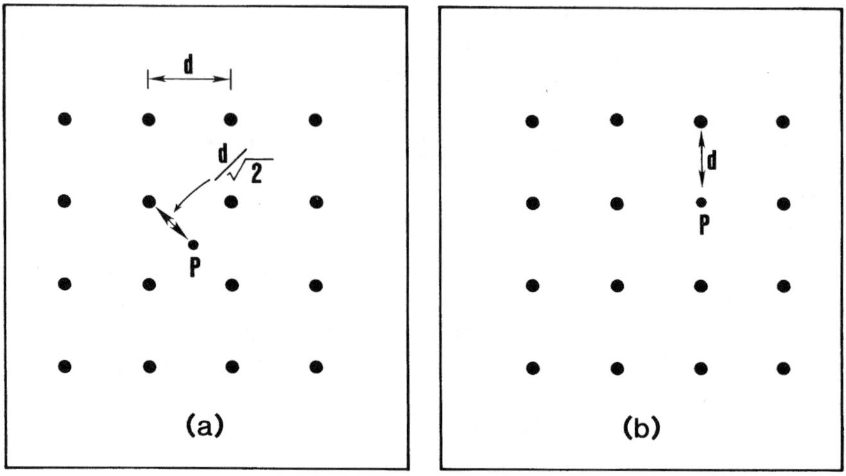

Figure 14-29. Problems in cross validation: the two-dimensional case. (From *Handbook of Applied Advanced Geostatistical Ore Reserve Estimation*, by M. David, Elsevier, 1988. Courtesy of Michel David.)

Figure 14-30. Problems in cross validation: the three-dimensional case. (From *Handbook of Applied Advanced Geostatistical Ore Reserve Estimation*, by M. David, Elsevier, 1988. Courtesy of Michel David.)

The geometry changes when a point is removed for cross validation. Figure 14-29b shows that the closest distance from point *P* to a sample is now *d* or 1. With all samples at a greater distance, it will be more difficult for any estimator to produce an accurate estimate or classification of the unsampled point.

14.5.7.2 Problems in Three Dimensions

Figure 14-30 shows the reverse problem from two dimensions. A vertical boring is shown with a central interval removed in anticipation of estimation by cross validation. Now, the nearest neighbors are directly above and below the point to be estimated. Regardless of the estimator of variogram model selected, the point estimate will be very close to the average of the two samples surrounding the interval.

Without additional sampling, the problem in two dimensions is not easily solved. In three dimensions, however, we may remove samples within a certain distance of the point to be estimated. This should be done with caution so as to avoid recreating the two-dimensional problem.

Conclusion

Successful implementation of kriging depends not only on a computer program that respects the equations developed in Chapter 13 but also on a practical understanding of the interactions between the various kriging parameters and block model constraints. These include topographic constraints, correct search procedures, and appropriate cross validation where typecasting error may be valuable. Computer kriging and contouring can often be accomplished with the push of a button, but the practitioner is responsible for determining whether the resultant map should be classified as "gospel" or "garbage."

PART 5 Data Quality Objectives

Chapter 15

Data Quality Objectives

SYNOPSIS: There are many opportunities during the life cycle of a hazardous waste site to introduce the techniques of sampling theory & practice (STP) and geostatistical appraisal (GA). The data quality objectives (DQO) process affords a regulatory-based and -approved structure through which STP and GA may be initiated to answer questions at the appropriate phase. This chapter briefly summarizes the key elements and steps that compose the DQO process (EPA 1994), and introduces spatial DQO concepts, including how DQOs change as site characterization progresses and the concept of exposure units (EUs).

KEY WORDS: Data Quality Objectives, Quality Assurance Project Plan, RI/FS, PA/SI, Decision Rule, Decision Errors, Design Performance Goal Diagram, False Positive, False Negative, Gray Region, Performance Curve, DQO Misclassification Ellipse, Exposure Units

15.1 Introduction

The GEM endeavors of STP and GA play an important role in site characterization and remediation. While important by themselves, it takes much more than sampling and statistics to shepherd a site through the entire characterization and remediation process. Whereas STP and GA may be "better mousetraps," other resources are necessary to harness the power of STP and GA for environmental studies. The DQO process serves this purpose and provides the interface to regulatory and legal areas.

To evaluate and remediate a hazardous waste site, the expertise of other professionals is needed, including those with regulatory, legal, chemical, managerial, administrative, engineering, and negotiating skills. Each skill is necessary but, as with STP and GA, no one of them is sufficient to complete the task.

Coordinating the skills of the professionals involved in the cleanup of the site can be a complex task. Interfacing and merging the ideas, needs, and approaches of the various professionals is a difficult balancing act, requiring thoughtful planning and execution.

DQOs offer a necessary series of planning guidelines that assist in integrating the STP and GA needs of a project with other site concerns. DQOs provide a structure through which many site issues can be addressed, especially those concerned with planning, collection, analysis, and interpretation of sample data. Strategic planning through DQOs can bring efficiency to the project and allow more effective and defensible decisions to be made.

DQOs provide a necessary link from STP and GA to the regulatory and legal processes associated with site investigations. DQOs consist of seven steps that serve to define many aspects of the data-collection process: when to collect samples, where to collect samples, acceptable levels of error, and the number of samples to collect. DQOs help to assure that the appropriate type, quantity, and quality of samples needed for decision making will be obtained at the appropriate phases of the project. This also acts as a safeguard to prevent allocation of resources to sampling efforts that are not cost-effective.

As with STP and GA, the DQO process should be used anytime a project requires data collection and must be implemented before the data are collected. While appropriate for all types of sample collection, the degree to which the DQO process is used may vary and depend on the specific nature and needs of the project. Outputs of the DQO process are often included in a quality assurance project plan (QAPP).

The purpose of DQOs is to develop qualitative and quantitative statements that are the resultant outputs of the first six DQO steps. According to EPA (1994), these statements are intended to:

- Clarify the study objective.
- Define the most appropriate data to collect.
- Determine the most appropriate conditions from which to collect the data.
- Specify tolerable limits on decision errors, which will be used as the basis for establishing the quantity and quality of data needed to support the decision.

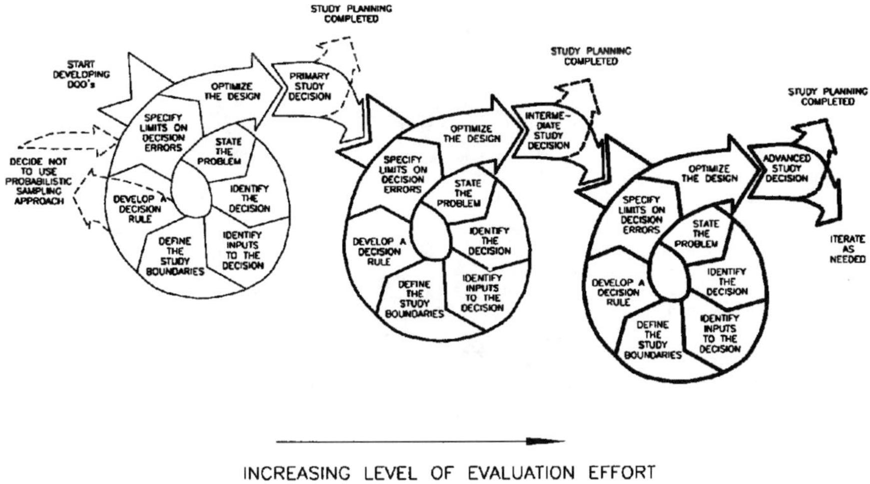

Figure 15-1. Illustration of the iterative nature of the DQO process. (From *Guidance for the Data Quality Objectives Process,* EPA QA/G4, USEPA, 1993.)

Different phases of site characterization typically require different data needs. For example, during the preliminary assessment/site investigation (PA/SI), it may only be necessary to determine the presence or absence of a suspected COC. Making such a determination may require only small amounts of data from relatively less accurate field-screening methods. During the remedial investigation/feasibility study (RI/FS), more data of better quality may be needed to determine the average COC concentration so that health risks may be determined. At the remedial stage, additional data representing the spatial nature of the contamination may be required in order to minimize remedial costs by surgical-type removals (Blacker and Goodman 1994b, Myers 1993).

The need for different data at different project phases implies that the DQO process is iterative. Figure 15-1 shows the seven-step DQO process and how it can be used in an iterative sequence until data collection is no longer demanded.

15.2 The Seven-Step Process

The DQO process consists of seven sequential steps, with each step impacting choices to be made in later steps. A diagram of these steps can be seen in Figure 15-2. In the first six steps, decision performance criteria are generated for use in the data collection design. The final step is to optimize the design under existing budget, manpower, and time constraints. This section will summarize the purpose, procedure, and outputs recommended by EPA.

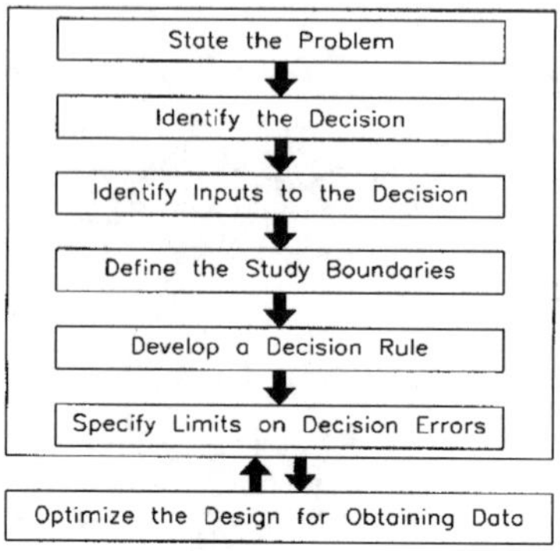

Figure 15-2. The seven steps in the DQO process. (From *Guidance for the Data Quality Objectives Process,* EPA QA/G4, USEPA, 1993.)

15.2.1 State the Problem

15.2.1.1 Purpose

By definition of EPA, the purpose of step 1 is to clearly define the problem so that the focus of the study will be unambiguous. Failure to complete this seemingly innocuous step can lead to severe problems. If the problem is not well defined, data incapable of answering the question of concern may be collected.

15.2.1.2 Activities

During this phase, four main activities are recommended:

1. Identify members of the planning team. This is a multidisciplinary group of technical experts that will develop the DQOs. All stakeholders should be represented by the team, with a typical team consisting of (but not limited to) samplers, chemists, scientists, engineers, modelers, technical project managers, administrative and executive managers, QA/QC experts, statisticians, data users, and decision makers.
2. Identify the primary decision maker and define each member's role and responsibility. The decision maker assumes ultimate authority for all final decisions once the team members have provided their recommendations.
3. Develop a concise description of the problem. Descriptions should include potential health risks or noncompliance situations and the conditions and circumstances surrounding them. Existing information should be reviewed and summarized to avoid potential duplication, and complex

Table 15-1. Summary of the elements to consider during the DQO problem description phase.

ELEMENTS OF THE PROBLEM DESCRIPTION

The following elements may be appropriate to include in the problem description. Note: This list only provides the basic elements of the problem description. Your elements may be slightly different.

- Study objectives/regulatory context.
- Persons or organizations involved in the study.
- Persons or organization that have an interest in the study.
- Political issues surrounding the study.
- Sources and amount of funding.
- Previous study results.
- Existing sampling design constraints (some aspects of sampling design may be specified in regulations or established through past planning efforts).

problems should be subdivided into smaller management units. Table 15-1 summarizes the elements of problem description.

4. Specify the available resources and relevant deadlines for the study. Items include budget, personnel, contract mechanisms, and intermediate and final deadlines.

15.2.1.3 Expected Outputs

In this phase, three outputs are expected:

- A list of the planning team members and identification of the decision maker
- A concise description of the problem
- A summary of available resources and relevant deadlines for the study

15.2.2 Identify the Decision

15.2.2.1 Purpose

The purpose is to define the decision statement that the study will attempt to resolve. This step should also identify the alternative actions that may be taken, based on the outcome of the study.

15.2.2.2 Activities

Four main activities occur at this DQO stage:

1. Identify the principal study question. A narrow and focused statement should be developed that identifies the unknown conditions or unresolved issues that the data will attempt to answer. Typical questions appear below:
   a. Is the contaminant concentration significantly above background levels, indicating that a release has occurred?
   b. Is the mean of the COC in soil at the site above a risk-based threshold?
   c. Has the contamination from the unlined storage lagoon infiltrated the aquifer below the site?
2. Define the alternative actions that could result from resolution of the principal study question. Generally, two courses of action will arise, one that will address the problem and one that requires no action. For example, if the answer to question 1a is yes, enforcement action is needed. If the answer is no, no action is required.
3. Combine the principal study question and the alternative actions into a decision statement, for example:

   Determine whether or not a release has occurred and requires enforcement action.
4. Organize multiple decisions in the proper sequence. Figure 15-3 shows a flowchart of how to organize and document multiple decisions.

15.2.2.3 Expected Outputs

The only expected output from this step is a decision statement that links the principal study question to possible actions that will solve the problem.

15.2.3 Identify the Inputs to the Decision

15.2.3.1 Purpose

The purpose is to identify the informational inputs that will be required to resolve the decision statement and to determine which inputs require environmental measurements.

15.2.3.2 Activities

The four activities for this step are summarized below:

1. Identify the information that will be required to resolve the decision statement. What type of samples or measurements will be needed: soil

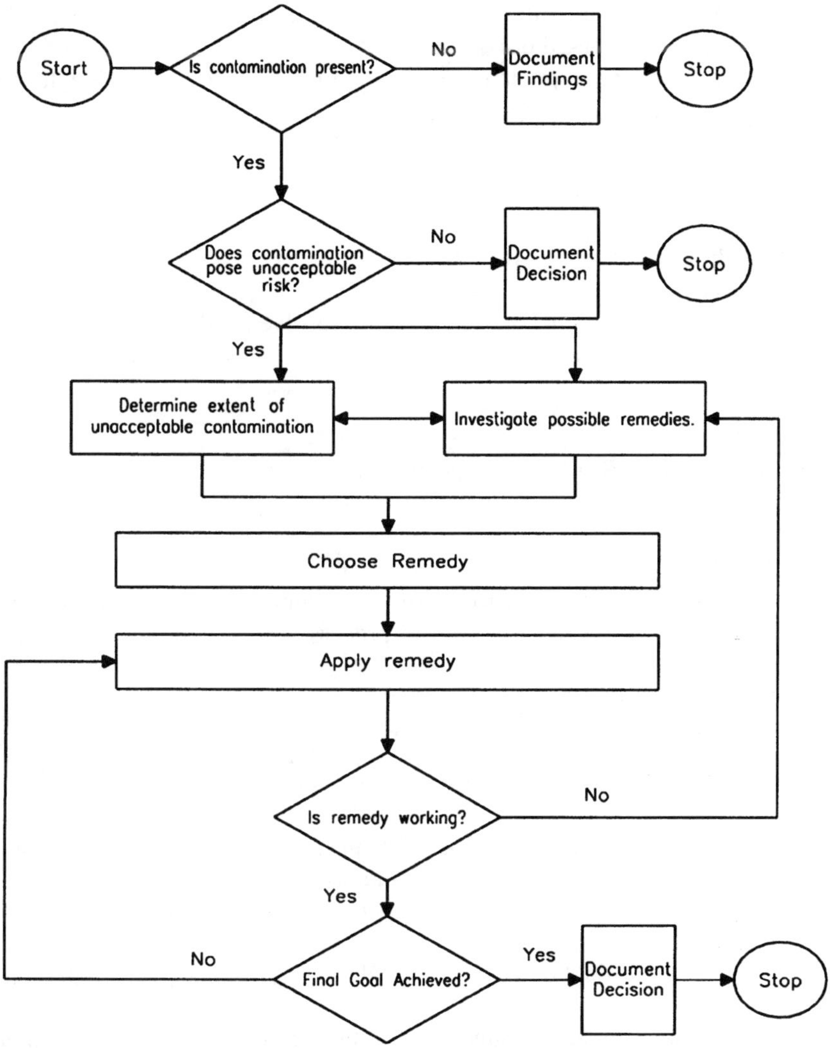

Figure 15-3. Flowchart for organizing and documenting multiple decisions. (From *Guidance for the Data Quality Objectives Process,* EPA QA/G4, USEPA, 1993.)

samples, aquifer thickness, depth to bedrock, groundwater elevation contour, soil thickness?

2. Determine the sources for each item of information identified above. Is information currently available, or is additional data gathering required?

3. Identify the information that is needed to establish the action level. Will it be threshold-based or risk-based?
4. Confirm that appropriate measurement methods exist to provide the necessary data.

15.2.3.3 Expected Outputs

Two outputs are expected from this step:

- A list of informational inputs needed to resolve the decision statement
- A list of environmental variables or characteristics that will be measured

15.2.4 Define the Boundaries of the Study

15.2.4.1 Purpose

Spatial and temporal boundaries of the problem are addressed in this step. More on how this step relates to the GEM process is presented in Section 15.3.

15.2.4.2 Activities

During this step, five primary activities take place:

1. Specify the characteristics that define the population of interest. Examples are hexavalent chromium concentrations in soil and TCE concentrations in groundwater. Reference to an appropriate support size should also be included.
2. Define the spatial boundaries of the decision statement and stratify when appropriate.
3. Define the temporal boundaries of the problem. This activity should include noting the appropriate time frames, any access restrictions for sampling, and determining appropriate times for data collection to ensure representative sampling of the target population, especially when temporal cycles may be present (e.g., Fig. 10-31).
4. Define the scale of decision making. The scale refers to the smallest and most appropriate areas for which decisions will be made. The scale is typically related to GA and risk concerns. Leaking PCBs from a transformer may have a "footprint" of 15 × 15 ft with decisions made on this scale. A risk model may integrate risk over a 1/2 acre area with decisions made on this scale.

5. Identify any practical constraints on data collection. These include seasonal or weather conditions, access restrictions, and equipment and manpower scheduling.

15.2.4.3 Expected Outputs

For this DQO step, two outputs are required:

- A detailed description of the spatial and temporal boundaries of the problem
- Any practical constraints that may interfere with the study

15.2.5 Develop a Decision Rule

15.2.5.1 Purpose

The purpose is to define the parameter of greatest interest, specify the action level, and integrate previous DQO outputs into a single statement that describes a logical basis for choosing among alternative actions. The decision rule must contain four main elements, including the parameter of interest (mean, median, proportion, etc.), the scale of decision making (support), the action level (risk- or threshold-based), and alternative actions.

15.2.5.2 Activities

Three activities occur during this step:

1. Specify the statistical parameter that characterizes the population (mean, median, proportion, etc.). Reference to an appropriate support should also be included.
2. Specify the action level for the study.
3. Develop a decision rule. This takes the format of an *if...then* statement, incorporating the four main elements described above. For example: *If* the mean of PCB soil concentrations of appropriate support size in any 1/2-acre risk unit exceeds 25.0 ppm, *then* that 1/2-acre unit will be remediated to a depth of 1 ft.

15.2.5.3 Expected Outputs

This step results in three outputs:

- The statistical parameter of interest that characterizes the population
- The action level
- An if...then statement that defines the conditions that would cause the decision maker to choose among alternative actions.

15.2.6 Specify Tolerable Limits on Decision Errors

15.2.6.1 Purpose

The purpose of this step is to specify the decision maker's tolerable limits on decision errors, which are used to establish performance goals for the data collection design. The goals generally relate to the rates of false positive and false negative error that will be allowed.

15.2.6.2 Activities

Four activities characterize this DQO step. They are listed below:

1. Determine the possible range of the parameter of interest. This activity is greatly aided by information from prior studies; lack of the same is likely to preclude determination of the range.
2. Identify the decision errors, and choose the null hypothesis. This procedure can be used provided that the relevant spatial scales are defined.
3. Specify a range of possible parameter values (gray region) where the consequences of decision errors are relatively minor. This activity is not appropriate for situations involving spatial data.
4. Assign probability limits to points above and below the gray region that reflect the tolerable probability for the occurrence of decision errors. Such limits can be shown graphically. Figure 15-4 shows a Design Performance

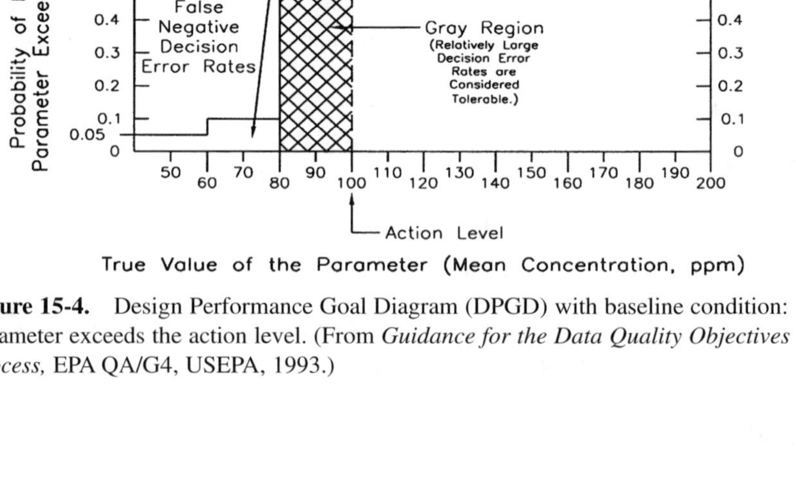

Figure 15-4. Design Performance Goal Diagram (DPGD) with baseline condition: Parameter exceeds the action level. (From *Guidance for the Data Quality Objectives Process,* EPA QA/G4, USEPA, 1993.)

Table 15-2. Summary of decision error limits for DQO design performance goal diagram in Figure 15-4.

True Concentration	Correct Decision	Type of Error	Tolerable Probability of Incorrect Decision
<60 PPM	NOT EXCEED	F(−)	5%
60 TO 80	NOT EXCEED	F(−)	10%
80 TO 100	NOT EXCEED	F(−)	GRAY REGION
100 TO 150	DOES EXCEED	F(+)	5%
>150	DOES EXCEED	F(+)	1%

Goal Diagram (DPGD) with the baseline condition: Parameter exceeds the action level; Table 15-2 presents the corresponding decision error limits shown in Figure 15-4. In a similar manner, Figure 15-5 shows the DPGD for which the baseline condition is: Parameter is less than the action level; Table 15-3 presents the corresponding decision error limits.

It should be noted that the DPGD makes some strong assumptions regarding the nature and symmetry of error loss functions. Thus, the DPGD can more appropriately answer certain types of DQO questions than others. A more detailed

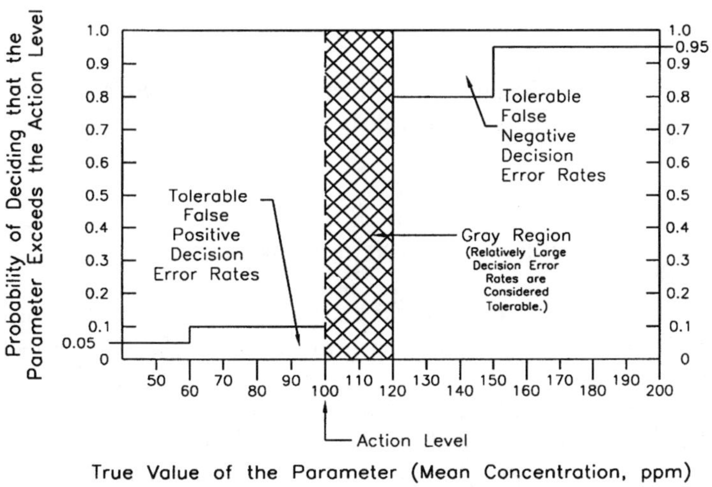

Figure 15-5. Design Performance Goal Diagram (DPGD) with baseline condition: Parameter is less than the action level. (From *Guidance for the Data Quality Objectives Process,* EPA QA/G4, USEPA, 1993.)

Table 15-3. Summary of decision error limits for DQO design performance goal diagram in Figure 15-5.

True Concentration	Correct Decision	Type of Error	Tolerable Probability of Incorrect Decision
<60 PPM	NOT EXCEED	F(+)	5%
60 TO 100	NOT EXCEED	F(+)	10%
100 TO 120	DOES EXCEED	F(−)	GRAY REGION
120 TO 150	DOES EXCEED	F(−)	20%
>150	DOES EXCEED	F(−)	5%

explanation of this loss function, along with a comparison to other types of loss functions, will be presented in Chapter 16.

15.2.6.3 Expected Outputs

The only output of this step is the decision maker's tolerable decision error rates based on a consideration of the consequences of making an incorrect decision.

15.2.7 Optimize the Design for Obtaining Data

15.2.7.1 Purpose

The purpose is to identify a resource-effective data collection design for generating data that are expected to satisfy the DQOs.

15.2.7.2 Activities

For this DQO step, six activities are recommended:

1. Review the DQO outputs and existing environmental data, and assess internal consistency. Determine any data gaps or deficiencies that should be addressed.
2. Develop general data collection design alternatives. Designs for spatial data situations will be addressed in Chapter 16.
3. Formulate the mathematical expressions needed to solve the design problem for each data collection design alternative. Cost functions for spatial data will be presented in Chapter 16.
4. Select the optimal sample size that satisfies the DQOs for each data

collection design alternative. If no design meets the specified limits, EPA recommends that some of the constraints be relaxed. For example:

- Increase the budget for sampling and analysis.
- Increase the width of the gray region.
- Increase the tolerable decision error rates.
- Relax other project constraints, such as schedule.
- Change the boundaries; it may be possible to reduce sampling and analysis costs by changing or eliminating subgroups that will require separate decisions.

5. Select the most resource-effective data collection design that satisfies all of the DQOs. Use of a power function on the DPGD can assist in this activity (Figure 15-6).
6. Document the operational details and theoretical assumptions of the selected design in the sampling and analysis plan.

Techniques to determine optimal sample sizes and number of samples for spatial data will be presented in Chapters 16 and 17.

15.2.7.3 Options for Improving Sampling

To improve sampling, the following options may be considered:

1. Increase the number of samples.
2. Increase the number of analyses of each sample.

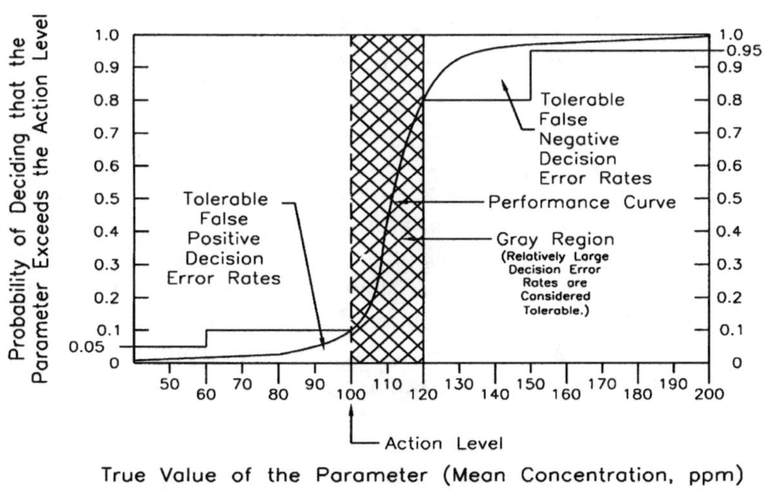

Figure 15-6. Power curve on a DPGD. (From *Guidance for the Data Quality Objectives Process,* EPA QA/G4, USEPA, 1993.)

3. Use a better analytical method.
4. Collect larger support volumes of material (or spatial composites).
5. Use correct sampling devices (does not necessarily increase cost).
6. Improve homogenization and subsampling procedures.

15.2.7.4 Expected Outputs

The only output is the most resource-effective design for the study that is expected to achieve the DQOs.

15.3 Defining Spatial Units for the DQO Process

15.3.1 Spatial DQO Themes

DQOs can be broadly summarized by three questions:

1. What is the question to be answered?
2. What is the statistical population of interest?
3. What is the acceptable level of uncertainty?

By answering these questions, much of what is required in DQO steps 1 through 6 can be accomplished, especially with regard to the GEM process. These questions are best answered in a context appropriate for spatial data. A discussion of some typical spatial contexts appears below.

15.3.1.1 What Is the Question?

This basic question is fundamental to designing an efficient and cost-effective sampling program. The question to be answered will dictate, to a large extent, the spatial or nonspatial nature of the sampling approach needed. Often, the question is related to the needs of the particular phase of the environmental investigation: PA/SI, RI/FS, RD/RA, and so forth, starting first with presence/absence determination, progressing to global average, then to local averages, and finally to long-term monitoring needs. Some examples are:

Early Phases

a. Is dioxin present in the soil at measurable concentrations in the surface soil anywhere at the site?
b. If sampling at a site indicates that no above-action limit COC concentra-

tions are present, what degree of certainty exists that no "hotspots" have been missed?

c. Does the average concentration of the COC in the top 6 in. of the soil within the property boundary of the site impose an increased cancer risk to residents or workers greater than 10^{-6}?
d. What is the statistical distribution of chromium values for 1 kg soil samples?
e. Do the hourly CO readings ever exceed federal standards?

Later Phases

f. What areas around the smelter contain soil concentrations of arsenic and lead above the action levels and to what depths?
g. Has the size of the TCE plume exceeding 0.05 ppm increased or decreased in the last year? Has the average concentration increased or decreased?

Table 15-4 summarizes some of the common objectives for sampling at various stages of site assessment.

15.3.1.2 What Is the Statistical Population?

This question combines the DQO needs of the study with the appropriate statistical domains and DQO boundaries. Along with these issues, the STP consideration of an appropriate sample support size must also be evaluated.

It may be appropriate in some cases to evaluate the site as a single unit for a given COC. In others, designating different OUs or domains may be necessary. Is the population a series of 1 kg samples that will estimate the average cadmium content, or is it the distribution of 10 × 10 × 1 ft estimated blocks? For COCs subject to significant daily or seasonal temporal variations, known or suspected

Table 15-4. Summary of common sampling plan objectives.

- Average or proportion
- Presence or absence
- Compare to threshold
- Search sampling
- Prove PRP
- Risk assessment
- Reportable quantity
- Cost recovery
- Determine extent of contamination
- Define/Locate action level contour
- Changes in plume size/shape

cycles must be addressed into the sampling design. Examples are ozone, CO, and particulates in air, wind patterns and directions, elevation of the groundwater table, and flow rates of streams.

15.3.1.3 What Is the Acceptable Level of Uncertainty?

A major outcome of both DQOs and GEM is the quantification of the error involved in the characterization process. Statistical models allow for the assessment of the error or loss expected for a given sampling design. These models, however, make certain assumptions as to the associated loss functions. In addition to knowing the level of confidence, it is also important to understand the assumptions, limitations, and implications of the statistical and loss models that are being used to characterize a site. Several of these models, along with their assumptions and sampling implications, are discussed in Chapter 17.

15.3.2 A DQO Misclassification Ellipse

The error tolerances developed and charted using the DPGD can also be displayed by means of the misclassification ellipse. An example of a DQO misclassification ellipse is shown in Figure 15-7 with the baseline condition that the parameter is less than the action level. As such, the gray region lies in the Type II error quadrant

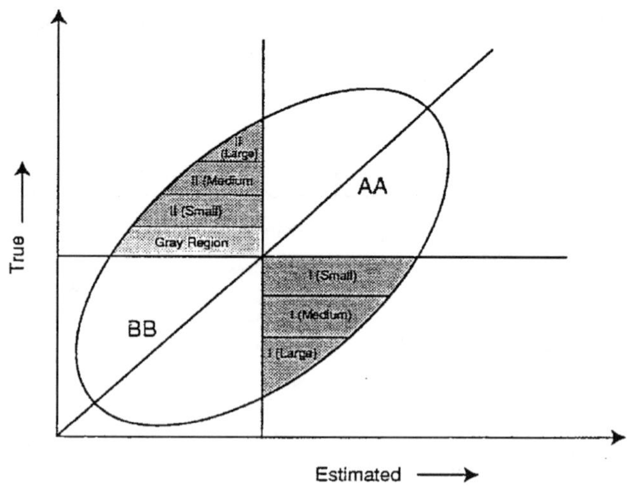

Figure 15-7. A DQO misclassification ellipse.

and indicates that areas with concentrations close to the action level that go unremediated are not of concern.

Both the Type II and Type I quadrants are divided into areas defining increasingly significant misclassification errors. In Figure 15-7, three different zones exist: small, medium, and large. The DPGDs shown previously used only two concentration ranges with associated error tolerances; Figure 15-7 indicates that three or more zones of concentration ranges may be defined for DQO error management.

15.3.3 Exposure Units

Because environmental remediation is largely risk-driven, the RPM frequently wants to know the smallest area on the site over which one can reasonably assume that workers or residents will be exposed to the contamination. The concept of an exposure unit (EU) has been proposed (Neptune et al. 1990, Ryti et al. 1991) as a spatial tool to combine DQOs and risk assessment. Figure 4-13 showed an updated version of Figure 1-3 with an EU definition stage added.

Neptune et al. (1990) define an EU as "the area over which people are expected to integrate exposure when routinely working at or visiting the site." EUs are site-specific, with their size depending on risk, exposure pathway, planned end use, and other factors. Sizes of EUs typically range between 5000 ft^2 and 1/2 acre (approximately 20,000 ft^2). An EU of 5000 ft^2 is often used to represent the size of a residential lot, where the end use will be residential housing (Blacker and Goodman 1994).

Cleanup decisions are made independently on an EU-by-EU basis. If an EU is determined to pose an unacceptable risk, remedial actions will be implemented; if not, no action is taken. For EUs having COC concentrations above the action level, at least two remedial alternatives are available. The first is to remediate the entire EU. The second is to remediate those portions of the EU that cause the average to exceed the action level (i.e., the highest concentrations). When the remaining areas that are more marginally above the action level are averaged with the areas below the action level, the average concentration in the EU (and, therefore, the average risk) will be within the tolerable limits.

Figure 15-8 shows an example of such an approach used by USEPA Region VII on a site contaminated with dioxin (Blacker and Goodman 1994). The initial area boundary planned for excavation is shown, along with groups of blocks that were actually excavated. The figure shows that a significant number of blocks were withdrawn from the remedial plan, resulting in a cost saving of more than 50% (Blacker and Goodman 1994). It should be noted that the cost saving required investing in additional sampling information; cost-benefit analysis techniques for sampling are discussed in Chapter 17.

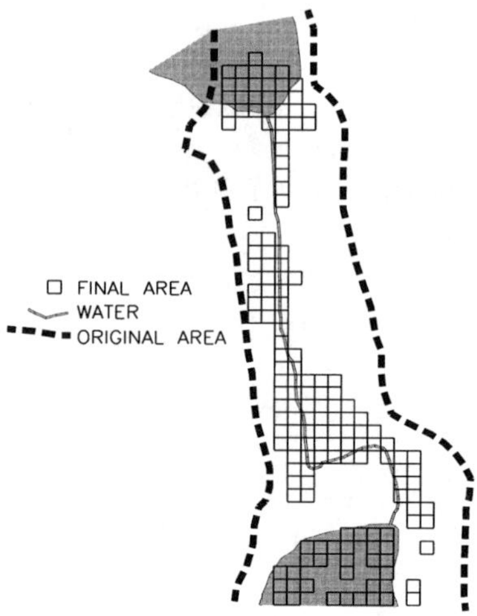

Figure 15-8. Example of how risk-based strategies can minimize remedial volumes, offering opportunities for "surgical" excavation. (Reprinted with permission from the American Chemical Society, from Case study: Application at a Superfund site, by S. Blacker and D. Goodman, *Environmental Science & Technology* 28 (11). Copyright © 1994 American Chemical Society.)

CONCLUSION

DQOs provide an Agency-approved seven-step sequential framework that can be used as an outline for almost any stage of site assessment. To be successful, the process requires input from several technical disciplines. Nonspatial DQOs are covered in depth in EPA guidance documents; this chapter has begun the process of putting DQOs in a spatial context. Chapters 16 and 17 will provide specific techniques and approaches that may be incorporated into the general process described in this chapter.

Chapter **16**

Integrating DQOs and STP: Development of Sampling Strategies

SYNOPSIS: Previous chapters have discussed the characteristics and components of the three GEM areas: DQOs, STP, and GA. Each has value separately, but they have even more value when used together to develop a unified and meaningful approach to site assessment and error management. This chapter will expand on the notions of spatial DQOs and DQOs in the context of project phase. It will focus on DQOs required early in the site characterization process and how STP plays a major role with regard to heterogeneities, dimension of the observation module, and equiprobability. STP concepts will be used to develop a generalized sampling strategy that is useful in a wide variety of sampling situations. In addition, spatial and equiprobability issues for sample placement will be discussed along with their relationship to exposure units (EUs). This chapter will describe techniques that address three questions commonly asked in the early phases of environmental investigation: (1) Can the presence or absence of a COC be established at a site? (2) If the COC is present, do concentrations exist at any locations above the action limit? (3) If concentrations exist above the action limit, is the average concentration of the COC at the site above the action limit, and is it posing a significant health risk?

KEY WORDS: Data Quality Objectives, Sampling Strategy, Sampling Protocol, Sampling Plan, Dimensional Mode of Observation, Heterogeneity, Delimitation Error, Extraction Error,

Zero-Dimensional Lots, One-Dimensional Lots, Two-Dimensional Lots, Three-Dimensional Lots, Optimal Sample Weight, Fundamental Error, Materialization, Sample Selection Mode, Random Sample, Systematic Sampling, Stratified Random Sampling, Area, Sector, Stratum, Substratum, Composite Sample, Increment Recombination, Exposure Units

16.1 Introduction

DQOs provide a general process with which environmental needs can be assessed and data collection planned in an organized manner. Under the current EPA guidance, specific examples generally refer to statistics that focus on problems related to nonspatial data. These assessments include means, medians, proportions, and percentiles that are intended to characterize the global situation, such as the concentration of cadmium in a fly ash container, whether discharge limits are exceeded for a point source, and if a COC exceeds the National Ambient Air Quality Standard.

Spatial data require a modified approach to contaminant assessment. The complete GEM process assumes that the COC is present at measurable levels at many locations at a site, some of which are above the action limit and some of which are below it. This implies that a local assessment, as well as a global assessment, of the concentrations and risks present must be undertaken.

Not all sites need the map-oriented analysis that is a part of GEM. This chapter will begin by describing how to incorporate STP in DQOs and site assessment, discriminating between nonspatial and spatial objectives. The approach developed can, therefore, be used in both spatial and nonspatial contexts. Later, initial elements of spatial analysis will be discussed. These will serve as a bridge to the more sophisticated GA techniques to be covered in Chapter 17.

16.2 Standardization of a Sampling Strategy

Among the important themes in the GEM process are maintaining equiprobability conditions, maintaining error tolerances, and incorporating appropriate spatial components. Starting with the techniques of STP, these goals can be achieved by establishing GEM processwide standards for sampling. Equiprobability was examined in earlier chapters dealing with STP, where it was shown to have a spatial component even at a very small scale. In this chapter, the concept will be expanded to a regional spatial context. This will serve as a link to spatial sampling for geostatistical appraisal.

16.2.1 Definition of a Sampling Strategy

It has often been proposed that a sampling strategy should be standardized for purposes of comparing the sampling at one hazardous waste site to the sampling at another. This is a commendable and legitimate recommendation. The problem is that people often confuse the idea of a sampling strategy with a sampling plan also called a *sampling protocol* (Pitard 1989). For GEM purposes, these definitions will apply:

> *Sampling strategy:* A standardized set of goals and conditions that provide for correct sample design, correct sample collection, and correct spatial assessment
>
> *Sampling plan:* A unique set of conditions and operations that are specific to a hazardous waste site and a COC at that site and that respect local characteristics and goals.

The desire to standardize a sampling strategy is a legitimate one as it respects the goals of each of the three GEM components. The same is not true, however, for sampling plans or protocols. As a result of the heterogeneity that is always present, it is an impossible task to standardize sampling plans. Whereas a sampling strategy can be generalized to all situations, a sampling protocol addresses the distinct and individual needs of a particular situation. A sampling protocol can satisfy only local conditions.

16.2.2 Components of a Sampling Strategy

As with DQOs, the basics of a sampling strategy are few, but the details quickly become complicated. Listed below are the key components of a standardized sampling strategy:

1. Select the module of observation.
2. Characterize the appropriate heterogeneities.
3. Develop a sampling plan.
4. Implement the sampling plan correctly.
5. Characterize the spatial variability.
6. Optimize within DQO constraints.

Despite a limited number of steps, standardization of a sampling strategy is not trivial. Elements of DQOs, STP, and GA must all be linked and applied in a proper manner. A model for the first four steps in a standardized sampling strategy will now be presented.

16.3 Select the Module of Observation

16.3.1 Introduction

The first step in the preparation of a sampling plan is to select the correct modules of observation, choosing between zero-, one-, two-, and three-dimensional models. The module selected prescribes the shape of the volume required to maintain equiprobability in each sample increment. The basic module of observation cannot be compromised: If it is not correct, sampling biases are necessarily introduced, with *DE* and *EE* increasing to an unknown degree.

In addition to module selection, the notion of heterogeneity and of its different types also comes into play. Heterogeneity and module dimension are so interdependent that it is difficult to talk about one without referring to and understanding the other. The following section will amplify the discussion on heterogeneity presented in Chapter 7.

16.3.2 Review of Heterogeneity

In Chapter 7, it was shown that heterogeneity h is equal to

$$h = h_1 + h_2 + h_3 \tag{16.1}$$

where:

h_1 = the short-range heterogeneity, typically random and discontinuous. It is responsible for two important sampling errors, *FE* and *GE*. Its variance is equal to $\sigma^2 CE_1$.

h_2 = the large-scale heterogeneity, usually nonrandom and noncyclic; reflects local trends. An important heterogeneity in terms of the GEM process, it can be quantified by the variogram. Its variance is equal to $\sigma^2 CE_2$.

h_3 = the periodic heterogeneity reflective of cyclic phenomena; most common in flowing streams of material, such as water, coal, and stack emissions. Its variance is written $\sigma^2 CE_3$.

While easier said than done, the total variance $\sigma^2 CE$ for correct and equiprobable sampling can be written:

$$\sigma_{CE}^2 = \sigma_{CE_1}^2 + \sigma_{CE_2}^2 + \sigma_{CE_3}^2 + \sigma_{AE}^2 \tag{16.2}$$

Note that this compares with the overall sampling error OE:

$$\sigma_{OE}^2 = \sigma_{CE}^2 + \sigma_{ME}^2 + \sigma_{PE}^2 + \sigma_{AE}^2 \tag{16.3}$$

where

$$\sigma^2 ME = \sigma^2 DE + \sigma^2 ME \tag{16.4}$$

where the materialization error ME is the sum of the delimitation error DE and the extraction error EE, and PE is the preparation error.

16.3.3 Dimension of the Lot

16.3.3.1 Zero-Dimensional Lots

A zero-dimensional lot is one that can be taken in its entirety for the purpose of sampling or subsampling. This implies a relatively small volume of material. One often encounters this situation at the laboratory for subsampling protocols. Conveniently, zero-dimensional problems are easy to solve. This in no way diminishes their importance, however. Many laboratory subsamples are incorrectly taken on one-dimensional lots, without regard to any STP conventions. Heterogeneity h_1, associated with zero-dimensional lots, will bias incorrect sampling in the laboratory.

16.3.3.2 One-Dimensional Lots

To ensure equiprobable sampling for one-dimensional lots, the correct isotropic module of observation is a slice of constant thickness across the entire stream (Fig. 8-29). For environmental sampling, a device producing such a result would be a rare exception. Two alternatives present themselves: (1) Composite several increments across the entire stream, or (2) assess the variability of the COC across the entire stream. These alternatives, however, place us in a two-dimensional observation module, described below. Along with h_1, heterogeneities h_2 and h_3 become issues for one-dimensional lots.

16.3.3.3 Two-Dimensional Lots

All three heterogeneities (h_1, h_2, h_3) play a role in two-dimensional lots. With increasing dimension, delimitation also becomes more difficult. Correct delimitation for two-dimensional lots requires that the isotropic module of observation be a cylinder of constant cross section (Fig. 8-15). This cylinder must be positioned at random within the entire mass and must be representative of the entire thickness of the unit to be sampled. Note that, although it is always possible to drill a two-dimensional unit correctly, many commonly available and commonly used sample devices are not correct (thiefs, augers, and triers).

Delimitation error *DE* is always present in the common technique of grab sampling. Thus, grab sampling is always affected by bias. This is not to say that grab sampling is always inappropriate. For characterization of large-scale trends, grab sampling may be adequate. But it is not a defensible method for attempting to quantify COC concentrations around an action limit threshold or to estimate an average value, as greater degrees of accuracy and precision are required in these situations.

16.3.3.4 Three-Dimensional Lots

The three-dimensional isotropic module of observation that ensures equiprobability is a sphere randomly placed within the entire mass of the sampling unit. For practical reasons, this is impossible to implement, as is the semi-isotropic module of a cube. To achieve correct sampling in three-dimensional situations, it is necessary to consider the unit as a series of layered two-dimensional units. It is always possible to do this and always possible to achieve correctness in two-dimensional situations.

16.4 Characterize the Appropriate Heterogeneities

The most important structural property of a COC is its heterogeneity *h*. Ironically, characterization of this parameter is rarely addressed. Fortunately, this lack of attention is being addressed by EPA and others (USEPA 1992, Pitard 1993).

16.4.1 *Zero-Dimensional Wastes*

Because there is no such thing as large-scale heterogeneity for zero-dimensional wastes, the heterogeneity h of the waste collapses to the simple formula:

$$h = h_1 \tag{16.5}$$

which can be characterized by calculating the constant factor of constitution heterogeneity IH_L.

Given the knowledge of IH_L, an appropriate increment weight, sample weight, and subsample weight may be calculated (Fig. 8-2), along with the optimal number of increments per sample. These calculations are based on the predicted variance of FE and the maximum variance of GE, which is itself related to FE by the following equality:

$$\sigma^2 GE_{\text{MAX}} = \sigma^2 FE \tag{16.6}$$

with

$$\sigma^2 GE \leq \sigma^2 FE \tag{16.7}$$

necessarily following since $\sigma^2 GE$ is always less than or equal to the variance of the fundamental error $\sigma^2 FE$.

16.4.1.1 *Determining Sample Weight Sensitivity*

Because of the cost and time constraints to which most environmental sampling is subject, Pitard (1993) recommends a cost-effective but legitimate method for characterizing IH_L and the associated error variance $\sigma^2 CE_1$, called a *10X experiment*.

As part of the process, two sample collection steps are undertaken, and the results are compared. The idea is to collect a number of random samples N_1 (1, 2,..., n) within a zero-dimensional lot, where the samples are composed of a single increment and are of equal weight M_{S1}, which is convenient for the analytical method of choice. The entire mass of all samples is then analyzed. Note that N_1 should be 10 or more samples.

In the second step, an equal number of samples N_2 should be collected from the lot at the same locations. The weight of each sample M_{S2} must be at least 10 times that of M_{S1}. For even better results, M_{S2} should be at least 100 times M_{S1}, which may not be possible for practical reasons. Again, samples of size M_{S2} must be composed of a single increment and analyzed in their entirety.

Using the following definitions:

M_{S1} = weight of samples of size $S1$
M_{S2} = weight of samples of size $S2$
$s^2{}_1$ = variance of the estimates from all samples of weight M_{S1}
$s^2{}_2$ = variance of the estimates from all samples of weight M_{S2}
$a^2{}_1$ = average impurity of interest content (COC) of all samples of weight M_{S1}
$a^2{}_2$ = average impurity of interest content (COC) of all samples of weight M_{S2}

the following relation holds:

$$0.5 \leq \left[\frac{\left(\dfrac{\sigma_1^2}{a_1^2}\right) M_{S1}}{\left(\dfrac{\sigma_2^2}{a_2^2}\right) M_{S2}} \right] \leq 2 \tag{16.8}$$

If the data fail to satisfy this inequality, a problem exists. A likely explanation is that the samples of mass M_{S1} are too small, possibly by more than one order of magnitude. The purpose of this type of experiment is to (1) determine an appropriate sample weight and (2) select a sampling protocol.

Before a sampling program begins, one has no idea as to the magnitude of the heterogeneities, even if the same COC has been evaluated at other sites. Possible sample weights range from 1 g or less to 1 ton or more. Even though a 1-ton sample will definitely cause consternation and, for practical reasons, be rejected, it is still useful to understand that a mass of this order is required. This fact can be recorded in the DQO documentation as a key issue and DQO constraints can be relaxed.

16.4.1.2 Determining Optimal Sample Weight

Fundamental Error Approach In Chapter 8, the constant factor of heterogeneity IH_L was shown to be

$$IH_L = clfgd^3 \tag{16.9}$$

and the variance of the fundamental error *FE* to be

$$\sigma_{FE}^2 = \frac{IH_L}{M_s} \qquad (16.10)$$

when the mass of the lot M_L is much greater than the mass of the sample M_S.

Using the sampling nomograph (Fig. 8-2), the sampling weight necessary to maintain an *FE* variance below a DQO specified level at each sampling and subsampling stage can be determined. Note that DQO error constraints may need to be relaxed if the sample weight for the desired *FE* variance is unmanageable.

Once the variance of *FE* is known, the grouping and segregation error *GE* can be addressed and minimized. This is done by including as many correctly delimited and extracted increments as possible as part of the sample. Keeping equation 16.10 in mind, each sampling substage should reflect the relationship:

$$S_{\text{TOTAL}}^2 = S_{FE}^2 + \frac{S_{GE}^2}{N} \qquad (16.11)$$

where *N* is the number of increments per sample.

Example

The optimum sample weight that restricts the standard deviation s_{FE} below 15% has been calculated to be 550 g. The DQO team leader wants to know the maximum standard deviation of the total sampling error resulting from heterogeneities if only a single 550 g increment is collected.

$$S_{\text{TOTAL MAX}}^2 = S_{FE}^2 + S_{GE\,\text{MAX}} \qquad (16.12)$$

$$S_{\text{TOTAL MAX}}^2 = (0.15)^2 + (0.15)^2 = 0.0450 \qquad (16.13)$$

$$S_{\text{TOTAL MAX}}^2 = \pm 21.2\% \qquad (16.14)$$

To what new level would the value of $s_{\text{TOTAL MAX}}$ decrease if the sample were composed of 16 equal random increments collected at very small intervals?

$$S^2_{\text{TOTAL MAX}} = (0.15)^2 + \frac{(0.15)^2}{16} = 0.0239 \tag{16.15}$$

$$s_{\text{TOTAL}} = \pm 15.5\% \tag{16.16}$$

Visman Constant Approach The procedure described in Section 16.4.1.1, which compares two sets of samples of different weights, can be expanded to include the calculation of an optimal sample weight. The concept of an optimal sample weight has been previously introduced by Ingamells and Switzer (1973), Ingamells (1974), Ingamells and Pitard (1986), and Pitard (1993).

Two sampling constants, A and B, were introduced by Visman (1969). The homogeneity constant is represented by A. The segregation constant is represented by B, which also accounts for the variance components for segregation, grouping, long-range fluctuations, and periodic fluctuations. Using some of the same parameters as those used for equation 16.8, the constants A and B may be calculated as follows:

$$A = \frac{M_{S1} M_{S2} (S_1^2 - S_2^2)}{(M_{S2} - M_{S1})} \tag{16.17}$$

$$B = s_1^2 - \left(\frac{A}{M_{S1}} \right) = s_2^2 - \left(\frac{A}{M_{S2}} \right) \tag{16.18}$$

The optimum sample weight can then be calculated as the ratio of A over B:

$$M_{s\,\text{OPT}} = \frac{A}{B} \tag{16.19}$$

As a general rule for environmental sampling, EPA recommends (USEPA 1992) that the sample weight should not be less that six times the minimum sample weight $M_{s\,\text{MIN}}$. The minimum sample weight is calculated:

$$M_{s\,\text{MIN}} = \frac{A}{(a_1 - a_{\text{BG}})^2} \tag{16.20}$$

where a_{BG} is the background concentration.

16.4.2 One-Dimensional Wastes

As with zero-dimensional wastes, calculation of h_1 and its variance $\sigma^2 CE_1$ provides useful information. In addition, it is necessary to calculate h_2 and h_3 along with their associated variances $\sigma^2 CE_2$ and $\sigma^2 CE_3$ through variogram analysis. The variogram can characterize one-dimensional phenomena (through space or time), according to their nature. Sources of COC fluctuations can be identified, allowing decisions to be made regarding increment weight, sample weight, subsample weight, optimum number of increments, spacing between increments, and sampling mode (Chaps. 7 and 8).

16.4.3 Two- or Three-Dimensional Wastes

The difficulty in obtaining sampling correctness grows significantly as one moves beyond zero and one dimensional wastes. It is always possible to characterize two-dimensional wastes correctly, but delimitation and extraction problems become less manageable. These problems only worsen for three-dimensional situations and, in fact, must be solved by two-dimensional approximation.

For two- and three-dimensional problems, the one-dimensional variogram analysis described above can be generalized into two and three dimensions. What can then be determined is whether the nature of the problem lies in (1) large-scale trends or cycles or (2) small-scale composition and distribution heterogeneities. Large-scale variations generalize heterogeneities h_2 and h_3 seen in one-dimensional wastes, whereas h_1 is characterized by zero- and one-dimensional wastes.

If h_1 is the problem, variography may be pointless because a pure nugget effect structure can result. On the other hand, if h_1 is small and a spatial component is present at the site, h_2 and h_3 may be of primary importance. The geostatistician can determine to what extent the nugget effect C_o is caused by h_1 versus regional short-range variability.

The focus of the sampling effort should be directed at whichever heterogeneities are contributing most to the error components. Pitard (1993) recommends a rapid testing procedure to distinguish the appropriate heterogeneity.

If the waste is three-dimensional, the first step is to stratify the area into a series of two-dimensional layered wastes to be analyzed separately. Within each of the arbitrarily layered strata (typically convenient to a drilling or sampling depth), the 10X experiment detailed in Section 16.5.1 should be performed using the selected sampling mode. Sampling modes (simple random, systematic random, or stratified) will be described in detail in Section 16.7. After the variances of test results are compared, one of four outcomes generally arises. These outcomes will now be described.

16.4.3.1 Outcome 1: $s^2{}_1 = s^2{}_2$ = Small Variance

This outcome is a welcome one but is unfortunately infrequent. A small variance for both sample weights indicates a homogeneous material on both large and small scales. In this case, the weight of the small samples is sufficient.

16.4.3.2 Outcome 2: $s^2{}_1 = s^2{}_2$ = Large Variance

This outcome implies that large-scale trends or cycles are the predominant variances present. This situation also indicates that the sample weight is not playing a significant role. Assuming a correct sampling plan and implementation, geostatistical methods may be used to optimize the sampling design.

16.4.3.3 Outcome 3: $s^2{}_1 \; s^2{}_2$

In this case, a general outcome of the volume-variance relationship is demonstrated: Small samples have a greater variance than large samples. This provides evidence that the sample weight is an important factor in variability. It is also the most common case. Pitard (1993) recommends estimating an optimum sample weight $M_{s\,\mathrm{OPT}}$, then sampling the waste using samples only of weight $M_S \geq M_{s\,\mathrm{OPT}}$. This may result in one of three situations:

1. $M_{s\,\mathrm{OPT}} < M_{S1}$: All samples from both series (with weights M_{S1} and M_{S2}) are valid.
2. $M_{S1} < M_{s\,\mathrm{OPT}} < M_{S2}$: The large samples from series M_{S2} are valid, but the samples from M_{S1} are not.
3. $M_{s\,\mathrm{OPT}} > M_{S2}$: This situation invalidates both sampling series (M_{S1} and M_{S2}). The solution is to repeat the sampling effort using samples where $M_S = M_{s\,\mathrm{OPT}}$.

16.4.3.4 Outcome 4: $s^2{}_2 \; s^2{}_1$

Despite the existence of a volume-variance relationship, this situation does occur and may be the result of factors other than analytical errors. It suggests the existence of low-frequency pollution clusters that cannot be reliably detected by small samples. This type of dangerous situation was discussed in Chapter 6 (Poisson distribution, Figs. 6-17 to 6-19) where the COC may be present at high levels, yet most sample results indicate low concentrations.

In this case, the samples may be too small by several orders of magnitude. Pitard (1993) recommends that the test be repeated using a series of larger samples where the weight $M_{S3} \geq 10\, M_{S2}$.

16.4.4 Summary: Heterogeneity Characterization

Equations 16.2 and 16.3 vividly point out that, while important, the analytical error *AE* constitutes only a portion of the overall error *OE* for sampling. In the words of Francis Pitard, "Many are those who are tempted to test an incorrect sampling system for its accuracy." This is the same principal stressed by TQM and QBD, where it is known that quality cannot be inspected in. A sampling plan must invoke a high-quality strategic design, implement it correctly, and analyze the results in the appropriate context.

16.5 Develop a Sampling Plan

Once the appropriate models and heterogeneities are understood, it is possible to proceed with the development of a sampling plan. This plan must, of course, honor the site DQOs and statistical requirements. Some of the tools necessary for developing a defensible sampling plan remain to be discussed. These tools will be presented in the remainder of this chapter and in Chapter 17.

16.6 Implement the Plan Correctly

The first few steps in the development of a standardized sampling strategy have focused on the characterization of heterogeneities and the selection of observation modules and sample weights. An important part of what remains is the actual extraction of the sample, along with the strategy for determining where it should be collected.

16.6.1 Materialization of the Sample

Correct implementation of a sample plan is crucial to its success and presents a significant challenge. Without correct implementation, significant biases are introduced to the sample in the form of delimitation error *DE* and extraction error *EE*. Environmental sampling is especially subject to these errors as little work has been done in the area of designing correct sampling devices. Among the correct devices, include the Coliwasa tube (Fig. 8-33) and the Kemmerer depth sampler (Fig. 8-34), which offer both correct delimitation and extraction.

Many devices remain in common use that neither delimit nor extract correctly. Among these are dippers, shovels, scoops, triers, thiefs, and augers. That these devices are approved by standards committees does not negate their incorrectness. Examples of correct and incorrect sampling devices can be seen in Chapter 8.

A correct sampling tool necessarily implies that the correct module of observation has been incorporated. Requirements vary significantly for zero-, one-, two-, and three-dimensional modules. These details are discussed in Section 16.4 and in Chapter 8.

16.6.2 Sample Location Selection and Sample Selection Modes

The sample, to this point in the chapter, has been carefully designed with regard to its initial weight, subweights, and number of increments so as to meet allowable DQO constraints and maintain equiprobability. What remains to be done is actually to locate the sample for spatial situations. Random sampling is generally recommended as a way by which valid sampling may be accomplished. Randomness, unfortunately, is only a necessary condition, not a sufficient one. Spatial factors regarding equiprobability must also be carefully considered.

For situations involving spatial sampling in the environment, it is only necessary to discuss sampling with regard to the two-dimensional module of observation for the following reasons:

- Zero-dimensional areas or lots have no spatial component.
- One-dimensional areas or lots lack proper sampling devices for environmental sampling; therefore, one-dimensional areas must be approximated by two-dimensional techniques, as discussed in Section 16.3.3.2.
- Three-dimensional areas or lots cannot be correctly delimitated or extracted and must be reduced to two-dimensional problems.

Three basic sample selection modes are available for spatial data. These are strict random sampling, systematic sampling, and stratified random sampling.

16.6.2.1 Strict Random Sampling

Strict random sampling works well in classical statistical situations, such as selecting black balls and white balls from the proverbial urn. In spatial data, this form of sampling is the least preferred and may be subject to biases (Matheron 1971). It has been demonstrated that:

- For situations in which no cyclic phenomena are present, systematic sampling is better than stratified random sampling, which is itself better than strict random sampling.

- For situations in which cyclic phenomena are present, stratified random sampling is better than strict random sampling, which is itself better than systematic sampling.

The conclusion to be derived is simple: Strict random sampling never provides the best option.

16.6.2.2 Systematic Sampling

Systematic sampling is defined as sampling that takes place at regularly spaced intervals in an organized way. Examples of systematic grids include square, rectangular, triangular, and radial grids. Examples of these grids are shown in Figures 16-1 through 16-3. Figure 16-4 shows six different systematic sampling patterns, each consisting of approximately 65 points within an equally sized area.

For square and rectangular grids, two categories exist:

1. Nonrandom origin: When the strata are composed of a systematic grid throughout the area of interest, sampling is referred to as simple systematic when the origin is selected in a nonrandom way. As shown in Figure 16-5a, the samples are taken in the center of each stratum.

   By using a nonrandom origin, the opportunity for human and other biases is increased. For example, one may wish to obtain the greatest coverage from a particular sample grid (most samples), position the grid to miss inappropriate physical structures (buildings, lakes, etc.), or obtain "good" coverage for hotspots or clean spots. All these objectives introduce a bias of unknown magnitude.
2. Random origin: When the strata are composed of a systematic grid throughout the area of interest, sampling is referred to as a random systematic mode when a random location selection is performed in one

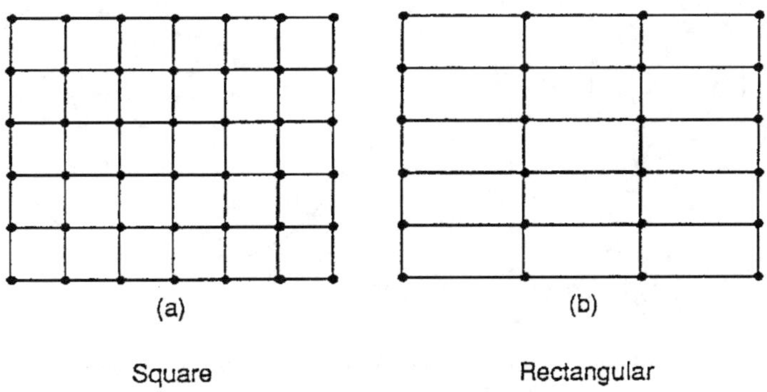

Figure 16-1. Systematic sampling: (a) Square and (b) rectangular grids.

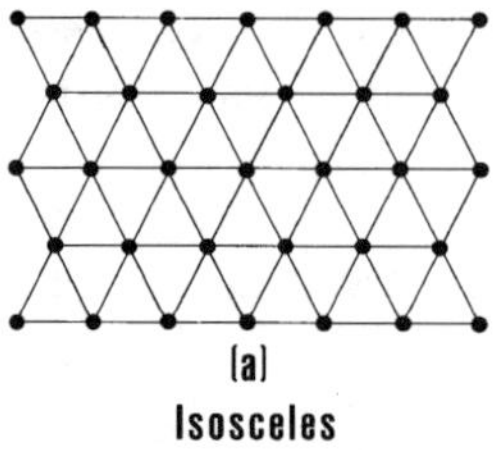

Figure 16-2. Systematic sampling: triangular grids. (a) Isosceles triangles and (b) equilateral triangles.

stratum. As shown in Figure 16-5b, the location of this random pick determines the location of all other locations throughout the strata.

It is often said that representative sampling is accomplished by introducing some form of randomness. Unfortunately, it is not that easy. Despite the random selection of the origin, the random systematic approach systematically excludes some areas as candidates for sampling, thus breaching the necessary component of equiprobability. The bias introduced is random, not human, but still of unknown magnitude.

The random systematic mode represents a small improvement over the simple systematic mode, but it is still biased. The simple systematic approach should not be used for any situation. As will be shown in Section 17.2.5 on indicator kriging, a square grid introduces a systematic bias in the interpolation process, prompting the development of probability kriging to reduce the influence of the bias.

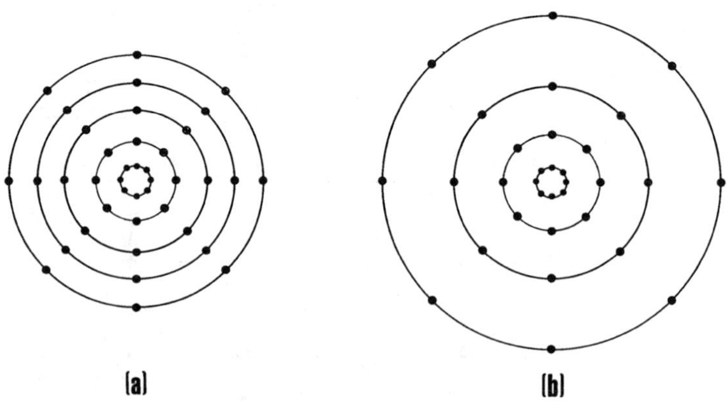

Figure 16-3. Systematic sampling: Examples of radial grids. (a) A pattern where the radius increases at a constant rate and (b) a pattern where the distance increases with each successive ring.

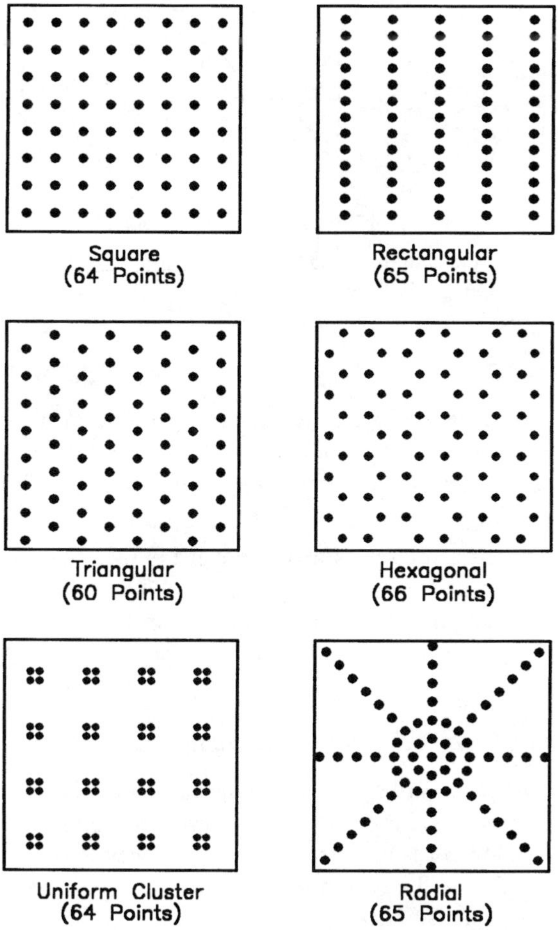

Figure 16-4. Illustration of six different sampling patterns that can be developed with approximately 65 sample locations. (From C. Kufs, unpublished course notes, 1988, with permission.)

16.6.2.3 Stratified Random Sampling

Stratified random sampling is defined as sampling that is random within defined strata. This approach maintains the necessary condition of equiprobability during a sampling event. Two categories of stratified random sampling exist:

1. When the strata are composed of a systematic grid throughout the area of interest, sampling is referred to as a *simple random stratified mode* when it is performed at random in each stratum.

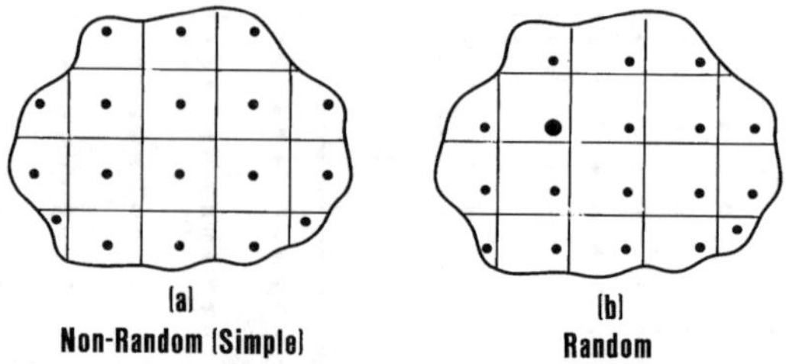

Figure 16-5. Systematic sampling modes. A non-random (simple) approach is shown in (a) and a random approach is shown in (b). (From F.F. Pitard, personal correspondence, 1991. Courtesy of Francis Pitard.)

2. When the strata are determined authoritatively in an effort to respect obvious domain or EU boundaries, sampling is referred to as *authoritative stratified random* when it is performed in the center of each stratum.

Stratified random sampling is a significant improvement on the systematic approach and is appropriate for many situations. Figures 16-6a and 16-6b show conceptual representations of systematic and authoritative strata definitions, respectively.

16.6.2.4 Definitions for Sampling Approaches

Before we proceed further, it is appropriate to define some of the terms introduced during the previous discussion on sample selection mode. Figures 16-5 and 16-6 took some liberties in order to show the general concepts. To minimize confusion, this section will develop specific graphic representations for each of the necessary subdivisions.

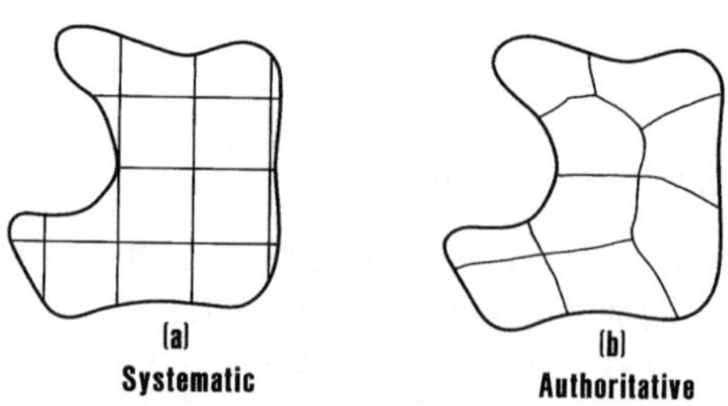

Figure 16-6. Stratified random sampling modes: (a) A systematic approach and (b) an authoritative approach. (From F.F. Pitard, personal correspondence, 1991. Courtesy of Francis Pitard.)

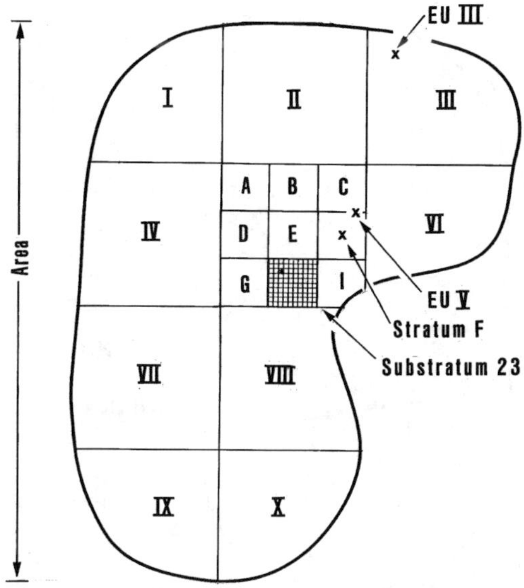

Figure 16-7. Area definition along with systematic EU definition, systematic strata definition, and systematic substratum definition. (From F.F. Pitard, personal correspondence, 1991. Courtesy of Francis Pitard.)

Area: This is the entire initial lot, typically the entire site. An example of an Area is shown in Figure 16-7.

EU: This is one portion of the Area. Rules governing the size of the EU should be established during the DQO process to meet local needs. Neptune et al. (1990) and Ryti et al. (1991) discuss EUs as large as 1/2 acre (approximately 20,000 ft^2) although no upper limits have been defined.

Figure 16-7 shows systematic sector definition, composed of a systematic grid. This is not to be confused with sampling on a systematic grid since, at the sector definition stage, no samples have been taken nor any sample locations excluded. Figure 16-8 illustrates the alternative method, showing sectors defined authoritatively to respect site domain boundaries.

Stratum: A stratum is a subunit of an EU. Each EU is divided into subunits using either (a) a systematic grid or (b) authoritative boundaries. Examples of systematic subdivision of strata are seen in Figure 16-7 (strata A–I); authoritative strata subdivision is demonstrated in Figure 16-8 (strata A–G). One sample must be collected and analyzed within each stratum. The number of strata depends on the goals, objectives, health hazards, and error limits allowed. A minimum of nine strata is recommended.

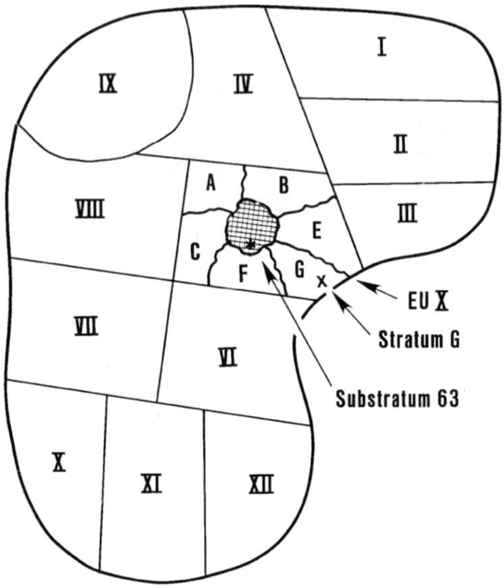

Figure 16-8. Area definition along with authoritative sector definition, authoritative strata definition, and substrata definition. (From F.F. Pitard, personal correspondence, 1991. Courtesy of Francis Pitard.)

> Substratum: Each stratum must be subdivided into smaller units so that either a random systematic or a simple stratified random mode can be implemented. The number of substrata necessary cannot be defined exactly. Pitard (1993) suggests that the number of substrata need not exceed the number of samples to be analyzed per EU. If equiprobability is to remain in effect until the sampling process is completed, many more substrata are required.

For the random systematic sampling mode, a single substratum is randomly selected from each stratum. The location of this selected substratum becomes the basis for sample location selection in all other strata. Examples are presented in Figure 16-9.

For the stratified random mode, a single substratum is selected randomly in each stratum. Which substratum is picked in a stratum does not influence which substratum is selected in the next or any other stratum. Examples of this mode can be seen in Figure 16-10.

16.6.2.5 Alternative Sampling Locations

While equiprobable and random selection for sample locations is statistically valid, the practical reality of obtaining access to the selected location may inter-

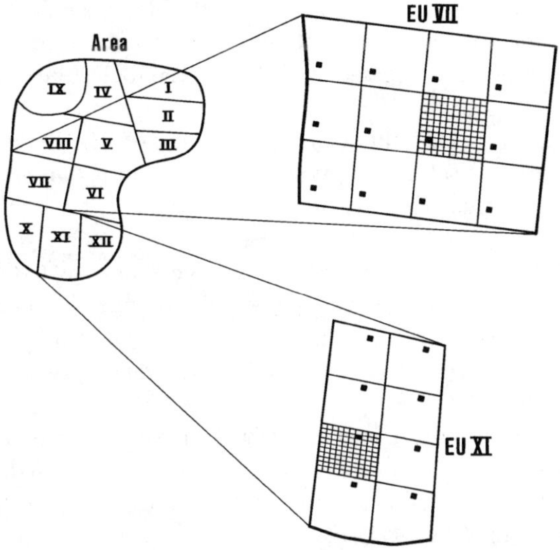

Figure 16-9. Substratum selection using a random systematic mode. (From F.F. Pitard, personal correspondence, 1991. Courtesy of Francis Pitard.)

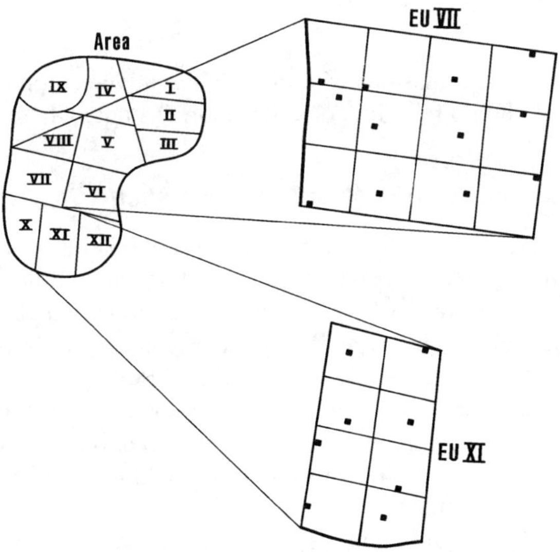

Figure 16-10. Substratum selection using a stratified random mode. (From F.F. Pitard, personal correspondence, 1991. Courtesy of Francis Pitard.)

fere. Obstructed access poses a potential for introducing a bias into the sampling program. Biases may be exacerbated if concentration and access are related: The highest concentrations are located under equipment that blocks access to the sample location.

A sampling protocol should decide on a policy of whether or not to relocate samples. If relocation is selected, a method for maintaining an unbiased approach is necessary. Some suggestions are offered:

- New substratum: During the planning stage, extra substrata can be selected within each stratum. This is a quick and effective procedure.
- Move a fixed distance: Starting from the original location, the sample will be located at a fixed distance in a direction randomly selected during the planning stage.
- Remain close to original point: A random systematic approach may specify a maximum fixed distance between sample points. In this case, samples should remain as near as possible to the original location. Note that several alternatives may be available with this relocation scheme and that they offer the potential for human bias. Random techniques should be employed for the final selection.
- Similarity to the original point: Historical usage, topography, soil type, and proximity to the original sample are some criteria for maintaining similarity. Random techniques should be used in situations in which multiple alternatives are available.
- Omit the sample: This may be necessary for practical or design considerations.

Note that all data omissions as well as alternative sampling locations and selection methods should be fully documented in QA and data reports.

16.6.2.6 Summary: Sampling Plan Implementation

Random systematic sampling offers a method of obtaining uniform coverage of an area or time period. However, its random element is limited to the random selection of the initial sampling point (with all other points based on this location) and the grid orientation. This type of approach is not equiprobabilistic and can lead to significant biases, especially when the objective is to characterize the average COC concentration and when relatively few samples may be taken. The effect of the bias is often mitigated by taking large numbers of samples, offering better coverage. Random systematic sampling should be avoided whenever contouring of data is anticipated.

Stratified random sampling is the safest form of sampling for environmental assessment, especially when small numbers of samples are planned. Most importantly, it is equiprobable, meaning that every part of each stratum in each sector has an equal chance of being selected. The location of one sample location does not influence the location of any other.

16.6.3 Compositing

16.6.3.1 General Definition

The process of compositing involves physically combining and mixing several volumes of material in an effort to form a single sample. This has traditionally been defined as combining several samples to create one composite sample. STP expands this definition to include the combination of several increments in order to make a sample. The distinction is important.

16.6.3.2 Increment Recombination

When several increments are combined to make a sample, the objective should be to minimize the small-scale segregation at a sample location. This will minimize heterogeneity h_1 as well as the grouping and segregation error *GE*. This type of compositing addresses local issues. Geostatistically, we are reducing the sample variance by increasing sample support. Because we cannot practically extract the larger volume, we can consider it a lot and subsample it with multiple increments.

16.6.3.3 Sample Recombination

When material from several sampling locations is combined, a single new sample is formed. This is the traditional definition of compositing. The objective of this approach is to smooth the effects of large-scale segregation resulting from heterogeneities h_2 and h_3. While effective as a money-saving device, it is also completely effective in wiping out any understanding of h_2 and h_3, the spatial variability components, leaving nothing for the geostatistician. This technique is useful for calculating the average COC concentration of an EU for risk-based DQOs, but it should be discouraged if spatial mapping is anticipated.

16.6.3.4 A Caveat for Decision Making

Often, several samples are composited and analyzed to determine the average concentration. If the average concentration is above a certain level, the existence of one or more hotspots is presumed. If the average falls below a specified level, we can presumably demonstrate that no hotspots exist. Note that the DQO decision rule has changed from that described in Section 16.6.3.3. Earlier, the objective was to establish average COC concentrations in EUs; now we are worried about hotspots.

The decision rule is commonly stated: If the average of the composite sample is less than the action limit threshold divided by the number of samples in the composite, the maximum COC concentration in the area is below the action limit.

This statement assumes that the COC is not a naturally occurring substance. In order to make a proper assessment, it is important first to establish the natural low background level (LBL). The decision rule should be modified to read as follows: If the average of the composite sample is less than the action limit threshold divided by the number of samples in the composite, plus the natural low background level (LBL), the maximum COC incremental contamination in the area is below the action limit.

Mathematically, this can be expressed as

$$\text{Average} < \frac{\text{action limit}}{N \text{ samples}} + LBL \tag{16.21}$$

Note that this formula does not apply to situations in which samples are composed of multiple increments collected at very short distances. It should be stressed that the need for, and the approach to, compositing is highly dependent on the DQOs, and any exercise in compositing should proceed with care.

16.6.4 Exposure Units, Sampling Selection Mode, and Compositing

For determining presence vs. absence, or average concentration, of a COC, the EU is often the smallest areal unit for decision making. In Chapter 15, EUs were described as a risk-based method of decision making for areal units, with typical EU sizes on the order of 5000 ft^2 to 1/2 acre.

Based on the results of the sampling effort, each of the EU areas may show an average COC concentration below the established risk threshold. Given that each EU is judged independently, this scenario would warrant no further action.

Figure 16-11 shows the danger in making such independent judgments. Two hotspots actually exist, each present in five separate EUs (i.e., 10 different EUs are impacted). Because the contamination is divided between several EUs, all with both clean and dirty zones, the clean zones act to dilute the concentration to below the action standards.

The theory of regionalized variables states that nearby areas are often alike. This can be true of any support unit: samples, strata, sectors, or EUs. Therefore, it is dangerous to make decisions on any support unit independently. This is one

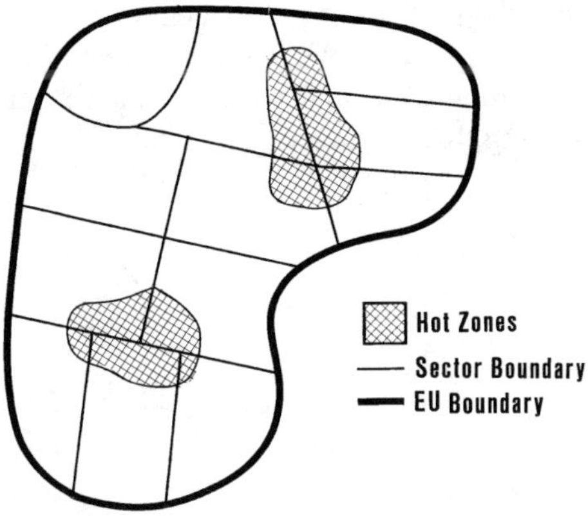

Figure 16-11. Undetected hotspots crossing EU boundaries because of a loss of spatial information during compositing

reason that compositing should proceed with caution: Important information regarding hotspots is lost that could be used later.

Figure 16-11 shows that the authoritative boundaries are not well set. This may be human error, or it may be due simply to chance. Systematic setting of sector boundaries could easily produce the same result. If spatial information is retained, cleanup decisions can be made based on the extent of the contamination with respect to the EU (Fig. 15-8).

Figure 16-11 is but one example of the problematic nature of environmental remediation and decision making in the face of uncertainty. However, DQOs are all about making tough choices and living with the uncertainty. If the defined objective is to obtain EU averages, we must accept the possibility of boundary hotspots. Otherwise, we must define the target hotspot and design the sampling plan accordingly.

Conclusion

The process of generating data that meets the specified DQOs begins with the definition of a sample: What dimensional module, what heterogeneities and subsequent errors apply? In addition, other key elements must be established to meet appropriate DQO requirements for any sampling strategy, including the quantification of sample support/optimum sample weight, the determination of correct delimitation and extraction procedures, and, finally, management of the error during subsampling protocols. Once a sample is defined, its placement within the site area of interest must follow strict guidelines to ensure that equiprobability conditions are maintained, no matter what the objective of the sampling may be. If economics allow, retention of spatial location and concentration information can be valuable in assuring that a maximum amount of the contaminated volume will be removed.

Chapter 17

Integrating DQOs and GA: Mapping and Appraisal

SYNOPSIS: In Chapter 16, DQOs and STP were used to develop sampling strategies and protocols for early phases of site characterization. When significant contamination is found at a site, later phases often require mapping of the contaminants of concern (COCs). DQOs for mapping and spatial variables require approaches that are consistent with the nature of the data. In this chapter, the techniques of geostatistical appraisal (GA) will be integrated with the DQO process to demonstrate some of the ways in which error, risk, and cost can be managed for spatial variables.

KEY WORDS: Ordinary Kriging, Polygonal Kriging, Decline Curve, Internal Grid Definition, Nonrectangular Area, Error Variance, Misclassification Ellipse, Kriging Error, Confidence Intervals, Type I Error, False Positive, Type II Error, False Negative, Action Line, Action Zone, Fat Contour, Cokriging, Indicator Kriging, Probability Kriging, Loss Function, Typecasting Error, Design Performance Goal Diagram, Symmetric Loss Function, Asymmetric Loss Function, Step Function, Asymmetric V Curve, Total Cost Curve, Cost of Errors, Probabilistic Targeting, Nomograph, Remedial Management Unit, Exposure Unit, Sampling Density, Sample Sufficiency, Phased Sampling, Conditional Simulation

17.1 Introduction

A remedial investigation (RI) usually gathers a sample data set sufficient in number to begin the process of geostatistical appraisal. With the initial data set, a variogram analysis can be performed to determine the spatial correlation structure present at the site for any COC. Given a variogram, kriging may then be implemented. The combination of variographic and kriging techniques can critically evaluate issues relating to data sufficiency and levels of statistical error. If error levels exceed the tolerances stipulated by the DQOs, these same tools can provide answers to how much additional sampling will be necessary to meet the goals and where to locate the samples.

The somewhat generic technique of ordinary kriging was examined in detail in Chapter 13. For specialized applications, many variations of the kriging system exist. The first section of this chapter will examine a few of the many kinds of kriging that are useful in the GEM process for environmental site characterization. In the second section, the assumptions and relationships between error and cost models will be discussed.

17.2 Kriging and Error Quantification for DQOs

17.2.1 Polygonal Kriging

17.2.1.1 Development of a Decline Curve

Whenever samples are collected from a site, two questions generally arise: (1) How many samples are necessary to reach a specified level of confidence? and (2) Where should these samples be located? If samples have already been taken during previous phases, the questions become: How many additional samples are needed? and Where should the new ones be put in relation to the existing ones?

A useful technique to answer these questions for RI and post-RI activities is polygonal kriging. In Chapter 13, it was shown that, when a block is estimated by kriging, an estimate of the error variance is also calculated. The examples given at the time suggested that blocks would always be rectangular and often on the order of 10 × 10 ft, 50 × 50 ft, 30 × 60 ft, or some other size that is an appropriate subportion of the site.

A "block" of particular interest may be the site itself. By kriging the site as a single block, an estimate of the global estimation variance on the mean of the

COC for the entire site can be obtained. This is a first step in answering question 1 posed above.

Intuition and the estimation variance both indicate that the confidence in the mean value grows as the number of samples taken at the site increases. This is also shown by the central limit theorem and the standard error of the mean for normally distributed data (Chap. 6). This concept can be displayed visually (Fig. 17-1) as a decline curve. Figure 17-1 plots the number of samples along the *X* axis and the error variance on the *Y* axis.

When the number of samples collected is low, the graph confirms that the error is high. As more samples are obtained, the error declines rapidly. As samples continue to be taken, the graph declines more slowly and asymptotically approaches the *X* axis. This demonstrates another intuitive notion: It is not appropriate to remediate by sampling!

Figure 17-2 shows the same decline curve, with additional decline curves positioned above the original. Each decline curve represents a specific sampling program geometry. Each of the new decline curves drops more slowly than the original. This indicates two things: (1) Not all sampling programs are created equal, and (2) the decline curve corresponding to the original sampling program is the most efficient.

Different decline curves can be the result of various factors. If an anisotropy is present, a square grid is a less efficient way to reduce error on the mean than a rectangular grid correctly oriented along the axis of anisotropy (Fig. 17-3). As shown in Chapter 12, clustering of data is an inefficient method for reducing the error variance. The family of decline curves shown in Figure 17-2 could also represent several COCs at a site, each with the same sampling geometry but with different variogram structures.

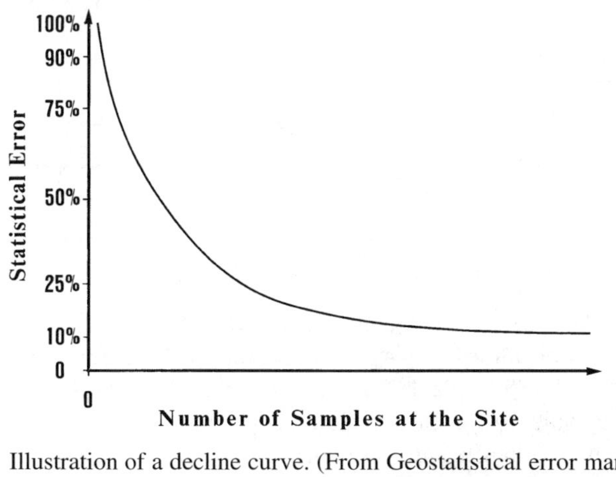

Figure 17-1. Illustration of a decline curve. (From Geostatistical error management for site assessment (GEM), by J.C. Myers, *Proceedings of HAZMAT West '91,* 1991.)

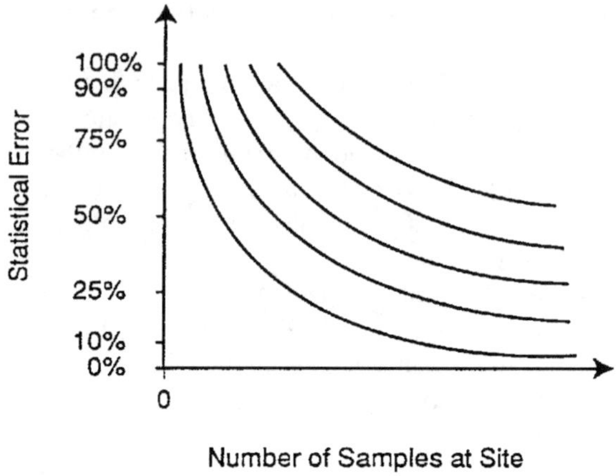

Figure 17-2. Multiple decline curves. (From Geostatistical error management for site assessment (GEM), by J.C. Myers, *Proceedings of HAZMAT West '91,* 1991.)

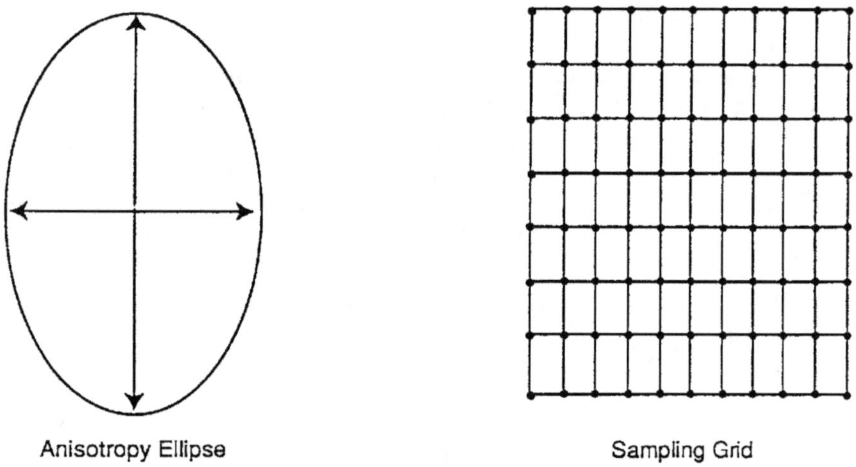

Figure 17-3. Illustration of a rectangular sampling grid reflecting spatial anisotropy.

17.2.1.2 Polygonal Boundary and Internal Grid Definition

It would be rare for a site boundary to be a perfect rectangle. Kriging programs, however, almost always assume that a perfect rectangle will be the object for estimation (Fig. 12-2). The reason for this is that the internal discretization grid can be quickly and easily calculated for rectangles. Nonrectangular areas (NRAs) present a more difficult problem.

To illustrate some of the practical considerations in polygonal kriging, Figure

17-4 shows an NRA representing an hypothetical site boundary. The kriging program was first requested to import a digitized file containing this site boundary. Note that not all kriging programs offer this function. The kriging program was then requested to insert a 10 × 10 internal grid. Because of the irregularities in the boundary, not all discretization points fall inside it. Discretization points outside the site boundary have been eliminated by the program. Only those falling inside will be used to calculate gamma bar.

Because of boundary irregularities, some areas may be underrepresented in the gamma bar calculation. It is recommended that a large number of internal grid points be used. Increasing the number of grid points increases calculation time but is offset by the fact that only one block is being kriged.

17.2.1.3 Interpretation of the Decline Curve

Assuming that data are available and a variogram has been calculated, the confidence on the mean for the site can be calculated by means of polygonal kriging. DQO goals to achieve a specified error level may be established before or after sampling begins. DQO goals may vary depending on the characterization phase or decision required: Less stringent goals may be acceptable in early phases than in final ones.

DQO goals may be set in either relative or absolute terms. For instance, the remedial project manager (RPM) may decide that a 90% reduction in the error variance is required for a particular site. At another site, the RPM may wish to

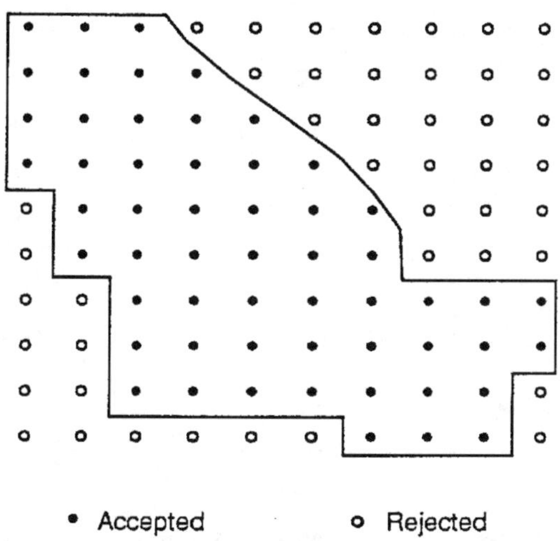

Figure 17-4. Hypothetical NRA site boundary with a truncated internal discretization grid.

reduce the error variance such that the corresponding 90% confidence interval on the mean does not exceed 50 ppm.

If existing sample data are insufficient to meet a DQO criterion, decline curve analysis and polygonal kriging can be used to determine the number and location of the additional sample required to meet compliance goals. Decline curves can be useful even if no additional sampling is planned. Figure 17-5 shows a decline curve from the Dallas Lead study. After the sampling was completed, questions arose as to how well the sampling had characterized the mean. The graph in Figure 17-5 shows that the 200 samples collected accomplished a sufficiently large reduction in the global estimation variance, with no additional samples needed.

17.2.1.4 Some Remarks on Error Reduction

In Chapter 12, it was shown that the location of samples, along with a variogram model, would allow kriging to determine the estimation variance. The same principle is applied in polygonal kriging. This property is especially powerful when the number and locations of additional samples are to be determined. As stated in the preceding section, it is not necessary to have knowledge of the sample concentrations at the proposed location; only the sample coordinates must be known. A review of the equation for the estimation variance reveals that nowhere are the sample concentrations used; only the distances between samples and from the samples to the block.

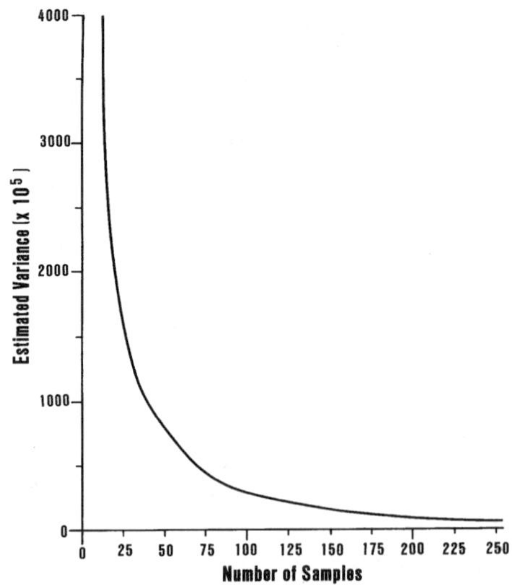

Figure 17-5. Decline curve from the Dallas Lead (Pb) study: DMC smelter. (From Geostatistical sampling and contamination analysis of lead values surrounding industrial smelters, by J.C. Myers, Master's thesis, Colorado School of Mines, 1985.)

The effect of additional sampling can also be seen on the misclassification ellipse. Figure 17-6a shows a "fat" ellipse where sampling is relatively sparse. As additional samples are taken, the width of the ellipse is reduced (Fig. 17-6b). More sample information allows for better estimation and fewer misclassifications.

17.2.1.5 A Remark on Sample Coordinates

The following situation is surprisingly and distressingly common: A geostatistician is asked what information is required as a basis for starting immediately on the GA. The geostatistician responds that, initially, only the sample IDs, sample values, and *XYZ* coordinates will be sufficient. This statement is often followed by a long pause, with those authorizing the GA looking nervously at each other. Finally, the bravest among them speaks up and asks sheepishly, "Coordinates?"

The "startling" fact is: Spatial data require coordinate locations. Use of spatial data techniques is relatively new in the environment, and not everyone is accustomed to obtaining the coordinates for samples during the collection. The moral of the story: Make sure all sample locations are surveyed.

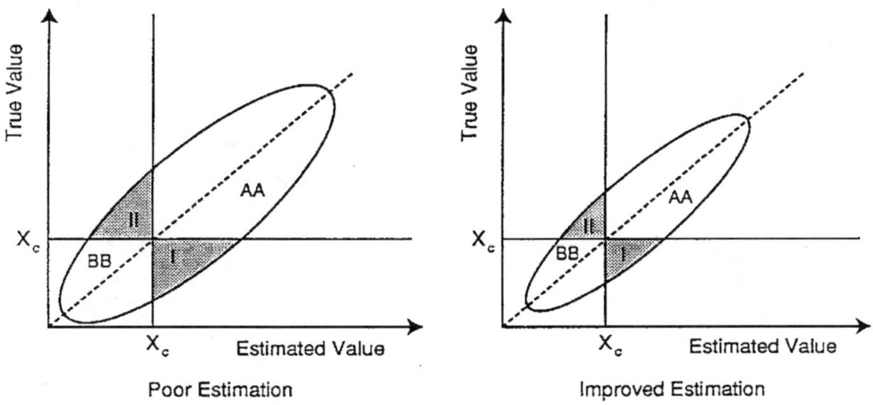

I = Type I Error = False Positive

II = Type II Error = False Negative

AA = Sample Above, Estimate Above

BB = Sample Below, Estimate Below

Figure 17-6. Reduction in the width of the misclassification ellipse resulting from additional sampling. (From *Optimization of Sampling Policies: A Geostatistical Approach*, by J-M. Rendu, MMIJ-AIME, 1980, with permission.)

17.2.2 Kriging Error Maps

In Chapter 13 it, was shown that the kriging process produces two results for each location: an estimated concentration and a kriging standard deviation. This is true for block model grids as well as single block or polygonal block estimation.

With most interpolators, only a concentration estimate is produced, with this estimate being transformed into a contour map. Whereas kriged concentrations can also be mapped, the implication of the kriging variance is that a second contour map can be produced. This "map of the second kind" contours the kriging variance for the area and shows the expected error for any location on the map.

Figure 17-7 is an example of a kriging error map from a portion of a dioxin

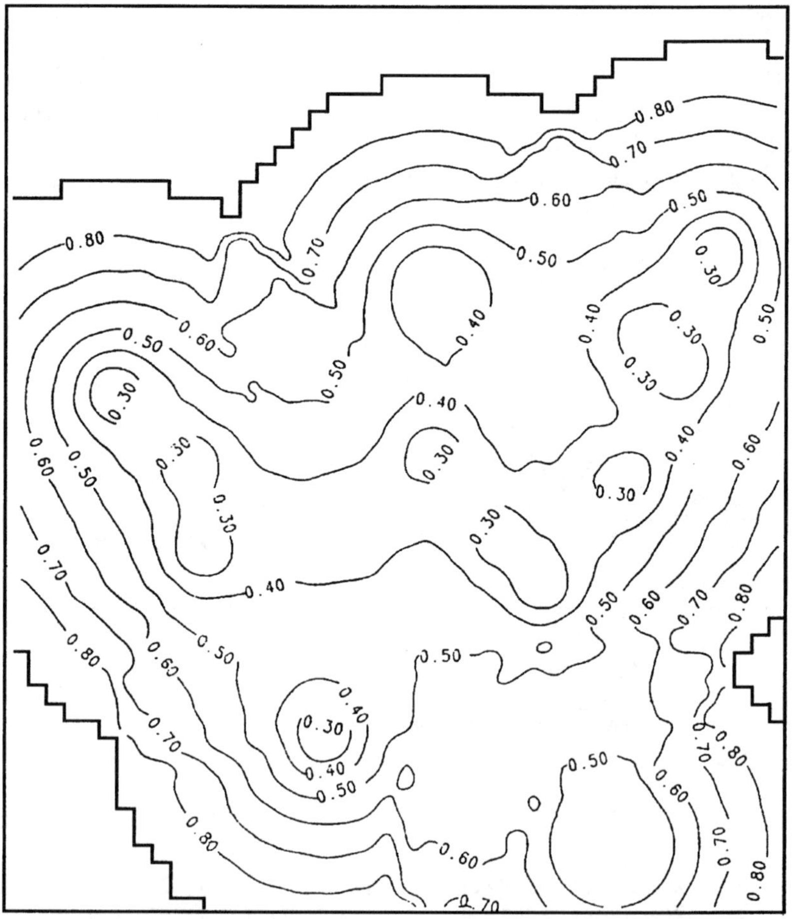

Figure 17-7. Kriging error contour map of a dioxin site.

site. A relative variogram has been used in kriging, and the resultant error values and contours show relative error. A contour of 0.50 indicates 50% error around the estimated concentration. As expected, the error is lowest near the sample data points and increases in unsampled areas. Outside the area of sampling, error rises rapidly to maximum values.

Figure 17-8 illustrates a commonly observed phenomenon in error mapping, in this case, sample data for lead (Pb) contamination in soil around the DMC smelter from the Dallas Lead study. A "swiss cheese" effect is noted, with error drawdown around each sample point on a semiregular grid. As in Figure 17-7, the error in Figure 17-8 increases rapidly outside the area of sampling. Interestingly, an area of relatively high error exists in the northeast portion of the map. This reflects a large unsampled area corresponding to the Trinity River and its flood plain.

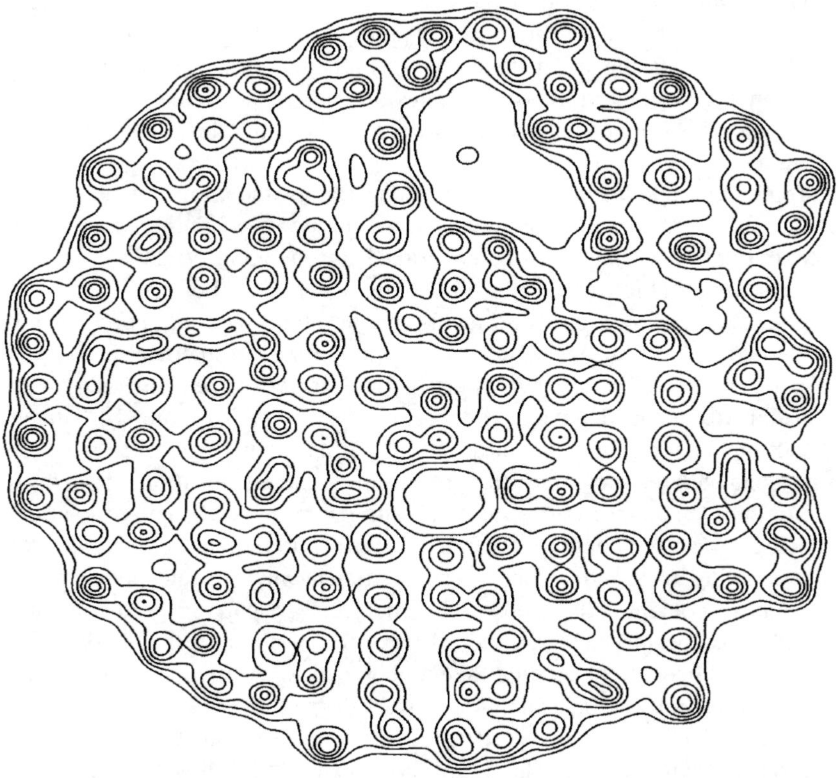

Figure 17-8. Kriging error map of the DMC smelter in Dallas showing a "swiss cheese" effect. (From Geostatistical sampling and contamination analysis of lead values surrounding industrial smelters, by J.C. Myers, Master's thesis, Colorado School of Mines, 1985.)

Error maps can be used to evaluate the levels of error occurring at specific locations at the site and at the site in general. Areas with high error values represent potential targets for additional sampling. If overall levels are too high, an increased sample density or tighter grid may be appropriate.

17.2.3 Confidence Intervals for Contour Lines

The functionality of both the kriging estimate grid and the kriging error grid may be merged to provide additional useful results for the design of sampling strategies. In environmental remediation, the contour of the COC concentration at the action level is of special significance. Areas enclosed by this line are often scheduled for remedial activities because of their unacceptable levels of associated risk.

With computers, this line can be sharply delineated and plotted. A crisp, color representation of the map on high-quality paper, delineating clean and dirty zones, is aesthetically pleasing, but it should not be immediately accepted as gospel. The accompanying kriging error map indicates that uncertainty exists at all locations of the map. This implies that the exact trace of the action level contour remains uncertain.

To quantify and visually display this uncertainty, the kriging standard deviations may be added and/or subtracted from the corresponding estimated COC concentrations at each point on the grid. This operation establishes a spatial confidence interval for the contour line at the selected level of confidence.

Figure 17-9 shows an action level contour, along with an upper and lower confidence level contour (± 1 standard deviation), for a hypothetical site. The crisp and precisely defined action level contour widens and becomes "fat" as it displays the spatial area included in the confidence interval around it. Figure 17-10 displays the 1000 ppm action line contour around the RSR smelter from the Dallas Lead study, along with the same line after one kriging standard deviation is added.

The area encompassed by the "fat" contour provides benefits to sampling strategy designs. Fat contours change an action line into an *action zone*. Additional sampling in the action zone offers an efficient way to locate more accurately the extent of contamination above the action level. Sampling in this manner is referred to as chasing the action line. Improved definition of the action line has the additional benefit of reducing the number of false positive and false negative blocks that result from the estimation process (Type I and II errors).

Figure 17-11 visualizes the efficiency of action zone sampling as it impacts the misclassification ellipse. The figure shows that, by sampling around the action line, the width of the misclassification ellipse is reduced around the action limit. This sampling approach squeezes the ellipse around the action threshold into an hourglass-like shape; *wasp-waisting,* as it is called, selectively minimizing the Type I and II errors. Note, however, that although a $+1\sigma$ action zone is efficient

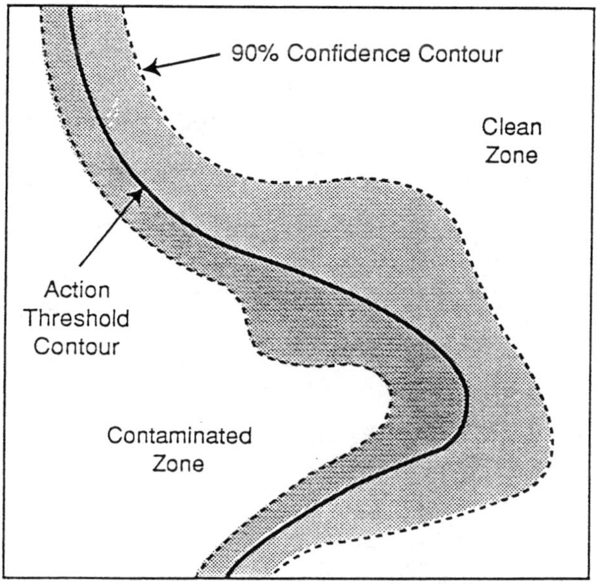

Figure 17-9. Example of a "fat" contour line from a hypothetical site. (From Geostatistical error management for site assessment (GEM), by J.C. Myers, *Proceedings of HAZMAT West '91,* 1991.)

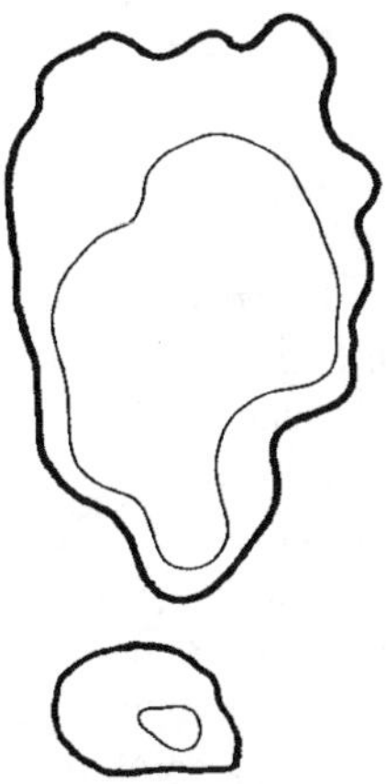

Figure 17-10. Action level contour (1000 ppm) and action level contour plus one standard deviation around the RSR smelter (Dallas Lead study). (From Geostatistical sampling and contamination analysis of lead values surrounding industrial smelters, by J.C. Myers, Master's thesis, Colorado School of Mines, 1985.)

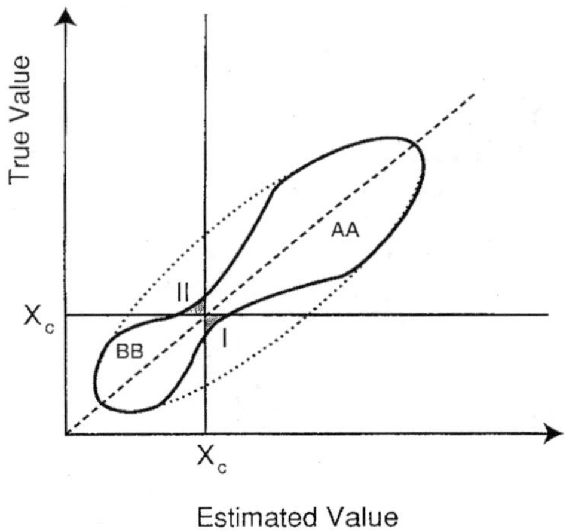

Figure 17-11. Impact of action zone sampling on the misclassification ellipse (wasp-waisting). (From Geostatistical error management for site assessment (GEM), by J.C. Myers, *Proceedings of HAZMAT West '91,* 1991.)

for sampling, it is not recommended as a remediation boundary. Remediating all material in the $+1\sigma$ region would be expected to greatly inflate the false positive rate.

17.2.4 Cokriging

In some instances, abundant data for one variable may be available while another variable may be significantly undersampled. Chapter 10 described a situation in which future soil pH was predicted based on the abundance of acid-forming materials in place. Sometimes, radiometric data are cheaply and abundantly available for radionuclides, whereas the more expensive wet assay results are sparse.

The precision of estimation can be improved by using the spatial correlations of both variables. In theory, cokriging is identical to ordinary kriging. From the practical and notational side, it is more complicated because more steps are involved and additional indexing is necessary to distinguish the two variables. Mathematical development of the cokriging system is given by Journel and Huijbregts (1978) and Myers (1982).

The process of cokriging involves two steps. First, three sets of variograms are developed, representing the primary variable (usually undersampled), the secondary variable, and the cross-correlations between the primary and secondary variable. Calculation of variograms and cross-variograms was described in Chapter 10.

The second step, while more complicated, follows the same procedure as that for ordinary kriging. A system of equations is developed and the error variance is minimized, subject to unbiasedness constraints. The result is an estimated value and an estimation variance.

17.2.5 Indicator Approaches to Kriging

17.2.5.1 Indicator Kriging

Chapter 5 introduced a nonparametric, distribution-free technique known as indicators. In this approach, a cutoff is selected and data are transformed into corresponding zero and one values, depending on their concentrations relative to the action limit. In Chapter 10, it was explained that these indicator-transformed data could be analyzed for their spatial correlation properties with the variogram. Indicator data may also be used in the kriging process as they offer some unique advantages. Note that Poisson-particle effects, described in Chapter 6, are assumed to be negligible.

As previously discussed, environmental data often span several orders of magnitude. To complicate the situation, these concentration transitions may take place over a very small distance either vertically or horizontally. Examples of rapid concentration changes for a PCB site are shown in Figure 17-12. Figure 17-12a shows the PCB concentration from 36,000 to 0.05 ppm over a horizontal distance of 35 ft; Figure 17-12b shows PCB values changing from 1920 to 0.6 ppm in 20 ft. Vertically, concentration changes at this site include 11,1000 to 1.05 ppm and 8600 to 1.00 ppm over a 1 ft interval.

Extreme concentration values create problems in estimation, especially when relatively low action limits have been established. For example, a concentration of 36,000 ppm, given a minuscule kriging weight of 0.001, will itself contribute an estimated concentration of 36 ppm, which is above the action limit at this site (25 ppm). Even if numerous other sample data are used and all other data have concentrations of 0.0 ppm, the estimated block value will be 36 ppm. In other words, this single sample can create block estimates above the action level in areas where all other samples are below the action level. While difficult to estimate, such a situation may reveal a larger problem: multiple but small and highly concentrated sources that are difficult to find and estimate.

Steep gradients also create problems in mapping. With a block size of 10 × 10 ft, approximately three blocks will be estimated between the locations exhibiting 36,000 and 0.05 ppm. At virtually any useful contour interval and scale, the contour lines will be so close together as to be unreadable. For block estimates,

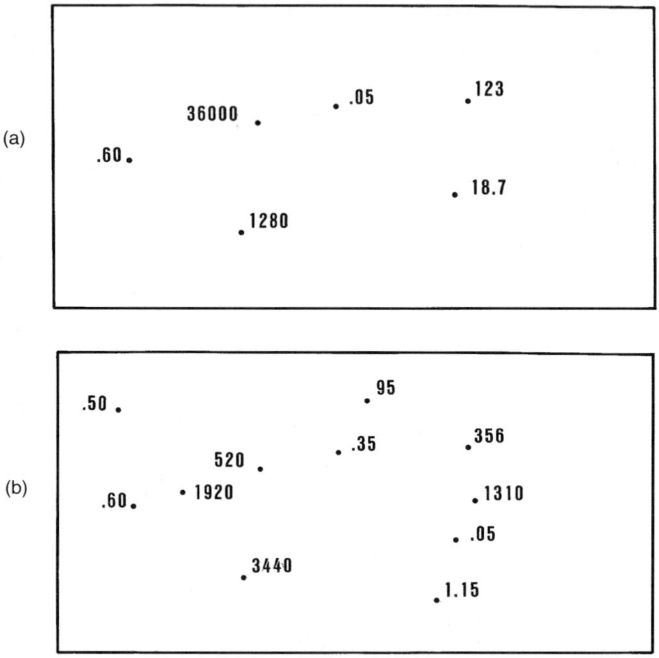

Figure 17-12. Examples of extreme data ranges and steep concentration gradients for environmental data (PCBs in soil). (From Geostatistical error management for site assessment (GEM), by J.C. Myers, *Proceedings of HAZMAT West '91,* 1991.)

the extreme value will most likely condemn blocks beyond the "clean" sample and well into the "clean" zone.

The use of indicator data offers a convenient solution to the problem of extreme concentrations and steep gradients. All indicator data are between zero and one and, correspondingly, the estimates are, as well. This transformation dramatically lessens the influence of extreme concentrations and reduces gradients between sample locations without resorting to logarithms or other methods requiring back transformations. The indicator transform is given by

$$i_j = \begin{pmatrix} 0 \text{ if } k_j \leq x_c \\ 1 \text{ if } x_j > x_c \end{pmatrix} \tag{17.1}$$

where x_c is the action limit concentration. In the transform, all samples below the action limit are given a value of zero, and all values above the action limit are set to one.

By performing the indicator transform, the resulting contour or block map must be interpreted in a different fashion than that for estimated concentrations. The indicator map displays the probability that a block or area exceeds the action limit. An estimated block value of .37 indicates that the probability that the PCB concentration in the block exceeds the action limit of 25 ppm is .37, or a 37% chance. Similarly, a contour line labeled as .90 encloses an area where PCB concentrations are estimated to have a .90 probability, or a 90% chance of exceeding the action limit. Figure 17-13 shows a block map of indicator-estimated probabilities. Figure 17-14 displays a contour map of indicator-estimated probabilities.

The technique of chasing the action line described in Section 17.4 can also,

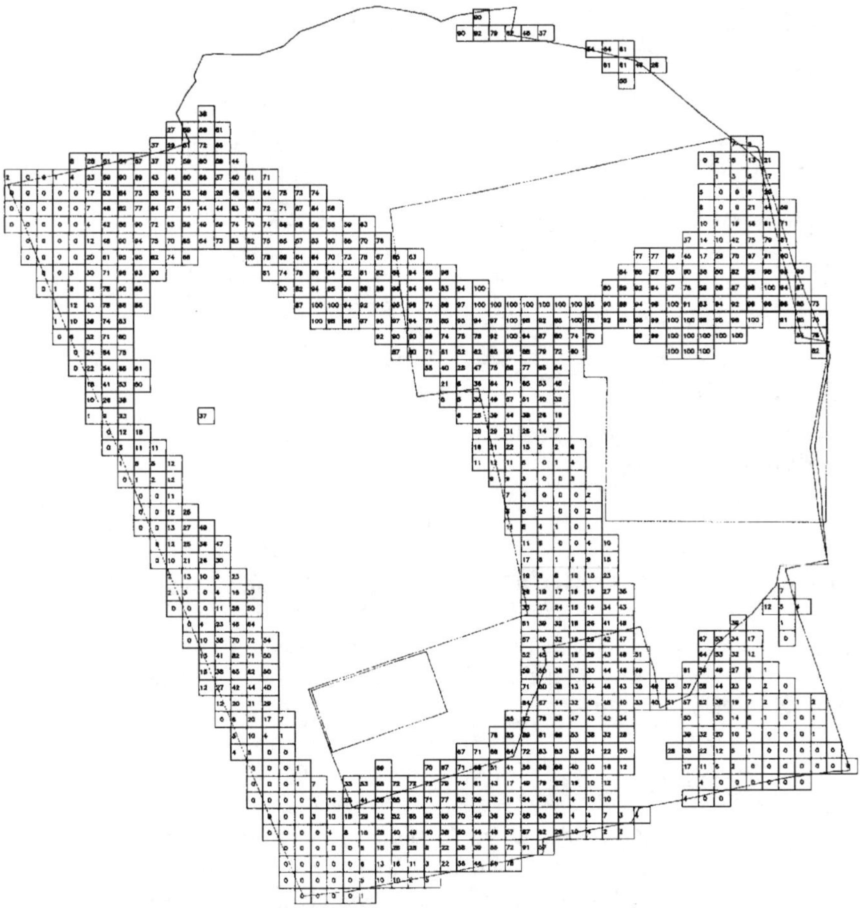

Figure 17-13. Block map of indicator-estimated probability values. (From A geostatistical analysis of PCB concentrations in soils at the ReSolve site, by J.C. Myers, 1992.)

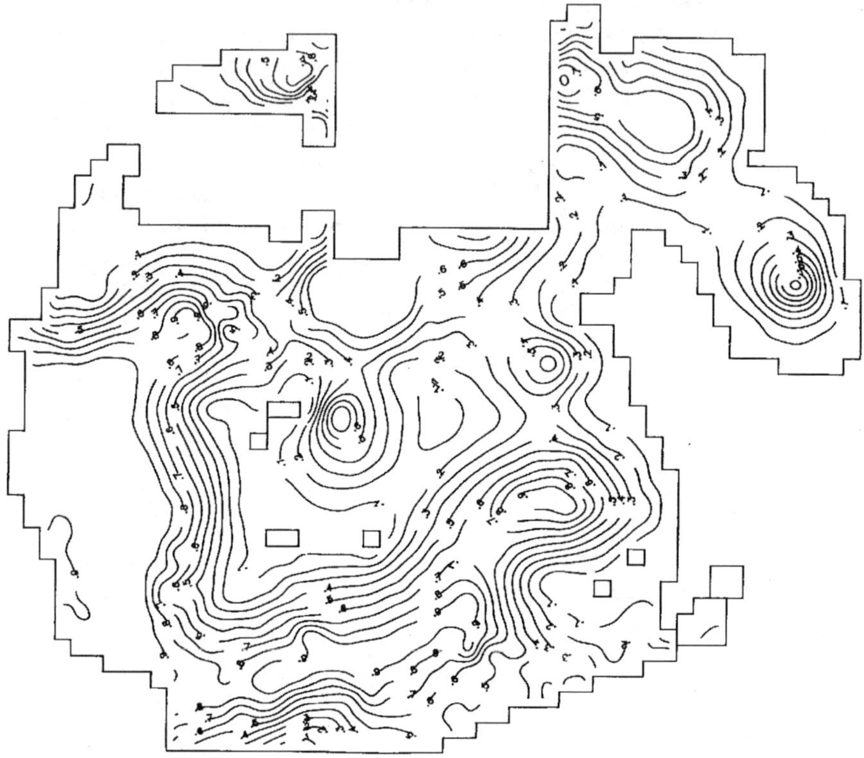

Figure 17-14. Contour map of indicator-estimated probability values. (From Geostatistical error management for site assessment (GEM), by J.C. Myers, *Proceedings of HAZMAT West '91,* 1991.)

with minor changes, be used effectively with indicator kriging approaches. Instead of following a concentration contour, a probability contour is used. The .50 or 50% contour line should be used as the action line because the probability of misclassifying estimates is at its highest. This also produces the wasp-waisting result on the misclassification ellipse.

17.2.5.2 Probability Kriging

Indicator kriging provides a simple and understandable approach to managing the extreme concentrations frequently encountered in environmental data. The transformation is straightforward and requires no inverse transformation.

The simplicity of the binary indicator approach can, at times, constrain its efficiency, especially with regard to sample configurations in which data are on a fixed grid (Sullivan 1984). Figure 17-15 shows a representative sampling and

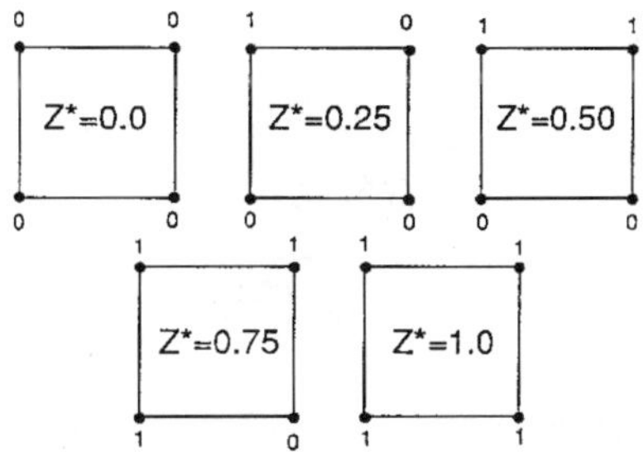

Figure 17-15. Example of a problematic sample and block configuration for indicator kriging.

kriging plan for a square grid sampling approach. A square block is to be estimated by means of four samples located at the corner of the block. The variogram model to be used is assumed to be isotropic.

The simple binary nature of the indicator data requires that each sample have an indicator value of either zero or one. It follows that, with the proposed variogram model, each of the four samples will receive the same weight. Thus, only five outcomes are possible for the block estimate: .0, .25, .5, .75, or 1.0 (Sullivan 1984). The resolving power of this type of estimator is severely limited and results in a large smoothing effect and large conditional biases.

Probability kriging incorporates some easily obtainable information from the data set to improve on indicator estimation. The concentration of the sample *Z*(x) can be simply replaced by the experimental cumulative distribution function (*cdf*) *F[Z(x)]*, the quantile value (Sullivan 1984). The estimate of the *cdf* always assumes a value between zero and one. This adds stability to the system as both the indicator and distribution data are of the same order. This is not necessarily true for *Z* (concentration) values, which may range upward of 10,000 ppm. This discrepancy in units could cause the estimate to become unstable.

The variogram of the quantile data is also required for probability kriging and is given by

$$\gamma_q(h) = \frac{\sum_{i=1}^{n} [q_{(x)} - q_{(x+h)}]^2}{2N(h)} \tag{17.2}$$

as well as the cross-variogram:

$$\gamma_{iq}(h) = \frac{\sum_{i=1}^{n} [(i_x - i_{(x+h)})(q_x - q_{(x+h)})]^2}{2N(h)} \tag{17.3}$$

and the indicator variogram:

$$\gamma_i(h) = \frac{\sum_{i=1}^{n} [(i_{(x)} - i_{(x+h)})]^2}{2N(h)} \tag{17.4}$$

Sullivan (1984) has shown that probability kriging is able to reduce the conditional bias of the estimated values and the smoothing effect. In addition, the probability kriging estimation variance is smaller than that for indicator kriging (Journel 1984).

Probability kriging output can be contoured or block-mapped for means,

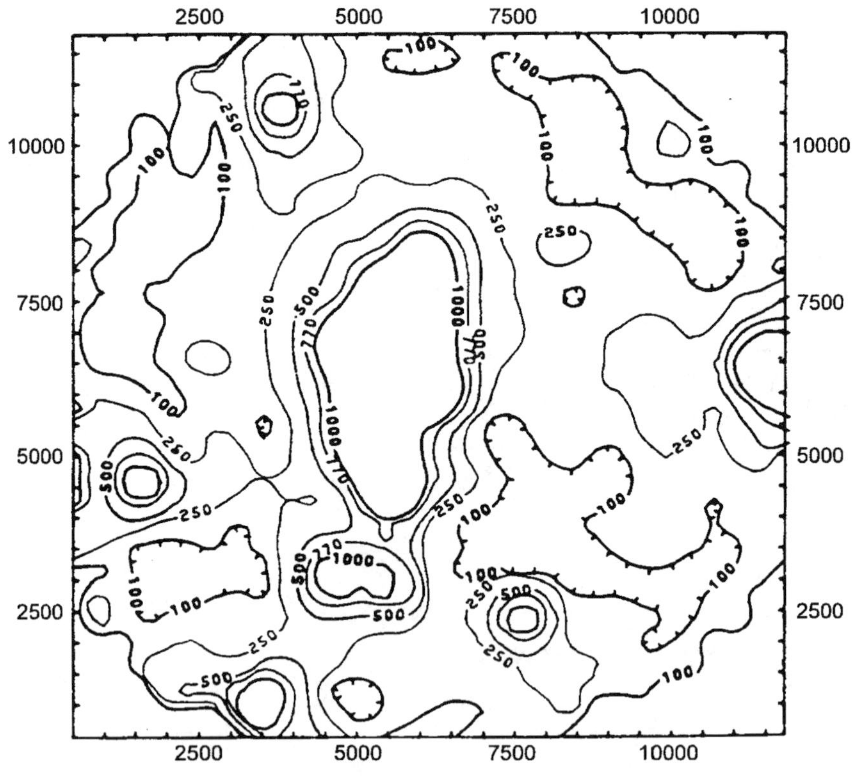

Figure 17-16. Contour map of mean lead (Pb) concentration data (ppm) around a smelter (probability kriging). (From Probabilistic spatial contouring of the plume around a lead smelter, by G. Flatman et al., *Proceedings of the Sixth National Conference on Management of Uncontrolled Hazardous Waste Sites,* 1985.)

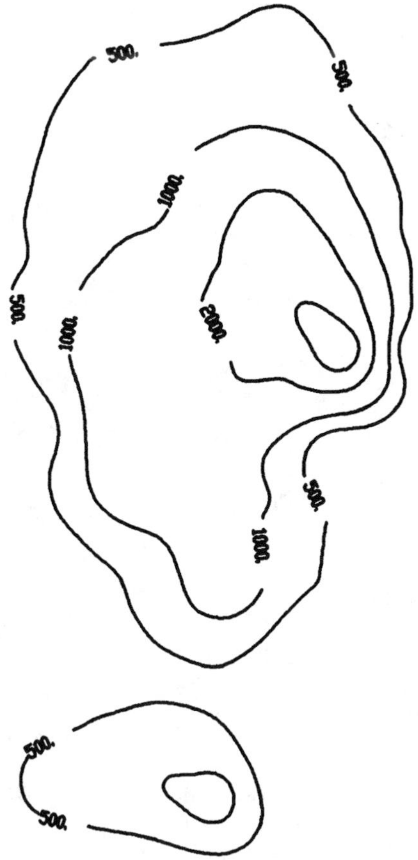

Figure 17-17. Contour map of mean lead (Pb) data (ppm) around a smelter developed by lognormal kriging. (From Geostatistical sampling and contamination analysis of lead values surrounding industrial smelters, by J.C. Myers, Master's thesis, Colorado School of Mines, 1985.)

medians, quantiles, or probabilities. Flatman et al. (1985) demonstrated the use of probability kriging for risk assessment involving lead (Pb) concentrations in soil around a smelter, shown in Figures 17-16 through 17-19. Figure 17-16 is a contour map of the mean lead (Pb) concentration in the region of the smelter. This compares well with a similar map (Fig. 17-17) developed using lognormal kriging (Myers 1983). Figure 17-18 shows a median/quantile concentration map of the same smelter region developed with probability kriging. The map in Figure 17-19 presents contours showing the probability of finding a concentration equal to or greater than 1000 ppm in a given area.

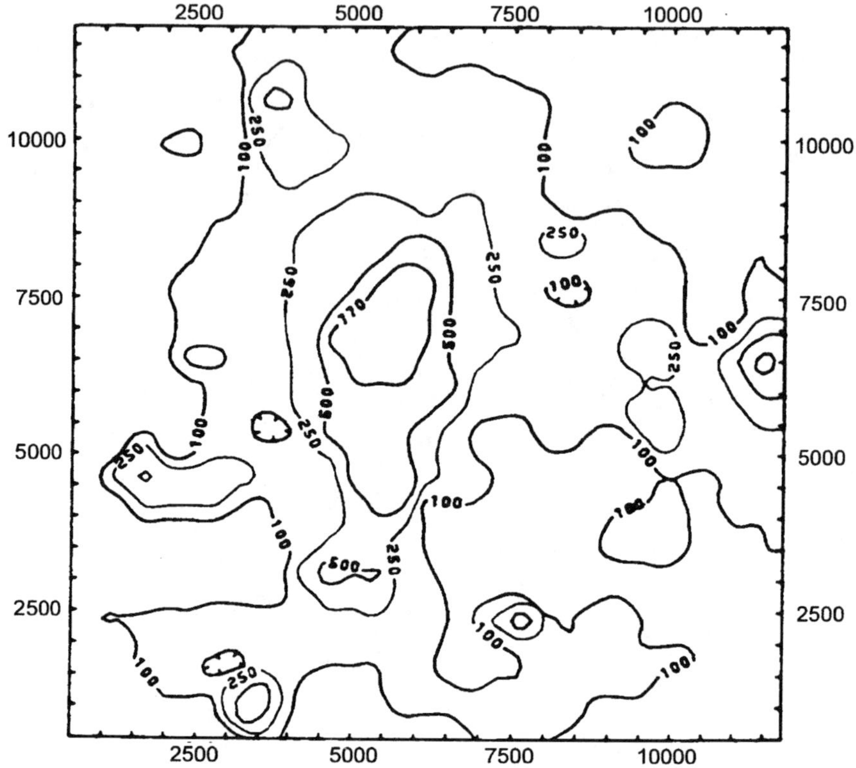

Figure 17-18. Contour map of median/quartile lead (Pb) data (ppm) around a smelter developed by probability kriging. (From Probabilistic spatial contouring of the plume around a lead smelter, by G. Flatman et al., *Proceedings of the Sixth National Conference on Management of Uncontrolled Hazardous Waste Sites,* 1985.)

17.3 Cost and Loss Functions

The tools of the misclassification ellipse, typecasting error, decline curve, fat contour line, and Design Performance Goal Diagram all provide a way of quantifying the error or uncertainty associated with environmental site characterization and decision making. As discussed, each is appropriate for use for different DQOs and for different phases of an environmental assessment. In addition to this distinction, each has an implied cost and loss function. It is important to understand the implications and assumptions surrounding the various cost and loss functions and the respective error quantification tools.

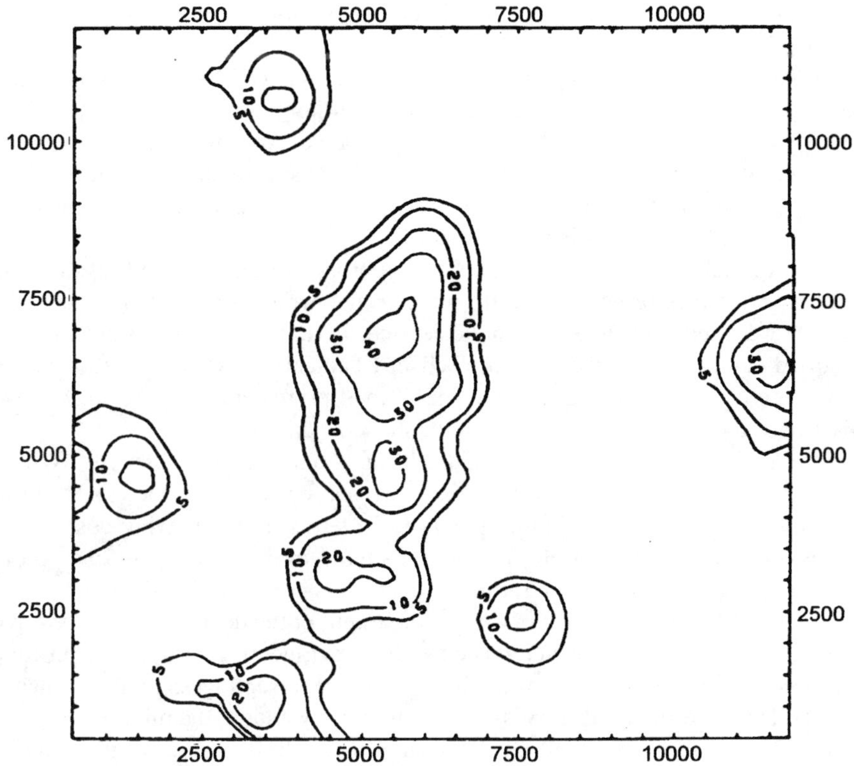

Figure 17-19. Contour map of the probability of finding a soil lead (Pb) concentration greater than or equal to 1000 ppm based on a probability kriging model. (From Probabilistic spatial contouring of the plume around a lead smelter, by G. Flatman et al., *Proceedings of the Sixth National Conference on Management of Uncontrolled Hazardous Waste Sites,* 1985.)

17.3.1 Symmetric Loss Functions

The misclassification ellipse and the decline curve are closely related as the width of the ellipse is proportional to the corresponding error variance in the decline curve. These approaches, as well as the typecasting of error technique, are all based on the kriging approach.

As described in Chapter 12, kriging is a form of regression analysis. All types of regression have a built-in loss function, namely, least squares, which is used to develop the normal equations. The least-squares approach is a symmetric loss function and works well for a variety of situations, including error minimization. An example of a symmetric least-squares loss function is pictured in Figure 17-20.

Figure 17-21 shows a square-grid sampling design (X's) around an industrial pollutant source. According to the criterion of minimizing the kriging standard deviation for the site as performed in a polygonal kriging exercise, the next locations (solid circles) are dictated strictly by the high estimation variance values associated with unsampled areas. The new locations are placed in the centers of the open areas left by existing grid points.

At early phases in site evaluation, this approach offers excellent results. The additional samples bring about a decrease in the error on the mean, as demonstrated by the reduced variances on the decline curve. It is also consistent with the stratified sampling approaches described in Chapter 16 that attempt to obtain a representative spatial coverage of the site in an effort to determine the presence or absence of a contaminant or a COC's average concentration.

Reductions in the error variance also bring about a reduced total cost. With a reduction in the error, misclassifications will also be reduced. The cost of false positives can be determined quite precisely in terms of engineering cost. Thus, "investing" in sampling can bring about a saving in remedial costs since these costs are generally higher than sampling costs.

Figure 17-22 shows a cost optimization graph for the decline curve. The error variance on the *Y* axis has been replaced with the engineering cost of false positive misclassification. In addition, a linear function representing sampling cost has been added since sampling cost is roughly linear. By adding the misclassification cost to the sampling cost at all levels of sampling, a third, total cost curve can be developed.

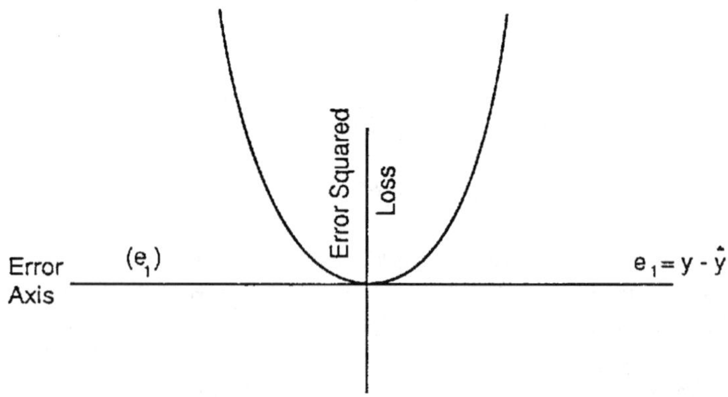

Figure 17-20. Illustration of a symmetric least-squares loss function. (From Asymmetric loss function for Superfund remediation decisions, by G. Flatman and E. Englund, *Proceedings of the Business and Economics Statistics Section of the American Statistical Association*, 1991.)

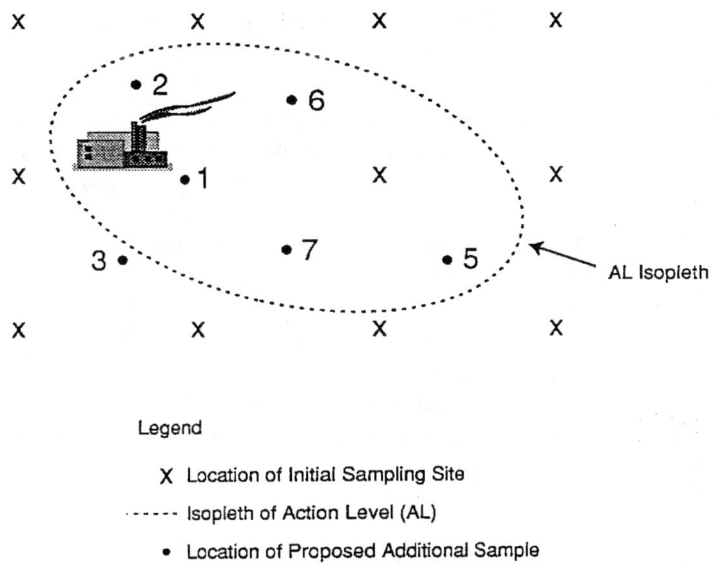

Figure 17-21. Example of a sampling design to optimize the least squares cost function. (From Asymmetric loss function for Superfund remediation decisions, by G. Flatman and E. Englund, *Proceedings of the Business and Economics Statistics Section of the American Statistical Association*, 1991.)

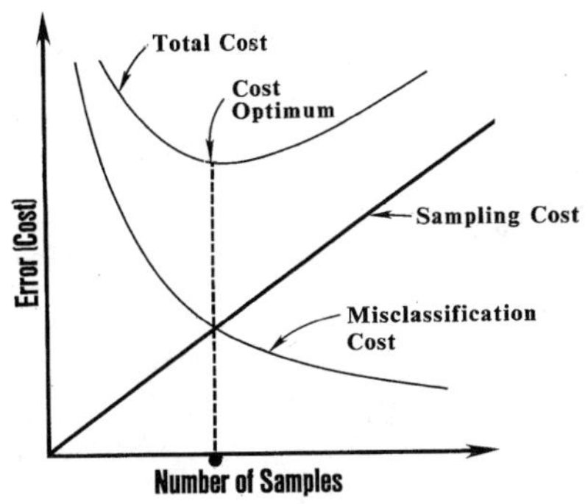

Figure 17-22. Decline curve with sampling cost curve and total cost curve. (From *Optimization of Sampling Policies: A Geostatistical Approach*, by J-M. Rendu, MMIJ-AIME, 1980, with permission.)

The total cost curve behaves in an expected manner. At low levels of sampling, the total cost declines as additional samples "purchase information" that serves to reduce misclassifications and, therefore, total costs. As more samples are taken, the "buying power" of additional samples is reduced; in other words, the marginal cost is increasing. Eventually, marginal costs equal marginal benefits, and the total cost curve reaches a minimum. Beyond this point, additional samples are no longer economical.

For later site characterization phases, symmetric loss functions are rather simplistic. The problem of environmental remediation focuses on classification, that is, deciding whether a block is clean or dirty based on estimates derived from sample information. For classification approaches, symmetric loss functions are not optimal as the loss is bounded for false positive classifications but unbounded for false negatives. This does not imply that the kriging approach should be abandoned; it is simply necessary to incorporate alternative, and more sophisticated, loss functions.

17.3.2 Asymmetric Loss Functions

17.3.2.1 Step Function

Figure 17-23 shows a commonly applied loss function known as a *step function*, where either type of decision error results in a fixed, constant loss. For a false positive, the loss equals the cost of unnecessarily remediating the block. For a false negative, the cost is associated with human health and safety and potential legal liability. The false negative loss is always set much higher than the false positive cost even though the exact cost of false negative errors is impossible to quantify.

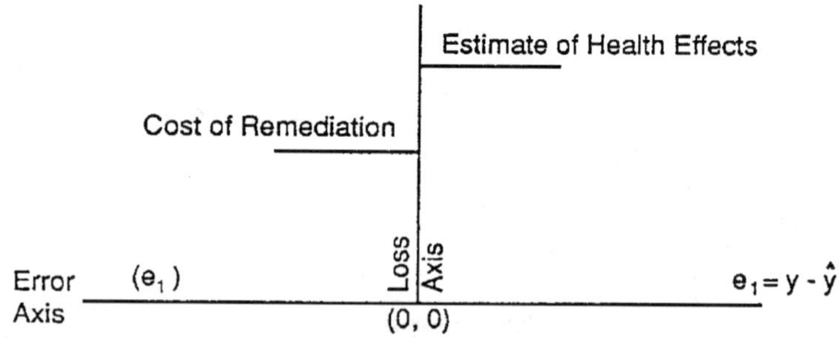

Figure 17-23. Step function for characterizing loss from incorrect decision making. (From Asymmetric loss function for Superfund remediation decisions, by G. Flatman and E. Englund, *Proceedings of the Business and Economics Statistics Section of the American Statistical Association*, 1991.)

The sampling implication of the step function loss curve can be seen in Figure 17-24. In this case, the "chasing-the-action-line" approach is implied since one wishes to minimize the number of false positives and false negatives. For this model, a false negative is viewed as extremely risky and costly, and a false positive is equal to the cost of remedial cleanup. The sampling design places new samples both inside and outside the action line contour in an attempt to minimize both false positives and false negatives. This approach to sampling can be used in conjunction with typecasting of error by placing new samples in regions of misclassification.

17.3.2.2 Asymmetric V Curve

An improvement on the step function loss curve can be seen in Figure 17-25. This model incorporates features of the risk-based perspective that drives many Superfund and other environmental remediations and specifically embod-

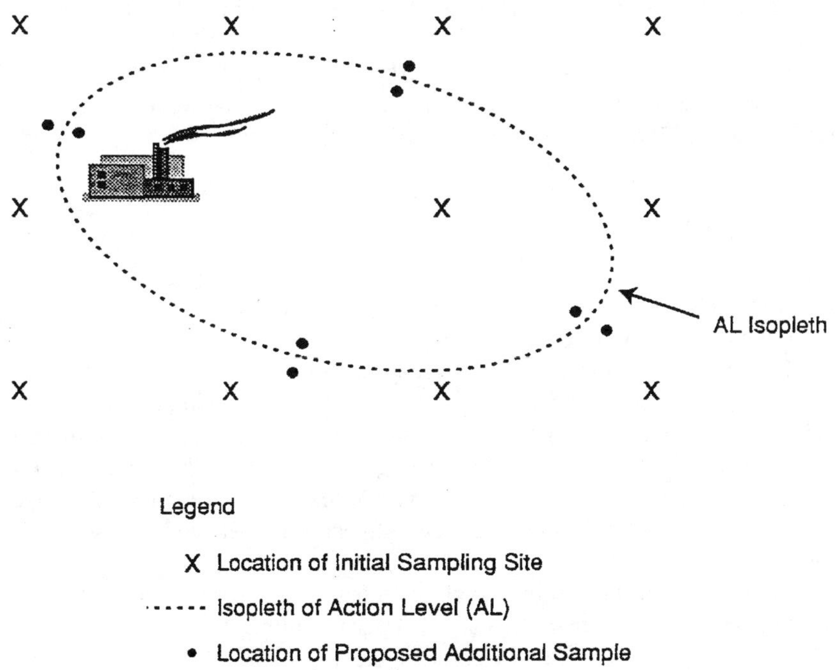

Figure 17-24. Sampling design to optimize the step function for loss. (From Asymmetric loss function for Superfund remediation decisions, by G. Flatman and E. Englund, *Proceedings of the Business and Economics Statistics Section of the American Statistical Association*, 1991.)

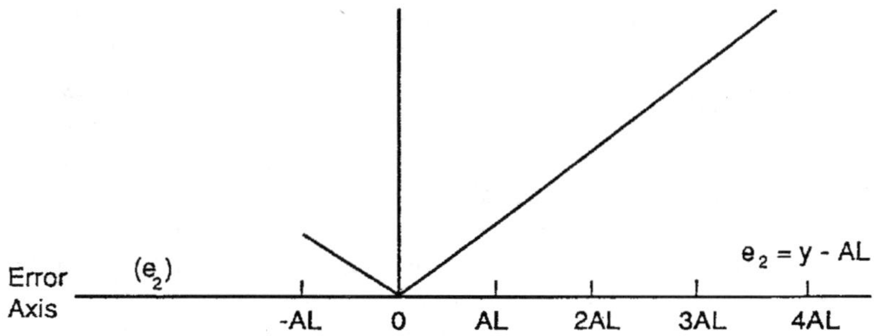

Figure 17-25. Illustration of the V curve loss function. (From Asymmetric loss function for Superfund remediation decisions, by G. Flatman and E. Englund, *Proceedings of the Business and Economics Statistics Section of the American Statistical Association*, 1991.)

ies the concept of "reasonable maximum exposure" to pollutants (Flatman and Englund 1991). It also closely parallels models used in EPA guidance documents.

In the V curve function, the *Y* axis graphs loss in units of remedial cost; the *X* axis graphs error. Error is defined as the difference between the true COC concentration (*TC*) and the action level (AL) concentration or

$$\text{Error} = TC - AL \tag{17.5}$$

and Error applies to both false positive and negative decisions. Loss is incurred only in the case of an incorrect decision.

The positive segment of the graph represents false negative decisions. In this region, the graph rises from zero (equal to the action level minus the action level) in a linear fashion, indicating that the greater the concentration left unremediated, the greater the health and safety risk. The graph in Figure 17-25 is linear with a slope of one. The toxicity of a given COC may increase or decrease the slope. A toxicity change could also produce a nonlinear function in the false negative region.

The negative portion of the graph reflects the false positive loss. It too rises linearly from zero but is truncated at $-1\,AL$ (zero minus *AL*) since COC concentrations are virtually never below zero. Figure 17-26 shows the sampling design implication for the asymmetric V curve function. For this model, all new sampling locations are placed outside the action line isopleth. Losses for false positives are limited in magnitude. The biggest losses occur for false negatives that are essen-

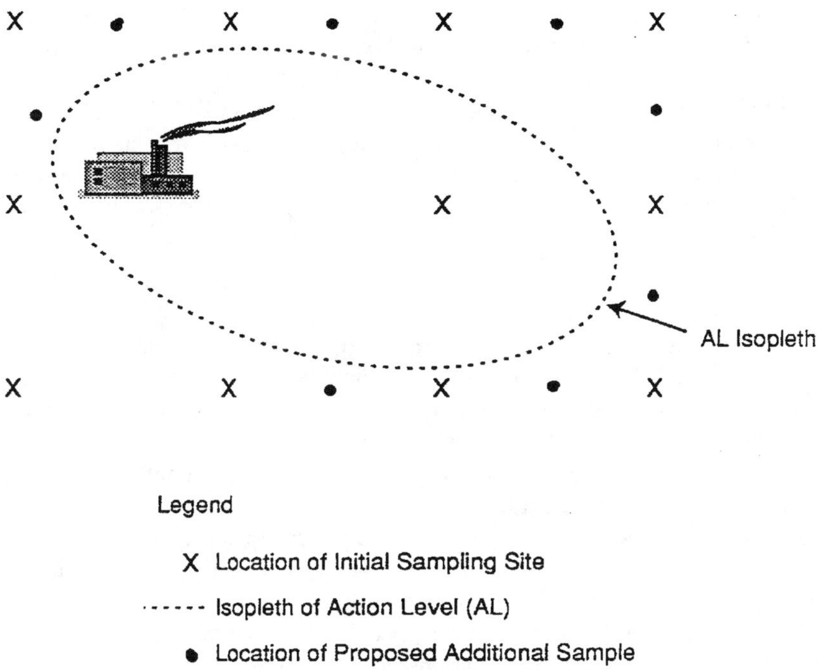

Figure 17-26. Sampling design to optimize the V curve loss function. (From Asymmetric loss function for Superfund remediation decisions, by G. Flatman and E. Englund, *Proceedings of the Business and Economics Statistics Section of the American Statistical Association*, 1991.)

tially unbounded, with a sampling design that is "searching" for undetected hot areas.

The V curve loss function closely parallels EPA's Design Performance Goal Diagram (DPGD) in its implicit assumptions. It is assumed that remedial cost equals the risk at the action level. This follows the logic that if a rational economic man will pay to remediate, then remediating a block with a true concentration slightly below the action level loses little (Flatman and Englund 1991). Conversely, failing to remediate a block slightly above the action level also loses little. This forces the loss lines to meet at the origin.

This logic can be seen in the DPGD (Fig. 15-4). The gray region indicates the concentration or risk range in which any decision is acceptable. Note that the hypothesis test is H_0: The COC exceeds the action level. With an action level of 100 ppm, decisions that a block is contaminated do not concern the RPM even if the true concentration is 80 ppm. For H_0: The COC is less than the action level (Fig. 15-5), the RPM is not concerned if a block with a true concentration of 120 ppm is declared to be clean.

While a step function type of approach is used on the DPGD, incorrect decisions in areas farther from the action level are less likely to be tolerated, implying a greater loss. This is true for either baseline H_0 condition. In any event, the RPM is willing to tolerate significant error around the action level.

17.3.2.3 Hybrid Step–V Curve

The logic used in the V curve loss function may not be embraced by all parties or stakeholders. Where the government is the sole PRP, the V curve may be an acceptable model. Industrial and other PRPs may not agree that the false positive function should be linear and increasing from the origin.

Figure 17-27 displays a hybrid alternative of the step function and V curve loss functions. For false positives, the function is fixed at the cost of remediation. This is typically how remedial costs are interpreted by corporations since this is generally how remedial costs impact them. The false negative portion remains an increasing function rising from zero and reflects the conventional wisdom for liability and health and safety. One can speculate that the tolerance of a gray area may reflect the fact that the true relationship between exposure to a toxic chemical and the associated risk to an individual is not exactly known and that risk safety factors have been built in.

The sampling design associated with the hybrid V curve loss function reverts to the design shown in Figure 17-24 for the step function where samples attempting to minimize both the Type I and Type II errors have been placed on both sides of the action line.

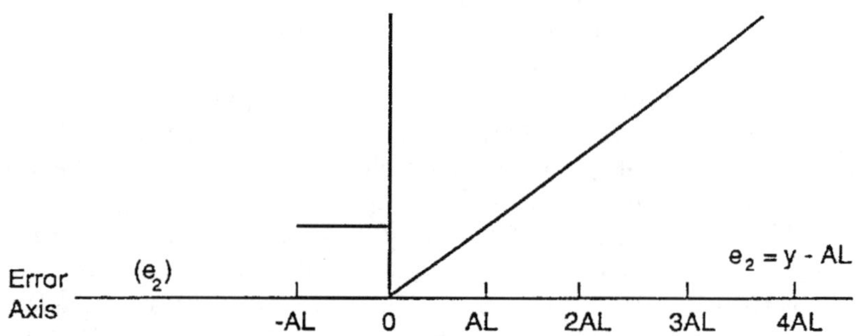

Figure 17-27. Illustration of the hybrid step–V curve loss function.

17.3.2.4 Probability-Based Cost Functions: Cost of Errors Approach

The total cost curve described in Section 17.3.1 serves well as a global indicator but does not directly address the issue of local misclassification. It would be desirable to develop a cost/loss optimization approach that incorporates the need to reduce global variance along with the objective of accurately defining the action line, thereby minimizing false positives and false negatives.

Aspie and Barnes (1991) propose an approach whereby these objectives can be achieved. The cost function can be written

$$E[\$_e] = \sum_{i=1}^{G} [\$_{FP} P_{FP}(x_i) + \$_{FN} P_{FN}(x_i)] dA_i \tag{17.6}$$

where:

G = number of cells (blocks)
x_i = center point of cell i
$P_{FP}(x_i) = Pr[Z(x_i) < \beta \cap Z^*(x_i)\ \ \beta(x)]$
$P_{FN}(x_i) = Pr[Z(x_i) > \beta \cap Z^*(x_i)\ \ \beta(x)]$
dA_i = area of cell i
β = action level threshold

The probabilities of a false positive or false negative can be obtained for each point in the block model by using ordinary, simple, indicator, or probability kriging.

Figure 17-28 shows a contour of true gold concentrations in parts per million within an area of interest (Aspie and Barnes 1991). The stipulated action level is 5 ppm. Figure 17-29 shows a contour map of gold concentrations developed using 10 initial sample points, along with the sample locations. A significant difference exists between the outline of the 5-ppm contour for the true vs. the estimated.

Figure 17-30 shows five new sample locations chosen using the cost classification approach with a step function loss model. Figure 17-30 also presents updated contours of gold concentrations. The 5-ppm contour is much closer to the true contour than before. In contrast, Figure 17-31 shows new sample locations and contours for samples selected based on the minimization of global estimation variance. The location of the 5-ppm action level contour is still quite inaccurate, and total costs of error are higher than for the optimal approach (Table 17-1).

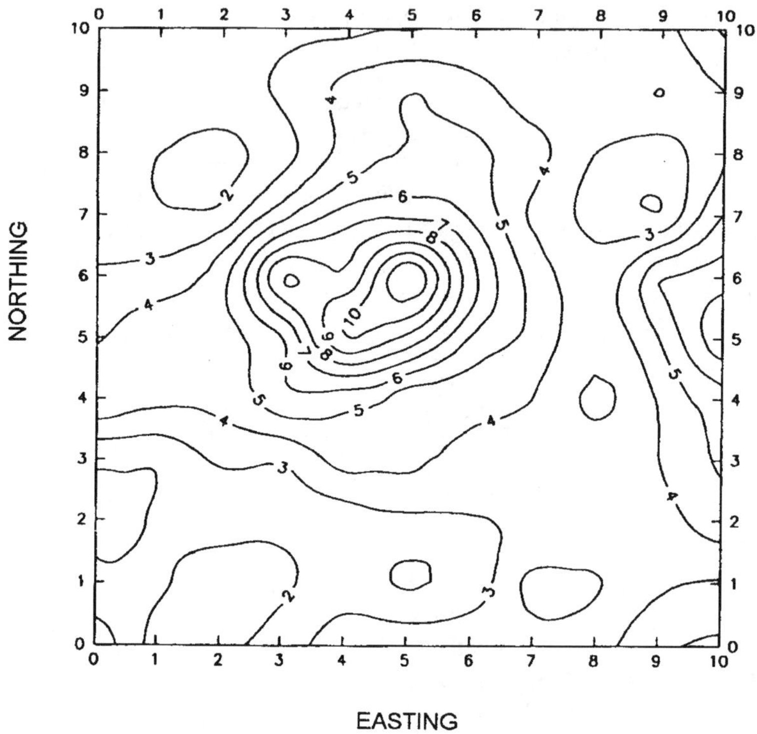

Figure 17-28. Contour map of true gold concentrations (ppm). (From Infill-sampling design and the cost of classification errors, by D. Aspie and R. J. Barnes, *Journal of Mathematical Geology*, 22:8. Reprinted with permission from Plenum Publishing Corporation.)

17.4 Considerations in Determining Sampling Density

17.4.1 Probabilistic Targeting

The approach of Aspie and Barnes blends the economic need of cost/loss minimization with the regulatory and sampling objectives of accurate plume definition and misclassification minimization. One implied objective is to find undetected hotspots, an important consideration in assuring health and safety.

Detecting hotspots can be an issue at any stage of site characterization. Occasionally, the existence of hotspots is suspected or even known, but their location remains unknown. An example is a site at which hazardous wastes are known to have been buried in a landfill.

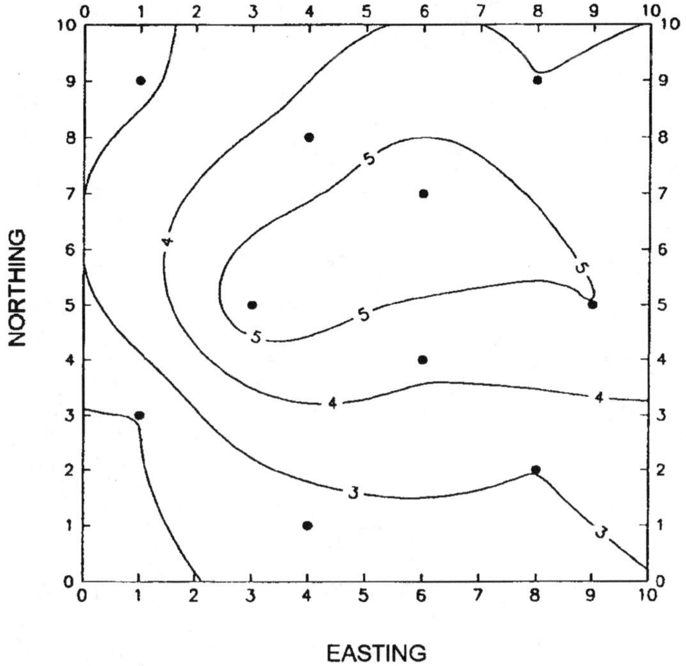

Figure 17-29. Contour map of estimated gold concentrations based on initial sample results taken at ten locations. (From Infill-sampling design and the cost of classification errors, by D. Aspie and R. J. Barnes, *Journal of Mathematical Geology*, 22:8. Reprinted with permission from Plenum Publishing Corporation.)

Singer (1972, 1975) developed the probabilistic targeting approach to locating hotspots as part of an effort to discover geologic ore deposits. Probabilistic targeting is a geometric approach and incorporates sampling on regular square, rectangular, or triangular grids. Zirschky and Gilbert (1984) developed nomographs (Figs. 17-32a to 17-32c) to answer questions such as the following posed by Gilbert (1987):

- What grid spacing is needed to locate a hotspot with a specified confidence?
- For a given grid spacing, what is the probability of hitting a hotspot of a specified size?
- What is the probability that a hotspot exists when no hotspots were found by the existing sample grid?

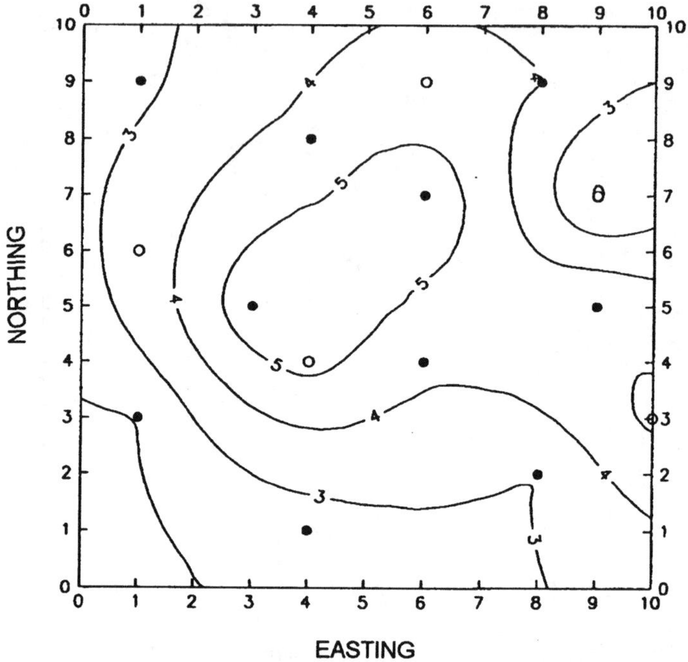

Figure 17-30. Updated contour map using the results from five new sample locations that provide the greatest reduction in the expected cost of errors. First-round samples are indicated by •; new sample locations are indicated by ○. (From Infill-sampling design and the cost of classification errors, by D. Aspie and R. J. Barnes, *Journal of Mathematical Geology*, 22:8. Reprinted with permission from Plenum Publishing Corporation.)

Parkhurst (1984) demonstrated that triangular grids are more reliable in providing an unbiased estimate of the density of randomly located hotspots at a site. If hotspots follow a regular but unknown pattern, however, samples should be taken randomly.

If undetected hotspots are a concern, probabilistic targeting offers a fast and understandable approach. The probability of detecting a hotspot is proportional to the ratio of the size of a hotspot vs. the area covered by a single grid square, rectangle, or square.

For example, when searching for a hotspot 100 ft in diameter with a square sampling grid spacing also equal to 100 ft, the probability of detection is .785, or an 78.5% chance. This is simply the ratio of the area of the circle (7854 ft^2) divided by the area of the grid (1000 ft^2). Note that sampling on a grid spacing equal to that of the diameter of the hotspot leaves more than a 20%, or 1 in 5,

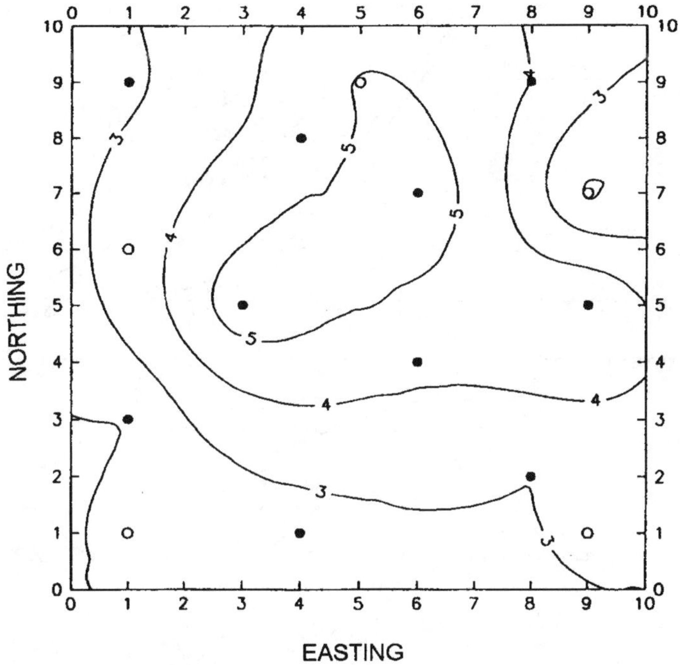

Figure 17-31. Map of five new sample locations selected on the basis of minimizing global estimation error along with updated contours. (From Infill-sampling design and the cost of classification errors, by D. Aspie and R. J. Barnes, *Journal of Mathematical Geology*, 22:8. Reprinted with permission from Plenum Publishing Corporation.)

Table 17-1. New sample locations, kriging variances, and expected costs for the cost of errors and kriging variance approaches.

Criterion	New Sample Locations	σ^2_k	*E* ($)
Initial		0.1201	43.61
Cost of errors	(1,6) (4,4) (6,9) (9,7) (10,3)	0.1083	35.16
Kriging variance	(1,1) (1,6) (5,9) (9,1) (9,7)	0.0994	36.35

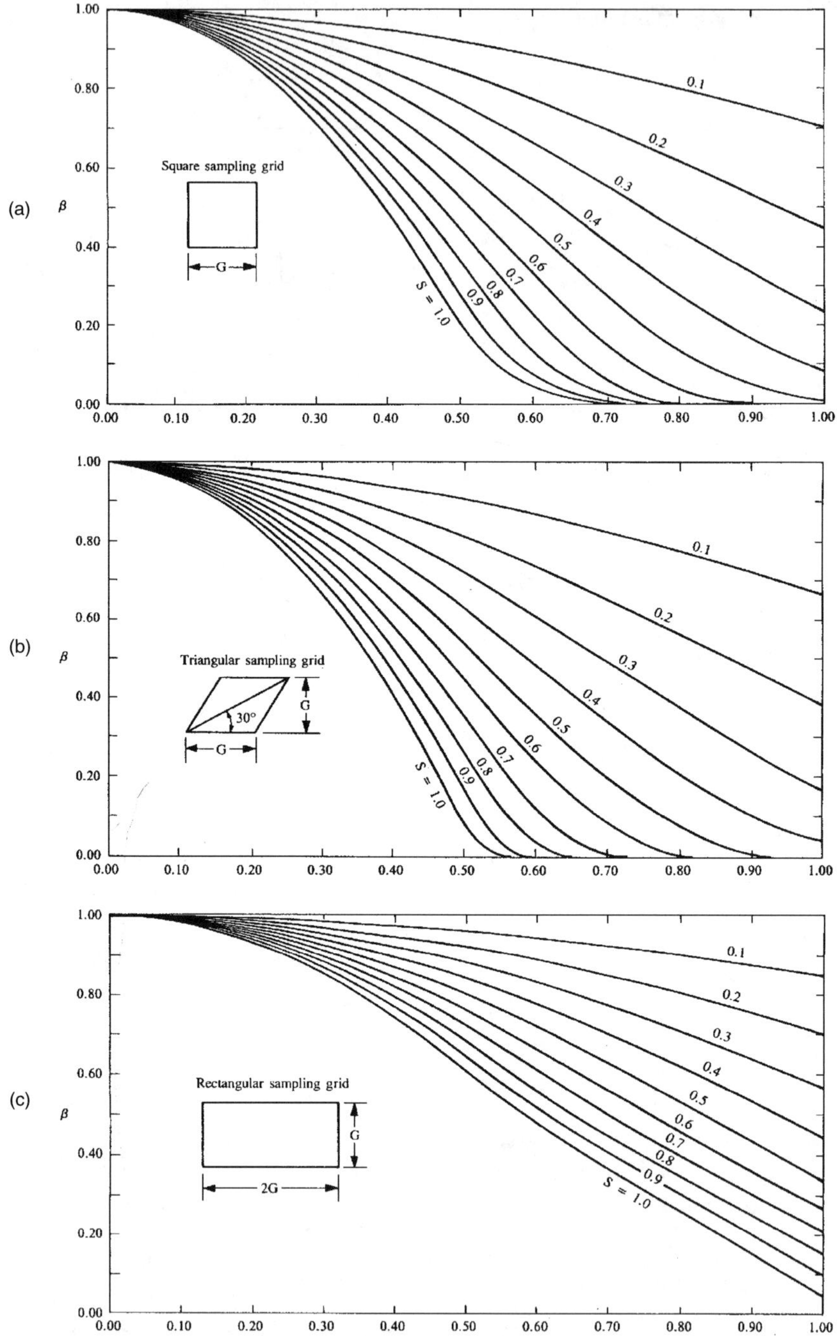

Figure 17-32. Illustration of sampling nomographs for probabilistic targeting: for (a) square, (b) rectangular, and (c) triangular grids. (From *Statistical Methods for Environmental Pollution Monitoring*, by R.O. Gilbert, Van Nostrand and Reinhold, 1987.)

chance (approximately) of missing the target. To limit the chance of not finding the target requires that the grid spacing be reduced to 89.3 ft between samples.

17.4.2 The Relationship between Exposure Units and Remedial Management Units

The EU approach described in Chapter 16 is an all-or-nothing binary method. An entire EU is subject either to remediation or to no action. Often multiple samples or multiple composite samples are used to determine the average COC concentration of an EU. Figure 16-11 illustrated how large areas of contaminated material could potentially go unremediated using this approach.

In geostatistical appraisal, a similar approach is used in that each block is considered independently for decision making but the scale is different. GA blocks, or *remedial management units* (RMUs), are smaller than EUs. Also, the sampling density for RMU estimation is markedly decreased. In GA, a grid is established for estimation by kriging or other interpolator. Assuming a square grid, typical RMUs are on the order of 10 ft to 50 ft, producing corresponding areas of between 100 ft^2 and 2500 ft^2 per RMU, respectively. These areas are significantly smaller than typical EUs, which measure 5000 ft^2 to more than 20,000 square ft^2.

Because a remedial decision for an EU is based on a global average for the area, a symmetric loss function can be used to determine error on the mean by using polygonal kriging. An alternative approach is to collect a single sample composed of *n* correctly materialized increments of an appropriate weight. The sampling nomograph can be used to assure DQO-specified tolerances for the error are respected.

By either method described, an EU is estimated by means of multiple samples within the EU boundary. RMUs differ in that generally only a few of them contain a sample within the RMU. Most RMUs are estimated from samples located outside the block. Although the estimation error is generally higher for individual RMUs, the advantage is the spatial map of the contamination plume, which is not subject to arbitrary EU boundaries. The action line for the COC can be mapped across any EU or stratum boundary. Surgical removal of the material above the threshold can then be implemented.

17.4.3 RMU Size vs. Sampling Density

A question that often arises is: How large should the RMU be? This is equivalent to determining the grid spacing for kriging. There is no universal answer for this simple question although a "universal" lower bound on RMU size is determined by the scale of the remediation technology. A frequently used rule of thumb is that the blocks should not be less than one-third or one-quarter of the sample spacing. For example, for samples taken on a regular grid spaced 100 ft apart, blocks should not be smaller than 25 × 25 ft.

Better "resolution" can be obtained by using smaller and smaller blocks, which yields a less jagged boundary between contaminated and uncontaminated areas. The resolution is artificial, however. While the boundary appears sharper, the estimation variance is simultaneously rising, increasing statistical "fuzziness" along the boundary. Smearing the same data to estimate more and more blocks is an artificial approach. If a crisp boundary is desired, a contouring option should be selected, with the reminder that the statistical "fuzziness" still remains.

17.4.4 Sample Sufficiency

The techniques described in this chapter offer several ways to determine the number of samples required at a site and their corresponding locations. Unfortunately, most of these tactics require computer support and a significant investment in time. As an alternative, a simple rule of thumb is proposed for obtaining a qualitative assessment of sample sufficiency.

The variogram gives a range of influence for a COC beyond which sample data are independent. Thus, local estimation needs sample data that are some fraction of the variogram range in order to be reliable. The smaller the fraction, the better the estimation. If a variogram cannot be developed for the site or yields a range almost equal to the sample spacing, more samples are needed. Note that this assumes that STP problems are not the cause of the pure nugget effect.

17.4.5 Phased Approaches to Sampling

In an attempt to improve data efficiency, to reduce costs, and to meet different DQOs, multiphase sampling programs are commonly implemented. During the planning phase(s) for multistage sampling programs, it is necessary to resolve how many stages are appropriate, how many samples should be collected during each stage, and what strategy for sample placement should be used for each phase. Objective answers to these questions are difficult to obtain.

Myers (1982) showed that it is possible in some situations to take a small initial number of samples in order to derive a variogram model with which subsequent planning may be performed. Using the 200-sample data surrounding the Dallas Lead smelters (DMC and RSR), data were selected at random to form multiple data subsets of the original, each with 100 samples. Variogram analysis of the data subsets produced a usable variogram structure.

The experiment was repeated to create multiple data subsets of 50, 25, and 16 samples. At all levels of subsampling, variogram structures were consistently obtained that closely approximated the original variogram model. While demonstrating the validity of using a geostatistically phased approach, the study did not address an optimal number of samples for each phase or how many phases should be implemented.

Englund and Heravi (1995) used phased repetitive sampling from "known" models to investigate these issues. The target site models were assumed to contain contaminated soils, were two-dimensional with a constant thickness, and be subject to the following DQO decision rule: Remediate any 100-m^2 area within the site if the estimated value of its average contaminant concentration exceeds a specified action level.

Two site models were selected for study (Walker Lake and Steamboat Falls), each containing 19,800 point values on a rectangular grid. The site models represented two different types of spatial contamination frequently encountered in environmental sampling. For each site model, 198 true remediation unit concentrations were calculated by averaging 10×10 arrays of points. After a sampling phase, the RMU concentrations at each site were estimated using ordinary kriging.

Given a data set of n values, 1,..., n phases can be constructed. Englund and Heravi (1994) implemented experiments with one-phase, two-phase, and n-phase design algorithms, with the cost-effectiveness of alternative methods assessed by means of a loss function (Fig. 17-33), and with an optimal sampling design defined as one that displays minimum total cost (or loss), including sampling.

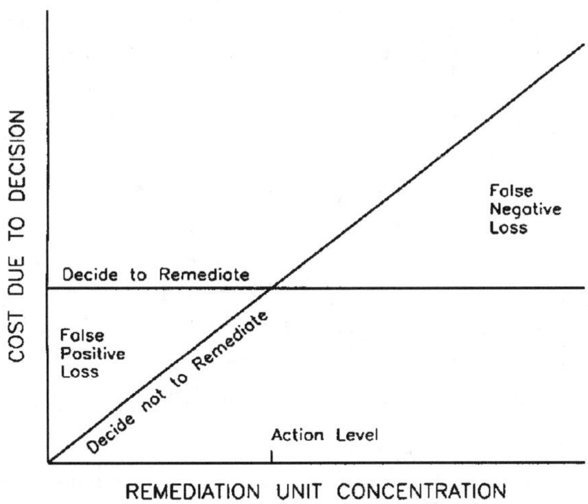

Figure 17-33. Loss function for multistage sampling analysis. (From Phased sampling for soil remediation, by E.J. Englund and N. Heravi, *Journal of Environmental and Ecological Statistics*, 1:247–263.)

Repetitive sampling outcomes in Figure 17-34 plot total cost vs. the number of samples. Solid lines connect the mean costs of multiple resampling, showing the expected total cost as a function of n. Dotted lines represent the upper 90th percentile, providing a "reasonable worst-case." Dashes represent individual resamplings and provide a visual measure of variability. While more pronounced in the Steamboat Falls graphs, both sites show the characteristics of the total cost curve (Fig. 17-22).

Cost curves in Figure 17-35 compare the one-phase, two-phase, and n-phase sampling programs. Table 17-2 details individual cost totals for samples of $n = 256$ and $n = 64$ for Walker Lake and Steamboat Falls, respectively. Two-phase designs are demonstrably better than one-phase designs, with n-phase designs better still.

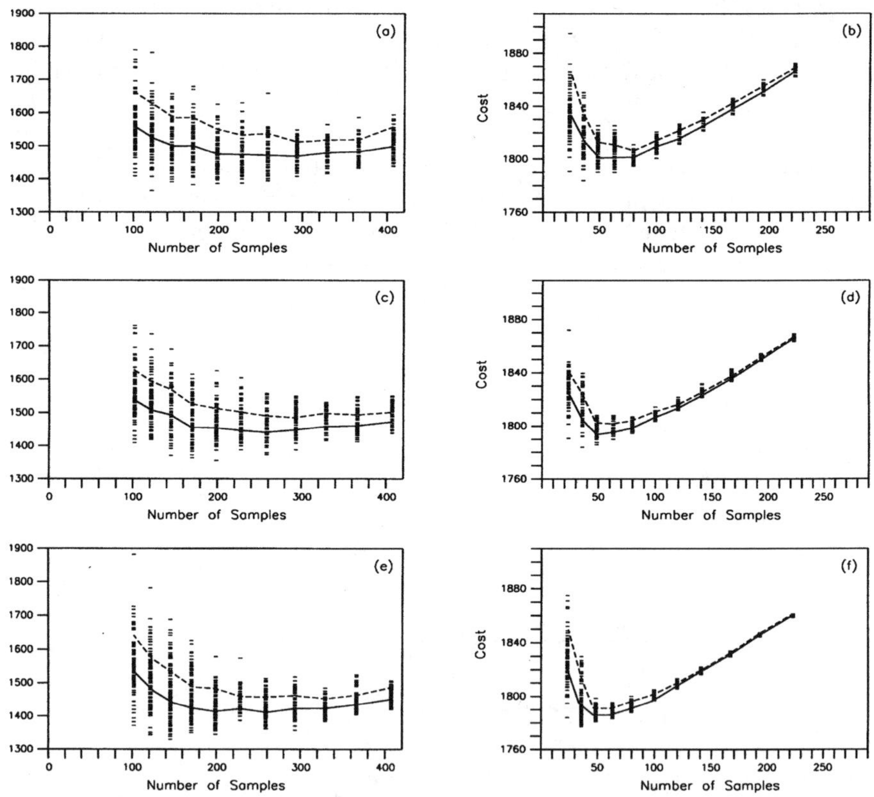

Figure 17-34. Total cost vs. number of samples for three sampling schemes. Left: Walker Lake; right: Steamboat Falls. A,B: One-phase; C,D: Two-phase; E,F: n-phase sampling. Solid lines show the mean cost of multiple resamplings; dotted lines show the 90th percentiles; dashes show the results of individual resampling events. (From Phased sampling for soil remediation, by E.J. Englund and N. Heravi, *Journal of Environmental and Ecological Statistics*, 1:247–263.)

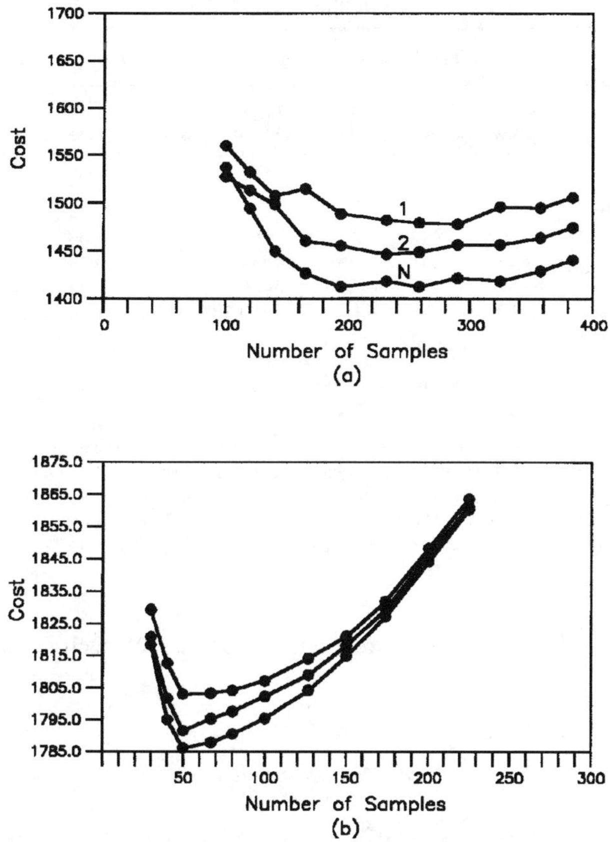

Figure 17-35. Cost curves comparing one-phase, two-phase, and *n*- phase sampling: (a) The Steamboat Falls model; (b) the Walker Lake model. (From Phased sampling for soil remediation, by E.J. Englund and N. Heravi, *Journal of Environmental and Ecological Statistics*, 1:247–263.)

For one-phase designs, the optimal number of samples is independent of the number of phases. Englund and Heravi found that, to optimize two-phase designs, 70 to 80% of the samples should be taken in the first phase. The usefulness of this result stems from the fact that n can be determined with relative ease by using a one-phase algorithm.

Optimally, n-phased designs should always be used. However, n-phased designs have practical problems. Potential logistical costs must be balanced against the relatively small improvements in total cost. Practical n-phase sampling implies the use of real-time field sampling where both data quality and unit costs are different. A likely choice might be between an n-phase design with inexpensive but less precise field-portable sampling equipment or a one- or two-phase

Table 17-2. Results from the three sampling design algorithms.

	Walker Lake: N = 256				*Steamboat Falls: N = 64*			
	One-phase	**Two-phase**[1]	**N-phase**	**One-phase**[2]	**One-phase**	**Two-phase**[3]	**N-phase**	**One-phase**[4]
Costs (× $1000):								
Sampling	128	128	128	128	32	32	32	32
Remediation	890	907	884	897	860	860	857	848
Non-remediation	449	404	398	441	907	899	897	911
Total	1467	1438	1410	1466	1798	1792	1786	1791
% False Positives	10.3	10.2	8.5	10.5	5.7	4.8	3.6	4.5
% False Negatives	7.3	6.3	5.8	7.1	5.7	4.8	3.8	5.1
Remediation								
% Units[5]	44.9	45.8	44.6	45.3	43.4	43.4	43.2	42.8
% Contamination[6]	80.5	82.5	82.7	80.9	53.8	54.2	54.3	53.6

[1]80% in the first phase. [2]Using the second-phase algorithm. [3]70% in the first phase. [4]Using the second-phase algorithm. [5]With perfect decisions, 41.9% of the units would be remediated for Walker Lake; 43.4% for Steamboat Falls. [6]With perfect decisions, 84.9% of the contamination would be remediated for Walker Lake; 54.9% for Steamboat Falls.

design using a laboratory. The correct will depend upon specific site conditions and site DQOs.

17.4.6 Power Curves for Spatial Sampling

The DQO process establishes maximum allowable rates for both false positives and false negatives, effectively bounding the family of acceptable power curves. Englund and Heravi (1995) offer ways to develop power curves for spatial data using Monte Carlo resampling techniques.

With a data set of 600 surface soil samples from a subportion pilot study at the Piazza Road dioxin site in Missouri (Ryti 1993), a conditional simulation procedure was used to create a regular grid of 20,000 dioxin values. The 600 original measurements plus the remaining 19,400 unestimated grid points are shown in Figure 17-36.

The technique of conditional simulation honors actual data measurements at the measured locations, as well as honoring the site variogram model that quantifies spatial variability (Deutsch and Journal, 1992). Thus, there is an order to the randomness of the simulations. Examples displaying grids of 20,000 conditionally simulated realizations for the Piazza Road site are shown in Figures 17-37a through 17-37e.

The site was partitioned into 25 rectangular RMU areas (Fig. 17-38). Each RMU was further subdivided into four quadrants, and a random sample was taken

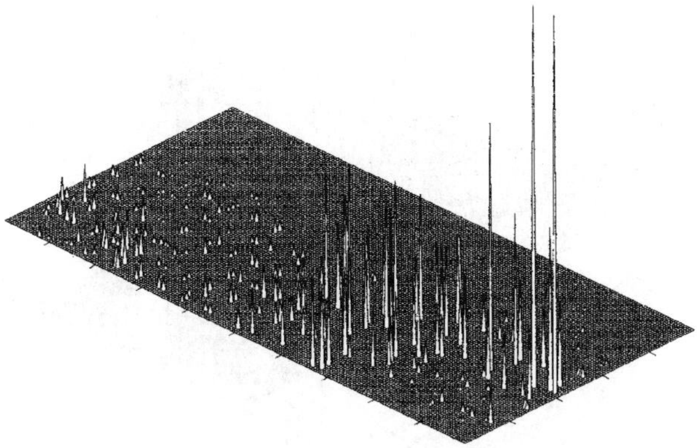

Figure 17-36. The original 600 data measurements from the Piazza Road site. (From Power curves for spatial data, E.J. Englund and N. Heravi, 1995. Reprinted with permission from E.J. Englund.)

in each quadrant. The average of the four samples served as the estimated average concentration for the RMU and applied to the decision rule that required remediation for RMUs having dioxin concentrations of 5 ppb and above. A decision to remediate is defined as a "positive" result. Therefore, power becomes the probability of the estimated mean exceeding the action level.

Two approaches to power curve construction are offered by Englund and Heravi (1995). The simple approach graphs the scatter plot of true vs. estimated RMU values. Figure 17-39a shows a scatter plot of a single sampling effort (25 samples), and Figure 17-39b shows the results of multiple resamplings.

Using the fraction of positive decisions as the power, power curves can be plotted (Fig. 17-40) that show the characteristic shape seen in the DPGD (Fig. 15-6). Unfortunately, this approach does not produce monotonically increasing graphs, which means that the power of the sampling design can differ for two units with identical means.

Englund and Heravi (1995) offer a second approach that tries to generalize the observed power values from the first approach. Using the scatter plot in Figure 17-39a, a new scatter plot can be created by projecting straight-line segments from the origin through each point on the scatter plot (Fig. 17-41). Now, each true concentration has 25 estimated values, and a power curve can be drawn as before. This time, however, the curve is monotonically increasing as it plots the cumulative frequency of the intercepts where the projected lines cross the horizontal line denoting the action level.

Figure 17-42 shows the power curves for 25 points (light line) and for 2500 points (dark line). These curves increase monotonically, with the 2500-point curve displaying the shape of a DPGD power curve. In a similar manner, 2500-point power curves were created for each of the other four simulated realizations. All

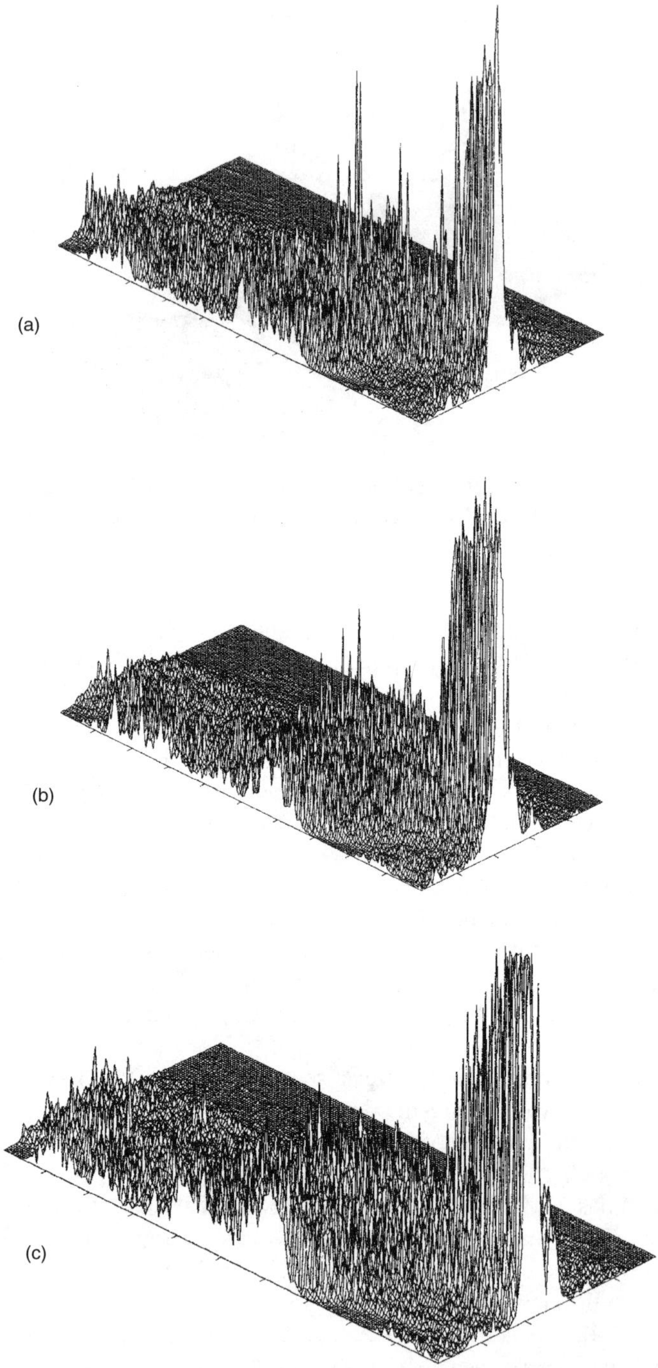

Figure 17-37. Alternative conditional simulations using the Piazza Road data. (From Power curves for spatial data, E.J. Englund and N. Heravi, 1995. Reprinted with permission from E.J. Englund.)

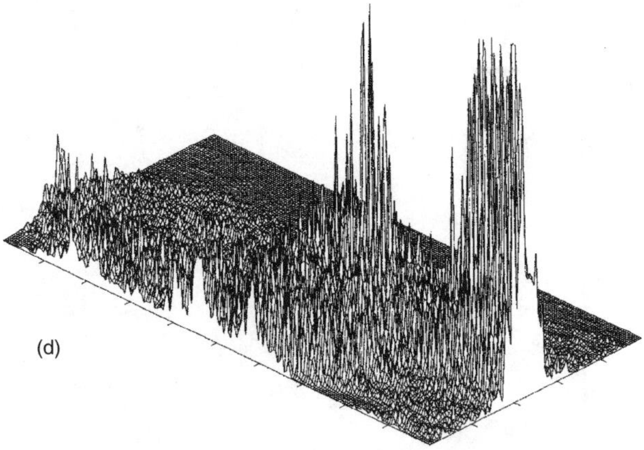

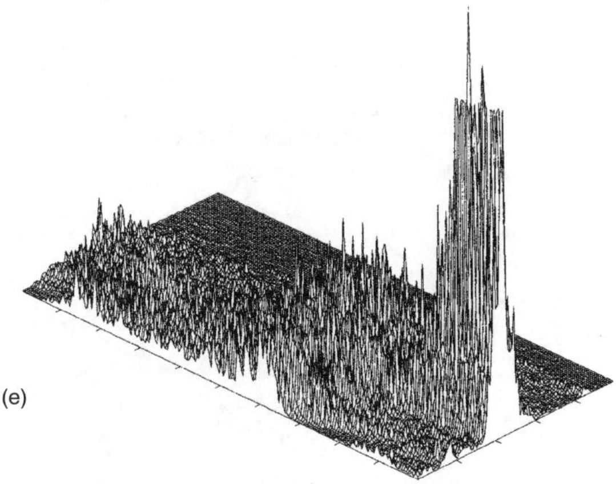

Figure 17-37. (*continued*)

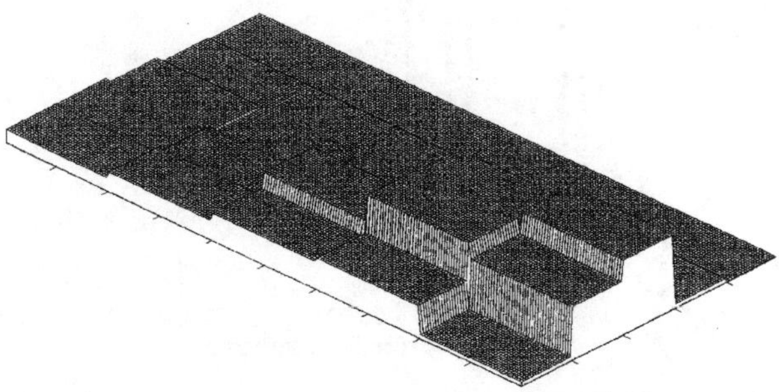

Figure 17-38. Twenty-five remediation unit means for the simulated model in Figure 17-37a.

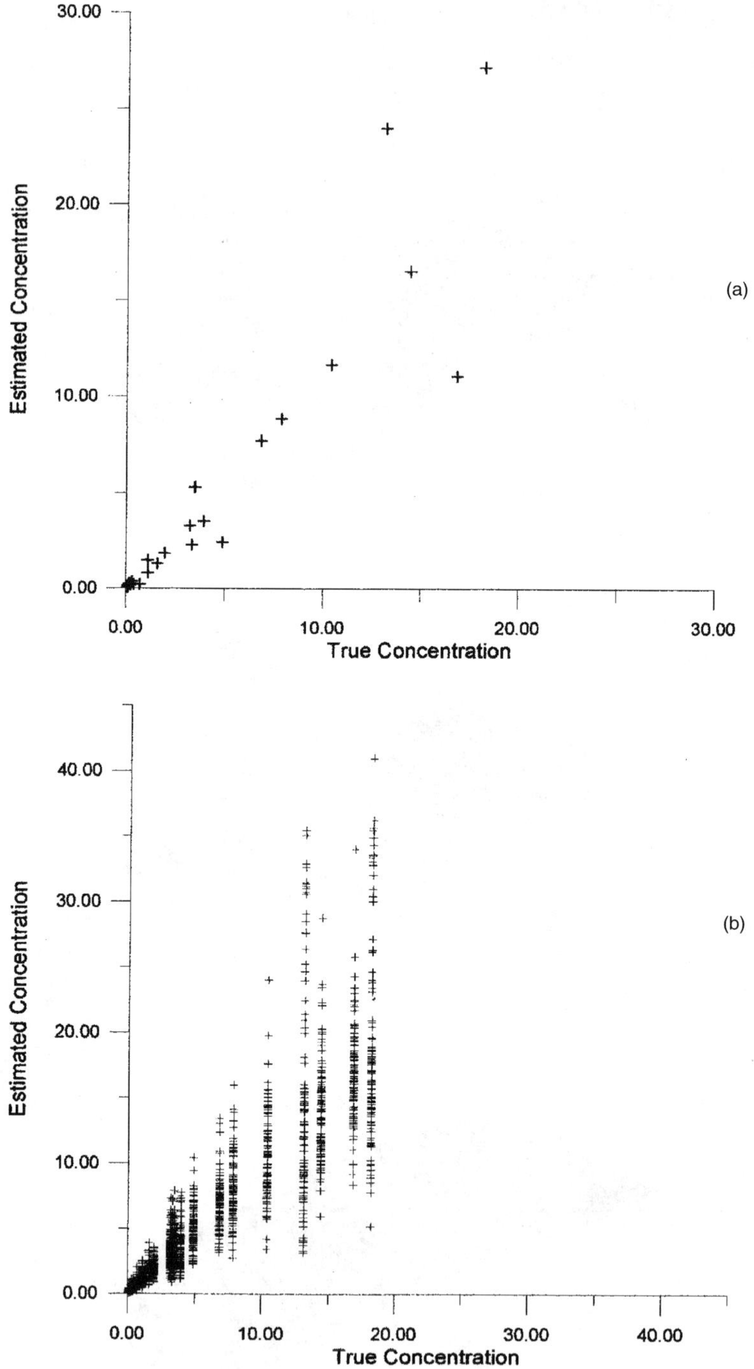

Figure 17-39. Scatter plots of (a) single and (b) multiple resamplings of simulated data for EU characterization at the Piazza Road site. (From Power curves for spatial data, E.J. Englund and N. Heravi, 1995. Reprinted with permission from E.J. Englund.)

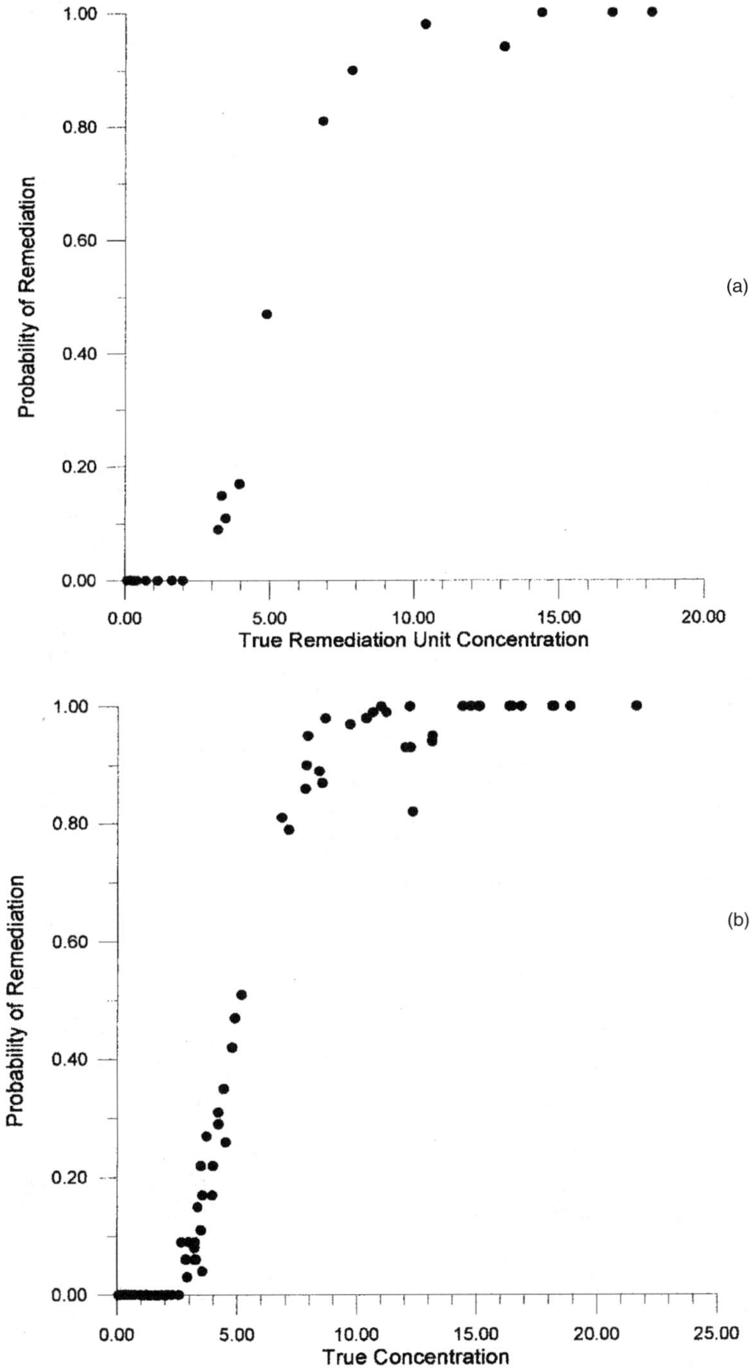

Figure 17-40. Simple approach to power curve generation for spatial data: (a) single sampling event; (b) results from multiple resamplings. (From Power curves for spatial data, E.J. Englund and N. Heravi, 1995. Reprinted with permission from E.J. Englund.)

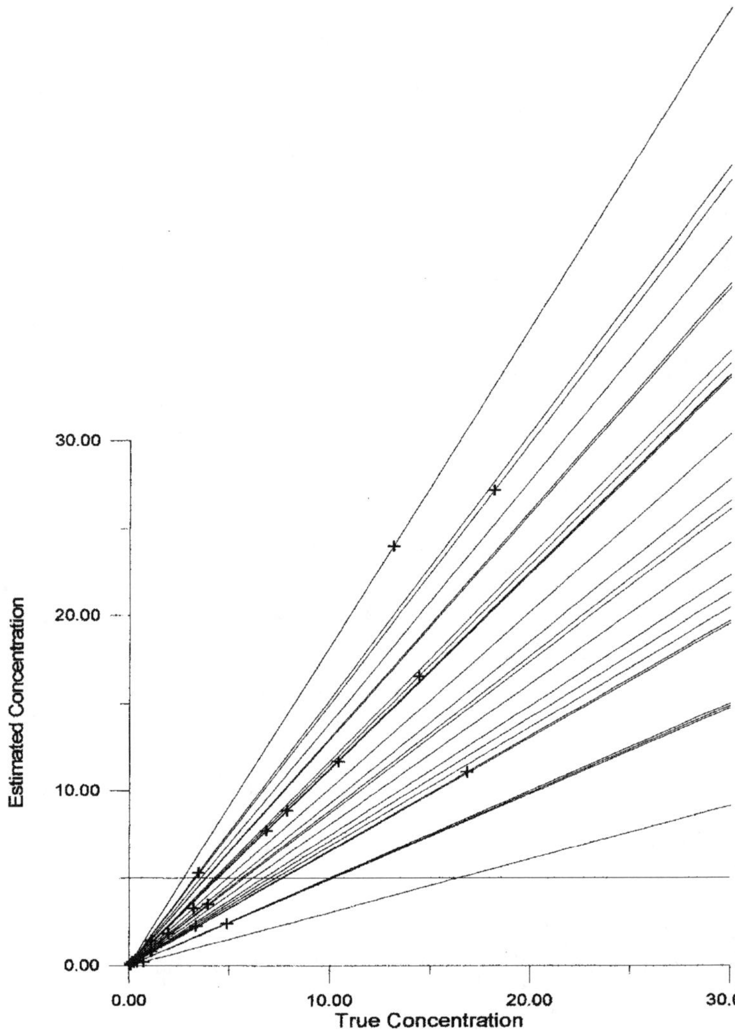

Figure 17-41. Generalized approach to power curves for spatial data. (From Power curves for spatial data, E.J. Englund and N. Heravi, 1995. Reprinted with permission from E.J. Englund.)

five 2500-point power curves are superimposed in Figure 17-43 and show a nearly identical trace.

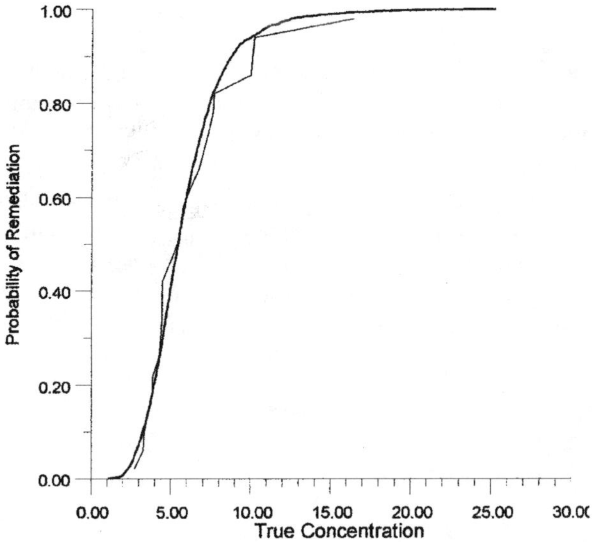

Figure 17-42. Power curves for 25 points (light line) and for 2500 points (dark line) . (From Power curves for spatial data, E.J. Englund and N. Heravi, 1995. Reprinted with permission from E.J. Englund.)

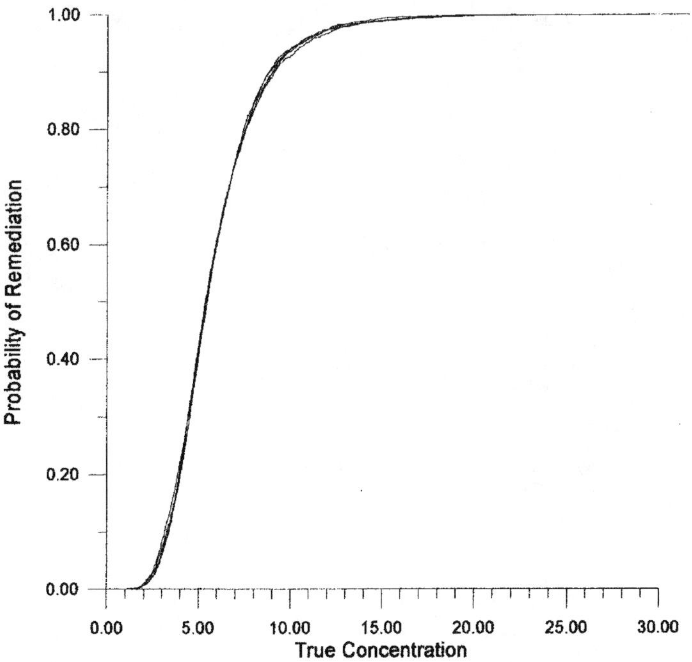

Figure 17-43. Five 2500-point power curves representing five conditionally simulated site realizations.

Conclusion

Error and uncertainty can be viewed in many contexts at a site: global estimation error, local estimation error, global and local misclassifications, uncertainties and misclassifications around an action level contour, probabilities of exceeding an action level, and probabilities of false positives and false negatives. Each of these perspectives has an associated cost/loss to it that needs to be understood so that appropriate DQOs can be developed. Spatially distributed data require spatial approaches to mapping and error quantification. This chapter has presented a few of the spatial techniques that are appropriate for integration into DQO plans for environmental site characterization and decision making.

PART 6 Appendixes

Appendix **A**

Definitions of Key Terms for Sampling Theory & Practice (STP)

A.1 Introduction

The ideas and concepts of sampling theory & practice (STP) are relatively new and, as such, may be unfamiliar to the reader. The theory brings along with it certain definitions from STP and from other fields as well. It is important that the reader become familiar with the definitions set forth below in order to follow the discussion of STP.

The definitions included have been taken, almost verbatim, from those of Pitard (1989), which, in turn, were based on the work of Pierre Gy.* The reason for doing this is to maintain some level of consistency within the field of STP. The reader will also notice that certain defined terms contain extended discussion. The elaboration is justified either by the importance of the term or by the potential conflict with similar terms in other fields.

It is also important to note that not all definitions provided by Pitard (1989) have been included in this chapter. A few definitions have been omitted as beyond the summary scope of STP presented in this book. It is recommended, however, that those wishing to implement STP techniques or to obtain a more complete understanding should review Pitard (1989).

A.2 Definition of Key Terms

Accuracy The definition of accuracy is often controversial, and it is incorrect to include the notion of precision with the notion of accuracy; consequently, definitions given by most current dictionaries are misleading, confusing, and incorrect. Accuracy has been given several definitions, and statisticians and engineers disagree completely on those definitions. A sampling bias is defined as the mean of the sampling error $m(SE)$ (also referred to as the selection error, SE). As far as sampling is concerned, a sample is said to be accurate when the absolute value of the bias $|m(SE)|$ is smaller than a certain standard of accuracy, $m_0(SE)$, which is regarded as acceptable for a given purpose. Accuracy is a property of the mean of a given error exclusively. It is very important to make a distinction between an accurate and unbiased sample. An unbiased sample implies that the mean of the selection error (SE) is equal to zero. Even when carried out in an ideal way, sampling is always biased as a result of the particulate structure of most materials. The bias is never strictly zero. It may be negligible or very small but always different from zero. One should speak of an accurate sample but not of an unbiased sample. In fact, one can forget about an unbiased sample because it is a limit case never encountered in practice.

Analysis Often when one speaks of analysis, one speaks, in fact, of estimation or determination of the content of a constituent of interest; this is an incorrect practice. Analysis is the opposite of synthesis. It is the separation, partial or complete, of a material into its constituents. How these constituents are determined after their separation is an entirely different matter. When one talks of analysis, analytical procedure, or analytical error, one should talk of determination, procedure for a determination, or determination error. One chooses to go along with the word *analysis*, although it is not correct, because it is so widely misused in the entire industry. One may speak of chemical analysis, moisture analysis, particle size distribution analysis, percent solid analysis, or variance analysis in statistics.

Analytical Error (*AE*) Error generated by analytical operations such as assaying, moisture determination, particle size analysis, estimation of the percent solids of a slurry, etc. This error does not include the last selection process from which the final analytical subsample is obtained; this stage shall be a part of the total sampling error, TE.

Bias *B* A bias is equal to the mean of the selection error $m(SE)$ when this mean is different from zero. The relative selection or sampling bias associated with the estimator a_s of the true unknown critical content a_L of a lot L can be written:

$$B(a_s) = m(SE) = \frac{m(a_s) - a_L}{a_L} \tag{A.1}$$

a_s being the critical content of a sample S selected from the lot L. A sampling bias is always introduced by incorrect sampling. In addition, we can state that correct sampling is not biased. This leads to two fundamental notions:

- Sampling correctness is defined as a structural primary property depending on specific conditions over which one has no control, but independent from the sampling problem itself over which one does have control.
- Sampling accuracy is defined as a circumstantial secondary property depending on the respect of the conditions of sampling correctness over which one has control, and on the intrinsic properties of the material to be sampled over which one has no control.

Circumstantial Property A property depending on the sampling problem itself on which one has no control. It is also a secondary property (e.g., accuracy is a circumstantial property).

Comminution A crushing, grinding, or pulverizing stage that diminishes the fragment size d of a lot, sample, or increment.

Component An elementary part or a constituent that can be separated and quantified by analysis. As far as sampling is concerned, there are several kinds of components:

- *Chemical components:* Usually assaying leads to the determination of the content of a chemical component; it can be an element, a molecule, or a mineral.
- *Physical components:* The domain occupied by the lot L can be divided into a set of subdomains respectively occupied by a single physical component of the material. The theory of sampling of particulate materials involves the proportions of physical components. There are two categories of physical components:
  Active physical components: A set of components the proportion of which is to be estimated by analysis; it could be a set of minerals, a set of particle size fractions, or a liquid such as water if the critical content is the moisture content.

Passive physical components: A set of components the proportion of which is irrelevant for a given analysis; for example, the interstitial air when estimating the moisture content of a solid material.

- *Critical component:* A chemical or physical component the proportion of which is to be estimated.

Composite Sample A sample made by the reunion of several distinct subsamples. Composite samples are often prepared when it is not economically feasible to analyze a large quantity of individual samples. Compositing may also be performed mathematically for two or more samples that have been analyzed, typically in the vertical direction.

Constitution Heterogeneity *CH* The constitution heterogeneity CH_L of a lot L is the heterogeneity that is inherent to the composition of each fragment or particle making up the lot. The greater the difference in composition between each fragment, the greater the constitution heterogeneity. The constitution heterogeneity *CH* could also have been called the composition heterogeneity.

Content Since it is always much easier to deal with dimensionless relative values, the word *content* shall be used for the proportion of a given active component:

$$\text{Content} = \frac{\text{weight of a given active component}}{\text{weight of all active components}} \tag{A.2}$$

Consequently, "critical content a" shall be called the proportion of a critical component: the critical content of a lot L is denoted a_L, the critical content of a sample S is denoted a_s, etc.

$$\text{Critical content } a_L = \frac{\text{weight of a critical component in lot } L}{\text{weight of all active components in lot } L} \tag{A.3}$$

Continuous Selection Error (*CE*) The continuous selection error is generated by the material selection process, and was regarded in former publications as the sum of two complementary errors:

- *The quality fluctuation error (QE):* Taking into consideration the fluctuation of the critical content a as a function of time t, *QE* can be estimated by investigating the properties of the quality function $a(t)$.
- *The quantity fluctuation or weighting error (WE):* Taking into consideration the fluctuations of the rate of flow μ of the active components as a

function of time t, WE can be estimated by investigating the properties of the quantity function $\mu(t)$.

In the new developments of Pierre Gy's sampling theory, this decomposition is no longer necessary since the continuous selection error CE is regarded as the result of the total heterogeneity contribution h. In other words, CE contains both the qualitative term QE and the quantitative term WE.

Correct Selection Sampling is correct when it gives all elements in the lot to be sampled a uniform probability of being selected. When this condition is not fulfilled, the sampling selection becomes necessarily incorrect, and one should remember that incorrect sampling is always a form of gambling responsible for large biases.

$$\text{Correct selection probability} = P = \text{constant} \tag{A.4}$$

Furthermore, sampling is correct when all elements that do not belong to the lot to be sampled have a zero probability of being selected.

Critical Component The critical component is a chemical or physical component, the proportion of which is to be estimated. There are several kinds of critical components. There are as many sampling problems as there are critical components, and these problems must be solved independently from one another. The critical component presenting the most difficult problem will determine the parameters of the sampling process for all critical components.

Critical Content The critical content is the proportion of a critical component, usually denoted as a_L for the true unknown critical content of the lot L, or as a_s for the true unknown critical content of the sample S.

Deterministic Sample A nonrandom sample, a sample selected for a given reason. The best example of a deterministic sample is a grab sample: a grab sample is usually deterministic and practically never probabilistic.

Discrete Model The discrete model identifies the lot L to be sampled with a discrete set of N_u units U_m $(m=1,2,3,...,N_u)$ that can be either individual particles or groups of particles. As far as sampling is concerned, a lot L is completely defined by the finite set of all particles or groups of particles of critical content a_m and weight M_m. The discrete model does not take into account the interstitial fluids and, as a consequence, this model is valid only if the interstitial fluid is a passive component.

Discrete Set A discrete set is a collection of separate and distinct particles or fragments, or units. Generally a discrete set shall be well defined in the sense that there are rules and properties that can be used to find out if one particle or unit is or is not a part of this discrete set.

Distribution Heterogeneity (*DH*) The distribution heterogeneity DH_L of a lot is the heterogeneity that is inherent to the manner in which separate and distinct particles or units are scattered or spread out within the lot *L*. The greater the difference in composition between each fragment, the greater the possible distribution heterogeneity; likewise, the greater the difference in density between each fragment, the greater the possible distribution heterogeneity.

Distribution Homogeneity It is an unstable stage that can be reached if one successfully mixes a segregated material.

Domain A domain *D* is a closed volume, the boundaries of which contain a lot *L*, a group of units, or a group of particles. Such a domain can be a three-, two-, or one-dimensional geometrical space. A domain *T* is a closed fraction of a one-dimensional temporal space, a notion one uses in the study of flowing streams.

Dynamic Phenomenon A phenomenon showing active changes; for example, sampling and homogenization are dynamic phenomena, contrary to heterogeneity, which is a static phenomenon.

Equitable A sampling operation is said to be equitable when the total sampling error *TE* becomes a random variable with an average $m(TE)=0$, which involves the unconditional respect of the rules of sampling correctness.

Errors Due to the fact that samples, and lots from which samples are collected, are not strictly equivalent, there are errors involved in any estimation procedure. The validity of a sampling method can be estimated by the magnitude of the sampling errors involved, and it is essential to clearly differentiate all the various sampling errors from one another because it is the key to a good understanding of the sampling theory. It is also important to discriminate analytical and preparation errors from sampling errors.

Estimate The result a'_s obtained from the analysis of a sample for a given constituent of interest is an estimate of the true unknown content a_s of the sample for the same constituent. The result a'_s is always affected by various errors, and there are no exact measurements or exact analyses. Note that an estimate is associated with a support unit. An estimate has the dual but related meaning of being an interpolated prediction of the average concentration of a critical component for a one-, two-, or three- dimensional unit such as a $25 \times 25 \times 1$ ft block.

Estimator By definition, the true content a_s of a sample is an estimator of the true content of a_L of the corresponding lot. Following a logical investigation an estimate a'_s can be selected for representing the true unknown content a_s of a sample and the true unknown content a_L of a lot. The estimate a'_{sp} can be selected for representing the true unknown content a_{sp} of a specimen; however, a_{sp} cannot and should not be selected for representing the true unknown content a_L of the corresponding lot. In other words, the true content a_{sp} of a specimen is not an estimator of the true content a_L of the corresponding lot because the hypothesis of equality between a_{sp} and a_L is not founded on any logical and scientific justification.

Exact Selection A selection, the error of which is identically zero, implies that the mean and the variance of an exact selection are both equal to zero. Therefore, an exact selection is characterized by a perfectly reproducible and unbiased mean. An exact selection is an inaccessible limit.

Extrapolation Empirical calculation made from values of one or several variables located beyond the known range of this or these variables. There is no legitimate method for making possible the prediction of a phenomenon beyond the known experimental domain: therefore, extrapolation should be performed with extreme caution, and only with the support of a theoretical analysis.

Fragment Broken pieces of a solid that form a compact unit belonging to the lot. During the selection process, this unit is assumed to be indivisible. As fragments become smaller, the term *particle* is preferred; however, both words have the same meaning irrespective of size.

Fundamental Sampling Error (*FE*) The fundamental error results from the constitution heterogeneity *CH* of the sampled material. The adjective *fundamental* is justified by the fact that, out of all the sampling errors, the fundamental error is the only one that can never cancel out. It is the error that remains when the sampling operation is perfect. The fundamental error is also the only sampling error that can be estimated beforehand.

Generalization One of the most frequent sources of problems generated by empirical approaches. For example, the sampling characteristics of a particular constituent are often generalized to the other constituents, which is nearly always a dangerous assumption.

Granulometric Factor It is the particle size distribution factor g. When calculating the variance of the fundamental error, we state that it is proportional to the third power of the maximum particle size d, defined as the opening of the square mesh retaining about 5% oversize. All particles do not have the same size, however, which is taken into account by a correcting factor g.

Group of Particles A subset of neighboring particles that belongs to the set of F_i particles making up the lot L.

Grouping and Segregation Error (*GE*) The grouping and segregation error results from the distribution heterogeneity *DH* of the sampled material, and its variance is directly proportional to three factors: (1) the constitution heterogeneity *CH*, (2) the grouping factor, and (3) the segregation factor. Under normal conditions, this error has a nonzero mean often negligible with major constituents; it could become important, however, with liberated minor constituents.

Grouping Factor The grouping factor is naturally introduced during the development of the notion of natural distribution homogeneity. Increments or groups making up a sample are, most of the time, composed of many neighboring particles. In other words, each particle of the lot does not have the same chance of being part of the sample, only each group does. Therefore, an error is generated that is taken into account by the grouping factor. The grouping factor is an increasing function for the average number of particles N_n making up each increment in a sample. This factor is equal to zero when each increment is made up of only one particle.

Heterogeneity The importance of a clear definition of heterogeneity is considerable. Since homogeneity is defined as the condition of a lot under which all elements of the lot are exactly identical, heterogeneity is also defined as the condition of a lot under which these elements are not strictly identical. Consequently, one can say that homogeneity is the zero of heterogeneity and is an inaccessible limit. All particulate materials are essentially heterogeneous. There are several kinds of heterogeneities, and one should always specify which kind of heterogeneity to which one is referring. In a first analysis, one can identify two different kinds of heterogeneity: (1) the constitution heterogeneity (*CH*), and (2) the distribution heterogeneity (*DH*). These two terms are defined separately. Constitution and distribution heterogeneities are responsible for the total heterogeneity contribution h of a given fraction toward the corresponding lot L. This fraction can be a single fragment or a group of fragments, such as an increment. The average heterogeneity contribution of all possible fractions within the lot is, by definition, nil. When one considers a lot as a sequential population, the total heterogeneity contribution h of a given fraction toward the corresponding lot L is the sum of only three components:

- The short-range heterogeneity contribution h_1 of h. It is a discontinuous term generated by the heterogeneity inside each fraction. Fluctuations of h_1 are essentially random.
- The long-range heterogeneity contribution h_2 of h. It is a continuous term generated by local trends. Fluctuations of h_2 are essentially nonrandom.
- The periodic heterogeneity contribution h_3 of h. It is a continuous term generated by handling, processing, reclaiming, and so on. Fluctuations of h_3 are essentially nonrandom.

Finally, the following formula can be written:

$$h = h_1 + h_2 + h_3 \tag{A.5}$$

Heterogeneity Fluctuation Error The heterogeneity fluctuation error is a synonym for the continuous selection error *CE*, which was, in former publications, the sum of the quality fluctuation error *QE* and the weighting error (*WE*)

$$CE = QE + WE \tag{A.6}$$

In recent texts, this decomposition is no longer necessary since one can directly investigate the total heterogeneity contribution *h* of a material. The total heterogeneity fluctuation error *CE* can be divided into three terms:

- CE_1: Short-range heterogeneity fluctuation error generated by h_1.
- CE_2: Long-range heterogeneity fluctuation error generated by h_2.
- CE_3: Periodic heterogeneity fluctuation error generated by h_3.

Incorrect Selection When one of the two conditions necessary in obtaining a correct selection are not fulfilled, an incorrect selection is always generated (see the definition of a correct selection).

Increment An increment *I* is a group of particles extracted from the lot *L* in a single operation of the sampling device. It is important to make a distinction between an increment and a sample that is obtained by the reunion of several increments.

Increment Delimitation Error (*DE*) The increment delimitation error *DE* results from an incorrect shape of the volume delimiting the increment, which renders nonuniform the probability for a given element of matter to fall between the boundaries of the model extended increment (ideal increment). The mean of this error is, most of the time, different from zero, which makes it an important source of sampling bias.

Increment Extraction Error The increment extraction error *EE* results from an incorrect extraction of the increment. The extraction is said to be correct if and only if the rule of the center of gravity is respected; all particles with their center of gravity inside the boundaries of the correctly delimited increment belong to the increment. The mean of this error is most often different from zero, which makes it an important source of sampling bias.

Increment Materialization Error The materialization error (ME) is the sum of three and only three errors not accounted for by the continuous selection model: the increment delimitation error DE, the increment extraction error EE, and the preparation error PE.

$$ME = DE + EE + PE \tag{A.7}$$

In this book, the preparation error PE will not be included in the materialization error ME because of its different nature:

- The variances of DE and EE are random variables with an average usually different from zero.
- Preparation errors are nonrandom variables, generated by accidents, that cannot be statistically analyzed.

Interpolation An insertion of fictive values between known values representing two consecutive experimental points. Having, in advance, a certain idea of what one is trying to demonstrate, an operator may often be tempted to alter the true meaning of some data by forcing them to fit a convenient model. Excessive interpolation is a common problem encountered when not enough experimental points are at one's disposal.

Liberation Factor When the mineralogical factor c is calculated, one assumes the constituent of interest to be completely liberated, which implies that one is dealing with a mixture of two pure constituents; the constituent of interest and the gangue. Of course, this is not often true, and the liberation factor ℓ is defined as the correcting factor, taking into account that the constituent of interest and the gangue are not perfectly separated from one another.

$$0 \leq \ell \leq 1 \tag{A.8}$$

Liberation Size Often called the *liberation diameter*, d_ℓ, the liberation size is defined as the particle size of a comminution that would ensure the complete liberation of the constituent of interest.

Lot A batch of material from which increments and samples are selected. A lot L of material shall have well-defined boundaries and can be represented by the content of a bag, truck, railroad car, ship, and so forth. Samples are used to estimate the content of various constituents of interest present in the lot.

Mass Quantity of matter in an object as measured in its relation to inertia, determined by dividing the weight of the object by the acceleration due to gravity. From a practical standpoint, the mass is a measure of the quantity of matter present in an object in terms of its heaviness referred to our specific environment. For all practical purposes *mass* is a synonym of *weight.*

Materialization Error See "Increment Materialization Error."

Mean Mean, arithmetic mean, and average are synonymous terms for the numerical value $m(x)$ obtained by dividing the sum of a set of x_i measurements by the total number N of individual measurements in the set. For the sake of clarity, when speaking of the mean, it is recommended that one qualify the type of mean to which one refers: arithmetic, quadratic, geometric, weighted, harmonic, logarithmic, or Sichel. All terms are defined within the text. Usually, when the word *mean* is used alone, without any specification, it is assumed to refer to the arithmetic mean.

Mean Square The mean square of a set of x_i measurements is nothing else than the square of their quadratic mean $r(x)$:

$$r^2(x) = m(x^2) = \frac{1}{N}\sum_{i=1}^{N} x_i^2 = m^2(x) + \sigma^2(x) \tag{A.9}$$

The mean square concept is useful as a measure of sample representativeness.

Mineralogical Factor The mineralogical factor c could be called more exactly the *mineralogical composition factor*. Since the constituent of interest does not have to be a mineral, one can generalize and call c "the constituent of interest composition factor." By definition, c is the maximum heterogeneity generated by the constituent of interest in the lot L. This maximum is reached when the constituent of interest is completely liberated.

Model A model is a more or less idealized representation of a complex reality. A model should provide an undistorted and unbiased picture of reality, and it should be simple enough to allow a straightforward mathematical processing. A model must link each of its subdivisions to one another. All subdivisions must cover all relevant aspects of the problem and answer all questions.

Overall Estimation Error (*OE*) The overall estimation error *OE* is defined as the sum of the total sampling error *TE*, which includes the sampling selection *SE* (i.e., proper sampling error) and the preparation error *PE*, and of the analytical error *AE*. It is the difference between the final analytical estimate and the unknown true value.

Particle See "Fragment."

Particle Size Distribution Factor See "Granulometric Factor."

Pearson's Coefficient Pearson's coefficient is the relative standard deviation. For comparing an experiment with another, it is often much easier to work with the relative standard deviation s_r, defined as the ratio of the standard deviation s of a population of values to the arithmetic average $\overline{x}$ of the same population.

$$s_r = \frac{s}{\overline{x}} \tag{A.10}$$

Precision Usually the term *precision* is used to describe the reproducibility of estimates. An estimate is precise or reproducible when the variance of the selection error $\sigma^2(SE)$ is not larger than a certain standard of precision $\sigma_o{}^2(SE)$ regarded as acceptable for a given purpose. Precision is a property of the variance of a given error exclusively.

Preparation Preparation is a sequence of nonselective operations, such as transfer, crushing, grinding, pulverizing, drying, and mixing, carried out on a batch of matter in order to bring it under a convenient form for the next processing stage, which can be another sampling stage or the ultimate analysis. Preparation shall not be confused with sampling, which is the selection process itself.

Preparation Error The preparation error *PE* is the sum of various errors: errors introduced by contamination, by loss, by chemical or physical alteration, by human mistakes, or by fraud. The preparation error *PE* should not be confused with the sampling selection error *SE*.

Probabilistic Selection A selection is said to be probabilistic when it is founded on the notion of selecting probability that necessarily includes the intervention of some random element. A probabilistic selection can be either correct or incorrect. It is correct when the selecting probability is uniformly distributed among all units making up the lot and nil for material that does not belong in the lot. A selection is incorrect when one of these conditions is not strictly fulfilled. A selection is said to be nonprobabilistic when the selection is not founded on the notion of probability, such as grab sampling, for example.

Probability A probability is the ratio of the number of possible elementary outcomes of a certain event to the total number of elementary outcomes, provided that all elementary outcomes are equally likely to occur. For instance, if one tosses a coin, the probability of showing one face instead of the other is equal to .5. Now, if one considers a lot of material, and if the probability of selecting a given elementary unit is a certainty, then the probability P of selecting the unit is equal to one. If the probability of not selecting a given elementary unit is a certainty, then the probability *P* of selecting this unit is equal to zero. Thus, all probabilities P_i encountered may be written:

$$0 \leq P_i \leq 1 \tag{A.11}$$

Purposive Selection A selection is said to be purposive when it is founded on the sampling operator's choice to decide which units of the lot *L* should be retained as part of the sample. This type of selection is often used by the exploration geologist or by the environmental regulatory officer to fulfill a given purpose. A purposive sample is nothing other than a specimen and should be labeled as such.

Random Variable A random variable *x* is a quantity that can take one of several possible values, each of these having a given probability of occurrence.

Representativeness The degree of representativeness $r^2(SE)$ is the mean square of the selection error *SE*, which is itself the sum of the square of the mean $m(SE)$ and of the variance $\sigma^2(SE)$ of the selection error:

$$r^2(SE) = m^2(SE) + \sigma^2(SE) \tag{A.12}$$

Thus, a sample is said to be representative when the mean square $r^2(SE)$ of the selection error is not larger than a certain standard of representatives $r_0^2(SE)$ regarded as acceptable. Representativeness is the exclusive property that synthesizes both accuracy and precision.

Reproducibility See the definition of "Precision."

Sample A sample *S* is a part of the lot *L* obtained by the reunion of several increments and meant to represent the lot *L* in further operations. A sample is not just any part of the lot: its positioning, delimitation, and extraction must respect certain rules that the theory of sampling intends to establish. Sampling is often carried out by progressive stages: a primary sample is extracted from the lot, then a secondary sample is extracted from the primary sample, and so on.

Sampling or Selection Error The relative sampling error SE occurs when replacing the true unknown critical content a_L of the lot by its estimator a_s. Thus, by definition:

$$SE = \frac{a_s - a_L}{a_L} \tag{A.13}$$

The sampling error shall not be confused with preparation errors, which have nothing to do with the selection process.

Sampling Protocol A record presented as a flowsheet of all sampling and preparation stages on which agreement has been reached after minimizing all the numerous sources of sampling error, with respect to the capital notion of sampling correctness. The term *sampling protocol* is used in this book rather than *sampling scheme*.

Sampling Ratio The sampling ratio τ is a random variable, the expected value of which is the selection probability. If the lot is one-dimensional, such as a flowing stream, the sampling ratio τ can be defined as the proportion of the time during which the stream is diverted to extract the sample.

Sampling Scheme A sampling scheme is a sequence of sampling and preparation stages carried out on a lot L and ending up with the extraction of one or several samples used to estimate the true unknown critical content a_L of the lot. It is a synonym of *sampling protocol*.

Sampling Stage A selective process implemented on a lot L or on a sample S in order to reduce its bulk without major alterations to its critical characteristics.

Segregation Error See "Grouping and Segregation Error."

Segregation Factor The distribution heterogeneity of a critical constituent has a natural range between the minimum distribution heterogeneity or distribution heterogeneity $(DH_L)_{\min}$ and the maximum distribution heterogeneity $(DH_L)_{\max}$ that is equal to the constitution heterogeneity CH_L:

- For $(DH_L)_{\min}$, the segregation factor $\xi=0$
- For $(DH_L)_{\max}$, the segregation factor $\xi=1$

Thus, the segregation factor ξ localizes the degree of distribution of heterogeneity of a critical constituent in a given material between a natural minimum and a maximum equal to the constitution heterogeneity of the material.

$$0 \leq \xi \leq 1 \tag{A.14}$$

Selecting Conditions Conditions controlling the selecting process; for example: cutter speed and cutter width when a cross stream sampler is used.

Selecting Results Results expressed in terms of sampling error and its moments.

Selection Error (*SE*) See "Sampling or Selection Error."

Set A set is a collection of elementary units, the domain of which is well defined in the sense that there are some rules or properties that can be used to determine whether a given elementary unit is or is not an element of the set.

Shape Factor The shape factor *f* is also called the coefficient of cubicity; it is a dimensionless factor that gives an idea of how different the shape of a particle is from the ideal cube. The shape factor of a cube is one.

Size Analysis The analysis and the estimation of the weight percentage of one or several size fractions of a particulate solid.

Specimen Part of a lot obtained without respecting the rules of delimitation and extraction correctness. A specimen should never be used for representing a lot *L* and should be labeled as such. Purposive sampling always leads to the collection of specimens.

Standard Deviation The standard deviation of a population measures the spread of values around the mean of the same population; it is the square root $\sigma(x)$ of the population variance $\sigma^2(x)$ of a distribution of x_i values. The standard deviation $s(x)$ of a sample is the square root of the sample variance $s^2(x)$.

Static Phenomenon Phenomenon showing very little or no change. For example, heterogeneity is a static phenomenon, contrary to sampling and homogenization, which are dynamic phenomena.

Structural Property An intrinsic property of the material to be sampled, independent of the sampling problem in which one has no control. For example, the heterogeneity of a material is a structural property; correctness is also a structural property.

Total Sampling Error Defined as the sum of the sampling error *SE* (i.e., proper sampling error) and of the preparation error *PE* for all sampling stages. If *n* is a given sampling stage ($n=1,2,3,...,N$) one can write:

$$TE = \sum_{N} (SE_n + PE_n) \tag{A.15}$$

Unit An element of a discrete set.

Variance The square of the standard deviation and the most common parameter used by statisticians because of its additive property. The reader should remember that standard deviations cannot be added, only variances can. The term σ^2 represents a population variance, often referred to as the true unknown variance; s^2

represents a sample variance. When referring to a variance for STP, the type of variance will be specified:

- Tolerated s_o^2: Variance of the distribution of n imaginary values x_i, regarded as acceptable, for example, when calculating an optimum sample weight.
- Estimated s^2: The variance s^2 is usually calculated from a limited series of N values x_i $(i=1,2,3,...,N)$.
- True unknown σ^2: All that is accessible is the estimator s^2 of σ^2. One must differentiate s, a statistical estimator of σ, which is an unknown probability distribution parameter.
- Relative: $\sigma^2(x)/m(x)^2$ is the population relative variance; $s^2(x)/\bar{x}^2$ is the sample relative variance. Its use facilitates comparisons between different experiments; it is also practical since it is dimensionless.

Appendix **B**

Mathematical Expressions of Heterogeneities CH_L and DH_L

B.1 Introduction

The concepts of constitution and distribution heterogeneity have been shown visually to be related. The next step is to develop the formal and mathematical relationship between these two entities and to link them together. The reader who is less mathematically oriented should note that formal definitions for the constitution heterogeneity (and its more practical form, the constant factor of constitution heterogeneity) and distribution heterogeneity have been developed. The more mathematically inclined reader is reminded that a complete description of the mathematics is set forth in Gy (1982) and Pitard (1989).

B.2 Constitution Heterogeneity

Before proceeding, it is necessary to define the terms that will be used in this section. Consider the following:

L = Lot of given material composed of discrete units
F_i = Individual fragment in lot L
h_i = Heterogeneity associated with fragment F_i
a_i = Critical content of fragment a_i

a_L = Critical content of the lot L
M_i = Weight of the active components in fragment F_i
$M_{\bar{i}}$ = Average weight of active components in fragment F_i
M_L = Weight of active components in lot L
N_F = Number of fragments in lot L

It is important to remember that heterogeneity always occurs in the context of a particular population (statistical vs. correlated). The heterogeneity of a fragment F_i is related to the set of all fragments constituting the lot L.

Given the above definitions, one can easily see that a homogeneous material can be defined as

$$a_i = a_L \tag{B.1}$$

as all fragments would be strictly identical. Building on this, the heterogeneity h_i of a fragment may be envisioned as proportional to the difference of these two quantities:

$$a_i - a_L \tag{B.1}$$

or, since it is generally more convenient to use relative and dimensionless units as opposed to absolute units, h_i is proportional to

$$\frac{(a_i - a_L)}{a_L} \tag{B.3}$$

It is also known that the fluctuation of a fragment F_i increases as a function of its weight M_i. This implies that h_i is proportional to the product:

$$(a_i - a_L)M_i \tag{B.4}$$

as well as its relative value:

$$\frac{(a_i - a_L)M_i}{a_L M_L} \tag{B.5}$$

As a result, the heterogeneity h_i of a fragment F_i in lot L is defined as

$$h_i = \frac{(a_i - a_L)}{a_L} \cdot \frac{M_i}{M_{\bar{i}}} = N_F \frac{(a_i - a_L)}{a_L} \cdot \frac{M_i}{M_L} \tag{B.6}$$

The average heterogeneities of the fragments and the variance of the heterogeneities carried by fragments in the lot L can be calculated, respectively, as follows:

$$m(h_i) = \frac{\sum_i h_i}{N_F} \tag{B.7}$$

and

$$s^2(h_i) = \frac{1}{N_F} \sum_i h_i^2 = N_F \sum_i \frac{(a_i - a_L)^2}{a_L^2} \cdot \frac{M_i^2}{M_L^2} \tag{B.8}$$

Since the constitution heterogeneity CH_L of a lot L is defined as the relative and dimensionless variance of the heterogeneities of the N_F fragments F_i composing lot L, the following equations are also true:

$$CH_L = s^2(h_i) \tag{B.9}$$

$$CH_L = s^2(h_i) = \frac{1}{N_F} \sum_i h_i^2 = N_F \sum_i \frac{(a_i - a_L)^2}{a_L^2} \cdot \frac{M_i^2}{M_L^2} \tag{B.10}$$

Equation B.10 presents certain problems for calculation in practical situations, for one reason, because it is normally extremely difficult to estimate the value of N_F, which is typically very large. To be able to calculate the CH_L in all cases, it is necessary to define a quantity independent of the size of the lot; thereby eliminating the need to calculate N_F.

The solution comes by multiplying CH_L by the quantity M_L/N_F. This quantity is simply the average weight $M_{\bar{i}}$ of a fragment. This produces the following result:

$$IH_L = \frac{CH_L M_L}{N_L} = CH_L M_{\bar{i}} = \sum_i \frac{(a_i - a_L)^2}{a_L^2} \cdot \frac{M_i^2}{M_L} \tag{B.11}$$

where IH_L is the constant factor of heterogeneity. IH_L has the dimension of a weight since CH_L is dimensionless and $M_{\bar{i}}$ has the dimension of a weight.

IH_L has the distinct advantage in practical situations over CH_L because IH_L can always be calculated. It is IH_L that is used in the calculation of the variance of the fundamental error *FE*.

B.3 Distribution Heterogeneity

During the sample selection process, it is rare that individual fragments are selected for inclusion in the sample. It is virtually always the case that groups of fragments are selected as incremental components.

The distribution heterogeneity DH_L addresses the "group dynamics" of sampling when a lot *L* is subdivided into groups. The mathematics for the DH_L is similar to that of the constitution heterogeneity CH_L but contains the grouping perspective.

For CH_L, the first step was to examine the heterogeneity h_i of a fragment F_i in the lot *L* followed by the variance $s^2(h_i)$ and the definition of CH_L. For DH_L, this process is expanded. First, the heterogeneity of a fragment within a group of fragments is examined, followed by the heterogeneity of a group of fragments within the lot *L*, then the variance of the heterogeneity, and, finally, a definition of the distribution heterogeneity DH_L.

As with CH_L, the notation used in this section will be specified in advance. The following definitions apply:

L = Lot of given material composed of discrete units
G_n = One of a certain number of groups in lot L
F_{nj} = Individual fragment in group G_n
h_{nj} = Heterogeneity associated with fragment F_{nj}
a_{nj} = Critical content of fragment F_{nj}
$a_{n\bar{j}}$ = Critical content of the average fragment a_{nj} ($a_{n\bar{i}} = a_n$)
a_n = Critical content of group G_n
a_L = Critical content of the lot L
M_{nj} = Weight of the fragment F_{nj}
$M_{n\bar{j}}$ = Average weight of fragment F_{nj}
M_n = Weight of active components in group G_n
N_n = Number of fragments in group G_n
N_g = Number of groups in lot L

When a lot L is partitioned into a given number of groups G_n, one may consider a certain fragment F_{nj} (where $j = 1,2,...,N_n$) inside each group G_n.

Associated with this single fragment in a group is a heterogeneity h_{nj} (Fig. B-1a). Using the same arguments as with CH_L, the heterogeneity h_{nj} can be written

$$h_{nj} = \frac{(a_{nj} - a_{n\bar{j}})}{a_{n\bar{j}}} \cdot \frac{M_{nj}}{M_{n\bar{j}}} \tag{B.12}$$

Given that

$$M_{n\bar{i}} = \frac{M_n}{N_n} \tag{B.13}$$

equation B.12 may be rewritten as

$$h_{nj} = \frac{(a_{nj} - a_n)}{a_n} \cdot \frac{M_{nj} N_n}{M_n} \tag{B.14}$$

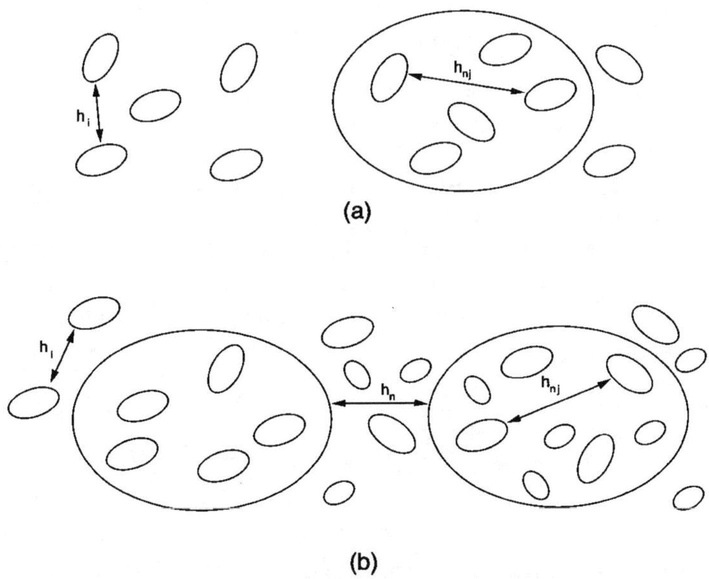

Figure B-1. Within a group G_n, a certain fragment F_{nj} generates a heterogeneity h_{nj}; (b) a group of fragments G_n generates the heterogeneity h_n.

In a fashion similar to CH_L, it follows that

$$m(h_{nj}) = \frac{\sum_i h_{nj}}{N_n} = 0 \tag{B.15}$$

and

$$s^2(h_{nj}) = \frac{1}{N_n} \sum_{nj} h_{nj}^2 = N_n \sum_{nj} \frac{(a_{nj} - a_n)^2}{a_n^2} \cdot \frac{M_{nj}^2}{M_n^2} \tag{B.16}$$

Next, one considers the heterogeneity h_n generated by a group G_n of fragments (Fig. B-1b). Using the same logic as above, it can be written:

$$h_n = \frac{(a_n - a_L)}{a_L} \cdot \frac{M_n}{M_{\bar{n}}} = N_G \frac{(a_n - a_L)^2}{a_L} \cdot \frac{M_n}{M_L} \tag{B.17}$$

It is also true that

$$h_n = h_{\bar{nj}} \tag{B.18}$$

indicating that the heterogeneity h_n is equal to the heterogeneity $h_{\bar{nj}}$ carried by the average fragment of the group.

This leads to the definition of the distribution heterogeneity DH_L of a lot. As with CH_L, the DH_L is a relative and dimensionless variance. It expresses the amount of heterogeneity h_n generated by N_G groups of fragments G_n composing the lot. It is written

$$DH_L = s^2(h_n) = \frac{1}{N_G} \sum_n h_n^2 = N_G \sum_n \frac{(a_n - a_L)^2}{a_L^2} \cdot \frac{M_n^2}{M_L^2} \tag{B.19}$$

As mentioned earlier, the distribution heterogeneity DH_L is a function of the size of the groups. Figure B-2 shows two situations where the support sizes differ. For support 1, the DH_L is greater than that for support 2.

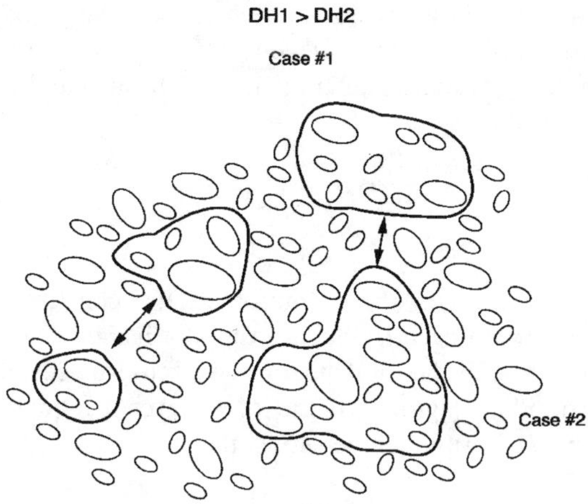

Figure B-2. Distribution heterogeneity is greater when support volumes are smaller.

One should also consider the limit case for DH_L. If the group is composed of only a single fragment, DH_L takes the same definition as the constitution heterogeneity CH_L.

In reality, DH_L never collapses to zero. A certain residual amount of distribution heterogeneity always remains. An inaccessible limit exists called the *natural distribution heterogeneity,* or $DH_{L\min}$.

This minimum quantity remains even after the best attempts at homogenization. When homogenization has been performed to its practical limit, say in a V-blender, the fragments are randomly distributed. Thus, a group of fragments is a random variable. In order to define a quantity $DH_{L\min}$, the mean of this random value must be obtained.

The mean of $DH_{L\min}$ is related to the constitution heterogeneity CH_L. The relationship can be written

$$m[(DH_L)_{\mathrm{MIN}}] = \frac{N_G - 1}{N_F - 1} \cdot CH_L \tag{B.20}$$

If the number of groups and the number of fragments considered are large, this formula converges to

$$m[(DH_L)_{\mathrm{MIN}}] = \frac{N_G}{N_F} \cdot CH_L \tag{B.21}$$

In contrast to the minimum distribution heterogeneity $DH_{L\min}$, there also exists a maximum distribution heterogeneity $DH_{L\max}$. This value turns out to be nothing else than the constitution heterogeneity CH_L and can be written

$$m[(DH_L)_{\text{MAX}}] = CH_L \tag{B.22}$$

The $DH_{L\max}$ is based on the definition of the constitution homogeneity. Homogeneity exists when each fragment in the lot has the same composition as the average composition of the lot. Under these circumstances, the quantity $(a_i - a_L)$ reduces to zero for all fragments in the lot. As a result, equation B.22 becomes equal to zero with $CH_L = 0$. It can be shown that

$$CH_L + CH_{\bar{n}} = DH_L \tag{B.23}$$

where $CH_{\bar{n}}$ is simply the average constitution heterogeneity of the various groups of fragments G_n in lot L.

All terms in equation B.23 are strictly positive. Thus, if $CH_L = 0$, it is implied that $CH_{\bar{L}} = 0$ and, since no negative terms can exist, DH_L must necessarily also be zero.

The nature of the terms also implies the following important inequality:

$$CH_L \geq DH_L > 0 \tag{B.24}$$

$DH_{L\max}$ is a limit case and should not be expected to occur. For $DH_{L\max}$ to occur, the following conditions are required:

- The constitution of the lot is homogeneous, thereby implying that

$$DH_L = CH_{\bar{n}} - CH_L \tag{B.25}$$

  and
- All groups G_n are composed of a single fragment ($G_n{=}F_i$).
- Multiple fragments compose each group G_n; however, all fragments are equivalent in composition ($a_{nj}{=}a_n$). This further implies complete liberation, complete separation, and grouping according to composition.

Having established a maximum and minimum value for DH_L, the practical domain for DH_L can be expressed by

$$DH_{L_{\text{MIN}}} < DH_L < DH_{L_{\text{MAX}}} \tag{B.26}$$

Because DH_L must fall within this range, the idea naturally follows that a factor may be developed to express where in the range DH_L actually occurs. A grouping factor γ can be defined as the ratio of the total number of fragments N_F to the total number of groups N_G. It can be written

$$\gamma_s = \frac{N_F - N_G}{N_G - 1} \tag{B.27}$$

or, equivalently,

$$\frac{1}{1+\gamma_s} = \frac{N_G - 1}{N_F - 1} \tag{B.28}$$

Substituting in equation B.21, the equation becomes

$$m\left[(DH_L)_{\text{MIN}}\right] = \frac{N_G}{N_F} CH_L = \frac{1}{1+\gamma_s} CH_L \tag{B.29}$$

The following relationships hold true for γ_s:

- γ_s is a positive and decreasing function of the number of groups G_n.
- γ_s is an increasing function of the average number of fragments $N_{\bar{n}}$ per group. This can also be stated as: γ_s is an increasing function of the size of the groups.
- γ_s attains a minimum value equal to zero if the number of groups equals the number of fragments; in other words, each group is composed of only one fragment (N_G=N_F).
- γ_s becomes very large and the minimum distribution heterogeneity correspondingly very small when the number of groups approaches one. This can be stated:

$$\gamma_s = +\infty \quad \text{and} \quad DH_{L_{\text{MIN}}} \qquad \text{when } N_G = 1 \tag{B.30}$$

Since 1 is very small in comparison to the number of groups N_G with a corresponding disparity between N_G and N_F written:

$$1 \ll N_G \ll N_F \tag{B.31}$$

the following can be stated:

$$\gamma_s = N_{\bar{n}} = \frac{N_F}{N_G} \tag{B.32}$$

where $N_{\bar{n}}$ represents the average number of fragments per group.

The practical domain of DH_L was established in equation B.25 as

$$DH_{L_{\mathrm{MIN}}} < DH_L < DH_{L_{\mathrm{MAX}}} \tag{B.33}$$

For purposes of convenient substitution, the following equality can also be defined:

$$1 = \frac{1+\gamma_s}{1+\gamma_s} \tag{B.34}$$

Borrowing from equations B.29 and B.34 and substituting into equation B.25, it can be stated that

$$\frac{1}{1+\gamma_s} CH_L < DH_L < \frac{1+\gamma_s}{1+\gamma_s} CH_L \tag{B.35}$$

As stated previously and implied by equation B.35, DH_L is a function of the constitution heterogeneity CH_L. In fact, according to equation B.35, all that needs to be done is to multiply CH_L by a factor whose value lies between 1 and $(1 + \gamma)$. The convention for writing this factor is

$$1 + \xi\gamma_s \tag{B.36}$$

where ξ is the segregation factor. DH_L may then be written

$$DH_L = \frac{1+\xi\gamma_s}{1+\gamma_s} CH_L \tag{B.37}$$

Substituting equation B.37 into equation B.35 and dividing all terms by $CH_L/(1 + \gamma)$ yields

$$0 < 1 \le (1+\xi\gamma_s) \le (1+\gamma_s) \tag{B.38}$$

Finally, if we subtract 1 from all terms and divide through by γ_s, the domain of γ_s is defined to be

$$0 \le \xi \le 1 \tag{B.39}$$

Now that the grouping and segregation factors have been developed, it is possible to conclude with the establishment of a general expression for distribution heterogeneity DH_L. Combining equations B.28 and B.37 yields the following equality:

$$DH_L = (1+\xi\gamma_s)\frac{N_G - 1}{N_F - 1} CH_L \tag{B.40}$$

Finally, the result of transposing equation B.10 into equation B.40 can be written

$$DH_L = (1+\xi\gamma_s)(N_G - 1)\frac{N_F}{N_F - 1}\sum_i \frac{(a_i - a_L)^2}{a_L^2}\frac{M_i^2}{M_L^2} \tag{B.41}$$

or, since $N_F/(N_F - 1) = 1$ because large numbers of fragments are usually present:

$$DH_L = (1+\xi\gamma_s)(N_G - 1)\sum_i \frac{(a_i - a_L)^2}{a_L^2}\frac{M_i^2}{M_L^2} \tag{B.42}$$

Equation B.42 is the general expression of the distribution heterogeneity DH_L. It demonstrates that DH_L is primarily a function of three factors:

1. The constitution heterogeneity CH_L, which, in turn, is an attribute of the population being investigated.
2. The grouping factor γ, an attribute of the module or scale of observation, which provides a measure of random selectivity.
3. The segregation factor ξ, which identifies the degree of heterogeneity of the distribution and is a measure of the spatial rearrangements.

Given these facts, some very significant conclusions may be drawn, which, ironically, do not coincide with many currently employed sampling practices.

1. To minimize the errors caused by the constitution heterogeneity CH_L, it is necessary to increase the number of fragments. In other words, the sample weight must be increased and comminution should be part of the subsampling process.
2. Minimizing the segregation factor ξ can be accomplished by homogenizing the material in the lot before sampling.
3. To minimize γ_s (grouping factor), reduce the size of the groups. For purposes of sampling, this means that as many increments as possible should be collected for each sample.
4. The constitution heterogeneity CH_L cannot be minimized as it is an intrinsic (structural) property of the lot. Only comminution can reduce CH_L.

The irony in sampling is that, generally, a lot of time and effort is spent on homogenization. Everyone does it, but it doesn't work! For example, it takes on the order of 10 min to achieve a state of homogeneity in a V-blender. Open the blender from the bottom and let the material pour out (1 sec), and it segregates by gravity.

Increasing the number of increments (step 3) is an effective way to better sampling but very few people do it. Increasing the sample weight (step 1) also works. Fortunately, this is sometimes done. In order to insure correct sampling, the appropriate order in which to perform these steps is 1, then 3, and then 2.

Appendix C

Mathematical Development of the Fundamental Error

C.1 The Fundamental Sampling Error *FE*

This appendix focuses on the development and derivation of the fundamental error *FE* of sampling. While the discussion is extensive, this appendix concentrates on the practical aspects of *FE*. A full mathematical development can be found in Gy (1982) and Pitard (1989, 1993).

The fundamental error stems from the constitution heterogeneity CH_L using the constant factor of constitution heterogeneity IH_L as a reasonable approximation. The idea of size-density fractions will first be presented and then introduced into the equation for the IH_L. This new form of the IH_L will then be partitioned into two convenient components for examination. These two components will then give rise to the final five essential parameters to be determined for the calculation of the fundamental error *FE*. Finally, a procedure for minimizing the fundamental error *FE* and its impact on the overall sampling error *OE* will be presented.

C.1.1 Equation of the Fundamental Error FE

Before developing the equation of the fundamental error, it is necessary to discuss some of the logic and assumptions that are part of the theory. These relate mostly to the size-density relationships. However, the reader is reminded that the constant factor of heterogeneity IH_L used the assumption of an average weight of a fragment.

It is a fact that all fragments of a lot belong to diverse size fractions, these size fractions to be called α. Similarly, fragments belong to assorted density fractions β. Often, a strong correlation exists between the critical content of a fragment and its density. On the other hand, the correlation between the critical content of a fragment and its size is not nearly as strong. Then, a size-density fraction $L_{\alpha\beta}$ of the lot L can be developed. Each fragment of $L_{\alpha\beta}$ can be represented by the average fragment $F_{\alpha\beta}$. By definition,

- The volume of $F_{\alpha\beta}$ is $v_\alpha = f_\alpha d_\alpha{}^3$, where f_α is a shape factor to be defined shortly.
- The density of $F_{\alpha\beta}$ is λ_β
- The weight of $F_{\alpha\beta}$ is $M_{F\alpha\beta} = v_\alpha \lambda_\lambda = f_\alpha d_\alpha{}^3 \lambda_\beta$
- The critical content of $F_{\alpha\beta}$ is $a_{\alpha\beta}$

Using this more complicated notation of size-density fractions, the variance of the fundamental error FE will be developed to a more simplified conclusion. Its equation can be written

$$S_{FE}^2 = \left[\frac{1}{M_s} - \frac{1}{M_L}\right] fgcld^3 \quad \text{(C.1)}$$

Although the development and notation are somewhat tedious, they provide a complete justification for this important equation.

The constant factor of constitution heterogeneity IH_L discussed previously is given by equation C.2:

$$IH_L = \frac{CH_L}{N_F} = CH_L M_{\bar{i}} = \sum_i \frac{(a_i - a_L)^2}{a_L^2} \frac{M_i^2}{M_L^2} \quad \text{(C.2)}$$

Beginning with the numerator (NUM) of equation C.2, defined as

$$NUM = \sum_i (a_i - a_L)^2 M_i^2 \quad \text{(C.3)}$$

the size-density fractions can be substituted to produce

$$NUM = \sum_\alpha \sum_\beta N_{\alpha\beta}(a_{\alpha\beta} - a_L) M_{F_{\alpha\beta}}^2 \quad \text{(C.4)}$$

By definition and substitution,

$$M_{F_{\alpha\beta}} = \frac{M_{L_{\alpha\beta}}}{N_{\alpha\beta}} \tag{C.5}$$

Substituting equation C.5 into equation C.4 yields the result

$$NUM = \sum_{\alpha}\sum_{\beta}(a_{\alpha\beta} - a_L)^2 M_{F_{\alpha\beta}} M_{L_{\alpha\beta}} \tag{C.6}$$

Again by definition, $M_{F\alpha\beta} = v_\alpha \lambda_\beta$. Substituting into equation C.6 yields

$$NUM = \sum_{\alpha} V_\alpha \sum_{\beta} \lambda_\beta (a_{\alpha\beta} - a_L)^2 M_{L_{\alpha\beta}} \tag{C.7}$$

Bringing back the denominator for IH_L and employing the numerator from equation C.7 results in an updated formula for IH_L equal to

$$IH_L = \sum_{\alpha} V_\alpha \sum_{\beta} \lambda_\beta \frac{(a_{\alpha\beta} - a_L)^2 M_{L_{\alpha\beta}}}{a_L^2 M_L} \tag{C.8}$$

In order to simplify the formula further, it is necessary to remove the terms expressed as αβ, specifically $a_{\alpha\beta}$ and $M_{F\alpha\beta}$. Pitard (1989) presents the following hypotheses to aid in the simplification process:

First Hypothesis: Experience shows that the critical content $a_{\alpha\beta}$ usually varies more from one density fraction to the next than from one size fraction to the next; therefore, all values of $a_{\alpha\beta}$ may be replaced by the average critical content a_β of the density fraction L_β.

Second Hypothesis: The study of a large number of real cases shows that the proportions $M_{L\alpha\beta}/M_{L\beta}$ usually vary little from one density fraction to the next; therefore, one can assume that all values $M_{L\alpha\beta}/M_{L\beta}$ can be replaced by their average $M_{L\alpha}/M_L$. It is then legitimate to write:

$$M_{L_{\alpha\beta}} = \frac{M_{L_\alpha} M_{L_\beta}}{M_L} \tag{C.9}$$

Substituting equation C.9 into the general equation C.8 and separating the α terms from the β terms results in the following formula:

$$IH_L = \left(\sum_{\alpha} V_{\alpha} \frac{M_{L_\alpha}}{M_L} \right) \left(\sum_{\beta} \lambda_{\beta} \frac{(a_\beta - a_L)^2}{a_L^2} \cdot \frac{M_{L_\beta}}{M_L} \right) \tag{C.10}$$

By grouping the terms with respect to size α and density β, two major components fall out. For convenience, the α components can be called X and the β components can be called Y. Thus, it follows that

$$IH_L = XY = fgcld^3 \tag{C.11}$$

Ways to estimate X and Y relative to the size and density fractions, respectively, can now be developed.

C.1.2 The Shape Factor f (Calculation of X)

By virtue of its volume term v_α, the calculation of X presupposes that one knows the volume of the average fragment in the size fraction L_α. As defined earlier, $v_\alpha = f_\alpha d_\alpha{}^2$, with f being a shape factor. The shape factor, which is dimensionless, adjusts the volume based on the fact that the fragments in the lot are not perfect cubes. If they were, the shape factor f_α would be equal to one. If all fragments were spheres, the shape factor f_α would be equal to 0.523.

Most minerals have a shape factor close to 0.5. Some examples are quartz, 0.474; pure pyrite, 0.470; and pure cassiterite, 0.530. Deviations from this rule include: (1) minerals occurring as flakes (mica, biotite, etc.), where the shape factor is approximately 0.1; (2) materials flattened by mechanical stress, such as small gold nuggets, where the shape factor is around 0.2; and (3) acicular minerals occurring as needles (asbestos, serpentine, tourmaline, etc.), where the shape factor is greater than 1 and can be as large as 10. For hazardous wastes, the shape factor is generally around 0.5.

C.1.3 The Granulometric Factor g (Calculation of X)

In factor X of equation C.11, the term d represents the particle size. Because v_α is proportional to the cube of the average diameter, one quickly realizes that the coarsest size fractions exert an overwhelming influence on the value of IH_L. Thus, the term d in equation C.11 represents the maximum fragment size.

The maximum fragment or particle size is defined as the opening of a square mesh that retains no more than 5% oversize material. Figure C-1 shows a series of *g(X)* curves for a number of different variables. There is a distinct convergence of all curves around the value 5% oversize and 0.25 for *g*. This brings a form of stability as *g* varies in only small amounts from one size fraction to the next, and generally occurs in the range of 0.20 to 0.30.

The factor *g* accounts for the fact that all fragments in the material are not identical in size *d*. The factor *g* is dimensionless. For perfectly calibrated material, *g* is equal to 1. More practical examples include:

- Noncalibrated materials: *g* is around 0.25.
- Calibrated materials: *g* is around 0.55. These are materials between two consecutive screen openings of a given series.
- Naturally calibrated materials: *g* is around 0.75. Such materials include cereals, beans, and wheat.

The factor *X* in equation C.11 is weighted by the proportion $M_{L\alpha}/M_L$ of the size fraction L_α. This leads to the equation

$$X = f\sum_{\alpha} d_\alpha^3 \frac{M_{L_\alpha}}{M_L} = fgd^3 \tag{C.12}$$

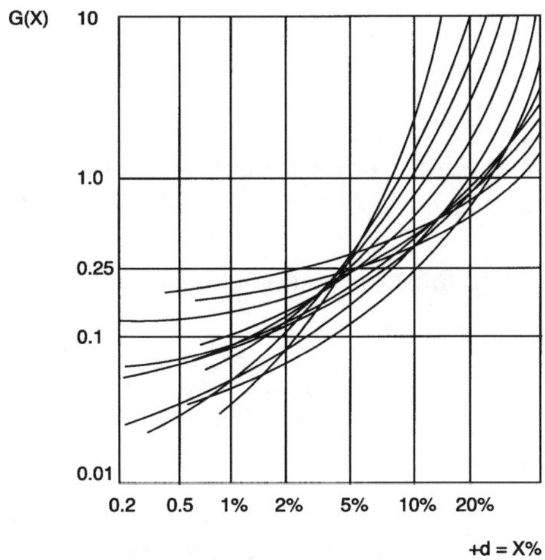

Figure C-1. Plotting *g(X)* curves for various materials shows a convergence of the graphs around 5% oversize and 2.5 for *g*.

This completes the derivation of the X term in the formula for IH_L and applies to the shape and granulometric factors.

C.1.4 The Mineralogical Factor c (Calculation of Y)

The remaining factors in the equation for IH_L, the mineralogical factor c and the liberation factor ℓ, are part of the term Y for density fractions. Y is written

$$Y = \sum_{\beta} \lambda_\beta \frac{(a_\beta - a_L)^2}{a_L^2} \cdot \frac{M_{L_\beta}}{M_L} \tag{C.13}$$

This sum is applied to all density fractions and reaches a minimum when the material of the lot is homogeneous. This implies that $a_\beta = a_L$ and, subsequently, $Y = 0$. To achieve homogeneity, complete liberation of the constituent of interest is required. In such a situation, only two density fractions exist, that for the constituent of interest and that for the host material.

Using this assumption, L_M can be called the density fraction for the constituent of interest of density λ_M, with the host material called L_g of density λ_g. From this, it can be said that $a_M = 1$, $a_g = 0$, and $M_M/M_L = a_L$. Substituting these values into equation C.13 results in

$$Y_{\mathrm{MAX}}\ C = \lambda_\beta \frac{(1 - a_L)^2}{a_L} + \lambda_g (1 - a_1) \tag{C.14}$$

$Y_{\max}$ shall serve as the definition of the mineralogical factor c. The term a_L should always be expressed as a proportion. For example, 5% is written as 0.05, 100 ppm as 0.0001, and so on.

For environmental applications, equation C.14 can be simplified to

$$c = \frac{\lambda_M}{a_L} \tag{C.15}$$

This formula is valid for situations where a_L is less than 0.1, which is the majority of the time. Note, however, that it is most important to have an accurate assessment of c around the regulatory threshold.

C.1.5 The Liberation Factor ℓ (Calculation of Y)

In practice, one cannot always expect complete liberation of the constituent of interest. However, Y was defined as the maximum liberation of the constituent of interest. Therefore, an additional factor is necessary to accommodate the reality that Y overestimates the actual liberation.

With $Y_{min} = 0$ and $Y_{max} = c$, the range of Y can be written

$$0 \leq Y \leq C \tag{C.16}$$

Since Y never reaches c, the dimensionless liberation factor ℓ can be applied as follows:

$$y = c\ell \tag{C.17}$$

where

$$0 \leq \ell \leq 1 \tag{C.18}$$

When the material is homogeneous, the liberation factor ℓ takes on the value of zero. With complete liberation (Y_{max}), $\ell = 1$.

Materials can be grouped according to the amount of heterogeneity:

- For very heterogeneous material: $\ell = 0.8$
- For heterogeneous material: $\ell = 0.4$
- For average material: $\ell = 0.2$
- For homogeneous material: $\ell = 0.1$
- For very homogeneous material: $\ell = 0.05$

It has been shown that the term $X = fgd^3$ and the term $Y = c\ell$. Combining X and Y again yields equation C.11:

$$IH_L = XY = c\ell fgd^3 \tag{C.19}$$

which serves as a practical formula for the calculation of the constant factor of constitution heterogeneity.

Appendix **D**

List of GEM Acronyms

AA Atomic Absorption

AE Analytical Error

ARAR Applicable, Relevant, and Appropriate Requirements

BRA Baseline Risk Assessment

BTEX Benzene, Toluene, Ethylbenzene & Xylene

CAA Clean Air Act

CWA Clean Water Act

CE Continuous Selection Error

CE_1 Short-Range Heterogeneity Fluctuation Error

CE_2 Long-Range Heterogeneity Fluctuation Error

CE_3 Periodic Heterogeneity Fluctuation Error

CERCLA Comprehensive Environmental Resource Conservation & Liability Act

CH Constitution Heterogeneity

CLP Contract Laboratory Program

CoC Chain of Custody

COC Contaminant of Concern

DCE 1,2 Dichloroethene

DE Increment Delimitation Error

DH Distribution Heterogeneity

DQA Data Quality Assessment

DQO Data Quality Objectives

ECD Electron Capture Detector

EE Increment Extraction Error

ELISA Enzyme Linked Immunosorbent Assay

EMSL Environmental Monitoring Systems Laboratory

EPA Environmental Protection Agency

FE Fundamental Error

FS Feasibility Study

GA Geostatistical Appraisal

GC Gas Chromatograph

GC/ECD Gas Chromatograph/Electron Capture Detector

GC/MS Gas Chromatograph/Mass Spectrometer

GE Grouping & Segregation Error

GEM Geostatistical Error Management

GSLIB Geostatistical Software Library

GUPI Guilty Until Proven Innocent

HRS Hazardous Ranking System

IHSS Individual Hazardous Substance Site

ISD Inverse Squared Distance

IPD Inverse Power Distance

ME Increment Materialization Error

MS Mass Spectrometer

NPL National Priorities List

OE Overall Estimation of Sampling

OU Operable Unit

PAH Polycyclic or Polynuclear Aromatic Hydrocarbons

PARCC Precision, Accuracy, Representativeness, Completeness & Comparability

PCB Polychlorinated Biphenyl

PCE Tetrachloroethene

PCP Pentachlorophenol

PE Preparation

PID Photo Ionization Detector

QAMS Quality Assurance Management Staff

QAPP Quality Assurance Project Plan

QA/QC Quality Assurance/Quality Control

QBD Quality By Design

QMP Quality Management Plan

RA Remedial Action

RAS Routine Analytical Service

RCRA Resource Conservation and Recovery Act

RD Remedial Design

RI Remedial Investigation

RI/FS Remedial Investigation/Feasibility Study

RFA RCRA Facilities Assessment

RFI RCRA Facilities Investigation

RV Regionalized Variable

SARA Superfund Amendments and Reauthorization Act

SAS Special Analytical Service

SD Standard Deviation

SOP Standard Operating Procedures

STP Sampling Theory & Practice

TAL Target Analyte List

TCA 1,1,1 Trichloroethane

TCE Trichloroethene

TCL Target Compound List

TDS Total Dissolved Solids

TE Total Sampling Error

TOC Total Organic Carbon

TOSCA Toxic Substances Control Act

TQM Total Quality Management

USEPA United States Environmental Protection Agency

VOC Volatile Organic Compound

XRF X-Ray Fluorescence

References

Agterberg, F.P. 1974. Geomathematics: Mathematical Background and Geoscience Applications. Amsterdam: Elsevier Scientific Publishing.

Agterberg, F.P. 1994. Fractals, multifractals, and change of support. In *Geostatistics for the Next Century*, edited by R. Dimitrakopoulos. Dordrecht, The Netherlands: Kluwer Academic Publishers, pp. 223–234.

Aspie, D., and R.J. Barnes. 1990. Infill-sampling design and the cost of classification errors. *Journal of Mathematical Geology* 22(8): 915–932.

Baker, W.H., J.S. McConnell, and J.C. Myers. 1991. Variability of nitrogen, phosphorus, potassium, and sulfur in cotton petiole monitoring. Poster Paper, Agronomy Society of America (ASA) Annual Meeting, Denver, CO.

Barnes, M.P. 1980. *Computer-Assisted Mineral Appraisal and Feasibility*. New York: Society of Mining Engineers of the American Institute of Mining, Metallurgical, and Petroleum Engineers.

Barnes, R.J. 1985. Estimation variance. Unpublished course notes.

Barnes, R.J. 1986. The cost of risk and the value of information in mine planning. *Proceedings of the 19th APCOM Symposium*. New York: Society of Mining Engineers of the American Institute of Mining, Metallurgical, and Petroleum Engineers.

Barnes, R.J. 1988. Sample design for geologic site characterization. Paper presented at the 3rd International Geostatistical Congress, Avignon, France.

Barnes, R.J. 1989. A partial history of spatial sampling design. *Newsletter of the North American Council on Geostatistics* 3(2): 10–13.

Barnes, R.J. 1989. Sampling design for geologic site characterization. In *Geostatistics*, edited by M. Armstrong. 2 vols. Dordrecht, The Netherlands: Kluwer Academic Publishers, pp. 809–822.

Barnes, R.J. 1991. The variogram sill and the sample variance. *Journal of Mathematical Geology* 23(4): 673–678.

Barnes, T.E. 1982. Orebody modeling: The transformation of coordinate systems to model continuity at Mt. Emmons. *Proceedings of the 17th APCOM Sympo-*

sium. New York: Society of Mining Engineers of the American Institute of Mining, Metallurgical, and Petroleum Engineers, pp. 765–770.

Barnsley, M.F. 1988. *Fractals Everywhere*. New York: Academic Press.

Barth, D., B. Mason, T. Sparks, and K. Brown. 1989. *Soil Sampling Quality Assurance User's Guide*, 2nd ed. Las Vegas: Environmental Monitoring and Support Laboratory, U.S. Environmental Protection Agency.

Bilonick, R.A. 1986. How far does Gy's theory go toward fulfilling the role of a general particulate material sampling theory? Presentation at the Society of Mining Engineers, St. Louis.

Blacker, S., and D. Goodman. 1994a. Case study: Application at a superfund site. *Environmental Science Technology* (American Chemical Society) 28(11).

Blacker, S., and D. Goodman. 1994b. An integrated approach for efficient site cleanup. *Environmental Science Technology* (American Chemical Society) 28(11).

Blank, L. 1980. *Statistical Procedures for Engineering, Management, and Science*. New York: McGraw-Hill.

Box, G. 1974. Statistics and the Environment. *Journal of the Washington Academy of Sciences* 64(2): 52–59.

Bresler, E. 1989. Estimation of statistical moments of spatial field averages for soil properties and crop yields. *Soil Science Society of American Journal* 53: 1645–1653.

Bresler, E., and A. Laufer. 1988. Statistical inferences of soil properties and crop yields as spatial random functions. Soil Science Society of American Journal 52: 1234–1244.

Brooker, P.I. 1978. Geostatistical calculations using true and estimated values from a simulated deposit. *Proceedings of the Australasian Institute of Mining and Metallurgy* 268: 63–69.

Bruno, R., and G. Raspa. 1988. Geostatistical characterization of fractal models of surfaces. In *Geostatistics*, edited by M. Armstrong. 2 vols. Dordrecht, The Netherlands: Kluwer Academic Publishers, pp. 77–90.

Bryan, R.C., and D. Splitstone. 1986. Risk qualified mapping of polychlorinated dibenzodioxin (PCDD) contamination. *Proceedings of the Symposium on Solving Hazardous Waste Problems*. New York: American Chemical Society.

Buffington, J.P. 1978. Developing recommendations to improve quality assurance for federal monitoring programs. *Proceedings of the National Conference on Quality Assurance of Environmental Measurements*. Denver.

Burgess, T.M., R. Webster, and A.B. McBratney. 1981. Optimal interpolation and isarithmic mapping of soil properties: IV. Sampling strategy. *Journal of Science* 32(4): 643–659.

Burrough, P.A. 1981. Fractal dimensions of landscapes and other environmental data. *Nature* 294: 240–242.

Buxton, B.E., and A. Pate. 1994. Joint temporal-spatial modeling of concentrations of hazardous pollutants in urban air. In *Geostatistics for the Next Century*,

Vol. 6, edited by R. Dimitrakopoulos. Dordrecht, The Netherlands: Kluwer Academic Publishers, pp. 75–87.

Chiles, J.P. 1988. Fractal and geostatistical methods for modeling of a fracture network. *Journal of Mathematical Geology* 20(8): 631–654.

Chou, D., and D.E. Schenk. 1984. *Selecting Optimum Drilling Locations by Groups*. New York: Society of Mining Engineers preprint 84–62.

Clark, I. 1979. *Practical Geostatistics*. London: Applied Science Publishers.

Copi, I.M. 1978. *Introduction to Logic*. New York: Macmillan.

Cressie, N. 1991. *Statistics for Spatial Data*. New York: John Wiley.

Dagbert, M., and J.C. Myers. 1981. Demonstration of 2D simulation of the orange marker bed. Report to the U.S. Geological Survey, Conditional Simulation Study. Denver.

Dagbert, M., and J.C. Myers. 1982. Precision of total reserve estimates: Old formulae applied to new situations. *Proceedings of the 17th APCOM (Applications of Computers in the Mineral Industry)*, edited by T. Johnson and R. Barnes. New York: Society of Mining Engineers of the American Institute of Mining, Metallurgical, and Petroleum Engineers, pp. 393–405.

David, M. 1977. *Geostatistical Ore Reserve Estimation*. Amsterdam: Elsevier Scientific Publishing.

David, M. 1988. *Handbook of Applied Advanced Geostatistical Ore Reserve Estimation*. Amsterdam: Elsevier Science Publishers.

Davis, D.R., L. Duckstein, and R. Krzysztofowicz. 1979. The worth of hydrologic data for nonoptimal decision making. *Water Resources Research*. 15: 1733–1742.

Davis, D.R., and W.M. Dvoranchik. 1971. Evaluation of the worth of additional data. *Water Resource Research*. 7: 700–707.

Davis, J. 1973. *Statistics and Data Analysis in Geology*. New York: John Wiley.

Dawdy, D.R. 1979. The worth of hydrologic data. *Water Resource Research*. 15: 1726–1732.

Dennison, J.M. 1972. Statistical meaning in geologic field work. Geological Society of America, Special Paper 146, pp. 25–37.

Deutsch, C., and Andre Journel. 1992. GSLIB: Geostatistical Software Library. New York: Oxford University Press.

Dimitrakopoulos, R. (Editor). 1994. *Geostatistics for the Next Century*. Dordrecht, The Netherlands: Kluwer Academic Publishers.

Duckstein, L., and C.C. Kisiel. 1971. Efficiency of hydrologic collection systems: Role of Type I and Type II errors. *Water Resources Research*. 7: 592–604.

Englund, E.J. 1993. Spatial simulation: Environmental applications. *Proceedings of the First International Conference/Workshop on Integrated GIS and Environmental Modeling*, edited by M.F. Goodchild, B. Parks, and L.P. Steyaert. New York: Oxford University Press, pp. 432–437.

Englund, E.J. 1989. A performance evaluation of interpolation methods. Las Vegas: U.S. Environmental Protection Agency EPA/600/X-89/049.

Englund, E.J. 1990. A variance of geostatisticians. *Journal of Mathematical Geology* 22: 417–456.

Englund, E.J., and N. Heravi. 1992. Conditional simulation: Practical application for sampling design optimization. *Proceedings of the Fourth International Geostatistical Congress, Troia*, edited by A. Soares. Dordrecht, The Netherlands: Kluwer Academic Press.

Englund, E.J., and N. Heravi. 1993. Conditional simulation: Practical application for sampling design optimization. In *Geostatistics*, Troia, edited by A. Soares. Dordrecht, The Netherlands: Kluwer Academic Press, pp. 613–624.

Englund, E.J., and N. Heravi. 1995. Phased sampling for soil remediation. *Journal of Environmental and Ecological Statistics* 1: 247–263. New York: Chapman & Hall.

Englund, E.J., and N. Heravi. 1995. Power curves for spatial data. Unpublished notes.

Englund, E.J., and A. Sparks. 1988. *GeoEAS (Geostatistical Environmental Assessment Software) User's Guide*. Las Vegas: U.S. Environmental Protection Agency EPA/600/4–88/033.

Englund, E.J., D. Weber, and N. Leviant. The effects of sampling design parameters on block selection. *Journal of Mathematical Geology* 24: 329–343.

Falconer, K. 1990. *Fractal Geometry*. New York: Wiley.

Feder, J. 1988. *Fractals*. New York: Plenum Press.

Flatman, G., K. Brown, and J. Mullins. 1985. Probabilistic spatial contouring of the plume around a lead smelter. *Proceedings of the Sixth National Conference on Management of Uncontrolled Hazardous Waste Sites*, pp. 442–448.

Flatman, G., and E. Englund. 1991. Asymmetric loss function for Superfund remediation decisions. *Proceedings of the Business and Economic Statistics Section of the American Statistical Association*.

Flatman, G., and A. Yfantis. 1984. Geostatistical strategy for soil sampling: The survey and the census. *Environmental Monitoring and Assessment* 4: 335–349.

Gershon, M. 1983. *Optimal Drillhole Location Using Geostatistics*. New York: Society of Mining Engineers preprint 83–63.

Gilbert, R.O. 1982. Some statistical aspects of finding hot spots and buried radioactivity. *TRAN-STAT: Statistics for Environmental Studies*. PNL-SA 10274. Richland, Washington: Battelle Memorial Institute, Pacific Northwest Laboratory.

Gilbert, R.O. 1987. *Statistical Methods for Environmental Pollution Monitoring*. New York: Van Nostrand Reinhold.

Gilbert, R.O., and P. Doctor. 1985. Determining the number and size of soil aliquots for assessing particulate contaminant concentrations. *Journal of Environmental Quality* 14(2): 286–292.

Gleick, J. 1987. *Chaos: Making a New Science*. New York: Penguin Books.

Guertin, K. 1987. A correction model for conditional bias in selective mining operations. *Journal of Mathematical Geology* 19: 407–423.

Gy, P.M. 1964. Variography. *International Symposium of Applied Statistics*. Golden, CO: Colorado School of Mines Q., 59, 693.

Gy, P.M. 1976. The sampling of particulate materials: A general theory. *Sampling Practices in the Mineral Industries Symposium*. Melbourne: Australasian Institute of Mining and Metallurgy.

Gy, P.M. 1979. *Sampling of Particulate Materials*. Amsterdam: Elsevier.

Gy, P.M. 1981. Sampling or gambling? *Coal Mining Process*. 18(9): 62.

Gy, P.M. 1982. *Sampling of Particulate Materials: Theory and Practice*, 2nd ed. Amsterdam: Elsevier.

Harnett, D.L., and J.L. Murphy. 1985. *Statistical Analysis for Business and Economics*, 3rd ed. Menlo Park: Addison Wesley.

Hughes, J.P., and D.P. Lettenmaier. 1981. Data requirements for kriging: Estimation and network design. *Water Resources Research*. 17(6): 1641–1650.

Hulse, D.E. The Consequence of block size decisions in ore body modeling. *Proceedings of the 23rd APCOM*. New York: Society of Mining Engineers of the American Institute of Mining, Metallurgical, and Petroleum Engineers.

Ingamells, C.O. 1974. Control of geochemical error through sampling and subsampling diagrams. *Geochim. Cosmochim. Acta* 38 (1974): 1225.

Ingamells, C.O. 1981. Evaluation of skewed exploration data: The nugget effect. *Geochim. Cosmochim. Acta* 45: 1209.

Ingamells, C.O., and F.F. Pitard. 1986. *Applied Geochemical Analysis*. New York: Wiley-Interscience 88.

Ingamells, C.O., and P. Switzer. 1973. A proposed sampling constant for use in geochemical analysis. *Talanta* 20: 547.

Isaaks, E.H., and R.M. Srivastava. 1989. *An Introduction to Applied Geostatistics*. New York: Oxford University Press.

Jablonski, J.R. 1992. Implementing TQM: *Competing in the Nineties Through Total Quality Management*. 2nd ed. Albuquerque: Technical Management Consortium.

Jessen, R.J. 1978. *Statistical Survey Techniques*. Wiley: New York.

Journel, A.G. 1984. Indicator approach to toxic chemical sites. *Annual Report on Project No. CR-811235–02–0*. Las Vegas: U.S. Environmental Protection Agency, Environmental Monitoring Systems Laboratory.

Journel, A.G. 1984. The place of nonparametric geostatistics. *Geostatistics for Natural Resource Characterization*, edited by G. Verly et al. 2 vols. Dordrecht, The Netherlands: Reidel, pp. 307–335.

Journel, A.G. 1989. Fundamentals of geostatistics in five easy lessons. *Short Course in Geology* 8. Washington, DC: American Geophysical Union.

Journel, A.G., and C. Deutsch. 1993. Entropy and Spatial Disorder. *Journal of Mathematical Geology* 25(3): 329–355.

Journel, A.G., and C.J. Huijbregts. 1978. *Mining Geostatistics*. New York: Academic Press.

Kleinbaum, D.G., and L.L. Kupper. 1978. *Applied Regression Analysis and Other Multivariable Methods*. North Scituate, MA: Duxbury Press.

Knudsen, H.P., and C.B. Fritz. 1989. Using IK to determine pod boundaries in a karst lignite deposit. *Proceedings of the 21st APCOM Symposium*. New York Society of Mining Engineers of the American Institute of Mining, Metallurgical, and Petroleum Engineers.

Koch, G., and R. Link. 1971. *Statistical Analysis of Geological Data*. Vols. 1 and 2. New York: Wiley.

Korvin, G. 1992. *Fractal Models in the Earth Sciences*. Amsterdam: Elsevier.

Krige, D.G. 1962. Statistical applications in mine valuation. *J. Inst. Mine Surv. S.A.* 12 (2, 3).

Krumbein, W.C., and F.A. Graybill. 1965. *An Introduction to Statistical Methods and Geology*. New York: McGraw-Hill.

Mandelbrot, B.B. *The Fractal Geometry of Nature*. San Francisco: Freeman.

Marcotte, D., and M. David. 1985. The Bi-Gaussian Approach: A Simple Method for Recovery Estimation. *Journal of Mathematical Geology* 17: 625–644.

Matheron, G. 1963. Principles of Geostatistics. *Economic Geology* 58: 1246–1266.

Matheron, G. 1965. *Les Variables regionalisee et leur estimation (Regionalized Variables and Their Estimation)*. Paris: Editions Masson.

Matheron, G. 1970. *La Theorie des variables regionalisees et ses applications (The Theory of Regionalized Variables and Their Applications)*. Paris: Cahiers du Centre de Morphologie Mathematique.

McBratney, A.B., R. Webster, and T.M. Burgess. 1981. Data design of optimal sampling scenes for local estimation and mapping of regionalized variables - I. *Computers and Geosciences* 7(4): 331–365.

Mood, A.M. 1950. *Introduction to the Theory of Statistics*. New York: McGraw-Hill.

Moss, M.E. 1979. Some basic considerations in the design of hydrologic data networks. *Water Resources. Res.* 15: 1673–1676.

Myers, D.E. 1982. Matrix formulation of Co-kriging. *Journal of Mathematical Geology* 14: 249–257.

Myers, J.C. 1983. *Geostatistical Soil Results for the Dallas Lead Study*. Las Vegas: U.S. Environmental Protection Agency, Environmental Systems Monitoring Laboratory, EPA Study 600/X-83-007.

Myers, J.C. 1983. *Sampling Sensitivity Study of the Dallas Lead Study*. Las Vegas: U.S. Environmental Protection Agency, Environmental Systems Monitoring Laboratory, EPA Study 600/X-83-008.

Myers, J.C. 1984. Geostatistics and ore reserve estimation on the microcomputer. *Proceedings of the GeoTech '84 Conference*. Denver.

Myers, J.C. 1985. Geostatistical sampling and contamination analysis of lead values surrounding industrial smelters. Master's thesis. Golden: Colorado School of Mines.

Myers, J.C. 1986. Objective quantification of sampling adequacy and soil contaminant levels around point sources using geostatistics. *Proceedings of the National Conference on Hazardous Wastes and Hazardous Materials*. Atlanta.

Myers, J.C. 1986. Calculation of kriging error for non-rectangular areas and global sampling patterns. *Proceedings of the Northwest Mining Association Convention*. Spokane, WA.

Myers, J.C. 1986. Sampling sufficiency using geostatistics. *Proceedings of the HazTech International Conference*. Denver.

Myers, J.C. 1987. Global risk assessment and optimum sampling geometries. *Proceedings of HAZTECH Canada*. Toronto.

Myers, J.C. 1987. SPACE: An interactive analysis of sample placement. *Proceedings of GeoTech '87*. Denver.

Myers, J.C. 1989. GeoEAS Software from EPA: A glimpse of the future. *HAZMAT World Magazine*. Chicago: Tower-Borner Publications.

Myers, J.C. 1990. The geostatistical renaissance: New applications for an old technology. *Proceedings of the HAZMAT Central '90 Conference*. Chicago.

Myers, J.C. 1990. Guilty until proven innocent (GUPI). *HAZMAT World Magazine*. Chicago: Tower-Borner Publications.

Myers, J.C. 1991. Geostatistical error management for site assessment (GEM). *Proceedings of HAZMAT West '91*. Long Beach.

Myers, J.C. 1992. A geostatistical analysis of PCB concentrations in soils at the ReSolve site. Unpublished report, Chemical Waste Management, Inc. Princeton, NJ.

Myers, J.C., and T.H. Brown. 1989. Geostatistical and sampling analysis of regraded spoil materials. *Proceedings of the 6th National RCRA/Superfund Conference & Exhibition, HWHM '89*. New Orleans.

Myers, J.C., and T.H. Brown. 1990. Geostatistical analysis of regraded spoil overburden materials. *Proceedings of the Billings Symposium on Disturbed Land Rehabilitation*. Billings.

Myers, J.C., and R.C. Bryan. 1983. Geostatistics applied to toxic waste: A case study. *Geostatistics for Natural Resource Characterization*, edited by G. Verly et al. 2 vols. Dordrecht, The Netherlands: Reidel, pp. 893–901.

Myers, J.C., and R.C. Bryan. 1987. Sampling strategies for site evaluation. Presented at HMCRI's 4th National RCRA Conference and Exhibition on Hazardous Wastes and Hazardous Materials. Washington, DC.

Myers, J.C., and R.C. Bryan. 1989. Applied sampling strategies for site assessment. *Proceedings of the HAZMAT Conference*. Atlantic City.

Myers, J.C., and G. Sergerie. 1984. *Downloading a Mainframe Package for Reserve Modeling and Contouring*. Toronto: Canadian Mining Journal.

Neptune, D., E. Brantly, M. Messner, and D. Michael. 1990. Quantitative decision making in Superfund: A data quality objectives case study. *Hazardous Materials Control* May–June.

Olea, R. 1975. *Optimum Mapping Techniques Using Regionalized Variable Theory*. Lawrence, Kansas: Kansas Geological Survey.

Olea, R. 1984. *Systematic Sampling of Spatial Functions*. Lawrence: Kansas Geological Survey.

Palmer, M.W. 1988. Fractal geometry: A tool for describing spatial patterns of plant communities. *Vegetation* 75: 91–102.

Parkhurst, D. 1984. Optimal sampling geometry for hazardous waste sites. *Environmental Science and Technology* 18(7): 521–523.

Pitard, F.F. 1989. Pierre Gy's sampling theory and sampling practice: Vol. 1 *Heterogeneity and Sampling*. Boca Raton: CRC Press.

Pitard, F.F. 1989. Pierre Gy's sampling theory and sampling practice: Vol. 2 *Sampling Correctness and Sampling Practice*. Boca Raton: CRC Press.

Pitard, F.F. 1993. *Pierre Gy's Sampling Theory and Sampling Practice: Heterogeneity, Sampling Correctness, and Statistical Process Control*, 2nd ed. CRC Press: Boca Raton, FL.

Pitard, F.F. 1994. Personal correspondence.

Puente, C.E. 1994. A fractal-multifractal approach to geostatistics. *Geostatistics for the Next Century*, edited by R. Dimitrakopoulos. Dordrecht, The Netherlands: Kluwer Academic Publishers, pp. 476–487.

Ramsey, C. 1989. The application of sampling theory and field analytical techniques to environmental investigation. Master's thesis. Golden: Colorado School of Mines.

Raymond, G.F. 1979. Ore estimation problems in an erratically mineralized orebody. June 1979. *Canadian Institute of Mining Bulletin*, pp. 90–98.

Rendu, J-M. 1978. *An Introduction to Geostatistical Methods of Mineral Evaluation*. Johannesburg: South African Institute of Mining and Metallurgy.

Rendu, J-M. 1979. Normal and lognormal estimation. *Journal of Mathematical Geology* 11(4): 213–221.

Rendu, J-M. 1980. *Optimization of Sampling Policies: A Geostatistical Approach*. Tokyo: MMIJ-AIME.

Rodriguez-Iturbe, I., and J.M. Mejia. 1974. The design of rainfall networks in time and space. *Water Resources Research*. 12: 1185–1195.

Rogers, J., P. Flyer, A. Chu, J. Braden, and E. Bryant. 1988. Statistical methods for evaluating the attainment of superfund cleanup standards. Vol. 1. *Soils and Solid Media*. Rockville, MD: Westat, Inc.

Rose, C.D. 1992. A fractal model for sampling coal, ores, and other natural populations. *Journal of Coal Quality* 11(1–2): 6.

Rose, C.D., and R.M. Srivastava. 1994. A fractal correlation function for sampling problems. *Geostatistics for the Next Century,* edited by R. Dimitrakopoulos. Dordrecht, The Netherlands: Kluwer Academic Publishers, pp. 142–155.

Royle, A.G. 1979. Why Geostatistics? *Engineering & Mining Journal*. May.

Royle, A.G., M.J. Newton, and V.K. Sarin. 1972. Geostatistical factors in design of mine sampling programmes. *Institute of Mining and Metallurgy*, April, pp. A82-A88.

Russo, D. 1984. Design of optimal sampling network for estimation of the variogram. *Soil Science Society of American Journal* 48: 708–715.

Russo, D., and E. Bresler. 1980. Scaling soil hydraulic properties of a heterogeneous field. *Soil Science Society of American Journal* 44: 681–684.

Russo, D., and E. Bresler. 1981. Soil hydraulic properties as stochastic processes: I. An analysis of field spatial variability. *Soil Science Society of American Journal* 45: 682–687.

Russo, D., and E. Bresler. 1982. Soil hydraulic properties as stochastic processes: II. Errors of estimates in a heterogeneous field. *Soil Science Society of American Journal* 46: 20–26.

Russo, D., and W. Jury. 1988. Effect of sampling network on estimates of the covariance function of stationary fields. *Soil Science Society of American Journal* 52: 1228–1234.

Ryti, R. 1993. Superfund soil cleanup: Developing the Piazza Road remedial design. *Journal of Air and Waste Management Association* 43: 197–202.

Ryti, R., and D. Neptune. 1991. Planning issues for Superfund site remediation. *Hazardous Materials Control*, Nov.–Dec.

Ryti, R., D. Neptune, and B. Groskinsky. 1991. Superfund soil cleanup. *Environmental Testing and Analysis*, Jan.–Feb.

Schroeder, M. 1991. *Fractals, Chaos, Power Laws*. New York: Freeman.

Sichel, H.S. 1966. *The Estimation of Means and Associated Confidence Limits for Small Samples from Lognormal Populations*. Johannesburg: South African Institute of Mining and Metallurgy. Discussed in Michel David, *Geostatistical Ore Reserve Estimation* (Amsterdam, Elsevier Scientific Publishing, 1977).

Singer, D.A. 1972. ELIPGRID: A FORTRAN IV program for calculating the probability of success in locating elliptical targets with square, rectangular, and hexagonal grids. *Geocom Programs* 4: 1–16.

Singer, D.A. 1975. Relative efficiencies of square and triangular grids in the search for elliptically shaped resource targets. *Journal of Research of the U.S. Geological Survey* 3(2): 163–167.

Singh, A., A.K. Singh, and G.T. Flatman. 1994. Estimation of background levels of contaminants. *Journal of Mathematical Geology* 26(3).

Skalski, J.R., and J.M. Thomas. 1984. *Improved Field Sampling Designs and Compositing Schemes for Cost Effective Detection of Migration and Spills at Commercial Low-Level Radioactive or Chemical Waste Sites*. PNL-4935. Richland, WA: Battelle Pacific Northwest Laboratory.

Solow, A.R. 1986. Mapping by simple indicator kriging. *Journal of Mathematical Geology* 18: 335–352.

Solow, A.R. 1993. On the efficiency of the indicator approach in geostatistics. *Mathematical Geology* 25(1).

Solow, A.R., and S.M. Gorelick. Estimating monthly streamflow values by cokriging. *Journal of Mathematical Geology* 18(8).

South, J. 1982. Selecting an acceptance sampling plan that minimizes expected error and sampling cost. *Quality Progress*, October, pp. 18–22.

Srivastava, R.M. 1987. Minimum variance or maximum profitability? *CIM Bulletin*, 80: 63–68.

Starks, T. 1986. Determination of support in soil sampling. *Journal of Mathematical Geology* 18(6): 529–537.

Starks, T., K. Brown, and N. Fisher. 1986. Preliminary monitoring design for metal pollution in Palmerton, Pennsylvania. In *Quality Control in Remedial Site Investigations: Hazardous and Industrial Solid Waste Testing*, edited by C.L. Peckert. ASTM STP 925. Philadelphia: American Society for Testing and Materials, pp. 57–66.

Starks, T., A. Sparks, and K. Brown. 1989. Geostatistical analysis of Palmerton soil survey data. *Environmental Monitoring and Assessment* 93: 239–261.

Sullivan, J. 1984. Conditional recovery estimation through probability kriging—theory and practice. *Geostatistics for Natural Resource Characterization*, edited by G. Verly et al. 2 vols. Dordrecht, The Netherlands: Reidel.

Switzer, P. 1979. Statistical considerations in network design. *Water Resources Research* 15(6).

U.S. Environmental Protection Agency. 1982. *Test Methods for Evaluation Solid Waste*, 2nd. ed. SW-846. Washington, DC. Office of Solid Waste and Emergency Response.

U.S. Environmental Protection Agency. 1987. *Data Quality Objectives: Development Process*, OSWER.

U.S. Environmental Protection Agency. 1987. *Data Quality Objectives: Example Scenario*, OSWER.

U.S. Environmental Protection Agency. 1989. *Documentation of EMSL-LV Contribution to the Palmerton Zinc Study*. Las Vegas: Environmental Monitoring and Support Laboratory.

U.S. Environmental Protection Agency. 1989. Risk assessment guidance for Superfund: Vol. 1, *Human Health Evaluation Manual* (Part A), EPA/540/1-89/002.

U.S. Environmental Protection Agency. 1990. *Guidance for Data Useability in Risk Assessment*, EPA/540/G-90/008, OSWER Directive 9285.7-05.

U.S. Environmental Protection Agency. 1991. Risk assessment guidance of Superfund (Part A, Baseline risk assessment) *Supplemental Guidance/Standard Exposure Factors*, OSWER Directive 9285.6-03.

U.S. Environmental Protection Agency. 1992. *Guidance on Implementation of the Superfund Accelerated Cleanup Model under CERCLA and the NCP*, OSWER Directive 9203.1-03.

U.S. Environmental Protection Agency. 1992. *Preparation of Soil Sampling Protocols: Sampling Techniques and Strategies*, B.J. Mason, Ph.D. Las Vegas:

Office of Research and Development, Environmental Systems Monitoring Laboratory. EPA/600/R-92/128.

U.S. Environmental Protection Agency. 1992. *Supplemental Guidance to RAGS: Calculating the Concentration Term*, OSWER Publication 9285.7-081.

U.S. Environmental Protection Agency. 1993. *Guidance for the Data Quality Objectives Process*, EPA QA/G4.

U.S. Environmental Protection Agency. 1994. *Guidance for the Data Quality Objectives Process*. EPA QA/G-4. Washington DC: Quality Assurance Management Staff, Office of Modeling Systems and Quality Assurance, Office of Research and Development.

Venziano, D., and Kitanidis, P.K. 1982. Sequential sampling to contour an uncertain function. *Journal of Mathematical Geology* 14: 387–404.

Verly, G. 1983. The multigaussian approach and its applications to the estimation of local reserves. *Journal of Mathematical Geology* 15: 263–290.

Verly, G., M. David, A. Journel, and A. Marechal. 1984. *Geostatistics for Natural Resource Characterization*. Vols. 1 and 2. Dordrecht, The Netherlands: Reidel.

Visman, J. 1962. Towards a common basis for the sampling of materials. *Tech. Surv. Res. Rep.* 98, Canadian Department of Mineralogy.

Visman, J. 1969. A general sampling theory. *Mater. Res. Stand.* 9(11): 8.

Walton, M. 1986. *The Deming Management Method*. New York: Putnam Publishing Group.

Watson, A.G., and R.J. Barnes. Infill sampling criteria to locate extremes. *Journal of Mathematical Geology* 27(5).

Weber, D., and E.J. Englund. 1992. Evaluation and comparison of spatial interpolators. *Journal of Mathematical Geology* 24(4): 381–391.

Weber, D., and E.J. Englund. 1994. Evaluation and comparison of spatial interpolators II. *Journal of Mathematical Geology* 26(5).

Wen, R., and Sinding-Larsen, R. 1994. Probability kriging of sub-seismic fault throws with multifractal distributions. *Geostatistics for the Next Century*, edited by R. Dimitrakopoulos. Dordrecht, The Netherlands: Kluwer Academic Publishers, pp. 488–497.

Yates, S.R., and M.V. Yates. 1988. Disjunctive kriging as an approach to management decision making. *Soil Science Society of America Journal* 52: 1554–1558.

Yfantis, A.A. 1990. A study before a nuclear dump site is operational. *Proceedings of the 7th International Conference on Systems Engineering*, pp. 647–652.

Yfantis, A.A. 1993. A new quadratic and biquadratic algorithm for curve and surface estimation. *Computer Aided Geometric Design* 10(6): 509–520.

Yfantis, A.A., G.T. Flatman, and J.V. Behar. 1987. Efficiency of kriging estimates for square, triangular, and hexagonal grids. *Journal of Mathematical Geology* 19: 183–205.

Yfantis, A.A., G.T. Flatman, and E.E. Englund. 1988. Simulation of geological surfaces using fractals. *Journal of Mathematical Geology* 20(6): 667–672.

Yfantis, A.A., G.T. Flatman, and F. Miller. 1993. Parametric surface estimation.

Proceedings of the Fourth International Geostatistical Congress, Troia, edited by A. Soares. Dordrecht, The Netherlands: Kluwer Academic Press.

Yfantis, A.A., G.M. Gallitano, and G.T. Flatman. 1992. A chaotic algorithm for surface estimation. *Journal of Landscape and Urban Planning*, Vol. 21: 309–311.

Yost, R.S., G. Uehara, and R.L. Fox. 1982. Geostatistical analysis of soil chemical properties of large land areas: I. Semi-Variograms. *Soil Science Society of America Journal* 46: 1028–1032.

Yost, R.S., G. Uehara, and R.L. Fox. 1982. Geostatistical analysis of soil chemical properties of large land areas. II. Kriging. *Soil Science Society of America Journal* 46: 1033–1037.

Zirschky, J., and R.O. Gilbert. 1984. Detecting hot spots at hazardous waste sites. *Chemical Engineering*, July, pp. 97–100.

Index